ELSEVIER'S ENCYCLOPAEDIC DICTIONARY OF MEDICINE

PART D

THERAPEUTIC SUBSTANCES

PART A. GENERAL MEDICINE

PART B. ANATOMY

PART C. BIOLOGY, GENETICS AND BIOCHEMISTRY

PART D. THERAPEUTIC SUBSTANCES

ELSEVIER'S ENCYCLOPAEDIC DICTIONARY OF MEDICINE

PART D
THERAPEUTIC SUBSTANCES

in five languages

English, French, German,
Italian and Spanish

compiled by

A. F. DORIAN

ELSEVIER

Amsterdam — Oxford — New York — Tokyo   1990

ELSEVIER SCIENCE PUBLISHERS B.V.
Sara Burgerhartstraat 25
P.O. Box 211, 1000 AE Amsterdam, The Netherlands

Distributors for the United States and Canada:

ELSEVIER SCIENCE PUBLISHING COMPANY INC.
655 Avenue of the Americas
New York, NY 10010, U.S.A.

ISBN 0-444-42826-7

Printed in The Netherlands

PREFACE

At the end of the fourth and last volume of my Encyclopaedic Dictionary of Medicine, I suddenly thought that to do justice to the subject of therapeutic products and pharmacology in general, I should start all over again and 'enlarge' it hugely. As it is, I have tried my best to choose and treat those items which represent the basic layer of pharmacological study either because of the active principles they are relating to or because of their particular therapeutic effects.

As I repeatedly wrote in the prefaces of other dictionaries, I can only hope that this book may prove useful as a humble contribution to the vast and unceasing work in the field of medicine.

GUIDE TO THE USE OF THE DICTIONARY

The dictionary is divided into two sections. The first section, the 'Basic Table', lists the English entries in alphabetical order. Each entry is followed by an English definition and by its French, German, Italian and Spanish equivalents. These equivalents are listed alphabetically in four separate indexes, which together form the second section of the dictionary. Each entry carries a number which refers to the relevant entry in the basic table.

ABBREVIATIONS

f	Français
d	Deutsch
i	Italiano
e	Español
s	see

BIBLIOGRAPHY

B. J. Large & I. E. Hughes, Learning Pharmacology. John Wiley & Sons.
Clark, Brater Johnson, Goth's Medical pharmacology. The C. V. Mosby Co.
Dukes' Side effects of Drugs Annual.
Johnson & Lawrence, Pocket Examiner in Therapeutics.
Orland and Saltman, Manual of Medical Therapeutics.
Faber's Medical Dictionary. Faber & Faber, London.
Dorland's Illustrated Medical Dictionary. W. B. Saunders, London.
Nöhring, J. Dictionary of Medicine (English/German). Elsevier, Amsterdam.
Dizionario Enciclopedico. Treccani Editore, Rome.
Cardinal's Diccionario Terminológico de Ciencias Médicas. Salvat Editores, Barcelona/Madrid.
Dorian, A.F., Elsevier's Encyclopaedic Dictionary of Medicine (Part C). Amsterdam.

A

1 abelmusk
Or abelmosk. Bushy herb found in tropical Asia and the East Indies. Its fruit and seed can be used for diuretic preparations.
f abelmosch; ambrette
d Moschus; Bisameibisch
i abelmosco
e abelmosco

2 abietate
The salt or ester of abietic acid, q.v.
f abiétate
d Abietat
i abietinato
e abietato

3 abietic acid
Acid containing two double bonds situated on two different rings. Owing to the presence of the two double bonds the acid is used for numerous reactions (isomerization, addition, oxidation, hydrogenation etc.) which are industrially exploited for the preparation of a large variety of drugs.
f acide abiétique; acide résinique; acide colophanique
d Abietinsäure; Resinolsäure; Harzsäure
i acido abietinico
e ácido abiético; ácido resínico

4 abietin
A glycoside extracted from some coniferous plants, in particular from the pitch of the silver fir. It is sometimes used to cure cough. Also known as laricine or coniferin.
f abiétine; coniféroside; laricine
d Abietin; Laricin
i abietina; coniferina; laricina
e abietina; laricina

5 abri semen
The seed of the abrus, a genus of tropical vines, whose aqueous extract may be used for the treatment of trachoma and some types of conjunctivitis, probably owing to its blood-agglutinant properties.
f abri semen
d abri semen; Abrussamen
i abri semen
e abri semen

6 absinthin
The bitter principle of wormwood (Artemisia absinthium). It is soluble in alcohol and used as a gastric, particularly as appetite-stimulating, tonic.
f absinthine
d Absinthin
i absintina
e absintina

7 absinthium
The dried leaves and the flowering tops of Artemisia absinthium used as a bitter tonic and anthelminthic.
f absinthe commune; armoise amère
d Absinth; Alsem; bitterer Beifuss
i assenzio
e ajenjo

8 absolute alcohol
Ethyl alcohol containing not more than one per cent weight of water.
f alcool absolu
d absoluter Alkohol
i alcool assoluto
e alcohol absoluto

9 acacia bark
The dried bark of Acacia arabica or Acacia decurrens which is used in the form of decoction as an astringent agent.
f écorce d'acacia
d Akazienrinde; Wattlerrinde
i corteccia di acacia
e corteza de acacia

* **acaciae gummi** s. **gum arabic**

10 aceaspargine
A neurotrophic and antiammonia agent capable of promoting the synthesis of urea in the liver and the transamination processes peripherically. It participates in the energy metabolism (the chemical changes releasing energy in the form available for work and biosyntheses) of the central nervous system. It also increases the yield of aspartic and glutamic acid in the brain. It is principally used in the treatment of hyperammoniaemia and of neurotoxic conditions due to inadequate ammoniacal detoxification.
f acéasparagine
d Azeasparagin
i acesparginina
e acesparginina

11 acebutolol
Adrenergic beta-receptor blocking agent. It is mainly used for the treatment of angina pectoris and tachyarrhythmia. The concomitant use of the drug with catecholamine depletion-causing therapeutic

agents should be avoided. Characteristic side effects are hypotension, bradycardia, gastrointestinal disorders and sometimes depression.

f acébutolole
d Acebutolol
i acetobutololo
e acetobutololo

12 acecarbromal

Tranquillizer and hypnotic used in neurotic conditions borne of psychic or physical stress, in cardiopathies (especially of a functional nature), insomnia and disorders caused by the menopause.

f acécarbromal
d Acecarbromal
i acecarbromale
e acecarbromalo

13 aceclidine

Parasympathomimetic agent whose action is particularly strong on the pupil. It is used to produce miosis (the contraction of the pupil) and in the treatment of glaucoma. It reduces intraocular pressure.

f acéclidine
d Aceklidin
i aceclidina
e aceclidina

14 acediasulfone sodium

Chemotherapeutic agent of the sulfonic group used in the treatment of infections caused by sulphamide-sensitive germs. It is frequently used in the treatment of acute and chronic otitis.

f acédiasulfone sodique
d Sodiumacediasulfon
i acediasolfone sodico
e acediasolfona sódica

15 acefylline piperazine

Agent inducing dilatation of the blood vessels frequently used in anginose conditions. Also employed to combat insomnia in various heart diseases and bronchopathies.

f acéphylline pipérazine
d Acephyllinpiperazin
i acefillina piperazina
e acefilinapiperacina

16 aceglatone

Antineoplastic agent capable of inhibiting beta-glucuronidase and relieving inflammations. It is recommended for the prevention of recidivation of vesical carcinoma following the surgical removal of the tumour.

f acéglatone
d Aceglaton
i aceglatone
e aceglatona

17 aceglutamide

Agent capable of exerting an antiasthenic activity in conditions of physical and mental fatigue. Its action is similar to that of a eutrophic substance as it sup-

ports metabolism at brain-cell level. Generally recommended in cases of psychic instability, excessive irritability, insomnia and to check or delay senile involution.

f acéglutamide
d Aceglutamid
i aceglutamide
e aceglutamida

18 acenaphthene

An intermediate derived from coal tar and used in the production of insecticides and pharmaceuticals.

f acénaphtène
d Acenaphthen
i acenaftene
e acenafteno

19 acenocoumarol

A coumarin-type anticoagulant (coumarin is the lactone of O-coumaric acid, a vasodilator, which is found in the common woodruff and in oil of lavender) capable of inhibiting prothrombin and other coagulation factors and used for the treatment of thromboembolic and phlebitic diseases.

f acénocoumarol
d Acenokumarol
i acenocumarolo
e acenocumarolo

20 acepromazine

A drug used in neuropsychiatry as an important tranquillizer in conditions associated with depression, recurrent or persistent insomnia and hyperexcitability. It is also used in surgical operations to enhance the action of narcotics and for artificial hibernation. It also finds some use in neurovegetative disorders.

f acépromazine
d Acepromazin
i acepromazina
e acepromacina

21 acerin

The active principle of a substance produced from the seeds of the Acer platanoides (the Norway maple). The substance itself is capable of inactivating the phage of several bacterial species as well as vaccinia virus.

f acérine
d Acerin
i acerina
e acerina

22 acetal

Diethylacetal. Highly flammable, colourless liquid obtained by the partial oxidation of ethyl alcohol and used in medicine as a hypnotic.

f acétal
d Acetal
i acetale
e acetal

23 acetamide

The white deliquescent crystalline amide

of acetic acid used as a plastifier and antacid substance, as well as a solvent in organic syntheses. Acetic acid amide.
f acétamide
d Acetamid; Azetamid; Essigsäureamid
i acetammide
e acetamida

24 acetamidophenol
Acetyl derivative of p-aminophenol used in medicine as an antipyretic agent.
f acétamidophénol
d Acetamidophenol
i acetamidofenolo
e acetamidofenolo

*** acetaminophen   s. paracetamol**

25 acetanilide
An acetyl derivative of aniline used as an antipyretic, analgesic and antirheumatic agent.
f acétanilide
d Acetanilid
i acetanilide
e acetanilida

26 acetarsol
A chemotherapeutic agent used in the treatment of amoebiasis, syphilis, trichomoniasis and pemphigus. It is now superseded. Acetarsol sodium, being a more soluble form, is sometimes injected for the treatment of yaws and general paralysis.
f acétarsol
d Acetarsol
i acetarsolo
e acetarsolo

27 acetazolamide
Oral diuretic mainly used in congestive heart failure. It is also recommended in glaucoma and epilepsy owing to its good results. The drug acts by eliminating carbonic anhydrase in the renal tubules. Owing to its teratogenic action, the drug should not be used during pregnancy.
f acétazolamide
d Acetazolamid
i acetazolamide
e acetazolamida

28 acetiamine
A derivative of vitamin B_1 characterized by a high degree of intestinal absorption and used in the treatment of neuralgia, cephalalgia, myalgia, hepatitis and alcoholic steatosis. Also used to improve the tolerance of glucose in diabetics and, topically, in the form of ointment.
f acétiamine
d Acetiamin
i acetiamina
e acetiamina

29 acetic acid
Saturated fatty acid produced by the oxidation of ethyl alcohol with aerobic bacteria or, in a more elementary form, during the destructive distillation of wood,

In variously diluted solutions the drug has an expectorant, eupeptic, antiseptic and astringent action and is used in cough and digestive preparations as well as in poisoning by alkalis. It is also used topically against inflammation of the joints, in contusions, in pruriginous affections and as a disinfectant for skin and mucous membranes. In concentrated solutions, e.g. glacial acetic acid, it has a strong caustic action.
f acide acétique
d Essigsäure
i acido acetico
e ácido acético

30 acetic anhydride
Mobile liquid characterized by a pungent smell and a lacrimatory and vesicant action and used in organic syntheses, in particular for the production of acetyl derivatives, e.g. aspirin.
f anhydride acétique
d Acetanhydrid; Essigsäureanhydrid
i anidride acetica
e anhídrido acético

31 acetidin
Ethyl acetate. Inhalant used in the treatment of laryngeal catarrh and, internally, as an antispasmodic and diaphoretic agent.
f acétidine
d Azetidin
i acetidina
e acetidina

32 acetin
Any of the three liquid acetates which are formed when glycerol and acetic acid are heated together, and occur naturally in cod-liver oil and butter fats.
f acétine; monacétine
d Acetin
i acetina
e acetina

33 acetiromate
Hypocholesterolaemic agent synthetically derived from the thyroid hormone and acting on the blood lipids. It is used in the treatment of hypercholesterolaemia associated with atherosclerosis, hypertension, diabetes mellitus and obesity.
f acétiromate
d Azetiromat
i acetiromato
e acetiromato

34 acetohexamide
Hypoglicaemic agent of the sulfonylurea group used in the treatment of stable non-ketonic diabetes. It cannot be regarded as a substitute for insulin whenever the endogenous source of this substance is absent. One of the principal side effects of the drug is a high alkaline phosphatase.
f acétohexamide
d Azetohexamid
i acetoesamide

e acetoesamida

35 acetomorphine
Diamorphine hydrochloride. A compound of morphine and acetoanhydride (heroin). It is more effective than morphine and is used in the relief of pain. The danger of addiction is very real.
f acétomorphine
d Acetomorphin
i acetomorfina
e acetomorfina

36 acetone
Dimethylketone. Anaesthetic and anthelmintic and sometimes used in dyspnoea. It constitutes the starting point of several important synthetic drugs.
f acétone
d Aceton; Dimethylketon
i acetone; dimetilchetone
e acetona; dimetilquetona

37 acetone oxime
Or acetoxime. The colourless crystalline volatile compound which is formed by acetone by the action of hydroxylamine.
f acétoxime
d Acetoxim
i acetossina
e acetoximo

38 acetonitryle
Methyl cyanide. Bad-smelling liquid present in the tar of coal. It is an important intermediate in the preparation of various organic compounds.
f acétonitrile; cyanure de méthyle
d Acetonitril; Methylcyanid
i acetonitrile; cianuro di metile
e acetonitrilo; cianuro de metilo

39 acetophenazine
A tranquillizer of the phenothiazine group acting on the central nervous system and cardiac innervation. It is used with good results in insomnia, tension due to stress, psychic hyperexcitability etc.
f acétophénazine
d Azetophenazin
i acetofenazina
e acetofenacina

40 acetophenetidin
Or phenacetin. Acetyl compound of phenetidin used as an analgesic and antipyretic.
f acétylphénétidine; phénacétine
d Acetylphenetidin; Phenacetin
i acetilfenetidina; fenacetina
e acetilfenetidina; fenacetina

41 acetophenone
An aromatic ketone occurring in small quantities in some essential oils. It is widely used as an intermediate in various organic syntheses and as a hypnotic.
f acétophénone
d Acetophenon
i acetofenone

e acetofenona

42 acetophenonephenetidin
A compound formed from acetophenone and phenetidin, used as an antipyretic and antineuralgic.
f acétophénonephénétidine
d Acetophenonephenetidin
i acetofenonefenetidina
e acetofenonafenetidina

43 acetopyrin
Antipyrin acetylsalicilate. An antipyrin or phenazone compound with acetylsalicylic acid. It is used as an analgesic and antirheumatic, particularly in sciatica and in the treatment of influenza.
f acétopyrine
d Acetopyrin
i acetopirina
e acetopirina

44 acetylaminosuccinic acid
A cerebral stimulant used in the treatment of mental and nervous stress and anxiety as well as in insomnia, depression and, more generally, neuropsychic disorders. It also finds some employment in behavioural disorders both in children and the elderly.
f acide acétylaminosuccinique
d Acetylaminobernsteinsäure
i acido acetilamminosuccinico
e ácido acetilaminosuccínico

* **acetylbetamethylcholine** s. **acetylcholine chloride**

* **acetylcarbromal** s. **acecarbromal**

45 acetylcholine
The acetic ester of choline found in the tissues of higher animals. It is the most powerful depressor base so far known. A sharp decrease in blood pressure results from an intravenous injection of the substance. It may also cause some fat accumulation in the liver. It is a chemical transmitter of nerve impulses at certain sites of the central nervous system, as well as at parasympathetic nerve terminals and at the neuromuscular junction.
f acétylcholine
d Acetylcholin; Vagusstoff
i acetilcolina
e acetilcolina

46 acetylcholine chloride
A parasympathomimetic drug more stable than acetylcholine having an action resembling that of muscarine. It is used to lower blood pressure, to enhance peristalsis and to promote the activity of the sudoriparous and salivary glands. It shows antagonism with atropine.
f chlorure d'acétylcholine
d Acetylcholinchlorid
i acetilcolina cloruro
e cloruro de acetilcolina

47 acetylcysteine
A secretolytic, mucolytic and fluidifying agent acting against the viscosity of the bronchial mucus owing to HS groups.
f acétylcystéine
d Acetylcystein
i acetilcisteina
e acetilcisteina

48 acetyldigitoxin
Digitoxin-like cardiac glycoside character-ized by a positive inotropic and brady-cardiac action. Mainly used for the treat-ment of cardiac failure, auricular flutter and paroxysmal tachycardia.
f acétyldigitoxine
d Acetyldigitoxin
i acetildigitossina
e acetildigitoxina

49 acetyldigoxin
Digoxin-like cardiac glycoside character-ized by complete absorption capacity and very low toxicity. Mainly used in the treatment of cardiac insufficiency, both acute and chronic.
f acétyldigoxine
d Acetyldigoxin
i acetildigossina
e acetildigoxina

50 acetyloxyphenol
A resorcinol monoacetate used in medicine for the treatment of acne.
f acétyloxyphénol
d Acetyloxyphenol
i acetilossifenolo
e acetiloxifenol

51 acetylphenylhydrazine
A compound easily soluble in alcohol and used as an antipyretic and analgesic as well as in the treatment of polycythaemia owing to its destructive affect when acting on erithrocytes. Phenylhydrazine, the parent substance, is more toxic.
f acétylphénylhydrazine
d Acetylphenylhydrazin
i acetilfenilidrazina
e acetilfenilidrazina

52 acetylsalicylic acid
Analgesic, antipyretic, antirheumatic, anti-inflammatory and uricosuric agent (aspirin). Also an antithrombotic agent and as such sometimes used for the pre-vention of myocardial infarction. It has a rapid and fairly long-lasting action. Also used topically for skin infections. It interacts with orally administered anti-coagulants and sulphonylureas.
f acide acétylsalicylique
d Acetylsalicylsäure
i acido acetilsalicilico
e ácido acetilsalicílico

53 acetylsulphadiazine
A compound which represents a proportion of the urinary excretion of sulphadiazine, q.v. It responds to colour tests only after hydrolysis.
f acétylsulfadiazine
d Acetylsulfadiazin
i acetilsolfadiazina
e acetilsulfadiacina; acetilsulfadiazina

54 acetylsulphadiazole
A compound which represents a proportion of the urinary excretion of sulphathiazole, q.v. It responds to colour tests only after hydrolysis.
f acétylsulfadiazole
d Acetylsulfadiazol
i acetilsolfadiazolo
e acetilsulfadiazolo

55 acetyl sulphaguanidin
A compound which represents a proportion of the urinary excretion of sulphaguani-dine, q.v. It responds to colour tests only after hydrolysis.
f acétylsulfaguanidine
d Acetylsulfaguanidin
i acetilsolfaguanidina
e acetilsulfaguanidina

56 acetyltannin
Odourless and tasteless grey powder which is soluble in ethyl acetate and used in medicine in the treatment of chronic diar-rhoea and intestinal catarrh.
f acétyltanin
d Acetyltannin
i acetiltannino
e acetiltanino

57 acexamic acid
Anti-inflammatory agent used as anti-cica-trizant, i.e. promoting the healing of a wound without leaving a scar, as it pro-tects connective and epithelial tissues. It also promotes the healing of fractures.
f acide acéxamique
d Acexamsäure
i acido acesamico
e ácido acesámico

58 acipimox
Agent chemically related to nicotinic acid and used to check the release of fatty acids from fat tissues thus reducing blood lipids.
f acipimox
d Acipimox
i acipimox
e acipimox

59 aconine
The alkaloid of aconite, which is virtual-ly non-poisonous.
f aconine
d Aconin
i aconina
e aconina

60 aconite
Also known as wolfsbane, friar's cap, blue rocket etc. Genus of plants contain-ing the important alkaloids of aconitine, picraconitine and aconine. It produces

local anaesthesia when applied to the tongue. It may be used externally for the treatment of neuralgia and internally, though hardly advisable, as an antipyretic.
f aconit; sabot; capuchon de moine; coqueluchon
d Akonit; blauer Eisenhut; blaue Wolfswurz
i aconito; malapelle; napello
e acónito; anapelo; uva verga

61 aconitine
Poisonous alkaloid derived from the common monkshood. It stimulates sensory nerve endings and is used in some liniments as it produces a tingling sensation.
f aconitine
d Aconitin
i aconitina
e aconitina

62 acridine
Heterocyclic compound resulting from the union of two benzene rings with one of pyridine. The diamino derivatives acriflavine and proflavine are antiseptics and bactericides widely used in surgery and in the treatment of burns, gangrene and gonorrhoea.
f acridine; dibenzopyridine
d Acridin; Dibenzopyridin
i acridina; dibenzopiridina
e acridina; dibenzopiridina

63 acriflavine
Chlormethyl derivative of the diamino-acridine. An antiseptic, also known as neutral trypaflavine, used against streptococci, staphylococci etc. It is not toxic. Its hydrochloride is used in ophthalmic surgery and ear, nose and throat surgery.
f acriflavine; tripaflavine
d Acriflavin; Trypaflavin
i acridina; tripaflavina
e acriflavina; tripaflavina

64 acriflavine chloride
Potent antiseptic agent marked by an activity that is greater than that of sublimate or phenol on cocci and on gonococci. There is little interference of blood and organic substances with the effects of the drug. It is mainly used for the disinfection of wounds and mucous membranes, particularly the oral and urinary mucosa.
f chlorure d'acriflavine
d Acriflavinchlorid
i acriflavina cloruro
e cloruro de acriflavina

65 acrisorcin
Fungicidal agent topically used in the form of ointment in the treatment of tinea versicolor.
f acrisorcine
d Acrisorcin
i acrisorcina
e acrisorcina

66 acrosoxacin
Antibiotic exclusively used in the treatment of gonorrhoea in case of allergy or resistance to penicillin.
f acrosoxacine
d Acrosoxacin
i acrosoxacina
e acrosoxacina

67 actinomycin
Antimicrobial substance with antineoplastic activity, which is obtained from cultures of Actinomyces antibioticus. It has a very considerable bacteriostatic effect on Gram-positive organisms but a reduced activity in the case of Gram-negative organisms. A tonic and antianaemic. It is employed in the treatment of neoplasias of any organ and in preoperative and postoperative phases.
f actinomycine
d Actinomyzin
i actinomicina
e actinomicina

68 activated charcoal
Strong adsorbant used orally in case of acute poisoning in order to diminish the absorption of toxins. The adsorbant capacity is increased by treatment at 800-900°C with steam or CO_2 to obtain a porous particle structure.
f charbon actif; charbon activé
d Aktivkohle; aktivierter Kohlenstoff
i carbone attivo; carbone attivato
e carbón activado

69 adenine
Vitamin B_4. Purine base extracted from several glandular organs and from tea and obtained by hydrolysis of nucleic acids. It has a protective action against granulocyte-damaging factors, and is used in the treatment of granulopenia caused by antibiotics, thyroid inhibitors, antineoplastic agents and psychotropic drugs.
f adénine
d Adenin
i adenina
e adenina

70 adenosine
The nucleoside (adenine D-ribose) that is obtained by the acid hydrolysis of the mononucleotide adenylic acid, with which it partly shares the pharmacodynamic properties. It is regarded as the principal source of ammonia in the blood.
f adénosine
d Adenosin
i adenosina
e adenosina

71 adenosine phosphate
A smooth-muscle relaxant and a vasodilator for coronary, cerebral and muscular vessels. Also used as a psychostimulant and a leukopoietic agent and in the treatment of angina pectoris and other myopathies, as well as in blood-coagulation disorders.

f phosphate d'adénosine
d Adenosinphosphat
i adenosina fosfato
e fosfato de adenosina

72 adenosine triphosphate
Or adenosine triphosphoric acid. The source of high-energy phosphate bonds for tissue metabolism. It has found some use in some cardiovascular diseases but its efficiency in this respect is still under examination. Also known as ATP.
f adénosine triphosphate
d Adenosintriphosphat
i adenosina trifosfato
e trifosfato de adenosina

73 adenosyl methionine
Physiological substance in which an adenosyl residue is attached to the sulphur atom of the amino acid methionine. It promotes transmethylation and transaminopropylation. It is used in some metabolic disorders, in hepatitis, both acute and chronic, biliary cirrhosis, gallbladder disorders and in case of damage to the liver caused by drugs.
f adénosylméthionine
d Adenosylmethionin
i adenosilmetionina; metionina attiva
e adenosilmetionina

* **adenylpyrophosphoric acid   s. adenosine triphosphate**

74 adionidine
A glycoside extracted from the Adonis vernalis. A yellow bitter water-soluble powder having cardiotonic and diuretic properties. It is occasionally used as an adjuvant of digitalis.
f adionine
d Adionin
i adionina
e adionina

75 adiphenine
Antispasmodic and myorelaxant agent marked by a slight inhibiting activity for muscarin and acetylcholine. It is similar to atropine in its parasympatholytic action which extends to the smooth muscle of the gastroenteric, biliary and urethral tracts and of the uterus, but not to the salivary and humoral glands and on the eye.
f adiphénine
d Adiphenin
i adifenina
e adifenina

76 adipiodone
A contrast medium which contains iodine in organic combination. It is used intravenously for cholecystography and cholangiography and is characterized by a choleretic action.
f adipiodone
d Adipiodon
i adipiodone
e adipiodona

77 adjuvant
The substance which is mixed and injected with an antigen which non-specifically supports or alters the antibody response to the antigen, e.g. aluminium hydroxide in diphtheria.
f adjuvant
d Adjuvans
i coadiuvante
e adyuvante

78 adonidis herba
Herbaceous plant containing specific enzymes and whose active principles are related to those of Strophantus. Although its action is similar to that of digitalis, it cannot be accurately graduated and is not constant.
f adonidis herba
d Adonidis Herba
i adonidis herba
e adonidis herba

* **adrenaline   s. epinephrine**

79 adrenalone
The ketone of adrenaline used as a vasoconstrictor and haemostatic. The drug is especially used in the treatment of glaucoma but is also valuable in the therapy of thrombopenic purpura and of hypotension.
f adrénalone
d Adrenalon
i adrenalone
e adrenalona

* **adstringent   s. astringent**

80 aerugo
Verdigris. A copper oxyacetate having the same medicinal properties of the copper sulphate which is used as an emetic or, externally, in various skin conditions.
f aerugo; vert-de-gris; acétate basique de cuivre
d Aerugo; Grünspan; Kupferacetat
i verderame; acetato basico di rame
e verdete; acetato básico de cobre

81 aesculin
A glycoside obtained from the horse chestnut. It has been used to protect the skin from ultraviolet rays and in bacteriological media.
f esculine
d Äsculin
i esculina
e esculina

82 agar
Or agar-agar. A purgative acting by a physical mechanism, i.e. by providing bulk to the faeces and thus stimulating peristalsis. Generally used with other laxatives. It also finds employment as emulsifier and as a basis for culture media in bacteriology.
f agar-agar
d Agar-Agar

ı agar-agar
e agar-agar

83 agmatine
Aminobutyl guanidine found in ergot and in protein decomposition as a breakdown product of arginine. It is used in medicine to induce uterine contractions.
f agmatine
d Agmatin
ı agmatina
e agmatina

84 ajmalicine
Or raubasine. Alkaloid of Rauwolfia serpentina capable of reducing the pulse rate and thus cardiac work load. The agent is also used as a coronary vasodilator and in the treatment of auriculo-ventricular extrasystoles as well as paroxysmal tach_y cardia and atrial fibrillation.
f ajmalicine; raubasine
d Ajmalicin; Raubasin
ı ajmalicina; rauwolfina
e rauwolfina; ajmalicina

85 alaninate
The salt of the amino acid alanine, used in medicine for the treatment of syphilis.
f alaninate
d Alaninat
ı alaninato
e alaninato

86 alantolactone
An antihelminthic agent used in the treatment of infestations by worms (Oxyuris, Ascarides, Trichocephalus etc.).
f alantolactone
d Alantolakton
ı alantolactone
e alantolactona

87 alclofenac
Analgesic and non-steroid antiphlogistic agent used in rheumatic and degenerative conditions, as well as in post-traumatic and postoperative pain relief.
f alclofénac
d Alclofenac
ı alclofenac
e alclofenac

88 alclometasone
A corticosteroid used topically for the treatment of eczema and non-infective inflammatory conditions.
f alclométasone
d Alclometason
ı alclometasone
e alclometasona

89 alcoholature
A tincture which is made with alcohol as solvent. Specifically, a pharmaceutical preparation obtained through the action of alcohol on fresh vegetables.
f alcoolature
d Alkoholatur; Auszug aus frischen Pflanzen
ı alcolatura

e alcoholatura

90 alcuronium chloride
A skeletal muscle relaxant used as adjuvant of anaesthesia in surgical operations relating to limbs, thorax and abdomen. It also finds employment in cardiosurgery, neurosurgery and obstetrics, as well as in ophthalmology and orthopaedics.
f chlorure d'alcuronium
d Alcuroniumchlorid
ı alcuronio cloruro
e clorure de alcuronio

91 aldosterone
A naturally-occurring corticosteroid hormone mainly acting on salt and water metabolism by increasing the retention of salt in the kidney. It is used in the treatment of adrenal insufficiency in Addison's disease or hypopituitarism, generally in combination with a glycocorticoid.
f aldostérone
d Aldosteron
ı aldosterone
e aldosterona

92 aleuronate
Yellowish powder obtained from the gluten of wheat. Added to flour (of wheat or rye) it is used to prepare a kind of bread which can be eaten by diabetics.
f aleuronat; gluten de blé purifié
d Aleuronat; Glidin
ı aleuronato
e aleuronato; gluten de trigo purificado

93 alexitol sodium
A compound of sodium polyhydroxyaluminium carbonate and hexitol (a hexahydroxy alcohol obtained by reduction of the corresponding hexoses and also occurring naturally) used as an antacid capable of neutralizing gastric acids (see also aluminium hydroxide).
f sodium d'alexitol
d Natriumalexitol
ı alexitol sodio
e sodio de alexitol

*** algedrate aluminium hydroxide gel   s. aluminium hydroxide**

94 algestone
A synthetic progestogen characterized by a very strong and prolonged activity. It is used to balance luteinic insufficiency of the menstrual cycle and for the hormone therapy of prostatic diseases in the male.
f algestone
d Algeston
ı algestone
e algestona

95 alginic acid
Polysaccharide that is obtained from algae and is sometimes used in the treatment of anaemia (the iron salt). The sodium salt may be used as a substitute for traga-

canth.
f acide alginique
d Algensäure; Alginsäure
i acido alginico
e ácido algínico

96 alimentazine
Antihistaminic and phenothiazine (a compound which is toxic to threadworm and tapeworm) neuroleptic marked by a weak adrenolytic and anticholinergic action). It is used to intensify anaesthesia and also in psychomotor disorders, neurovegetative diseases and bronchial cough and/or asthma.
d alimentazine
d Alimentazin
i alimentazina
e alimentacina

97 alkavervir
An alkaloid obtained from Veratrum viride and used in medicine for its antihypertensive action.
f alkavervir
d Alkavervir
i alcavervir
e alcavervir

*** all-heal   s. valerian**

98 allicin
A colourless liquid, oily and irritant substance used as an antibacterial agent. It is extracted from garlic.
f allicine
d Allisin; Allicin
i allicina
e alicina

*** allocinnamic acid   s. cinnamic acid**

99 allopurinol
A drug used in the treatment of gout as it inhibits the endogenous synthesis of uric acid by suppressing the action of xanthinoxidase, the enzyme which oxidizes both xanthine and hypoxanthine into uric acid.
f allopurinol
d Allopurinol
i allopurinolo
e alopurinolo

100 alloxan
Mesoxalyl urea, produced by the oxidation of uric acid. It yields alloxantin by reduction. It is capable of producing diabetes in various animals, e.g. rabbit, cat, dog, rat and monkey by selectively destroying the insulin-producing cells of the pancreas and is used for the study of diabetes.
f alloxane; mésoxalylurée
d Alloxan; Mesoxalyl-Harnstoff
i allossana; mesosallilurea
e aloxana; mesoxalilurea

101 alloxantin
A compound obtained from uric acid by the action of the nitric acid. It yields alloxan by oxidation and has an equal diabetogen action.
f alloxantine
d Alloxantin
i allossantina
e aloxantina

102 allyl alcohol
Colourless liquid characterized by a penetrating smell used in dilute solution as a disinfectant.
f alcool allylique
d Allylalkohol
i alcool allilico
e alcohol alílico

103 allyl sulphide
Yellowish oil having a garlic smell, generally regarded as a germicide and used in the past in the treatment of tuberculosis. Also known as oil garlic.
f sulfure d'allyle
d Allylsulfid
i allile solfuro
e alilo sulfuro

104 aloes
A cathartic agent acting in a short time by causing a considerable increase of tone and peristalsis in particular in the large intestine. It has a cholagogic action and is used in cases of obstinate constipation.
f aloès
d Aloe
i aloe
e áloe; acibar

105 aloin
A glycoside of aloe. Eupeptic, cholagogic and very mildly purgative action when administered in small doses. A high dosage brings about a drastic and emmenagogic action. The drug may exercise an irritant action on the uterine muscle and should therefore be avoided during pregnancy.
f aloïne; barbaloïne
d Aloin; Barbaloin
i aloina; barbaloina
e aloina; barbaloina

106 alpha amylase
Amylolytic (capable of transforming starch into sugar) enzyme isolated from saliva and pancreas bacteria. Its action is antiphlogistic and antioedematous. It is used in the treatment of pathological conditions affecting the ear, nose, mouth, larynx etc. It also finds some use in traumatology, obstetrics, proctology etc.
f alpha amylase
d Alpha-Amylase
i alfamilasi; amilopsina
e alfamilasa

107 alpha-calcidol
The alpha-hydroxyvitamin D_3 which is converted in the liver in dihydroxyvita-

min D_3 (the metabolite of vitamin D char
acterized by its optimum effect on calcium and phosphate balance). It is used
in the treatment of renal bone disease,
rickets, osteomalacia etc. whenever these
conditions prove resistant to vitamin D.
f alpha calcidol
d Alphacalcidol
i alfa-calcidol
e alfa-calcidol

108 alphadolone
An anaesthetic used to promote general
anaesthesia for short surgical interven-
tions.
f alphadolone
d Alphadolon
i alfadolone
e alfadolona

109 alphamethadol
The alcohol of which methadone (a narcotic drug) is the compounding ketone.
Its properties are similar to those of
morphine.
f alphaméthadol
d Alphamethadol
i alfametadolo
e alfametadol

110 alphaprodine
Analgesic related to pethidine, both chemically and pharmacologically. It is wide
ly used in preoperative treatment and in
minor surgical operations.
f alphaprodine
d Alphaprodin
i alfaprodina
e alfaprodina

111 alphaxalone
Anaesthetic agent used to promote general
anaesthesia for short surgical operations.
f alphaxalone
d Alphaxalone
i alfaxalone
e alfaxalona

112 alprenolol
Beta-receptor adrenergic blocking agent
whose selective action is balanced by a
mild intrinsic sympathomimetic capacity.
It is used in the treatment of angina
pectoris and disorders of the cardiac
rhythm, as well as hypersympathicotonia
accompanied by cardiovascular disorders.
f alprenolol
d Alprenolol
i alprenololo
e alprenololo

113 alstonia
A genus of plants whose bark, in particular the dita bark is used as a tonic,
a febrifuge etc.
f alstonia; dita
d Dita
i alstonia; dita
e alstonia; dita

114 alum
Potassium aluminium sulphate used to
stop bleeding (in its solid form) and to
apply to the umbilical cord (in its powder form). It is a potent astringent.
f alum
d Alaun
i allume
e alumbre

115 aluminium aceglutamide
A chemotherapeutic agent used in the
treatment of gastric ulcers.
f acéglutamide d'aluminium
d Alluminiumaceglutamid
i alluminio aceglutamide
e aceglutamida de aluminio

116 aluminium carbonate
Non-absorbable antacid used in the treatment of peptic ulcers owing to its capacity of neutralizing acids. It also finds
employment in the prevention of urinary
phosphate stones.
f carbonate d'aluminium
d Aluminiumcarbonat
i carbonato di alluminio
e carbonato de aluminio

117 aluminium chloride
An astringent used to prevent hyperhi-
drosis (excessive sweating).
f chlorure d'aluminium
d Aluminiumchlorid
i cloruro di alluminio
e cloruro de aluminio

*** aluminium clofibrate   s. clofibrate**

118 aluminium formate
A topical antiseptic used in the treatment
of wounds, chilblains, itch and, generally, lesions.
f formiate d'aluminium
d Aluminiumformiat
i alluminio formiato
e formiato de aluminio

119 aluminium gluconate
A drug used in the treatment of retarded
mental development in children, in mongo
lism and the consequences of vaccinal
encephalitis.
f gluconate d'aluminium
d Aluminiumglukonat
i alluminio gluconato
e gluconato de aluminio

120 aluminium glycinate
A delayed-action antacid used for the
treatment of hyperacidity, gastritis and
as an adjuvant in gastroduodenal ulcer.
f glycinate d'aluminium
d Aluminiumglycinat
i alluminio glicinato
e glicinato de aluminio

121 aluminium hydroxide
Antacid used in the treatment of peptic
ulcer and more generally of gastrointesti-

nal disorders accompanied by diarrhoea and abdominal pains. It may also be used topically for the treatment of some dermatoses.

f hydroxyde d'aluminium
d Aluminiumhydroxyd
i idrossido di alluminio
e hidróxido de aluminio

122 aluminium nicotinate
Therapeutic agent used in the treatment of hypercholesterolaemia and hyperproteinaemia.

f nicotinate d'aluminium
d Alluminiumnikotinat
i alluminio nicotinato
e nicotinato de aluminio

123 aluminium phosphate
An antacid agent used in the treatment of dyspepsia, gastritis, peptic ulcers, a sensation of uneasiness in the stomach during pregnancy, hiatal hernia, hyperacidity etc.

f phosphate d'aluminium
d Aluminiumphosphat
i alluminio fosfato
e fosfato de aluminio

124 aluminium sulphate
A stronger than alum astringent which is used in dilute solutions for topical appli_cations in the treatment of ulcers, chronically recurrent sores, pus-producing excretions from mucous surfaces and as a mild caustic agent. Also known as cake alum or pickle alum.

f sulfate d'aluminium
d Aluminiumsulfat
i alluminio solfato
e sulfato de aluminio

125 aluminium trilactate
Aluminii lactas. Antiseptic agent for the upper respiratory tract, i.e. oral cavity and pharynx. It is characterized by a fungistatic activity against yeast and pseudoyeasts.

f trilactate d'aluminium
d Aluminiumtrilactat
i alluminio trilattato; alluminio lattato
e trilactato de aluminio

126 alveloz milk
The milky sap of a Brazilian plant (the alveloz) which is used for the treatment of cancerous ulcers.

f lait d'alveloz
d Alvelozbalsam
i alvelos; latte d'alvelos
e balsamo de alveloz; leche de alveloz

127 alverine
A strong synthetic antispasmodic acting partially as peripheral anticholinergic agent by a process which is similar to that of atropine and papaverine. It is used for the treatment of spastic gastrointestinal and genitourinary disorders.

f alvérine

d Alverin
i alverina
e alverina

128 amantadine
Chemotherapeutic agent characterized by antiviral action. It has been used in prophylaxis against influenza A virus. It also displays some action on Parkinsonism. It may just possibly increase the dopaminergic activity in the brain. Extra-pyramidal syndromes caused by neuroleptic drugs have also been treated with amantadine.

f amantadine
d Amantadin
i amantadina
e amantadina

129 amaranth
The red dye which is synthesized from naphthionic acid and is used to colour some medicines.

f amarante
d Amarant
i amaranto
e amaranto

130 ambazone
An antiseptic particularly effective against Streptococcus pneumoniae, viridans and pyogenes and used in the treatment of infections affecting mouth and pharynx.

f ambazone
d Ambazon
i ambazone
e ambazona

131 ambenonium chloride
A cholinesterase inhibitor used in the treatment of myastenia gravis and as a stimulant of gastrointestinal motility.

f chlorure d'ambénonium
d Ambenoniumchlorid
i ambenonio cloruro
e cloruro de ambenonio

132 ambergris
Opaque, mostly grey, waxlike mass of intestinal origin. It is obtained from the sperm whale and was used in the past as a stimulant in slight fevers and in nervous disorders.

f ambre gris
d Ambergries; Amber
i ambra grigia
e ámbar gris

133 ambucaine
Local anaesthetic used in ophthalmology mainly for subconjunctival injections, the probing of the tear duct, electroretinography etc.

f ambucaïne
d Ambucain
i ambucaina
e ambucaina

134 ambucetamide

Antispasmodic agent used in drugs prescribed for dysmenorrhoea.
f ambucétamide
d Ambucetamid
i ambucetamide
e ambucetamida

135 ambuphylline
A derivative of theophylline characterized by diuretic and cardiostimulant action. It is widely used in the treatment of bronchial asthma, chronic bronchitis and pulmonary oedema.
f ambuphylline
d Ambuphyllin
i ambufillina
e ambufilina

136 ambuside
Thiazide-type diuretic carrying out its action by inhibiting the reabsorption of the sodium at the level of the renal tubules thus increasing its elimination. Used in the treatment of some cardiopathies accompanied by visceral or peripheral oedemas, of disorders of the left ventricle, of pulmonary heart, arterial hypertension, oedema-ascites syndrome etc.
f ambuside
d Ambusid
i ambuside
e ambusida

* **amethocaine s. tetracaine**

137 amfepramone
A drug causing the loss of appetite for food. It acts at the level of the hypophysis and peripheral tissues, and used in psychogenic obesity and sometimes against the effects of menopausa.
f amfépramone
d Amfepramon
i amfepramone
e amfepramona

138 amidefrine mesylate
A vasoconstrictor capable of decongesting the nasal mucous membrane by a sympathomimetic mechanism. It is mainly used in nasal congestions due to allergy or to some infectious disease of the upper respiratory tract.
f amidéfrine mésylate
d Amidefrinmesylat
i amidefrina mesilato
e mesilato de amidefrina

139 amikacin
Aminoglycoside antibiotic active against Gram-positive and Gram-negative bacteria used for the treatment of some infections of the respiratory tract.
f amikacine
d Amikacin
i amicacina
e amicacina

140 amikhellin
A derivative of khellin acting as a smooth-muscle relaxant and coronary dilator. It is mainly used in the treatment of essential hypertension and angiospastic conditions.
f amikelline
d Amikhellin
i amikellina
e amiquelina

141 amiloride
A diuretic which acts by inhibiting the exchange of sodium for potassium in the distal tubule of the kidney. It is frequently combined with a thiazide.
f amiloride
d Amilorid
i amiloride
e amilorido

142 aminacrine
An antiseptic and disinfectant agent derived from acridine and used in the treatment of stomatitis, tonsillitis, angina, dermatosis etc.
f aminoacridine
d Aminoakridin
i aminoacridina
e aminoacridina

143 aminaphthon
Haemostatic agent capable of reducing capillary bleeding time by increasing the resistance and diminishing the permeability.
f aminaphtone
d Aminaphthon
i aminaftone
e aminaftona

144 amineptine
Quick-action thymoanaleptic agent used in the therapy of manic-depressive psychoses, neuroses and general conditions borne of involution.
f amineptine
d Amineptin
i amineptina
e amineptina

145 aminitrozole
A chemotherapeutic agent capable of acting against Trichomonas vaginalis and used in the treatment of male and female trichomoniasis.
f aminitrozole
d Aminitrozol
i aminitrozolo
e aminitrozolo

146 aminoacetic acid
Antiacid agent used in the treatment of gastric hyperacidity and irritation and capable of activating the functional tone of muscles and blood vessels. It is therefore used in the therapy of myopathies and vascular disorders, e.g. coronary diseases and muscular dystrophy.
f acide aminoacétique; glycine; glycocolle
d Aminoessigsäure; Glycin; Glycocoll
i acido amminoacetico; glicina; glicocolla

e ácido aminoacético; glicina; glicocola

147 aminobenzoic acid
Member of the vitamin complex present in various compound vitamin preparations and mainly used in the form of ointment to protect the skin from ultraviolet radiations.
f acide aminobenzoïque
d Aminobenzoesäure
i acido amminobenzoico
e ácido aminobenzóico

148 aminobutyric acid
A derivative of butyric acid obtained by the substitution of an atom of hydrogen with an aminic group. The gamma-amino butyric acid occurs in some plants, in various microorganisms and in the brain. It probably has an important role in the metabolism and in the function of the nervous tissue, particularly in the transmission of inhibitory nervous impulses.
f acide aminobutyrique
d Aminobuttersäure
i acido amminobutirrico
e ácido aminobutírico

149 aminocaproic acid
Antifibrinolytic agent acting by specific inhibition of plasminogen activation and used against haemorrhages and anaemia.
f acide aminocaproïque
d Aminocapronsäure
i acido aminocaproico
e ácido aminocapróico

150 aminochlorthenoxazin
A salt of tetracycline used as antibiotic, analgesic and antipyretic against various infections caused by strains of Gram-positive and Gram-negative organisms. It is mainly used for the prevention of respiratory disorders following influenza and of septic and phlogistic disorders following surgical operations.
f aminochlorténoxicycline
d Aminochlortenoxicyclin
i aminoclortenossazina
e aminoclortenoxacina

151 aminoglutethimide
An agent having an inhibitory action on adrenal cortex and used to reduce or eliminate adrenal activity in cases of hyperadrenalism, especially when caused by adrenal carcinoma. Also used in prostatic carcinoma and to reduce oedema due to hyperaldosteronism.
f aminoglutéthimide
d Aminoglutethimid
i aminoglutetimide
e aminoglutetimido

152 aminohippuric acid
A diagnostic agent used to measure the functional capacity of the kidneys.
f acide aminohippurique
d Aminohipursäure
i acido amminoippurico

e ácido aminohipúrico

153 aminomercury chloride
Or mercury ammonium chloride; white precipitate. The chloride of ammonium and mercury, soluble in water and less poisonous than mercury chloride. It is used for antipruriginous drugs.
f chloramidure de mercure; lait de mercure; sel d'Alembroth
d Amidoquecksilberchlorid; weisser Quecksilberpräzipitat
i cloramiduro di mercurio; sale di Alembroth; precipitato bianco
e cloroamiduro de mercurio; sal de Alembroth; precipitado blanco

154 aminomethylthiazole
Synthetic drug used in thyroid disorders by inhibiting the organic binding of iodine and biosynthesis of iodine-containing thyroid hormone.
f aminométhylthiazole
d Aminomethylthiazol
i amminometiltiazolo
e aminometiltiazolo

155 aminometradine
Non-mercurial diuretic agent especially used whenever mercurials prove unsuitable or harmful.
f aminométradine
d Aminometradin
i amminometradina
e aminometradina

156 aminophenazone
Analgesic and antipyretic agent used in symptomatic treatment of fever-causing diseases and as antirheumatic and antiphlogistic.
f aminophénazone
d Aminophenazon
i aminofenazone
e aminofenazona

157 aminophenylstibinic acid
An antimony compound which is used in the treatment of kala-azar (an infectious disease caused by a flagellate) in combination with urea.
f acide aminophénylstibinique
d Aminophenylstibinsäure
i acido aminofenilstibinico
e ácido aminofenilstibínico

158 aminophylline
Smooth-muscle relaxant and dilator of bronchial alveoli. It has a stimulating and tonic action on the cardiovascular system and a diuretic action. It is used as an antispasmodic agent for bronchial and cardiac asthma. Its action is short-lasting.
f aminophylline
d Aminophyllin
i aminofillina
e aminofilina

159 aminopicoline

A sympathomimetic and a cardiac, respiratory and circulatory analeptic agent. As it increases the blood pressure and stimulates the heart muscle without provoking tachycardia, it is generally used in the treatment of cardiac and respiratory asthenia, hypotension and respiratory depression.

f aminopicoline
d Aminopikolin
i aminopicolina
e aminopicolina

160 **aminopromazine**
A phenothiazine derivative capable of parasympathetic action without atropine-
-like effects. Used as a spasmolytic, particularly in post-operative visceral spastic pains, in colitis and dysmenorrhoea.

f aminopromazine
d Aminopromazin
i aminopromazina
e aminopromazina

161 **aminopropionic acid**
Open-chain amino acid occurring in various proteins. It may be considered the parent of the polypeptides. It acts as a glycogenic agent in diabetic animals. It actively participates in the metabolism although its presence in aliments is not indispensable as the organism is capable of synthesizing it from other compounds.

f alanine; acide aminopropionique
d Aminopropionsäure; Alanin
i acido amminopropionico; alanina
e ácido aminopropiónico; alanina

162 **aminopterin**
Synthetic derivative of folic acid capable of reversing the effects of the latter by competitive antagonism. Owing to its action on leukaemic bone, it has been used in acute leukaemia as it diminishes the leukaemic cells and brings about a reversal to normal of the peripheral white-cell count.

f aminoptérine
d Aminopterin
i amminopterina
e aminopterina

163 **aminosalicylic acid**
Chemotherapeutic agent characterized by a bacteriostatic action against the tuberculosis mycobacterium, generally used together with streptomycin.

f acide aminosalicylique
d Aminosalicylsäure
i acido aminosalicilico
e ácido aminosalicílico

* **aminosuccinic acid   s. aspartic acid**

164 **amiodarone**
Anti-anginal and antiarrhythmic agent whose principal action is that of supporting the work of the heart. It is used in the prophylaxis and the treatment of chronic anginous conditions borne of coronary disorders, i.e. ischaemic cardiopathies and infarctions.

f amiodarone
d Amiodaron
i amiodarone
e amiodarona

165 **amiphenazole**
Analeptic agent used in the treatment of acute respiratory disorders, specifically in respiratory depression caused by barbiturates and narcotics.

f amiphénazol
d Amiphenazol
i amifenazolo
e amifenazol

166 **amitriptyline**
Antidepressant agent characterized by an anticholinergic mechanism. It is used to correct behavioural disorders and acts directly on the central nervous system to fight mental depression.

f amitriptyline
d Amitriptylin
i amitriptilina
e amitriptilina

167 **amixetrine**
Chemotherapeutic agent capable of antihistaminic, spasmolytic, anticholinergic and antidepressant action. It is principally used in the therapy of gastroenteropathies characterized by inflammatory conditions, in gastritis, duodenitis, gastroduodenal ulcers, colitis, generally together with antispastics and antacids.

f amixétrine
d Amixetrin
i amixetrina
e amixetrina

168 **ammonium camphocarboxylate**
Analeptic agent acting on the heart and on the circulation. Though readily absorbed and of a short duration, the drug is used in the treatment of heart failure, circulation disorders, lypothymia, toxicosis, senescence etc.

f camphocarboxylate d'ammonium
d Ammoniumcamphocarboxylat
i ammonio canfocarbossilato
e canfocarboxilato de amonio

169 **ammonium carbonate**
An analeptic agent for the central nervous and the respiratory system used as an expectorant in chronic bronchitis, as well as in the treatment of depressive states. It has a mild diuretic action.

f carbonate d'ammonium
d Ammoniumcarbonat
i ammonio carbonato
e carbonato amónico

170 **ammonium chloride**
Analeptic agent acting on the central nervous system and on the respiratory system. It has an expectorant and transitory diuretic action. Used mainly to treat

poisoning accompanied by a depression
of the respiration, and neuropsychic disorders. Also used in acute and chronic
bronchitis.
f chlorure d'ammonium
d Ammoniumchlorid
i ammonio cloruro
e cloruro amónico

171 ammonium ferrous sulfate
Antianaemic agent used in the therapy of
acute and chronic hypoferric anaemia and
sideropenia in the course of pregnancy or
in states of undernourishment.
f sulfate de fer et d'ammonium
d Ferro-Ammoniumsulfat
i ammonio ferroso solfato
e sulfato ferroso amoniacal

172 amnion
A fragment of human amnion prepared according to Filatov's method: the tissue is
kept for seven or eight days in perfectly
aseptic vital-process-inhibiting conditions
with consequent formation of defence substances or biogenous stimulants. A non-
-specific opotherapeutic (relating to a
treatment of some disease by the extracts
of ductless glands) is used in tissular
therapy if a local action is required, as
it stimulates cellular metabolism, rigene-
rative and defence processes. Also recommended for the therapy of sores, varicose
ulcers, arteritis obliterans, pains following amputation etc.
f amnios
d Amnion
i amnio
e amnios

173 amobarbital
A barbiturate used as a sedative and
hypnotic as well as anticonvulsant. Also
used for pre- and post-operative sedation.
f amobarbitale
d Amobarbital
i amobarbital
e amobarbital

174 amodiaquine
Antimalarian agent, particularly active
against Plasmodium falciparum. Its action
is similar to that of Chloroquinine, q.v.
It is also used in the treatment of hepatic amoebiasis, giardiasis etc.
f amodiaquine
d Amodiaquin
i amodiaquina
e amodiaquina

175 amoxycillin
Semisynthetic penicillin having a wide
ampicillin-like spectrum comprising Gram-
-positive (with exception of penicillase-
-producing staphylococci) and Gram-nega
tive germs. It is used in the treatment
of bacterial infections of the respiratory
system, urinary, biliary and gastrointestinal infections, septicaemia, meningitis
and venereal diseases.

f amoxicilline
d Amoxycillin
i amossicillina
e amoxicilina

176 amphetamine
ß-phenylisopropylamine. A sympathicomimetic agent marked by a highly stimulant
effect on the central nervous system, in
particular the cerebral cortex and the
respiratory and vasomotor centres. It
also displays an alpha- and beta-adrener
gic activity. It relieves fatigue and appreciably increases mental awareness and
activity promoting a general feeling of
elation. Among its uses, of which there
is a great variety, behaviour disorders
in children, post-encephalitic parkinsonism, obesity, enuresis, nausea in the
course of pregnancy and hyperkinetic
states are noteworthy.
f amphétamine
d Amphetamin
i anfetamina
e anfetamina

177 amphotalide
Therapeutic agent of the phthalimide
group used against infections caused by
Schistosoma Haematobium, e.g. urinary
bilharziosis.
f amphotalide
d Amphotalid
i amfotalide
e amfotalida

178 amphotericin B
Antifungal used for skin infections. It
has no antibacterial, antiviral or antiprotozoal activity. Its side effects are
systemic and frequently severe, e.g.
fever, dyspepsia, anorexia, nausea,
vomiting and epigastric spasms, as well
as kidney damage.
f amphotéricine B
d Amphotericin B
i anfotericina B
e anfotericina B

179 ampicillin
Semisynthetic penicillin with a remarka-
bly broad spectrum of activity against
Gram-positive organisms and Gram-negative bacteria. It should be noted that its
activity is neutralized by penicillase-pro
ducing bacterial strains. It is used to
fight infections caused by susceptible
bacteria of the urinary, respiratory and
gastrointestinal tracts, as well as in
otitis and meningitis in children. It is
particularly used against typhoid fever.
f ampicilline
d Ampizillin
i ampicillina
e ampicilina

180 amsacrine
A cytotoxic agent whose activity is similar to that of doxorubicin, q.v.
f amsacrine

d Amsakrin
i amsacrina
e amsacrina

181 amygdalin
Glucose occurring in the oil of bitter almonds, in the kernels of apricot, peach and cherry and in the leaves of cherry-
-laurel and used in bronchial catarrh as an expectorant.
f amygdaline
d Amygdalin
i amigdalina
e amigdalina

182 amyleine
A local anaesthetic agent used in the past for spinal anaesthesia and presently as a component for drugs for tonsillitis, hoarseness, spastic conditions of the stomach and sometimes in ophthalmology.
f amyléine
d Amylein
i amileina
e amileina

183 amyl nitrite
Vasodilator mainly used as a spasmolytic agent in the acute phase of angina pectoris and asthma. Also used to treat epileptic fits, tetanus and other intoxications.
f nitrite d'amyle
d Amylnitrit
i isoamile nitrito
e nitrito de amilo

184 amylobarbitone
A barbiturate and general depressant of the central nervous system used in insom̲nia and in states of anxiety. As it is̲ metabolized by the liver it must be used with great caution in liver diseases.
f amylobarbitone
d Amylobarbiton
i amilobarbitone
e amilobarbitona

* **amylocaine s. amyleine**

185 amyl valerate
Apple oil. Synthetic flavouring used in medicine as a solvent for gallstones.
f valérianate d'amyle; essence de pomme
d Amylvalerianat; Apfelöl
i valerianato di amile; olio di mela
e valerianato de amilo; aceite de manzana

186 anaesthetic
Any drug inducing anaesthesia. The anaesthetics form a group of drugs of a varying chemical composition but having the common capacity to abolish, albeit by a differing mechanism, sensitivity, particularly dolorific sensitivity.
f anesthésique
d Anästhetikum
i anestetico
e anestésico

187 analeptic
Drug acting as a stimulant on the central nervous system. Specifically, any drug having a tonic action and capable of improving, rapidly but transiently, the cardio-circulatory conditions.
f analeptique
d Analeptikum
i analettico
e analéptico

188 analgesic
Any drug capable of eliminating or mitigating a pain. The most widely used analgesic agents are substances prevailingly acting at a mesencephalic and cephalic level and capable of depressing both the thermoregulating centres and the dolorific paths.
f analgésique; antalgique
d Analgetikum; schmerzlinderndes Mittel
i analgesico
e analgésico

189 anchoic acid
Or azelainic acid. An acid of the oxalic acid series used as a standard in the analysis of fats.
f acide anchoïque; acide azélaïque
d Azelainsäure
i acido azelaico
e ácido anchóico; ácido azeláico

190 ancitabine
An oncolytic therapeutic agent used in the treatment of myeloid leukaemia, lymphatic leukaemia and some carcinomata, in particular the adenocarcinoma.
f ancitabine
d Ancitabin
i ancitabina
e ancitabina

191 ancrod
The active principle that is obtained from the poison of a Malayan viper and is specifically active on the fibrinogen. In fact, it is an anticoagulant having an enzymatic activity on the fibrinogen which is transformed into an unstable fibrin that is rapidly eliminated from the circulation by fibrinolysis or phagocytosis. It is generally used in the therapy of arteriosclerosis obliterans, diabetic microangiopathy, vein thrombosis, pulmonary hypertension of thrombotic origin, embolias etc.
f ancrod
d Ankrod
i ancrod
e ancrod

192 anemone
Or pulsatilla, windflower. Genus of ranunculaceous plants growing in subarctic and temperate regions, which are alteratives and depressants and used in the treatment of inflammation.
f anémone pulsatile; coquerelle; fleur de Pâques

d Anemone; Kuhschelle
i anemone; pulsatilla comune
e anémona; pulsatila

193 anemonin
The active principle of the anemone; yellow crystals, insoluble in water, characterized by an acrid taste. It is an antispastic and sedative agent.
f anémonine; camphre d'anémone
d Anemonin; Anemonenkampfer
i anemonina; canfora di pulsatilla
e anemonina; alcanfor de anémona

194 anethole
An essential oil generally prescribed as a carminative and expectorant agent. It is a constituent of the essential oils of anise. Also known as anise camphor.
f anéthol
d Anethol; Aniskampfer
i anetolo
e anetol

195 aneurine
Or thiamine, vitamin B_1. A vitamin prescribed against cardiac failure, peripheral neuritis and Wernicke's encephalopathy. It is found in barm, eggs, cereals and fruit.
f aneurine; thiamine; vitamine B_1
d Aneurin; Thiamin; Vitamin B_1
i aneurina; tiamina; vitamina B_1
e aneurina; tiamina; vitamina B_1

196 angelic acid
Or angelicic acid. Unsaturated acrylic acid present with isomeric acids as esters in chamomile oil and angelica root. It is used in medicine as a diuretic and a diaphoretic.
f acide angélique
d Angeliksäure
i acido angelico
e ácido angélico

197 angelica oil
The oil distilled from angelica and used in medicine for its tonic, stimulant and diaphoretic properties.
f essence de racines d'angélique
d Angelikaöl
i olio di angelica; essenza della radice di angelica
e esencia de raíz de angélica

198 angiotensin
Peptide pressor agent having a biological mechanism. It is produced by the kidneys in inactive form and subsequently activated in the circulating blood. Being a potent hypertensive agent, it is used in states of shock and collapse.
f angiotensine
d Angiotensin
i angiotensina
e angiotensina

199 angiotensinamide
The drug commonly used to obtain the effects of the angiotensin, q.v.
f angiotensinamide
d Angiotensinamid
i angiotensinamide
e angiotensinamida

200 angostura
The bark of the Galipea officinalis, which contains a bitter substance, i.e. angosturin, and various bitter alkaloids. It is used as an aromatic bitter tonic.
f angusture
d Angostur
i angostura
e angostura

201 anileridine
Synthetic analgesic of the pethidine group marked by a central mechanism of action and used to relieve severe pains and in hepatic and renal colics. Also used to produce anaesthesia in the course of labour.
f aniléridine
d Anileridin
i anileridina
e anileridina

202 aniline
Colourless liquid derived from coal tar and used for the preparation of several drugs. It has a strong antipyretic action but, owing to its tendency to cause anilinism (a condition of poisoning) it has been replaced by derivatives, e.g. phenacetin and p-phenetidin.
f aniline
d Anilin
i anilina
e anilina

203 anisate
Any of the salts of anisic acid, which are characterized by antiseptic and antipyretic action.
f anisate
d Anisat; anissäures Salz
i anisato
e anisato

204 anise
The dried ripe fruit of Pimpinella anisum used in medicine for its carminative and expectorant properties due to the volatile oil present in it.
f anis vert
d Anis; süsser Kümmel
i anice; anice verde
e anis

205 anisic acid
Acid obtained from anethole, the phenolic ether contained in anise and fennel oils, and used in medicine as antiseptic, carminative and antipyretic.
f acide anisique
d Anissäure
i acido anisico
e ácido anísico

206 anisindione
Anticoagulant marked by an antivitamine-
-K action and acting by reducing the
prothrombotic rate. It is used in the ther
apy of thrombophlebitis, lung embolism,
in the prophylaxis of chronic arteritis
and traumatic lesions of blood vessels.
f anisindione
d Anisindion
i anisindione
e anisindiona

207 anorectic
Relating to any drug capable of causing
loss of appetite.
f anorexigène
d Appetitzügler; Anorecticum
i anoressigeno
e anoréxico

208 antazoline
Antihistaminic mainly used in ophthalmology for the treatment of conjunctivitis,
blepharitis, keratitis. Also used in ear,
nose and throat diseases and in dermatology.
f antazoline
d Antazolin
i antazolina
e antazolina

209 anthelmintic
Drug capable of destroying parasiting
worms or of causing their expulsion. The
term refers particularly to those drugs
which are active on intestinal parasites,
e.g. piperazine and its derivatives.
f vermifuge
d Anthelmintikum; Wurmmittel
i antielmintico
e vermífugo

210 anthemis oil
Or Roman chamomile oil. Volatile oil chief
ly containing n-butyl and isoamyl esters
of angelic acid and coumarine, as well
as flavone derivatives. It is used to
relieve spastic conditions in the digestive
tract and in dysmenorrhoea.
f essence de camomille romaine
d römisches Kamillenöl
i essenza di camomilla romana
e esencia de manzanilla romana

211 anthiolimine
An antiprotozoal agent used in intestinal
and urinary schistosomiasis, leishmaniasis, subacute lymphogranulomatosis in-
guinalis, kala-azar and Aleppo boil (tropical ulcer). Chemically, antimony lithium thiomalate.
f anthiolimine
d Anthiolimin
i antiolimina
e antiolimina

* **anti-anxiety agent   s. anxiolytic**

212 antibiotic
Substance produced by microorganisms
and capable of acting on other micro-
organisms by inhibiting their growth or
destroying them. Antibiotics also act on
living cells.
f antibiotique
d Antibiotikum
i antibiotico
e antibiótico

* **antidiuretic hormone   s. vasopressin**

213 antidote
Any drug capable of neutralizing the effects of a poison on the organism. There
are chemical and chemico-physical antidotes which act directly on the poison by
means of respectively chemical and chemico-physical reactions, and pharmacodynamic (or antagonist) antidotes which
determine a physiological reaction that is
opposed to that caused by the poison.
f antidote; contrepoison
d Antidot; Gegengift
i antidoto
e antídoto

214 antifibrinolysin
A substance which is regarded as being
produced as an immune allergic reaction
to infection by a streptococcus and as
being capable of preventing fibrinolysis
by streptococcal cultures.
f antifibrinolysine
d Antifibrinolysin
i antifibrinolisina
e antifibrinolisina

215 antihistamine
Antagonist of histamine. The action of
the antihistaminics antagonizes the phar-
maco-dynamic action of the histamine
either destroying it or by exerting a spe
cific antagonism in relation to its physiological effects. Antihistaminics find employment in the treatment of allergies,
yet they represent a mere symptomatic
rather than a causal remedy.
f antihistaminique
d Antihistamin...
i antistaminico
e antihistamínico

216 antimony
Chemical element (Sb) whose organic compounds find a useful employment in medicine for the treatment of protozoal diseases, e.g. potassium or sodium antimonyl
tartrate, antimony thioglycollamide, antimony sodium thioglycollate and stibophen
which contain trivalent antimony have
been usefully employed in the treatment
of schistosomiasis, filariasis, leishmaniasis etc. but require very great precaution as they can produce violent vomit
ing, a dry cough leading to pneumonia,
arthritis and various pains. The compounds of pentavalent antimony are less
toxic and have proved very useful in the
treatment of leishmaniasis and trypanosomiasis.

f antimoine
d Antimon
i antimonio
e antimonio

217 antimony aniline tartrate
Yellow antimony compound used against
trypanosomes. It is less toxic that tartar
emetic.
f tartrate d'aniline-antimoine
d Anilin-Antimontartrat
i antimonio anilina tartrato
e tartrato de anilina-antimonio

218 antimony chloride
Or butter of antimony. Antimony compound
forming a fuming solution which can be
used as a caustic.
f trichlorure d'antimoine; beurre d'antimoine
d Antimontrichlorid ; Antimonbutter
i tricloruro antimonico; burro di antimonio
e antimonio tricloruro; manteca de antimonio

219 antimony potassium tartrate
Water-soluble antimony compound having
a very irritant effect on the cells sur-
rounding cutaneous glands. It is used
orally to stimulate the vomiting centre in
the medulla.
f tartre émétique
d Brechweinstein
i tartaro emetico; tartaro stibiato
e tártaro emético; tártaro estibiado

220 antipyretic
Relating to any drug acting against fever, e.g. derivatives of salicylic acid,
quinine etc. The mechanism of their action is very complex and must be seen in
several directions: the lowering of the
production of heat through their action
on the thermoregulating nervous centres
or the decrease of the processes of combustion of the tissues, or again through
the heightening of the caloric dispersion
caused by the increase of the peripheral
circulation and perspiration.
f antipirétique; fébrifuge
d Antipyretikum
i antipiretico
e antipirético

221 antiseptic
Any of a group of organic and inorganic
substances capable of inhibiting the development and the multiplication of the
microorganisms.
f antiseptique
d Antiseptikum
i antisettico
e antiséptico

222 antispasmodic
Any drug which, by its action on the
central nervous system or on the peri-
phery, eliminates the spasms of the musculature, e.g. papaverine, benzyl benzoate as well as some sympatholytic and

parasympatholytic agents. Also called
spasmolytic.
f antispasmodique; spasmolytique
d Antispasmodikum; Spasmolytikum
i antispasmodico; antispastico; spasmolitico
e antiespasmódico; espasmolítico

223 antistreptolysin
Any agent or substance capable of inhibit
ing the action of the haemolytic strepto-
lysin.
f antistreptolysine
d Antistreptolysin
i antistreptolisina
e antistreptolisina

224 antistrumous
Term relating to all those drugs and
agents which are effectively used in the
treatment of tuberculous adenitis and of
goitre.
f antistrumeux
d antistrumös; strumalindernd
i antistrumoso
e antiestrumoso

225 anxiolytic
Any agent capable of dissolving or reliev
ing anxiety, e.g. mental agitation. A
good anxiolytic action is offered by barbiturates, some tranquillizers and neuroleptic drugs.
f anxiolytique
d angstlösendes Mittel
i ansioliti co
e anxiolítico

226 apomorphine hydrochloride
An emetic agent used parenterically for
the treatment of acute intoxication.
f chlorhydrate d'apomorphine
d Apomorphinhydrochlorid
i apomorfina cloridrato
e clorhidrato de apomorfina

227 aprindine
Antiarrhythmic agent used in the treatment of extrasystoles, ventricular and
10 · D2. supraventricular tachycardia,
both idiomatic or due to cardiac disorder,
and arrhythmias caused by digitalis
intoxication.
f aprindine
d Aprindin
i aprindina
e aprindina

228 aprobarbital
A barbiturate used for sedation and sleep
induction in narcosis prior to a surgical
operation. It is also used in the treatment of insomnia.
f aprobarbital
d Aprobarbital
i aprobarbitale
e aprobarbital

229 aprotinin
An extract of bovine lung or parotid
gland used for the therapy of pancreati-

tis and in states of shock.
f aprotinine
d Aprotinin
i aprotinina
e aprotinina

230 arginine
Essential amino acid protecting the liver.
It is used in the treatment of liver diseases and haepatic coma, as well as in
acute pancreatitis. It also finds employment in the physiological therapy of
oligospermia and male impotence.
f arginine
d Arginin
i arginina
e arginina

231 argyrol
A combination of silver and proteinic
substances containing 19 to 23% of metallic silver. It is an antiseptic and astrin
gent agent mainly used in ophthalmology
for the treatment of mucopurulent conjunc
tivitis, as well as for the prophylaxis of
neonatal ophthalmias.
f argyrol
d Argyrol
i argirolo
e argirolo

232 aristic acid
Or aristolochic acid, aristidic acid. An
acid derived from the roots of various
plants and used as an activator of phago
citosis in the treatment of chronic purulent infections.
f acide aristolochique
d Aristolochiasäure
i acido aristolochico
e ácido aristolóquico

233 arnica
Genus of flowering plants, whose dried
flower heads and rhizome roots may be
used in the form of tincture for the treat
ment of sprains and bruises. Also known
as wolfsbane or mountain tobacco.
f arnica; herbe aux chutes; plantain des
Alpes
d Arnika; Bergwohlverleih; Wundkraut
i arnica; piantaggine di montagna
e árnica; tabaco de montaña

234 arnicina
The glycoside that is contained in the
flower heads and in the rhizome of arnica (arnica montana).
f arnicine
d Arnicin
i arnicina
e arnicina

235 aromatic
Having the nature, smell and taste of an
aroma. Specifically, relating to those
chemical compounds whose atoms (mostly
carbon atoms) are arranged in rings, i.e
have a benzene or quinonoid ring thus
differing from an aliphatic or open-chain

compound.
f aromatique
d aromatisch
i aromatico
e aromático

236 aromatic spirit of ammonia
A solution of ammonia and ammonium carbonate in alcohol and distilled water,
which is used as a stimulant, carminative and antacid.
f alcoolat aromatique ammoniacal
d aromatischer Ammoniumspiritus
i alcoolato ammoniacale aromatico
e espíritu amoniacal aromático

237 aromatic sulfuric acid
A mixture of sulfuric acid, tincture of
ginger, cinnamon oil or spirit and cinnamon used in the past as a tonic and
astringent.
f élixir acide aromatique; teinture aromatique
d saure aromatische Tinktur
i acido solforico aromatizzato
e ácido sulfúrico aromático; tintura aromática ácida

238 aromatic water
Saturated aqueous solution of a volatile
substance obtained by distillation or by
the dissolution of the substance itself.
f eau aromatique; hydrolat
d Aromatisches Wasser; Hydrolat
i acqua aromatica; idrolato
e agua aromática; hidrolato

239 arsanilic acid
p-aminobenzenarsonic acid obtained by
the reaction of the arsenic with aniline.
It is a white, crystalline, ether-soluble
powder used for the preparation of the
arsanilates (the salts of the acid) and
as a starting product for the manufacture
of medicinal arsenical compounds, e.g.
salvarsan.
f acide anilarsinique; acide atoxylique
d Arsanilsäure
i acido arsanilico
e ácido arsanílico

240 ascorbic acid
Vitamin C. The only natural antiscorbutic
product, widely diffused in both the animal and the vegetable kingdoms. It is an
essential factor in the formation of collagen and intercellular cementing substances and thus in the development of
bone, cartilage and dental tissue. It further promotes the assimilation of folic
acid and absorption of iron, as well as
the formation of haemoglobin and the full
development of the erythrocytes. In both
natural and synthetic forms, it is essential to combat various diseases and/or
disorders, e.g. hypovitaminosis, bone
fractures, haemorrhagic diatesis, wounds
etc. and is recommended in pregnancy,
lactation etc.
f acide ascorbique; vitamine C

d Ascorbinsäure; Vitamin C
i acido ascorbico; vitamina C
e ácido ascórbico, vitamina C

241 asparagine
Substance originally isolated from asparagus and subsequently from several other plants. Chemically, the monoamide of the aspartic acid. It functions as a reserve of ammonia in plants and participates in the constitution of many proteins.
f asparagine
d Asparagin
i asparagina
e asparagina

242 asparagus officinalis
Edible asparagine-containing vegetable used in medicine as a diuretic.
f asperge; asparagus officinalis
d Spargel; Asparagus officinalis
i asparago; asparagus officinalis
e espárrago; asparagus officinalis

243 aspartic acid
Amino acid capable of regulating the metabolism of nitrogen and used in the treatment of hyperammoniaemia and poisoning due to ammonia. It is also used in cardiac insufficiency. Also known as asparaginic acid, asparagic acid and asparamic acid.
f acide aspartique
d Asparaginsäure
i acido aspartico
e ácido aspártico

244 aspidinol
A dihydroxy alcohol obtained from the extract of a male fern (Dryopteris) and used as taenicide.
f aspidinol
d Aspidinol
i aspidinolo
e aspidinol

* **aspirin   s. acetylsalicylic acid**

245 astemizole
Antihistaminic agent used for sedation.
f astémizole
d Astemizol
i astemizolo
e astemizolo

246 astringent
Relating to any drug capable of arresting a discharge, e.g. diarrhoea.
f astringent
d stopfendes Mittel; astringent
i astringente
e astringente

247 atenolol
Beta-adrenergic blocking drug used in the treatment of hypertension and renal hypertension. It has a limited cardioselectivity.
f atenolole
d Atenolol

i atenololo
e atenololo

248 atropine
Alkaloid extracted from the roots of Atropa belladonna where it is present with the laevorotatory isomer, hyoscyamine. Atropine has an inhibiting action on the parasympathetic vegetative nervous system and thus determines a dilatation of the pupils, a decrease or stoppage of the salivary, gastric, bronchial and sudoral secretions, an increase of the pulse frequency and, in cases of intoxication, some physical disorders.
f atropine
d Atropin
i atropina
e atropina

249 atropine methobromide
Or atropine methylbromide. Parasympatholytic agent whose action is similar to that of the atropine sulphate. It is chiefly used to dilate the pupil in the examination of the eye.
f atropine méthylbromure
d Atropinmethylbromid
i atropina metilbromuro
e bromuro de metilbromuro

250 atropine sulphate
A sulphate of the alkaloid atropine. A natural cholinergic agent having both central and peripheral activity. It is a cholinergic receptor blocking agent displaying a considerable antispastic activity on smooth muscles, antisecretive and antivagal activity, as well as cycloplegic and mydriatic activity. Atropine sulphate controls gastric motility and reduces salivary, intestinal and bronchial secretions. It also relieves muscular rigidity and tremor. It is chiefly used in the symptomatic treatment of hyperkinetic states in parkinsonism, paralysis agitans, cardiac spasms, hyperkinesis of the gastro-intestinal tract and convulsions following organo-phosphoric poisoning.
f atropine sulfate
d Atropinsulfat
i atropina solfato
e sulfato de atropina

251 attapulgite
A form of magnesium or aluminium silicate which is used as an adsorbent agent in acute poisoning, diarrhoea and deodorants.
f attapulgite
d Attapulgit
i argilla del Texas
e atapulguita

252 auranofin
Active gold compound used for the treatment of rheumatoid arthritis, chronic polyarthritis and infantile polyarthritis.
f auranofine

d Auranofin
i auranofina
e auranofina

253 auric sulphide
A colloidal compound used in antirheumat
ic chrysotherapy, in particular for the
treatment of chronic polyarthritis, spon-
dyloarthritis, arthropathic psoriasis etc.
f sulfure aurique
d Goldsulfid
i solfuro aurico
e sulfuro áurico

254 aurothioglucose
A water-soluble derivative of gold used
in the treatment of chronic polyarthritis,
juvenile chronic polyarthritis and in
psoriatic arthritis.
f aurothioglucose
d Aurothioglukose
i aurotioglucosio
e aurotioglucosio

255 aurothiomalate sodium
Active gold compound used in the treatment of rheumatoid arthritis. Jaundice
and kidney disorders are frequently identified following intramuscular injection of
the drug.
f aurothiomalate sodium
d Aurothiomalatsodium
i aurotiomalato sodio
e aurotiomalato sodio

256 azacyclonol
A psychotropic drug which can be chemically defined as diphenyl-piperidine-me-
thanol. It is marked by an antilysergic
action and has been used, with varying
results, in the treatment of schizophrenia,
hallucinatory states and, more generally,
psychotic disorders.
f azacyclonol
d Azacyclonol
i azaciclonolo
e azaciclonolo

257 azadirachta
The bark of Melia azadirachta (cultivated
in India) which contains a bitter alkaloid, the margosine, and can be used in
the form of an infusion as a tonic.
f azedarach; margousier
d Zedarach; Margosa
i acedarach; margosa
e acedarac; margosa

258 azalomycin
Antibiotic active against Gram-positive
bacteria, antimycotic and antiprotozoan,
used in the treatment of infections caused
by Trichomonas vaginalis and Candida
albicans.
f azalomycine
d Azalomycin
i azalomicina
e azalomicina

259 azapetin

Vasodilator and adrenergic blocking a-
gent used in the treatment of peripheral
vascular disorders caused by diabetes,
endocarditis and, above all, atherosclerosis. The drug is also used against
varicose ulcers, Sudeck's atrophy (an
acute atrophy of the bone appearing as
an area of osteoporosis), chilblains etc.
f azapétine
d Azapetin
i azapetina
e azapetina

260 azapropazone
Non-steroid anti-inflammatory and analgesic agent used in the treatment of acute
and chronic phlogistic processes of rheumatic and traumatic origin.
f azapropazone
d Azapropazon
i azapropazone
e azapropazona

261 azaribine
Systemic antipsoriatic agent used in symp
tomatic treatment of psoriasis that is al-
ready widespread and does not respond to
local treatment. It is also a cytostatic
agent used in the treatment of leukaemia
and in some neoplasms which do not respond to other cytostatics.
f azaribine
d Azaribin
i azaribina
e azaribina

262 azatadine
Antihistamine and antiserotonin agent
used in the treatment of allergic itching,
as well as allergies affecting the respiratory system, occupational diseases,
itching at the anal orifice and of the
vulva, and in the prophylaxis of anaphylactic reactions.
f azatadine
d Azatadin
i azatadina
e azatadina

263 azathioprine
A derivative of mercaptopurine, into
which the drug reverts in the body. It
is used in the treatment of leukaemia. It
inhibits the synthesis of nucleic acids by
interfering with the utilization of purine.
It is also used in the therapy of reject
in transplantation.
f azathioprine
d Azathioprin
i azatioprina
e azatioprina

264 azidamfenicol
Broad-spectrum antibiotic acting on Gram-
-positive and Gram-negative germs and
on some large viruses. It is used for the
treatment of eye infections, e.g. styes,
blepharitis, chronic conjunctivitis, kerati
tis and, more generally, lesions.
f azidamfénicol

d Azidamfenikol
i azidamfenicolo
e azidamfenicolo

265 azidocillin

Or azlocillin. Broad-spectrum antibiotic -a semisynthetic penicillin- whose action is effective against penicillin-sensitive Gram-positive and Gram-negative bacteria. It is thus used both to combat Haemophilus influenzae and infections of the gastrointestinal system, urinary and respiratory tracts and of the skin. Should an allergic diathesis be observed, great caution is required in the administration of the drug.

f azidocilline
d Azidocillin
i azidocillina
e azidocilina

266 azintamide

A true choleretic whose action brings about an increase of the amount of excreted bilirubin which corresponds to that of the volume of duodenal juice. The effects of the drug are very rapid, generally 20 minutes, and last for about one hour. It is also used as a substitute therapy when bile secretion is somehow inhibited.

f azintamide
d Azintamid
i azintamide
e azintamida

267 azostamine

Histamine azoprotein. A histamine-protein antigen causing anti-histaminic immunization. It is used in the treatment of various skin diseases, specifically eczema, urticaria, as well as in vasomotor rhinitis, bronchial asthma, heat and cold allergy etc.

f azostamine
d Azostamin
i azostamina
e azostamina

B

268 bacampicillin
Ampicillin derivative. Broad-spectrum antibiotic active against Gram-positive and Gram-negative germs but not against penicillase-producing bacteria. It is easily absorbed in the gastro-intestinal tract and metabolized to ampicillin (q.v.) without any delay. It is used to treat infections of the respiratory system, sinusitis, otitis and also infections of the genitourinary tract.
f bacampicilline
d Bacampicillin
i bacampicillina
e bacampicilina

269 bacitracin
Antibiotic produced by a bacterium resembling the Bacillus subtilis that is often present in infected wounds. It acts against a large number of Gram-positive organisms, e.g. haemolytic streptococci, and a few Gram-negative ones. It is used to combat spirochetae and protozoa. Its mechanism of action is similar to that of penicillin. Skin infections are often treated with the drug.
f bacitracine
d Bacitracin
i bacitracina
e bacitracina

270 baclofen
A muscle relaxant acting on the spinal cord and inhibiting the monosynaptic and polysynaptic transmission of reflexes. It is used in the treatment of spasticity of skeletal muscle, especially in multiple sclerosis, as well as in muscle spasticity occurring in diseases affecting the spinal cord and in cerebral disorders.
f baclofène
d Baclofen
i baclofene
e baclofeno

271 bactericide
Any agent capable of destroying bacteria, in particular pathogenic bacteria. The bactericide activity of the various agents displays a different effectiveness on the various types of bacteria.
f substance bactéricide
d Bakterizid
i battericida
e bactericida

272 bacteriocine
Any of naturally-occurring substances displaying an antimicrobial activity particularly against other strains of the same species but also against members of a different genus.
f bactériocine
d Bakteriozin; Bacteriocin
i batteriocina
e bacteriocina

273 bacteriolysin
Term denoting an immunoglobulin capable of acting with non-specific and thermolabile complement and of killing bacteria by lysis.
f bactériolysine
d Bakteriolysin
i batteriolisina
e bacteriolisina

274 bacteriolysis
The fragmentation into granules and the dissolution of the bacterial body that is commonly obtained by using a mixture of fresh antiserum and a bacterial suspension. It may be carried out in vivo and in vitro.
f bactériolyse
d Bakteriolyse
i batteriolisi
e bacteriólisis

275 badian
Or star anise, stellate anise. The ripe and dried fruit of the Illicium verum Hook of the Magnoliaceae family used to treat gastro-intestinal disorders.
f badiane; anis étoilé
d Badian; Sternanis
i badiana; anice stellato
e badiana; anís estellado

* **baking soda   s. sodium bicarbonate**

276 balsam of Tolu
Antiseptic and anti-inflammatory agent used externally for skin lesions, in particular chilblains, and internally as an expectorant in bronchitis.
f baume de Tolu
d Tolubalsam
i balsamo del tolù
e bálsamo del tolù

277 bamethan
A vasodilator causing a direct myolytic

action on the smooth vasal muscles, an action which is intensified by the nicotin component in the drug, and an indirect action consisting of the stimulation of ß-adrenergic terminations. It is used in disorders of the peripheral circulation, vasospasticity and cerebrovascular insufficiency, as well as in post-traumatic syndromes.

f baméthan
d Bamethan
i bametano
e bametano

278 bamifylline
Spasmolytic agent characterized by myotropic action. It is a derivative of theophylline and its main properties are the capacity of acting as a bronchodilator and a bronchospasmolytic and that of so doing without affecting the central nervous system. It is mainly used in respiratory and cardiorespiratory insufficiencies, coronary disorders, chronic bronchitis and asthma.

f bamifylline
d Bamifyllin
i bamifillina
e bamifilina

279 bamipine
An antihistamine and anticholinergic agent. Also an antispasmodic, sedative and anaesthetic. The drug is used in the treatment of allergies and of itching rashes.

f bamipine
d Bamipin
i bamipina
e bamipina

280 baptisia
Or wild indigo. Genus of plants of the Leguminosae family whose dried root contains an alkaloid, glycosides and a cathartic substance that is used, in the form of a decoction, as a laxative.

f baptisie sauvage; indigotier sauvage
d Baptisia; wilder Indigo
i baptisia
e baptisia; indigo silvestre

281 barberry
The dried bark of Berberis vulgaris which is indigenous to the British isles. It contains alkaloids, specifically the berberine, i.e. a bitter tonic.

f berbéris; épine vinette; chivafou
d Berberitze; Sauerdorn
i berberis; crespino; trespino; uva vinetta
e agracero; bérbero; arleta

282 barbexaclone
An anticonvulsant agent mainly used in the treatment of epilepsy independently of the occurrence of crises.

f barbexaclone
d Barbexaclon
i barbesaclone
e barbesaclona

283 barbital
Sedative and hypnotic agent chiefly acting at the level of the thalamus and the reticular formation and thus affecting the transmission of impulses to the cortex. It is used in the treatment of insomnia, epilepsy, some forms of intoxication and in preliminary narcosis. Together with antipyretics it acts as an analgesic. The term is used to denote barbitone.

f barbital
d Barbital
i barbitale
e barbital

284 barbital sodium
Practically immediate yet prolongued-action sedative and hypnotic. Its slow excretion may cause accumulation. It is used in the treatment of epilepsy and of analeptic poisoning. It finds also use together with antipyretics and analgesics.

f barbital sodique
d Natriumbarbital
i barbitale sodico; barbitone sodico
e barbital sódico; barbitona sódica

285 barbiturate
The salt of barbituric acid. Any of a large number of alphyl radicals, e.g. barbitone, phenobarbitone, thiopentone etc. used as a strong sedatives and hypnotics in insomnia, hyperthyroidism, hysteria and more generally in all those states in which a depression of the nervous system is indicated.

f barbiturate
d Barbiturat
i barbiturato
e barbiturato

286 barium
Element of the alkaline earth group, whose salt is very poisonous. The barium ion increases the tone of plain muscles and cardiac contractions. It also lengthens the contraction of voluntary muscle.

f barium
d Barium
i bario
e bario

287 barium sulphate
A contrast medium used in the radiography of the gastrointestinal tract, in particular the stomach. It is also used for bronchographies in the form of a suspension with carboxymethylcellulose and mucin.

f sulfate de barium
d Bariumsulfat
i bario solfato
e sulfato de bario

288 batroxobin
A biologic coagulant agent involving thrombin and thromboplastin in its action. It is used in the treatment of many haemorrhages, haemophilia, delayed coagulation etc. It is also used occasionally

in dental extractions whenever the danger of haemorrhage is suspected.
f batroxobine
d Batroxobin
i batroxobina
e batroxobina

289 beclamide
N-benzyl-ß-chloropropionamide. An anticonvulsant agent used in the treatment of epilepsy as well as in various states of psychomotor disorders, cramps, enuresis etc.
f béclamide
d Beklamid
i beclamide
e beclamida

290 beclomethasone
A corticosteroid used topically in the treatment of skin diseases and allergic diseases. Also used as an aerosol in the treatment of bronchial asthma. The drug should not be applied on large body surfaces covered by occlusive dressings in order to avoid a massive absorption which may cause systemic side effects.
f béclométasone
d Beklomethason
i beclometasone
e beclometasona

291 behenic acid
A glyceride present in some vegetable oils. It can be obtained by catalytic reduction of the erucic acid. Various salts used in medicine are prepared from this acid.
f acide béhénique
d Behensäure
i acido beenico
e ácido behénico

292 belladonna
The dried leaves or roots of Atropa bella donna (the deadly nightshade) or Atropa acuminata containing the alkaloids L-hyoscyamine and L-hyoscine, the first of which racemizes rapidly and yields the mixture called atropine. The belladonna leaves produce a parasympatholytic, antispasmodic, antisecretory and mydriatic action and is used in the treatment of gastrointestinal disorders, hypersecretory phenomena and poisoning due to toxic cholinergic substances. The roots produce a parasympatholytic agent acting on the central nervous system, the heart, smooth muscles, metabolism etc. It is used for the symptomatic treatment of spastic disorders of the gastrointestinal tract and in postencephalitic parkinsonism.
f belladone
d Dollkraut ; Tollkirsche ; Atropa belladonna
i belladonna
e belladona

293 bemegride
Analeptic agent, chemically defined as ß-ethyl-ß-methylglutarimide, characterized by a central stimulating action. It is a strong antagonist in barbiturate poisoning as it increases the muscular tone, improves reflex activity as well as the frequency and amplitude of the respiration. It is also used to accelerate the reawakening after narcosis.
f bémégride
d Bemegrid
i bemegride
e bemegrida

294 bemetizide
A benzothiodiazine preparation and diuretic agent.
f bémétizide
d Bemetizid
i bemetizide
e bemetizida

295 benactyzine
2-diethylaminoethyl benzilate. A tranquillizer used in treatment of psychoneuroses for its antispasmodic and an anticholinergic action. It also proves useful for the relief of states of anxiety and excessive emotivity.
f bénactyzine
d Benaktizin
i benactizina
e benactizina

296 benaprizine
2-(ethylpropylamino)benzilate. A parasympatholytic agent used in the treatment of idiopathic and postencephalitic parkinsonism and in extrapyramidal syndromes (in neurolepsy). The drug is generally used in combination with levodopa.
f bénaprizine
d Benaprizin
i benaprizina
e benaprizina

297 bencyclane
Therapeutic agent capable of arresting a smooth-muscle spasm by acting on the cells of arteries and arterioles. The drug has the added value of inhibiting platelet aggregation. It is recommended for the treatment of cerebral ischaemia as well as of peripheral ischaemic conditions, circulatory disorders affecting the limbs in arteriosclerosis, in acrocyanosis, diabete-related angiopathies, Raynaud's disease (cyanosis of the extremities accompanied by intermittent pallor and brought about by a degree of coldness which would not affect a healthy individual), arteriopathy obliterans etc.
f bencyclane
d Benzyklan
i benciclano
e benciclano

298 bendazac
Topically administered antiphlogistic agent having no antiseptic, anaesthetic and anticholinergic action. It finds employment in various types of dermatosis

and ulcerations.
f bendazac
d Bendazac
i bendazac
e bendazac

299 bendazol
Vasodilator, hypotensive and spasmolytic agent characterized by its capacity to stimulate the reticuloendothelial system, as well as lymphopoiesis and antibody formation. Its action also promotes the synaptic transmission of the nervous impulse. It is used both to prevent and to treat the conditions generally following myocardial infarction, coronaropathies, neuritis, the consequences of polio myelitis etc.
f bendazol
d Bendazol
i bendazolo
e bendazolo

300 bendroflumethiazide
Very effective thiazide diuretic capable of inhibiting the reabsorption of the electrolytes in the proximal renal tubules and consequently promoting their excretion. It is used in the treatment of cardiac oedema, anasarca, (particularly in cases of ascites caused by an excess of water in the blood) and in gravidic intoxication.
f bendrofluméthiazide
d Bendroflumethiazid
i bendroflumetiazide
e bendroflumetiazida

301 benethamine penicillin
Long-acting antibiotic (a derivative of benzylpenicillin, q.v.) used in the treatment of infections caused by penicillin-sensitive microorganisms so as to avoid complications caused by infectious diseases. It is also used to prevent relapses in acute articular rheumatism.
f bénéthamine pénicilline
d Benethaminpenicillin
i benetamina penicillina
e benetamina penicilina

302 benfotiamine
Therapeutic agent derived from thiamine and used in cases of deficiency of vitamin B_1. It is easily absorbed by the intestine and acts on the central nervous system. Used also as an analgesic against neuralgias, sciatica etc.
f benfotiamine
d Benfotiamin
i benfotiamina
e benfotiamina

303 benfurodil hemisuccinate
Vasodilator used in the treatment of degenerative diabetic arteriopathies, of thromboangiitis, peripheral arteritis, angina pectoris, inflammation of the coronary arteries, pathological conditions following a myocardial infarction and

cerebral arteriopathy. It is frequently used in combination with a cardiotonic and/or anticoagulant.
f benfurodil hémisuccinate
d Benfurodilhämisuccinat
i benfurodil emisuccinato
e hemisuccinato de benfurodil

304 benmoxin
An antidepressant and inhibitor of monoamine oxidase used in the therapy of chronic asthenia accompanied by depression, mild senile dementia or presenile depression and more generally to combat states of psychoneurotic depression.
f benmoxine
d Benmoxin
i benmoxina
e benmoxina

305 benorylate
4-acetamidophenyl 0-acetylsalicylate. Analgesic, antipyretic and antiinflammatory agent used in cases of rheumatoid arthritis, osteoarthritis and muscular pains.
f bénorilate
d Benorylat
i benorilato
e benorilato

306 benperazine bromide
A muscle relaxant used in the treatment of muscular spasms, particularly in orthopaedics, neurology and in psychiatric disorders.
f bromure de benpérazine
d Benperazinbromid
i bromuro di benperazina
e bromuro de benperazina

307 benperidol
Butyrophenone derivative. A tranquillizer characterized by a depressant action on the central nervous system and used in the treatment of various psychomotor conditions, e.g. delirium, manic agitation, extreme anxiety, mild senile dementia, aggressivity etc.
f benpéridol
d Benperidol
i benperidolo
e benperidolo

308 benproperine
Therapeutic agent capable of relieving or checking cough. It is particularly used in cases of dry cough, pertussis (whooping cough), pharyngitis, laryngitis and tracheitis.
f benpropérine
d Benproperin
i benproperina
e benproperina

309 benserazide
Specific inhibitor of the peripheral decarboxylase, capable of checking the decarboxylation of levodopa at the extracerebral plane and thus of enabling the latter to reach the central nervous system

in adequate quantities. It is used in the treatment of the Parkinson's syndrome.
f bensérazide
d Benserazid
i benserazide
e benserazida

310 bentonite
Colloidal hydrated aluminium silicate used as an adsorbent in the treatment of poisoning. Also used as suspending agent in pharmaceutical preparations.
f bentonite
d Bentonit
i bentonite
e bentonita

311 benzalamide
A cholesterol reducer and an antihyperlipidaemic used in the treatment of cerebral-circulation, coronary and peripheral vascular disorders. The drug finds also use in cases of hyperlipaemia, cholesterosis of the liver and spleen.
f benzalamide
d Benzalamid
i benzalamide
e benzalamida

312 benzaldehyde
A constituent of oil of bitter almond used in the manufacture of several drugs.
f benzaldéhyde; essence d'amandes amères
d Benzaldehyd; Bittermandelöl
i benzaldeide; essenza di mandorle amare
e benzaldehido; esencia de almendras amargas

313 benzalkonium chloride
A mixture of alkyl chlorides of dimethylbenzyl ammonium. Being a cationic detergent with bactericidal properties it is used for irrigation of the bladder and more generally for the disinfection of surgical instruments.
f chlorure de benzalkonium
d Benzalkoniumchlorid
i benzalconio cloruro
e cloruro de benzalconio

314 benzamine
Beta-eucaine. Synthetic substitute for cocaine to which it resembles in its chemical structure. It is practically no longer used in medicine.
f eucaïne; benzamine
d Eucain; Benzamin
i eucaina; benzammina
e eucaina; benzamina

315 benzarone
Vascular tonic capable of regulating the permeability and checking the fragility of capillary vessels and thus used in the treatment of oedemas, varicose veins and haemorrhoids in all those forms which are not related to organic cardiopathies and inflammation of the blood vessels.
f benzarone
d Benzaron

i benzarone
e benzarona

316 benzathine benzylpenicillin
Or benzathine penicillin. Penicillin generally active on all sensitive germs. It is used for the treatment of infections caused by susceptible organisms and above all to combat meningitis, pneumonia (by Diplococcus), septicaemia, syphilis etc.
f benzathine benzylpénicilline
d Benzathinbenzylpenicillin
i benzatina benzilpenicillina
e benzatina benzilpenicilina

317 benzatropine
An atropine derivative used in the treatment of idiopathic and postencephalitic parkinsonism, as well as of other parkinsonism-related syndromes, in particular when caused by neuroleptic agents.
f benzatropine
d Benzatropin
i benzatropina
e benzatropina

318 benzbromarone
A hypouricacidaemic (relating to a subnormal amount of uric acid in the blood) agent capable of promoting the clearance and the secretion of uric acid.
f benzbromarone
d Benzbromaron
i benzbromarone
e benzbromarona

319 benzene
Colourless, volatile, toxic aromatic hydrocarbon generally obtained from petroleum fractions by catalytic hydrogenation and regarded as a raw material of various pharmaceutical products. It has an antiseptic action and may be used, though very rarely, as gastro-intestinal disinfectant. Also known as benzol.
f benzène; benzol
d Benzol
i benzene; benzolo
e benceno

320 benzestrol
Non-steroid oestrogen used for the treatment of menopausal syndromes, as well as in senile vaginitis. It is also sometimes used to inhibit lactation.
f benzestrol
d Benzestrol
i benzestrolo
e benzestrolo

321 benzethidine
Ethyl 1 - (2 - benzyloxyethyl) - phenilpiperidine - 4 - carboxylate. A narcotic and analgesic agent.
f benzéthidine
d Benzethidin
i benzetidina
e bencetidina

322 benzethonium chloride

Antiseptic and disinfectant agent used in medicine for the treatment of blepharitis, conjunctivitis, keratitis etc. It may be used by itself or in combination with a corticosteroid.

f chlorure de benzéthonium
d Benzethoniumchlorid
i benzetonio cloruro
e cloruro de bencetonio

323 benzidine

Diaminodiphenyl. Aminic derivative of the diphenyl obtained from the nitrobenzene by reduction in alkaline medium and subsequent benzidinic transposition or by electrolytic reduction. Used in medicine because it forms a very sensitive test for blood. Also employed in the detection of organic substrates of enzymes having a peroxidase action.

f benzidine; diaminophényle
d Benzidin; Diphenylamin
i benzidina; diamminodifenile
e bencidina; difenildiamina

324 benzilonium

A parasympatholytic agent characterized by peripheral and toxic effects. For its action, see benzilonium bromide.

f benzilonium
d Benzilonium
i benzilonio
e bencilonio

325 benzilonium bromide

A parasympatholytic agent used in the treatment of peptic ulcers. Also used as antacid and inhibitor of the gastric secretion. It relieves gastrointestinal spasms in gastric hyperacidity.

f bromure de benzilonium
d Benziloniumbromid
i benzilonio bromuro
e bromuro de bencilonio

326 benziodarone

Spasmolytic and anti-anginous agent capable of improving cardiac metabolism and protecting against the danger of infarction.

f benziodarone
d Benziodaron
i benziodarone
e benziodarona

327 benzocaine

Ethylic ester of the p-aminobenzoic acid. A local anaesthetic drug which is not irritant and does not display any systemic toxicity. It is effective against pains caused by gastric disorders and to prevent nausea and vomiting in the course of dental operations.

f benzocaïne
d Benzocain
i benzocaina
e benzocaina

328 benzoctamine

A tranquillizer and anxiolytic used in the treatment of anxiety states, aggressiveness and more generally behaviour disorders. The drug promotes socialization and smooths inner unrest without causing any neurovegetative disorder.

f benzoctamine
d Benzoktamin
i benzottamina
e benzoctamina

329 benzododecinium

Quaternary ammonium disinfectant agent displaying bacteriostatic and fungicide properties and used in the treatment of acute and chronic rhinitis and rhinusitis.

f benzododécinium
d Benzododecinium
i benzododecinio
e benzododecinio

330 benzohydrol

Or benzhydrol. Secondary aromatic monovalent alcohol generally obtained by reduction of benzophenone. It is used in organic synthesis.

f benzohydrol
d Benzhydrol
i benzidrolo
e benzhidrol

331 benzoic acid

Mild disinfectant marked by keratolytic action and the capacity of inhibiting seborrhoea. It is used for mild fungus infections of the skin and in the treatment of pulmonary abscess and gangrene. Also used in acute bronchitis in conjunction with an expectorant. It finds occasionally employment as a conservative agent.

f acide benzoïque
d Benzoesäure
i acido benzoico
e ácido benzóico

332 benzoin

Short for gum benzoin; masses of balsamic resin obtained from various species of Styrax (Styrax benzoin) in the form of "tears" or almonds. It consists of resin, benzoic acid, cinnamic acid and vanillin. It is used in medicine to combat catarrh (internal use) and as an antiseptic and disinfectant (externally).

f benjoin
d Benzoeharz; süsser Azant
i benzoino
e benjui

333 benzonatate

Chemotherapeutic agent capable of relieving and checking cough by acting without any delay on the sensory receptors in the chest and in the medulla.

f benzonatate
d Benzonatat
i benzonatato; benzononatina
e benzonatato

334 benzonitrile

Aromatic nitrile. An oily transparent liquid, alcohol- and ether-soluble, used as an intermediate for the preparation of various pharmaceutical products.
f nitrile benzoïque
d Benzonitril
i benzonitrile
e benzonitrilo

335 benzosulphimide
A sweetening agent. White crystalline powder whose degree of sweetness is approximately 500 times that of sugar. It is prepared by oxidation of o-toluenesulphonamide and is generally used as a substitute for sugar, particularly in the diet of diabetics.
f benzosulfimide
d Benzosulfimid
i benzosulfimide
e benzosulfimido

336 benzotherapy
Treatment by means of a benzoate, in particular the intravenous injection of sodium benzoate to treat abscesses of the lung.
f benzothérapie
d Benzotherapie
i benzoterapia
e benzoterapia

337 benzoxyethylphenylpiperidine
An ester of carbonic acid used as an analgesic. It belongs to the pethidine group.
f benzoxyéthylphénylpipéridine
d Benzoxyethylphenylpiperidine
i benzossietilfenilpiperidina
e benzoxietilfenilpiperidina

338 benzoyl
Monovalent radical occurring in benzoic acid, the basis of a large number of organic compounds some of which important in medicine.
f benzoyle
d Benzoyl
i benzoile
e benzoilo

339 benzoyl disulfide
An antipruritic and antimycotic agent used in the treatment of itch-producing eczemas, mycosis, epidermophytosis etc.
f dibenzoyl disulfide
d Benzoyldisulfid
i dibenzoile disolfuro
e disolfuro di dibenzoilo

340 benzoyl peroxide
A keratolytic and antibacterial agent capable of promoting the desquamation and dryness of the skin and thus used in practically all forms of acne vulgaris. Its use may cause the discoloration of the hair and eyebrows.
f peroxyde de benzoyle
d Benzoylperoxyd
i perossido di benzoile
e peróxido de benzoilo

341 benzphetamine
(4) - N - benzyl - Na -dimethylphenethyl amine, belonging to the amphetamine group. It is used against obesity by displaying a strong anorexigenic property, an action which does not involve psychic, vasopressor and cardiac effect to any notable extent.
f benzphétamine
d Benzphetamin
i benzfetamina
e benzfetamina

342 benzpiperylone
An antirheumatic and antiinflammatory agent used as analgesic and antipyretic. Its action is practically the same as that of aminopyrine.
f benzpipérylone
d Benzpiperylon
i benzopiperilone
e benzopiperilona

343 benzquinamide
Or benzoinamide. Reserpine-related antiemetic, antihistaminic, anticholinergic and sedative agent used in the prophylaxis and the therapy of nausea and vomiting due to anaesthesia.
f benzquinamide
d Benzoinamid
i benzchinamide; benzoinamide
e benzoinamida

344 benzthiazide
Very effective diuretic causing an increase of the urinary excretion of water, sodium and chloride. The loss of potassium is limited. Its action is prompt but of short duration. The drug is used as an antioedemic and antihypertensive.
f benzthiazide
d Benzthiazid
i benztiazide
e benztiazida

*** benztropine   s. benzatropine**

345 benzydamine
An antiinflammatory, analgesic and antipyretic agent used both in medicine and surgical ginecology. It also finds use as a local anaesthetic in odontoiatry, in the treatment of pneumonia, in urology. It also has a diuretic action and may be employed with antibiotics, particularly tetracyclines.
f benzydamine
d Benzidamin
i benzidamina
e benzidamina

346 benzylamine
Primary aromatic amine, isomer of the toluidine, obtained from benzyl chloride by reaction with ammonia. It has a strongly alkaline reaction, is soluble in water, ancohol and ether and used as a

local analgesic.
f benzylamine
d Benzylamin
i benzilamina
e benzilamina

347 **benzylaniline**
Secondary aromatic amine obtained by treating benzylchloride with aniline; colourless crystals insoluble in water but soluble in alcohol and ether. It is an intermediate in various organic syntheses.
f benzylaniline
d Benzylanilin
i benzilanilina
e benzilanilina

348 **benzyl benzoate**
Volatile oil present in various balsams and also prepared synthetically. Though possessing antispasmodic properties it is mainly used in the form of emulsion (25%) in the treatment of scabies. It is applied topically.
f benzyl benzoate
d Benzylbenzoat; benzylbenzoesäures Salz
i benzil benzoato
e benzoato de bencilo

349 **benzyl morphine**
Colourless bitter powder whose action is similar to that of codeine and ethylmorphine.
f benzyl morphine
d Benzylmorphin
i benzil morfina
e bencil morfina

350 **benzyl nicotinate**
Phenylmethylic ester of the 3-pyridincarboxylic acid used as a vasodilator related to nicotinic acid.
f benzyl nicotinate
d Benzylnikotinat
i benzile nicotinato
e nicotinato de bencilo

351 **benzylpenicillin**
Bactericidal antibiotic obtained by the fermentation of the Penicillum Notatum. It is active on Gram-positive and on some Gram-negative organisms. The drug is successfully used against meningococci, diplococci, gonococci etc.
f benzylpénicilline
d Benzylpenicillin
i benzilpenicillina (potassica)
e bencilpenicilina

* **benzylpenicillinum procainum   s. procaine benzylpenicillin**

352 **benzylphenol**
A derivative of phenol by substitution of a hydrogen atom of the benzenic ring with a benzylic radical. Three isomers are possible, but only the ortho- and para- isomers are important as organic solvents with antiseptic and germicide properties.
f benzylphénol
d Benzylphenol
i benzilfenolo
e benzilfenolo

353 **benzyl salicylate**
Ester of the salicylic acid used in medicine as a local analgesic and antirheumatic agent.
f salicylate de benzyle
d Benzylsalicylat; Salicylsäurebenzylester
i salicilato di benzile
e salicilato de bencilo

354 **benzyl succinate**
Ester of the succinic acid; the sodium salt forms a colourless, tasteless, water-soluble powder having spasmolytic properties.
f succinate de benzyle
d Benzylsuccinat
i succinato di benzile
e succinato de bencilo

355 **benzylthiouracile**
Chemotherapeutic agent capable of regulating the thyroid secretion and used in the treatment of the Basedow's disease, hyperthyroidism, thyrotoxicosis and hyperthyroidism. Also combined in various preparations for the treatment of angina pectoris and, generally, cardiac insufficiency.
f benzylthiouracile
d Benzylthiouracil
i benziltiouracile
e benziltiouracilo

356 **bephenium hydroxynaphthoate**
An anthelmintic agent particularly used in the treatment of hookworm. It is only slightly absorbed in the gastrointestinal tract.
f hydroxynaphtoate de béphénium
d Bepheniumhydroxynaphthoat
i befenio idrossinaftoato
e hidroxinaftoato de befenio

357 **betacaine**
Local anaesthetic characterized by a toxicity which is about half that of cocaine and by an activity in producing anaesthesia of mucous membranes which is roughly a quarter of that obtained by it. When injected, the solution of the drug contains some adrenaline to avoid possible hyperaemia.
f bétacaïne
d Betacain
i betacaina
e betacaina

358 **betacarotene**
A provitamin A used in the treatment of photo-sensitivity (the stimulation by the influence of light) in patients suffering from erythropoietic protoporphyria.
f bétacarotène
d Betakarotin
i betacarotene

e betacaroteno

359 betahistine
A diamine oxidase inhibitor; anti-vertigo drug displaying a relaxant activity on the precapillary sphincters of the inner ear. Such activity promotes a better microcirculation in the internal auditory arteries and reduces the pressure of the liquid in the labyrinth. It is especially used in the Ménière's syndrome.
f bétahistine
d Betahistin
i betaistina
e betahistina

360 betahypophamine
The pressor constituent of posterior pituitary extract used for its power to constrict capillary vessels and to stimulate peristalsis. It is also accompanied by antidiuretic action.
f bétahypophamine
d Betahypophamin
i betaipofamina
e betahipofamina

361 betaine
A trimethylic derivative of the glycocoll and a natural constituent of animal and vegetable tissue. It is considered an internal (or intramolecular) salt originating from one basic and one acid group. In fact, once in solution, it is an ampho̲teric ion as it contains a positive and a̲ negative charge. Betaine is a lipotropic agent and is employed in functional liver insufficiency and dyslipidosis.
f bétaïne
d Betain
i betaina
e betaina

362 betaine ascorbate
A form of betaine capable of promoting a higher and more durable vitamin level in the organism.
f ascorbate de bétaïne
d Betainascorbat
i betaina ascorbato
e ascorbato de betaina

363 betaine aspartate
A form of betaine capable of regulating ammonaemia and promoting the exploitation of lipides and glycides.
f aspartate de bétaïne
d Betainaspartat
i betaina aspartato
e aspartato de betaina

364 betaine citrate
A form of betaine having an antacid action.
f citrate de bétaïne
d Betaincitrat
i citrato di betaina
e citrato de betaina

365 betaine nicotinate
A form of betaine which proves less toxic and more effective when used in nicotinic therapy.
f nicotinate de bétaïne
d Betainnicotinat
i betaina nicotinato
e nicotinato de betaina

366 betalanine
Or ß-alanine. Inhibitor (non-hormonal) of postmenopausal vasodilatation mainly used to treat menopausal flush, i.e. the mounting of the blood to face, head and neck and even, though rarely, to the whole body.
f bétalanine
d Betalanin
i betalanina
e betalanina

367 betamethadol
Beta analogue of alphamethadol displaying morphine-like properties.
f bétaméthadol
d Betamethadol
i betametadolo
e betametadolo

368 betamethasone
A corticosteroid hormone displaying a good glucocorticoid action. It is used in the therapy of rheumatism and rheumatoid arthritis, inflammatory and allergic dermatopathies, traumatic ophthalmopathies, autoimmune haemolytic anaemia and acute and/or chronic leukaemia. It also finds employment in multiple sclerosis.
f bétaméthasone
d Betamethason
i betametasone
e betametasona

369 betanaphthol
A compound characterized by a faint phenolic odour present in coal tar and used medicinally as an internal antiseptic and anthelmintic, as well as externally in skin diseases in the form of ointment.
f bétanaphtol
d Betanaphthol
i betanaftolo
e betanaftol

370 betanidine
A long-duration adrenergic blocking agent used as an effective anti-hypertensive agent. It is rapidly absorbed and does not involve the parasympathetic and central nervous system in its action. Moderate and fairly severe conditions of hypertension may be treated with it, especially when resistant to other types of hypotensive agents.
f bétanidine
d Betanidin
i betanidina
e betanidina

371 betaxolol
A beta-adrenoreceptor blocking agent

marked by cardioselectivity and used as an antihypertensive and also in ophthalmology in cases of glaucoma.

f bétaxolol
d Betaxolol
i betassololo
e betaxololo

372 betazole

Histamine-related agent used as a diagnostic agent particularly for testing gastric secretion and pheochromocytoma. Possible side effects are very similar to those caused by histamine.

f bétazole
d Betazol
i betazolo
e betazolo

373 bethanecol chloride

Acetylcholine-like parasympathomimetic which is inactivated by cholinesterase. The drug promotes both the tone and the motility of the urinary bladder and is therefore recommended in all cases of urinary retention.

f chlorure de béthanécole
d Bethanekolchlorid
i cloruro di betanecolo; carmecolina cloruro
e cloruro de carmecolina

374 betoxycaine

Local anaesthetic characterized by a rapid and long-duration action. It is mostly used in surgery.

f bétoxycaïne
d Betoxycain
i betossicaina
e betoxicaina

375 betulin

The bitter principle of the outer bark of the European birch, from which it is extracted with benzene. It is constituted by five aromatic nuclei with one phenolic and one alcoholic group. It yields betulinamaric and betulinic acid by oxidation. Also known as betulinol.

f bétuline
d Betulin; Birkenkampfer
i betulina
e betulina

376 bevonium methylsulfate

Anticholinergic-type spasmolytic agent displaying musculotropic papaverine-like properties. It is characterized by a selective myolytic action in the smooth muscles of the hollow viscera and excretory tracts. It somewhat reduces the vermiform movement of the intestine (peristalsis) and acidity. The drug is also used as a bronchodilator and in the treatment of spastic visceral conditions.

f méthylsulfate de bévonium
d Bevoniummethylsulfat
i metilsolfato di bevonio
e metilsulfato de bevonium

377 bezitramide

A narcotic and analgesic drug. Bezitramide is also used to relieve respiratory disorders, mainly against cough, and in the treatment of acute chronic pains deriving from them.

f bézitramide
d Bezitramid
i bezitramide
e bezitramida

378 bibenzonium bromide

2-(1,2-diphenylethoxy)ethyltrimethylammonium bromide. Antitussive agent marked by a central action. It may be used orally, in the form of syrup, drops or pills for the relief of violent, spasmodic cough. It is more potent (approximately 3 times) than codein.

f bromure de bibenzonium
d Bibenzoniumbromid
i bibenzonio bromuro
e bromuro de bibenzonio

379 bibrocathol

Bismuth derivative of the tetrabromopyrocathecol. An insoluble salt of bismuth capable of a protective action on mucous membranes and displaying astringent and antiseptic properties. It finds employment as an eye ointment in the treatment of blepharitis.

f bibrocathol
d Bibrocathol
i bibrocatolo
e bibrocatolo

380 bicarbonate

Any salt of carbonic acid in which one only of the hydrogen atoms has been replaced by a metal or radical. More generally, any salt containing the acid group HCO_3, e.g. sodium carbonate.

f bicarbonate
d Bicarbonat
i bicarbonato
e bicarbonato

381 bichromate

Or dichromate. A bivalent-radical containing salt. Bichromates are soluble and find use in the production of drugs for their oxidizing properties.

f bichromate
d Bichromat; doppeltchromsaures Salz
i bicromato
e bicromato

382 bietamiverin

Atropine- and papaverine-like parasympatholitic agent used as a muscle relaxant and in the treatment of hepatic and renal colics, gastritis, colitis etc. Being a secretion-reducing agent it is also used in the therapy of dysmenorrhoea.

f biétamivérine
d Bietamiverin
i bietamiverina
e bietamiverina

383 bietaserpine

Water-soluble derivative of reserpine (a drug capable of reducing sympathetic tone by noradrenalin depletion) having an hypotensive action that does not involve a neurodepressive one. It is used in cases of hypertension including those forms which are related to renal insufficiency.

f biétaserpine
d Bietaserpin
i bietaserpina
e bietaserpina

384 bilberry

Or blueberry. Term denoting various species of Vaccinium and above all the Vaccinium myrtillus. Its berries are used to prepare astringent and antidiarrhoic infusions.

f airelle; myrtille; brimbelle
d Blaubeere; Heidelbeere
i mirtillo
e arándano

385 bile salts

Salts extracted from animal bile and used to stimulate the bile flow without increasing its bile salts and pigment contents. They are administered in some compound preparations for the treatment of biliary insufficiency.

f sels biliaires
d Gallensalze
i sali biliari
e sales biliares

386 bioflavonoids

Any agent belonging to the group of antihaemorrhagic so-called vitamin P. Flavonoids are widely diffused in various plants, particularly in citrus, some Greek grapes etc. Chemically, they are flavone derivatives and they contain the structure of the phenylbenzopyrone often with OH groups in the benzene ring. They are biologically interesting as they contribute to the maintenance in normal conditions of the walls of the blood vessels. They are therefore used in the prophylaxis and the treatment of vascular disorders in cases of hypertension, arteriosclerosis, diabetes, nephropathies etc. as well as of some forms of retinitis, post-traumatic oedema, haemorrhoids etc.

f bioflavonoïdes
d Bioflavonoide
i bioflavonoidi
e bioflavonoides

387 biotin

Vitamin of the B group (vitamin B_7, also known as vitamin H or coenzyme R). Chemically a sulphur-containing organic acid. Biotin is present in all cells of animal and vegetable tissues, generally in a combined form. The most abundant sources are the liver, yeast, kidneys, eggs and milk. No deficiency of biotin is ever found in man, as it is synthesized by intestinal bacteria. In addition to

participating in the fixation of carbon dioxide in the metabolic processes as a transcarboxylase coenzyme, biotin, as a drug, protects the skin and acts against seborrhoea. It is therefore used in the treatment of skin lesions, e.g. acne, psoriasis, alopecia etc.

f biotine
d Biotin
i biotina
e biotina

388 biperiden

Anticholinergic agent whose action is very similar to that of atropine. It is mainly used in the symptomatic treatment of paralysis agitans (parkinsonism or shaking palsy) of arteriosclerotic, idiopathic or post-encephalitic nature. It is also employed as a sedative in the extra pyramidal syndrome induced by phenothiazine drugs and reserpine (q.v.), as it inhibits rigidity, salivation, perspiration, hypokinesis and, more generally, the frequency of the crises.

f bipéridène
d Biperiden
i biperidene
e biperideno

389 birch tar oil

Brown toxic phenolic oil which is obtained by destructive distillation of the outer bark and wood of the European white birch and is used in the preparation of ointments for skin diseases, e.g. some forms of acne or eczema.

f essence de bouleau; goudron de bouleau; huile de bouleau
d Birkenteer
i catrame di betulla; olio di Rusco
e aceite de abedul; alquitrán de abedul

390 bisacodyl

Purgative acting on the colon by stimulating the peristalsis without causing spasms. Generally used in the treatment of chronic constipation and during pregnancy.

f bisacodile; bisacodyl
d Bisacodyl
i bisacodile
e bisacodilo

391 bismuth

An element most of whose inorganic salts are used as protective agents, in particular of the mucous membranes of the alimentary canal, from the action of irritants. Some of them, e.g. subnitrate, are astringent and antiseptic. The metal in suspension, the inorganic salts and the organic compounds, e.g. tartrate, salicylate etc. as well as combinations of bismuth and arsenic, have been used, albeit with very limited results, in the treatment of trypanosomiasis, amoebiasis, syphilis, yaws etc. The acid bismuth sodium tartrate finds frequent employment in the treatment of digestive disorders.

f bismuth
d Wismut
i bismuto
e bismuto

392 bismuth aluminate
Insoluble bismuth salt acting as antacid and adsorbent agent capable of protecting the gastric mucosa and mainly used in the symptomatic treatment of gastric ulcers.
f aluminate de bismuth
d Wismutaluminat
i bismuto alluminato
e aluminado de bismuto

393 bismuth camphocarbonate
Therapeutic agent capable of acting against spirilla (abacterial genus) and used for the treatment of tonsillitis, infectious angina etc. It is only used in cases of allergy to antibiotics.
f camphocarbonate de bismuth
d Wismuthcamphocarbonat
i bismuto canfocarbonato
e canfocarbonato de bismuto

394 bismuth camphorate
Therapeutic agent capable of acting against spirilla (bacterial genus) and used to treat tonsillitis and tonsillar abscesses. It also finds employment in preoperative treatment of acute appendicitis and in the prevention of possible complications of exanthematic diseases.
f camphorate de bismuth
d Wismutkamforat; Wismuthcamphorat
i bismuto canforato; bismuto canfocarbossilato
e canforato de bismuto

395 bismuth carbonate
Insoluble substance marked by weak antacid effect. It reacts slowly with the hydrochloric acid of the stomach and inhibits the secretion by lining the mucous membrane. It is used, in combination with carbonates of magnesium and calcium in gastric or duodenal ulcers and in some cases of diarrhoea. It may also be incorporated in ointments and used as a protecting and sedative agent on inflamed skin.
f carbonate de bismuth
d Wismuthkarbonat
i carbonato di bismuto
e carbonato de bismuto

396 bismuth dipropylacetate
Therapeutic drug having almost exactly the same action of bismuth camphorate.
f dipropylacétate de bismuth
d Wismuthdipropylacetat
i bismuto dipropilacetato; bismuto valproato
e dipropilacetato de bismuto

397 bismuth heptadiencarbonate
Chemotherapeutic drug acting against spirilla, quickly absorbed and slowly excreted. It is employed parenterally for the treatment of syphilis and yaws and by means of pessaries for the treatment of acute and chronic tonsillitis of spirillary and fusospirillary origin. It is also used for the preoperative treatment of appendicitis.
f heptadiencarbonate de bismuth
d Wismutheptadiencarbonat
i bismuto eptadiencarbonato
e heptadiencarbonato de bismuto

398 bismuth magma
Or milk of bismuth. Bismuth carbonate and bismuth hydroxide suspended in water and used in the treatment of enteritis in children.
f lait de bismuth
d Wismuthydroxyd
i latte di bismuto
e lechada de bismuto; magma de bismuto

399 bismuth oxide
A compound, characterized by a yellowish colour and having the same properties of bismuth subnitrate, q.v.
f oxyde de bismuth
d Wismutoxyd
i ossido di bismuto
e óxido de bismuto

400 bismuth pectate
Therapeutic agent capable of protecting the mucosa of the gastroenteric tract and used in the treatment of hiatal hernia, dyspepsia, gastritis, oesophageal ulcer, colonopathies, biliary reflux, diarrhoea etc.
f pectate de bismuth
d Wismuthpektat
i pectato di bismuto; bismuto pectato
e pectato de bismuto

401 bismuth phosphate
Therapeutic agent promoting the gastric secretion of mucus and thus acting as an antacid and astringent. It is particularly used in functional disorders of the colon accompanied by either constipation or diarrhoea, in acute and chronic colitis, in diverticolitis, peptic ulcers etc. It also finds employment in combination with other drugs in the treatment of gastroenteritis marked by dehydration and to relieve the consequences of a gastrectomy.
f phosphate de bismuth
d Wismutphosphat
i bismuto fosfato
e fosfato de bismuto

402 bismuth salicylate
A disinfectant, adsorbent and protective agent in various infectious gastroenteric disorders. In the form of an oily solution, it may be used parenterally for the treatment of syphilis. It is also employed against throat diseases, e.g. tonsillitis.
f salicylate de bismuth
d Wismutsalicylat
i bismuto salicilato

e salicilato de bismuto

403 bismuth subgallate
Or bismuth oxygallate. Internal astringent and antiseptic as well as absorbent
agent. It is used for the treatment of
gastroenteritis, toxic diarrhoea and, in
the form of ointment, in dermatology. It
is also used in the treatment of syphilis
and, in the form of suppositories, for
haemorrhoids.
f bismuth oxygallate
d Wismuthoxygallat
i bismuto gallato (basico)
e galato de bismuto

404 bismuth subnitrate
Insoluble compound used in cases of gastric ulcers and, more generally, for the
protection of gastrointestinal membrane
in the event of infectious and toxic enteritis. The drug may be used in combination with magnesium oxide, antispasmod-
ics, carminatives and disinfectants. Also
known as basic bismuth nitrate.
f nitrate basique de bismuth
d Wismutsubnitrate
i bismuto nitrato basico; bismuto sottoni-
trato
e bismuto subnitrato básico

405 bismuth succinate
A compound used against spirilla-caused
disorders of the throat, e.g. soreness,
tonsillar abscesses, in cases of surgery
in the tonsillary region, in otitis and
sinusitis. It should not be used in diphtheria.
f succinate de bismuth
d Wismutsuccinat
i bismuto succinato
e succinato de bismuto

406 bisoprolol
Beta-adrenoceptor blocking agent (cardioselective). It finds employment in the
treatment of angina pectoris and of hyper
tension.
f bisopropole
d Bisopropol
i bisopropolo
e bisopropolo

407 bisoxatin
2,3-dihydro-2,2-di-(4-hydroxyphenyl)-1,4-
-bensoxazin-3-one. A purgative which is
stronger than phenolphthalein and other
antraquinone derivatives. Its action in
the large intestine is carried out in
three to four hours. It is generally used
in chronic constipation.
f bisoxatine
d Bisoxatin
i bisossatina
e bisoxatina

408 bistort
Snakeweed. A herbaceous plant with large
oval-oblong leaves and pink flowers.
Widely diffused in European countries.

Its rhizome is rich in tannin and is used
as an astringent.
f bistorte; coulevrine
d Drachenwurzpflanze; Schlangenknöterich
i bistorta
e bistorta

409 bisulfate
An acid sulfate, i.e. a salt containing
the monovalent radical $-HSO_4$ derived
from the sulphuric acid. The sodium bisulfate is used in the production of artificial yeasts.
f bisulfate
d Bisulfat
i bisolfato
e bisulfato

410 bisulfite
Acid sulfite, a salt containing the monovalent radical $-HSO_3$ derived from sulphurous acid. The most important bisul-
fites are the sodium bisulfite used for its
reducing properties, and the calcium bisulfite used to extract cellulose from
plants.
f bisulfite
d Bisulfit
i bisolfito
e bisulfito

411 bitter almond
The seed of Prunus amygdalus from which
an oil is obtained which is laxative.
Bitter almonds, if ingested in excessive
quantities, can prove toxic: sixty of
them may well represent a lethal dosis
for an adult.
f amande amère
d bittere Mandel
i mandorla amara
e almendra amarga

*** bitter almond oil   s. benzaldehyde**

412 bittern
The bitter mother liquor found in salt-
works after the salt has crystallized and
containing other salts, e.g. magnesium
chloride, magnesium sulfate, bromides
and iodides.
f eau mère (amère)
d Salzmutterlauge (bittere)
i acque madri (amare)
e agua madre (amargua)

413 bitter principle
The active principle of those drugs (bitters) which are used to stimulate appetite. Quassia and calumba are the most
important simple bitters; they contain no
tannic acid and may thus be dispensed
with iron. Aromatic bitters, e.g. gentian
and bitter-orange peels, have a more
potent effect; they contain tannic acid
and are incompatible with iron. The aromatic bitters contain volatile oils together with the bitter principle. Bitters are
astringent when they contain tannic acid.
f principe amer

d Bitterstoff; Extraktivstoff
i principio amaro
e principio amargo

414 bleomycin polypeptide
Cytotoxic antibiotic acting by inhibiting DNA biosynthesis and used to treat lymphomas and solid tumours, as well as lung cancer and cutaneous spinocellular carcinomata.
f bléomycine
d Bleomyzin
i bleomicina
e bleomicina

415 blue ointment
A mercurial ointment containing 10% of mercury and used as disinfectant and against infestation.
f onguent mercuriel double
d Läusesalbe; Merkurialsalbe
i pomata mercuriale mite; unguento napoletano
e pomada mercurial simple

416 boldine
Alkaloid obtained from a South American evergreen shrub (boldo) and used to lower blood pressure. The drug may cause clonic convulsions.
f boldine
d Boldin
i boldina
e boldina

417 boric acid
Or boracic acid. A natural chemical compound whose solution has a slightly disinfectant action. It inhibits the growth of bacteria but is incapable of destroying them.
f acide borique
d Borsäure
i acido borico
e ácido bórico

418 bornaprine
An anticholinergic drug used in the treatment of Parkinson's disease.
f bornaprine
d Bornaprin
i bornaprina
e bornaprina

419 borneol
Bornyl alcohol, Borneo camphor. Secondary terpene alcohol obtained by the reduction of common camphor. Its laevorotatory isomer has the same properties of camphor for therapeutic purposes but its esters show properties that are associated with the particular acid radical, e.g. salicylate, that is used to relieve rheumatic pains, and valerate that is used in nervous disorders.
f bornéol
d Borneol
i borneolo
e borneol

420 bran
The coat of a cereal grain suitably milled and processed. It is a non-irritant purgative agent. Also used as a breakfast cereal. Its indigestible cellulose increases the intestinal bulk.
f son
d Kleie
i crusca
e afrecho; salvado

421 bretylium
Adrenergic neurone blocking drug used in cardiac arrythmias. Its side effects, e.g. nausea, diarrhoea, orthostatic hypotension etc., limit its use as an antihypertensive agent acting by selectively blocking transmission in post-ganglionic adrenergic nerves.
f brétylium
d Bretylium
i bretilio
e bretilio

422 brinase
Fibrinolytic proteolytic enzyme used in the treatment of occlusive peripheral vascular diseases. The drug is also used to release sticking or clogged-up cannulae in haemodialysis. Among its side effects, the drug presents the danger of thrombosis at the site of the injection.
f brinase
d Brinase
i brinase
e brinasa

423 bromazepam
Benzodiazepine tranquillizer characterized by an anxiolytic, hypnotic and muscle-relaxant action. It is used in the treatment of anxiety, tension and irritation states, as well as of functional disorders of the cardiovascular, gastrointestinal and genitourinary systems and in psychosomatic conditions.
f bromazépam
d Bromazepam
i bromazepam
e bromazepam

424 bromazine
Antihistamine agent marked by a mild sedative action and devoid of spasmolytic action. It is mainly used in the treatment of skin-, food-, drug- and cosmetics-related allergies and of vasomotor rhinitis.
f bromazine
d Bromazin
i bromazina
e bromazina

425 bromelains
Proteolytic enzymes obtained from the Ananas comosus. An antiphlogistic and antioedematous agent used for the treatment of cutaneous and ocular inflammatory disorders. It is frequently used in conjunction with antibiotics.

f bromélaïne
d Bromelaine
i bromelina
e bromelina

426 brometenamine
Therapeutic agent used as a sedative, especially for the relief of cough independently of the nature of the latter.
f brométénamine
d Brometenamin
i brometenamina
e brometenamina

427 bromethol
Concentrated solution of tribromoethyl alcohol in amylene hydrate, which is diluted and instilled into the rectum as a preliminary to anaesthesia.
f brométhol
d Bromethol
i brometolo
e brometolo

428 bromethyl
Volatile liquid used to produce anaesthesia but seldom employed as it easily causes depression of the respiratory centres.
f brométhyle
d Bromethyl
i brometile
e brometilo

429 bromhexine
A bronchial mucolytic used for the treatment of diseases of the respiratory tract, e.g. tracheobronchitis, pneumoconiosis, emphysema and emphysematous bronchitis. It also displays an antiphlogistic action.
f bromhexine
d Bromhexin
i bromesina
e bromhexina

430 bromide
Salt of hydrobromic acid resulting from bromine and a metal (or an organic radical). The most commonly used are the ammonium, potassium and sodium bromides which display a sedative action as well as a depressive one on the nervous system and particularly cerebral cortex and bulbar centres. They are used in the treatment of insomnia, psychic agitation, convulsive and neurotic syndromes etc.
f bromure
d Bromid
i bromuro
e bromuro

431 bromidione
An anticoagulant agent used in the treatment of coronary thrombosis, myocardial infarction, angina pectoris and, more generally, cardiac insufficiency.
f bromidione
d Bromidion
i bromidione
e bromidiona

432 bromisoval
A sedative and hypnotic agent used in the treatment of nervous disorders, insomnia, irritability and excitability also when accompanied by irregular cardiac rhythm.
f bromisoval
d Bromisoval
i bromisoval
e bromisoval

433 bromocalcenamine
Calcium bromide and hexamethylentetramine. A condensation product capable of balancing the action of calcium on the neurovegetative system and displaying at the same time the sedative action of bromine on the cortical centres. It further includes the mildly diuretic action of the metenamine. It is used in the treatment of excessive nervous irritability, insomnia etc.
f méthionine et bromure de calcium
d Calciumbromid-Methionin
i bromocalcenamina; metionina e calciobromuro
e metionina y bromuro cálcico

434 bromocamphor
Camphor-smelling crystalline substance used as a hypnotic and antispasmodic in chorea, asthma, whooping cough and hysteria.
f camphre monobromé
d Bromkampfer
i canfora monobromata; bromuro di canfora
e alcanfor monobromado

435 bromocriptine
A compound capable of stimulating dopamine receptors and used in the treatment of acromegaly, inhibition or suppression of lactation. It is also capable of reducing prolactin levels and thus used in galactorrhoea, amenorrhoea and some cases of infertility or impotence, as well as in patients suffering from Parkinson's disease.
f bromocriptine
d Bromokriptin
i bromocriptina
e bromocriptina

436 bromocyclodipentene
An organic compound of liposoluble bromine used for its marked predilection for nervous tissues. It is generally prescribed in all cases requiring a bromine treatment, e.g. nervous excitation, cardiac neurosis etc.
f bromocyclodipentène
d Bromozyclodipenten
i bromociclodipentene
e bromociclodipenteno

437 bromoform
Tribromomethane. The bromine analogue of chloroform and iodoform. A sedative agent used in mental disorders and especially in the form of syrup against whooping

cough. Its action lasts for a comparative‑
ly short time.
f bromoforme; tribromométhane; tribromure
   de formyle
d Bromoform; Tribrommethan
i bromoformo; tribromometano; bromuro di
   formile
e bromoformo; metano tribromado; tribromu‑
   ro de formile

438 bromoiodocasein

Synthetic brominated thyroid hormone
displaying a specific thyromimetic action
on the basal metabolism together with
bromine neuro-sedative effects. It is used
in thyroid insufficiency, in obesity and
various other conditions borne of thyroid
disorders. Its effect is gradual and fair‑
ly prolongued.
f bromoiodocaséine
d Bromoiodocasein
i bromoiodocaseina
e bromoiodocaseina

439 bromophenol

Dark-yellow oily liquid (phenol monobro‑
mate) characterized by a strong odour
and used, in mild solution, for the treat‑
ment of erysipelas.
f bromophénol
d Bromophenol
i bromofenolo
e bromofenol

440 brompheniramine

Antihistaminic agent acting like chlorphe‑
niramine. It occupies the receptor sites
in the effector cells causing the exclu‑
sion of histamine. The drug is mainly
used in the treatment of various allergies,
asthma, rash, rhinitis etc.
f bromphéniramine
d Brompheniramin
i bromfeniramina
e bromfeniramina

441 bromsulphthalein

Phenoltetrabromphthalein sodium sulpho‑
nate, a non-toxic, non-irritant dye used
to diagnose liver functions. Injected in‑
travenously, the remains of the drug in
the circulation allow for a measure of
hepatic sufficiency after a given interval
of time.
f bromosulfophtaléine; sulfobromophta‑
   léine
d Bromosulfophthalein; Sulfobromphthalein
i bromosolfoftaleina; bromosulfaneina
e bromosulfoftaleina

442 broparestrol

Oestrogenous agent mainly used in the
treatment of menopausal disorders, galac‑
torrhoea, premenstrual mammal tension,
some forms of seborrhoea and juvenile
acne.
f broparestrol
d Broparestrol
i broparestrolo
e broparestrol

443 brovanexine

A mucolytic agent generally used as an
expectorant in cases of acute and chronic
bronchitis.
f brovanexine
d Brovanexin
i brovanessina
e brovanexina

444 broxaldine

A bacteriostatic and disinfectant agent
used to combat Gram-positive bacteria
and mycetes, in cases of intestinal infec‑
tions and in the prophylaxis of the lat‑
ter.
f broxaldine
d Broxaldin
i broxaldina
e broxaldina

445 broxuridine

An antineoplastic agent capable of promot‑
ing the radiosensitivity of the tumoral
cells and used together with other drugs
for the radiotherapeutic treatment of
brain, neck and head tumours.
f broxuridine
d Broxuridin
i broxuridina
e broxuridina

446 broxyquinoline

A compound derived from oxyquinoline, q.
v., and used as a potent antiseptic and
disinfectant in the treatment of disorders
affecting the intestinal tract.
f broxyquinoline
d Broxyquinolin
i brossichinolina
e broxiquinolina

447 buclizine

Antihistamine and antiemetic drug used
in the treatment of allergic disorders,
seasickness, some forms of dermatosis and
in particular allergies caused by drugs.
f buclizine
d Buclizin
i buclizina
e buclizina

448 buclosamide

Topical antimycotic agent especially used
to combat Candida albicans and, more
generally, for the treatment of dermato‑
mycosis and onichomycosis.
f buclosamide
d Buclosamid
i buclosamide
e buclosamida

449 bucloxic acid

Antiphlogistic, analgesic and antipyretic
agent mainly used in the treatment of
painful and fever-inducing inflammations,
e.g. in arthritis and rheumatism causing
musculo-skeletal, dental and respiratory-
-infection-related pains. It is also used
in lung and vein inflammations.
f acide bucloxique

d Bucloxinsäure
i acido buclossico
e ácido buclóxico

450 bucolome
Antiphlogistic and antilithic (preventing the formation of a stone or calculus) agent with a mild analgesic and antipyretic action. It is used in the treatment of inflammation-causing diseases and conditions, as well as in tumefactions.
f bucolome
d Bukolom
i bucolomo
e bucolomo

451 budesonide
A corticosteroid used by inhalation in the treatment of asthma and, intranasally, of allergic disorders.
f budésonide
d Budesonid
i budesonide
e budesonido

452 bufeniode
An alpha-blocking agent used in the treatment of hypertension, particularly in old age, when accompanied by a weak condition of the heart, coronary insufficiency, as well as marked psychic disorders.
f buféniode
d Bufenjod
i bufeniodo
e bufeniodo

453 bufetolol
A beta-blocking agent used in the treatment of sinus arrhythmia and tachycardia as well as angina pectoris.
f bufétolol
d Bufetolol
i bufetololo
e bufetololo

454 bufexamac
Non-steroid anti-inflammatory topical agent used in the treatment of acute and chronic skin diseases, as well as in proctology, phlebology, in rheumatic and traumatic diseases.
f bufexamac
d Bufexamac
i bufessamac
e bufexamac

455 buflomedil
Peripheral vasodilator used in the treatment of disorders of the vascular system of the limbs accompanied by a deficient or defective circulation.
f buflomédil
d Buflomedil
i buflomedile
e buflomedilo

456 bufogenine
Organic substance isolated in a Chinese

drug obtained from the secretion of the skin glands of the toad. It is a derivative of the bufotaline ($C_{24}H_{34}O_5$).
f bufogénine
d Bufogenin
i bufogenina
e bufogenina

457 buformin
Biguanide hypoglycaemic agent used in the treatment of diabetes mellitus in adults. It may be prescribed by itself or in conjunction with insulin or sulfanilureas. In young people, however, the drug must always be administered with insulin.
f buformine
d Buformin
i buformina
e buformina

458 bufotalin
Chemical compound; genin obtained from the poisonous secretion of the skin glands of the common toad ($C_{26}H_{36}O_6$), which presents analogy of composition with the biliary acids. Bufotalin contains an oxhydrile which is esterified with suberilarginine to yield bufotoxin, q.v.
f bufotaline; bufotalol
d Bufotalin
i bufotalina
e bufotalina

459 bufotenedine
Organic substance isolated from the secretion of the skin glands of the common European toad. It is the betaine of the bufotenin and displays an action which is similar to that of adrenaline.
f bufoténédine
d Bufotenedin
i bufotenedina
e bufotenedina

460 bufotenin
Alkaloid base, ethyldimethylaminine derived from the oxindole, contained in the secretion of the skin gland of the common European toad; an oily liquid, little soluble in water. It displays an action that is similar to that of adrenaline. It is also used as a hallucinogenic agent in the experimental production of temporary psychoses.
f bufoténine
d Bufotenin
i bufotenina
e bufotenina

461 bufotoxin
Poison secreted by the skin glands of the common European toad ($C_{40}H_{60}O_{10}$). It decomposes by hydrolysis into a nitrogen base derived from arginine and into bufotalin. The drug displays an analogy of composition and action with the digitalin.
f bufotoxine
d Bufotoxin
i bufotossina

e bufotoxina

462 bufylline
Theofilline 2-amino-2-methylpropan-1-ol.
Therapeutic agent used as a bronchodilator characterized by its stimulating
and tonic action on the cardiovascular
system.
f bufylline
d Bufillin
i bufillina
e bufilina

463 bumadizone
Antiphlogistic, antipyretic and analgesic
agent used in the treatment of articular
and spinal rheumatism, inflammatory
rheumatism, periarthritis, bursitis, myalgias etc.
f bumadizone
d Bumadizon
i bumadizone
e bumadizona

464 bumetanide
A diuretic agent carrying out its action
by inhibiting the reabsorption of the sodium at the level of the renal tubules
and thus increasing its elimination. It is
mainly used in the treatment of oedemas
of cardiac, renal and hepatic origin and
whenever it is necessary to induce a forc
ed diuresis, e.g. in cases of poisoning.
f bumétanide
d Bumetanid
i bumetanide
e bumetanida

465 bunaftine
Drug capable of depressing the excitability of the cardiac muscle to electrical
stimulation and lengthening the refracto-
ry period, thus acting as an antiarrhyth
mic agent. It is also used for the prophylaxis and the treatment of extrasystoles, tachyarrhythmias and sinus tachycardia, as well as for the maintenance
therapy following cardioversion, i.e. the
process of converting the heart from an
abnormal sinus rhythm.
f bunaftine
d Bunaftin
i bunaftina
e bunaftina

466 bunamiodyl
Iodine-containing radiopaque contrast
medium which is fully absorbed by the
intestines and eliminated by the bile after rendering the bile ducts opaque. It
is generally used in cholecistography and
cholangiography.
f bunamiodyl
d Bunamiodyl
i bunamiodil
e bunamiodil

467 bunitrolol
A beta-adrenergic blocking agent used in
the treatment of angina pectoris, tachycardia and paroxysmal tachycardia, and
arrhythmia.
f bunitrolol
d Bunitrolol
i bunitrolol
e bunitrolol

468 buphenine
Chemotherapeutic agent causing peripheral dilatation by its direct action on the
arterioles. By mobilizing blood from the
blood depots, it increases cardiac output.
It is particularly used in the treatment
of intermittent claudication.
f buphénine
d Buphenin
i bufenina
e bufenina

469 bupivacaine
Local anaesthetic of the amide type, long
acting and fairly potent. It is used in
regional nerve and epidural block when
a prolongued effect is required. Whenever
the patient is subject to some psychotropic therapy, great caution is necessary as side effects can be very serious
and include the loss of perineal sensation, paralysis of the lower limbs and a
grave weakening of sexual functions.
f bupivacaïne
d Bupivakain
i bupivacaina
e bupivacaina

470 bupranolol
A beta-blocking agent used in the treatment of disorders of the coronary arteries
(angina pectoris), arrhythmias of organic
or vegetative origin and neurovegetative
cardiac disorders. Particular caution
should be taken in the case of diabetic
patients.
f bupranolol
d Bupranolol
i bupranololo
e bupranololo

471 buprenorphine
Pure analgesic used to relieve the pain
which is due to myocardial infarction,
neoplasias and postoperative conditions.
The drug may easily cause drowsiness
and respiratory depression, as well as
nausea.
f buprénorphine
d Buprenorphine
i buprenorfina
e buprenorfina

472 burodiline
Spasmolytic agent also capable of relieving convulsions.
f burodiline
d Burodiline
i burodilina
e burodilina

473 buspirone
Anxiolytic agent used for the relief of

anxiety symptoms without having muscle relaxant or anticonvulsant effects. Its action is slower than that of the diazepines but causes less sedation.

f buspirone
d Buspiron
i buspirone
e buspirona

474 busulphan
1,4-Dimethanesulphonyloxybutane. A drug characterized by its specific depressant action on granulocytes. It can thus be considered an oncolytic agent whose mechanism resembles that of alkylating agents. It is mainly used in the treatment of chronic myeloid leukaemia. Effects are rather slow to appear (generally some weeks) and remission may occur after almost two years from the beginning of the therapy.

f busulfan
d Busulfan
i busulfano
e busulfano

475 butacaine
Local anaesthetic agent, similar to cocaine but with an action which is more rapid and prolonged. Unlike cocaine, it does not cause mydriasis or accommodation difficulties. It is used by injection or spray.

f butacaïne
d Butacain
i butacaina
e butacaina

476 butalamine
Chemotherapeutic agent acting on the vascular system particularly in relation to the peripheral circulation. It augments the flow of blood and promotes the cellular metabolism thus supporting the oxygenation of body tissues and the exploitation of glucose. It is mainly used for the treatment of cerebral vascular diseases, arteriopathies of the lower limbs and labyrinth disorders.

f butalamine
d Butalamin
i butalamina
e butalamina

477 butalbital
Sedative and hypnotic of the barbituric series, acting intermediately. It is used, in combination with aminophenazone, for its selective analgesic and antineuralgic action.

f butalbital
d Butalbital
i butalbitale
e butalbitalo

478 butamben
Butyl aminobenzoate. Local anaesthetic used diluted with sterile talc and applied as a dusting powder. It is soluble in oil and may therefore be incorporated in ointments and suppositories. Being capable of reacting with picric acid it forms a yellowish amorphous picrate which combines local anaesthetic and antiseptic action.

f aminobenzoate de butyle
d Butylaminobenzoat
i butile amminobenzoato
e benzoato de aminobutilo

479 butamirate
Cough-relieving drug marked by a central mechanism of action. It is used in the treatment of acute respiratory cough and catarrh, dry bronchitis and in some cases of irritation caused by smoke.

f butamirate
d Butamirat
i butamirato
e butamirato

480 butanilicaine
Amide-related local anaesthetic mainly used in dentistry. The duration of the drug is short as it is very rapidly transformed in the liver by the metabolic action. It also finds employment in infiltration anaesthesia.

f butanilicaïne
d Butanilicain
i butanilicaina
e butanilicaina

481 butaperizine
A potent tranquillizer mainly used in the treatment of schizophrenic conditions, e. g. a difficult-to-control excitement, depression etc. It does not act on the neurovegetative system, nor does it affect cerebral circulation. It may also be used as antiemetic agent.

f butapérizine
d Butaperizin
i butaperizina
e butaperizina

482 butaprobenz
A local anaesthetic whose action is similar to that of cocaine. It does not promote the action of adrenaline and does not dilate the pupil when applied to attain the anaesthesia of the eye. It is administered in combination with adrenaline when given by subcutaneous injection so as produce vasoconstriction and localize with exactitude the anaesthetic.

f butaprobenz
d Butaprobenz
i butaprobenz
e butaprobens

483 butazopyridine
Therapeutic agent used in the treatment of urogenital disorders.

f butazopyridine
d Butazopyridin
i butassampiridina; butazopiridina
e butazopiridina

* **butethal   s. butobarbitone**

484 butethamate
2-diethylaminoethyl 2-phenylbutyrate. A spasmolytic agent used in gastrointestinal disorders.
f butétamate
d Butethamat
i butetamato
e butetamato

485 butidrine
A beta adrenergic blocking agent used in the prophylaxis and treatment of disorders affecting the coronary arteries and for the relief of anginose pains.
f butidrine
d Butidrin
i butidrina
e butidrina

486 butixirate
Topical antiphlogistic and analgesic agent mainly used to relieve the painful conditions borne of phlebitis, thrombophlebitis, rheumatism affecting muscles, contusions, tendinitis and, more generally, any skin inflammation. It is also used in the treatment of post-traumatic oedemata.
f butixirate
d Butixirat
i butixirato; butisirato
e butixirato

487 butizide
Chlorothiazide-related diuretic agent capable of inhibiting the reabsorption of chlorine and sodium and, to a smaller measure, potassium in the proximal renal tubules. It is used in the treatment of oedemata of various nature, as well as for the treatment of hypertension.
f butizide
d Butizid
i butizide
e butizida

488 butobarbitone
A barbiturate hypnotic acting on the sub cortex. It induces a form of sleep that is very similar to the physiological or normal one and used in idiopathic insomnia.
f butobarbitale; butétal
d Butobarbiton
i butobarbitone; butobarbitale
e butobarbitona; butobarbital

489 butopiprine
A mucolytic agent used to relieve cough and in disorders of the respiratory tract, as well as in cases of tracheobronchitis, pneumonia, bronchopneumonia, acute laryngitis etc.
f butopiprine
d Butopiprin
i butopiprina
e butopiprina

490 butriptyline
Antidepressant agent used to relieve states of depression that are intrinsically not psychotic and not necessary accompanied by a state of anxiety.
f butitriptyline
d Butitriptylin
i butitriptilina
e butitriptilina

491 butropium bromide
Atropine-derivative parasympatholytic agent with antisecretory and antispastic properties in relation to the gastrointestinal smooth muscles and used for the relief of the spastic pains which are caused by gastritis, spastic colitis, gasstroduodenal ulcer etc.
f bromure de butropium
d Butropiumbromid
i butropio bromuro
e bromuro de butropio

*** butyl aminobenzoato   s. benzonatate**

492 butylchloral hydrate
Trichlorobutylene glycol. A compound having the same action of chloral hydrate as hypnotic and sometimes used as an analgesic.
f butylchloral hydraté
d Butylchloralhydrat
i butilcloralo idrato
e hidrato de butilcloral

*** butylethylbarbituric acid   s. butobarbital**

493 butylscopolamine bromide
Anticholinergic spasmolytic agent capable of blocking the transmission of impulses at the parasympathetic ganglionic synapses of the smooth muscles of pelvic and abdominal organs. It is characterized by the total absence of atropine-like effects affecting the central nervous system, the heart, eye and salivary glands. The drug is used for the treatment of spastic conditions of the gastrointestinal tract, gastrointestinal ulcers, spasms and dyskinesia of the urinary and biliary tracts and cervical spasms.
f bromure de butylscopolamine
d Skopolaminbromid
i scopolamina bromuro; ioscina butobromuro
e bromuro de scopolamina

494 buxamine
A physiological substance having a regulating action on the nervous excitability and used in some states of anxiety and tension, subjective disorders due to hypertension, idiopathic and symptomatic epilepsy and convulsion affecting children.
f buxamine
d Buxamin
i bussamina
e buxamina

C

495 **cabbage tree**
Genus of plants found in tropical America whose bark and seeds are used locally as a substance capable of expelling parasitic worms from the intestine. Also known as worm bark tree or yellow bark tree.
f geoffré; geoffrier jaune
d gelber Wurmrindenbaum; Geoffroy-Baum
i andira; andira gialla
e angelin amarillo

496 **cabbage-tree bark**
The bark of the cabbage tree, from which a substance is extracted that is capable of expelling parasitic worms from the intestine.
f écorce de geoffré
d Geoffroy-Rinde
i corteccia di andira
e corteza de andira

* **cacaine**   s. theobromine

497 **cacao**
Term denoting both the plant and the seed of some species of the genus Theobroma (of the Sterculaceae family) cultivated in tropical countries. The cacao seeds contain two alkaloids, i.e. theobromine and caffeine and a fat known as cocoa butter. The theobromine that is obtained from the husks is used as a diuretic and a mild stimulant.
f cacao
d Kakao
i cacao
e cacao

498 **cacodylate**
Salt of the cacodylic acid, one of the least toxic and irritant of the arsenical preparations. The most important cacodylates are those of iron, sodium, calcium and manganese, which are used in the treatment of anaemias, leukaemias, dermatoses, asthenias etc. In particular, sodium cacodylate is administered by injection in those cases in which an arsenical therapy is indicated.
f cacodylate
d Kakodylat
i cacodilato
e cacodilato

499 **cade oil**
The oil obtained from the juniper tar; an essential, cadinene-containing yellowish liquid used for pharmaceutical products and as a parasiticide in some cutaneous diseases. (Cadinene is a dicyclic sesquiterpene which contains a reduced naphthalene skeleton) Also known as juniper tar.
f huile de cade; goudron de cade
d Kadeöl; Wacholderteer; Kaddigöl
i olio cadino; olio nero di ginepro
e aceite de enebro; alquitrán de cade

500 **cadmium sulphide**
A yellow precipitate used in medicine to treat skin diseases. Also called cadmium yellow.
f sulfure de cadmium; jaune de cadmium
d Cadmiumsulfid; Schwefelcadmium; Cadmium gelb
i solfuro di cadmio; giallo di cadmio
e sulfuro de cadmio; amarillo de cadmio

501 **cafaminol**
A xanthine derivative capable of promoting the secretion of nasal mucosa and used in the therapy of common colds, in particular when accompanied by respiratory difficulties, as well as of inflammation and congestion of nasopharyngeal mucosa. It is also used in the postoperative treatment of the nasal septum.
f cafaminol
d Cafaminol
i cafaminolo
e cafaminolo

502 **cafcycline**
Chloramphenicol succinate of tetracycline. A broad-spectrum antibiotic formed by a complex salt of chloramphenicol succinate and tetracycline. It is frequently used to combat Gram-positive and Gram-negative microorganisms that are sensitive to the above mentioned salt.
f cafcycline
d Cafzyklin
i cafciclina
e cafciclina

503 **cafedrine**
Analeptic agent used in the therapy of cardiovascular disorders.
f cafédrine
d Cafedrin
i cafedrina
e cafedrina

504 caffeine
Purine-related alkaloid naturally present in the leaves of the tea plant (5%), in coffee seeds, in maté and in the kola nut. It may also be prepared synthetically starting from the theobromine of the cocoa or from the cyanacetic acid and urea, or from uric acid. Caffeine has a strong tonic action on the central nervous system by speeding up intellective processes and promoting a greater muscular and intellectual power, both by diminishing the sense of somnolence and tired ness and improving the capacity of coordination of all movements. It is also used to combat poisoning by narcotics and to improve cardiac tone as it has a stimulating effect on the myocardium. Caffeine is also a mild diuretic and vaso dilator.
f caféine; théine; triméthylxanthine
d Kaffein; Thein; Trimethylxanthin
i caffeina; teina; trimetilxantina
e cafeína; teína; trimetilxantina

505 cainca
The common term of the two species of Chiococca (anguifera and racemosa), the small trees of the Rubiaceae found in South America, whose roots have diuretic, powerfully purgative and other properties, e.g. the capacity to neutralize the poison from a snake bite. They contain, as active principles, tannic acid, a glucoside (caincic acid) and an alkaloid. Also known as David's root tree.
f caïnca; chiocoque
d Kainka; Schneebeerenbaum
i cainca
e cainca

506 caincic acid
Or cahinic acid, cahincic acid. A glucoside that is obtained from cahinca roots and used as a purgative and a diuretic.
f acide caïncique; caïncine
d Kainkasäure
i acido caincinico; caincina
e ácido caincínico; caincina

507 cajeput
Or cajuput, swamp tree tea, paperbark. An East Indian tree (Malaleuca leucadendron) from which, by the distillation of the leaves in a stream of steam, a camphor-smelling liquid is obtained, which is constituted by a mixture of terpenes. It is used in medicine as a pulmonary antiseptic and antispasmodic and also, externally, as analgesic in otalgia.
f cajeput; cajeputier
d Kajeputbaum
i caieput
e cayeput

508 cajeputene
A terpene ($C_{10}H_{16}$) contained in the cajeput oil and identical with depentene, a liquid terpene hydrocarbon. Also known as cinene or limonene.
f cinène; limonène
d Kajeputen; Cinen; Kautschin
i caieputene
e cayeputeno; cineno

509 cajeput oil
A camphor-smelling liquid, colourless when pure, consisting of a mixture of terpenes (cineole, pinene, terpineol etc.) enriched with turpentine, lavender etc. and used as a pulmonary antiseptic and antispasmodic, as well as for external use to mitigate earache.
f essence de cajeput
d Kajeputöl
i olio di caieput
e esencia de cayeput

510 Calabar bean
Or chapnut, esere nut. The dark-brown potently poisonous seed of a tropical African woody wine, used in pharmacology as a source of physostigmine from which it differs in its action.
f fève de Calabar; éseré
d Kalabarbohne; Kalabarsamen
i fave del Calabar
e haba del Calabar

511 calamine
Native carbonate of zinc generally found in stalactic aggregates or concretions. Pharmacologically, an amorphous, pink or reddish powder composed of basic silicate of zinc with the addition of a variable amount of ferric oxide and used as dusting powder, cream or ointment in the treatment of some forms of dermatitis.
f calamine
d Kalamin
i calamina
e calamina

512 calamus
The dried rhizome of the Acorus calamus used as an agent promoting digestion by stimulating gastric action and as a tonic capable of improving appetite.
f acore
d Kalmus
i calamo
e cálamo

513 calamus oil
Yellow liquid characterized by an aromatic smell and containing various terpenes. It is used for stimulating baths and anti-rheumatic ointments.
f essence d'acore
d Kalmusöl
i olio di calamo
e esencia de cálamo

514 calcifediol
First active metabolite of vitamin D_3. It is used in osteomalacia, renal osteodystrophia, idiopathic or postoperative hypoparathyroidism, spasmophilia, hypocalcaemia, rickets etc.

f calcifédiol
d Calcifediol
i calcifediolo
e calcifediolo

515 calciferol
Vitamin D_2, ergocalciferol. A colourless crystalline substance obtained by the ultraviolet irradiation of ergosterol. It is an antirachitic agent and promotes the absorption of calcium and phosphorus from the intestine thus proving essential for the normal development of bones and teeth. In therapeutical medicine, the drug is indicated for the treatment of rickets, spasmophilia and tetany, as well as for those osteopathies which are characterized by a reduced amount of normally calcified bone. Also used to treat tuberculosis and in the course of pregnancy and lactation.
f calciferol; ergostérine irradiée; vitamine D_2
d Calciferol; Vitamin D_2
i calciferolo; ergosterolo; ergocalciferolo; vitamina D_2
e calciferol; vitamina D_2

516 calcitonin
Hormone from the thyroid glands capable of controlling and reducing calcium metabolism. It also lowers the levels of plasma phosphates by inhibiting osseous reabsorption. It is indicated in hypercalcaemia, the Paget's disease and osteoporosis.
f calcitonine
d Calcitonin
i calcitonina
e calcitonina

517 calcitriol
A potent drug indicated in order to improve the metabolism of vitamin D.
f calcitriol
d Calcitriol
i calcitriolo
e calcitriolo

518 calcium
Reactive metal of the alkaline-earth group copiously distributed in nature as compound, mainly the carbonates, the phosphates and the silicates. In the human body it is present both in the blood and in the extracellular fluid either ionized or bound to the plasma protein. In addition to being essential for a normal functioning of cardiac muscle, blood coagulation and the regulation of cell permeability, it plays a rather important role in nerve-impulse transmission and in the mechanism of the neuromuscular system.
f calcium
d Calcium; Kalzium
i calcio
e calcio

519 calcium alginate

Precipitate which is formed when sodium alginate solution is applied to raw surfaces. It is slowly absorbed and is used as a haemostatic agent.
f alginate de calcium
d Calciumalginat
i calcio alginato
e alginato de calcio

520 calcium aminosalicylate
Antituberculous agent used in the treatment of pulmonary tuberculosis either by itself or in association with isoniazid and streptomycin. It has proved capable of acting by inhibiting tuberculosis activators, e.g. benzoic acid and salicylic acid.
f aminosalicylate de calcium
d Calciumaminosalicylat
i calcio aminosalicilato
e calcio aminosalicilato

521 calcium benzamidosalicylate
Antituberculous agent, both potent and long-acting, derived from para-aminosalicylic acid (the essential metabolite for certain bacteria). Calcium benzamidosalicylate combines in the liver with glycocoll or undergoes acetylation. Peak serum levels are quickly reached and the excretion takes place slowly. The drug, unless there is resistance to para-aminosalicylic acid (PAS), is mainly used for the treatment of pulmonary or urogenital tuberculosis.
f benzamidosalicylate de calcium
d Calciumbenzamidosalicylat
i calcio benzamidosalicilato
e calcio benzamidosalicilato

522 calcium bromide
A salt of hydrobromic acid which was used in the past in the treatment of epilepsy. It is a granular, white, odourless, saline-tasting powder and finds employment as a sedative as well as for its action on the central nervous system, on the heart and bones, on the blood coagulation and on the muscles. More generally, it presents the sedative action of bromine.
f bromure de calcium
d Calciumbromid
i calcio bromuro
e bromuro de calcio

523 calcium bromolaevulinate
An organic compound combining the neuro vegetative sedative, allergy-inhibiting and mineralizing action of the calcium with the sedative action of bromine on the cortical centres. It is used in states of excessive nervous irritability, in dystonias affecting the neuro-vegetative system, some allergic states, headache, convulsions and spasmophilia (spasmophilic diathesis), i.e. a morbid state marked by an abnormal response of the motor nerves to electrical or mechanical excitation, shown by a tendency to teta-

ny, convulsions and spasm.
f bromolévulinate de calcium
d Calciumbromolävulinat
i calcio bromolevulinato
e bromolevulinato de calcio

524 calcium carbaspartate
Chemotherapeutic agent acting as tranquil‐
lizer and capable of causing stimulation.
It is used in the treatment of asthenia,
anxiety, insomnia and, in children, in
behaviour anomalies and anorexia.
f carbaspartate de calcium
d Calciumcarbaspartat
i calcio carbaspartato
e carbaspartato de calcio

525 calcium carbimide
Therapeutic agent used in the treatment
of alcoholism as it inhibits the acetal-
dehyde-oxidizing enzymes.
f carbimide de calcium
d Carbimidcalcium
i calcio carbimide
e carbimido de calcio

526 calcium caseinate
Yellow-white powder used as nourishment
in cases of diarrhoea, particularly in
children. It further serves as a supplement protein intake.
f caséinate de calcium
d Calcium‑caseinat
i calcio caseinato
e caseinato de calcio

527 calcium chloride
Therapeutic agent used in combination
with vitamin D and phosphorus in the
treatment in diseases affecting the bones
and the central nervous system. It is
also indicated in the therapy of tetany
due to calcium deficiency. The drug is
essential for the coagulation of the blood
and for the strengthening of the tone of
the myocardium.
f chlorure de calcium
d Calciumchlorid
i cloruro di calcio; calcio cloruro
e cloruro de calcio

528 calcium cyclamate
The dihydrate of calcium cyclohexylsul-
phamate used as a sweetening agent as
an alternative to sodium cyclamate whenever it is advisable to avoid an excess
of sodium in the diet. In a dilute solution, the sweetening power of calcium
cyclamate is about 30 times that of sugar.
f cyclamate de calcium
d Calciumcyclamat
i calcio ciclamato
e ciclamato de calcio

529 calcium dobesilate
Angioprotective drug especially indicated
for the treatment of patients requiring a
systemic control of capillary bleeding
and in order to correct an abnormal permeability and fragility of the capillary

vessels. It therefore has an antihaemorrhagic effect and is used to check bleeding caused by diabetes (in particular
diabetic retinopathy), in eye surgery and
in the therapy of venous inadequacy.
f dobésilate de calcium
d Calciumdobesilat
i calcio dobesilato
e dobesilato de calcio

530 calcium folinate
Therapeutic agent capable of exerting an
antianaemic action by improving biosynthesis in the course of the maturation of
the erithroblast. It is thus used in the
therapy of macrocytic non-megaloblastic
anaemia caused by wrong or deficient
diet.
f folinate de calcium
d Calciumfolinat
i calcio folinato
e folinato de calcio

531 calcium glubionate
A drug which is used to improve the process of ossification and blood coagulation.
It further shows a considerable trophic
action on the central nervous system. It
should not be administered in cases of
renal insufficiency.
f glubionate de calcium
d Calciumglubionat
i calcio glubionato
e glubionato de calcio

532 calcium gluconate
Tasteless white powder used in medicine
for its antiinflammatory and antianaphylactic action, as well as in cases of
calcium deficiency. It is highly advisable after massive blood transfusions.
f gluconate de calcium
d Calciumgluconat
i calcio gluconato
e gluconato de calcio

533 calcium glycerophosphate
An anabolic and antianaemic agent capable of promoting calcification. It also
supports the metabolism of cerebral phosphorus and muscular activity. It is
mainly used in mental states associated
with advancing age, in phosphaturia,
osteoporosis and in convalescence.
f glycérophosphate de calcium
d Calciumglycerophosphat
i calcio glicerofosfato
e glicerofosfato de calcio

534 calcium glycolate
Mild diuretic agent used in the treatment
of functional disorders of the liver, of
arteriosclerosis and oliguria.
f glycolate de calcium
d Calciumglykolat
i calcio glicolato
e glicolato de calcio

535 calcium hydroxide
White powder characterized by a slightly

bitter taste. It is used as solution of lime water in order to activate the chymo sin (rennin or chymase) and thus facilitate the digestion of milk in infants. It is also used to treat diarrhoea in babies.
f hydroxyde de calcium; chaux éteinte
d Calciumhydroxyd; gelöschter Kalk
i calcio idrossido
e hidróxido cálcico

536 calcium hydroxide solution
The solution of lime water which is used as an alkali and as an antidote.
f solution d'hydroxyde de calcium
d Calciumhydroxydlösung
i soluzione d'idrossido di calcio
e solución de hidróxido cálcico

537 calcium iodide
A salt of hydriodic acid found in nature in sea water and in marine algae, and prepared industrially by the reaction of iodine with calcium sulphide or by the neutralization of hydriodic acid with lime or calcium carbonate. It is used in medicine as an antisyphilitic agent.
f iodure de calcium
d Calciumjodid
i ioduro di calcio
e yoduro cálcico

538 calcium lactate
Salt of the lactic acid, a white, crystalline, odourless and tasteless powder, soluble in water but insoluble in alcohol and ether. It is used in medicine like calcium chloride, i.e. mainly as a trophic agent for tissues and bones and as an antiallergic.
f lactate de calcium
d Calciumlactat
i calcio lattato
e lactato de calcio

539 calcium lactophosphate
Mixture of lactate, acid lactate and acid phosphate, a white granular powder in appearance, which is rarely used in cases of calcium deficiency.
f lactophosphate de calcium
d Calciumlactophosphat
i calcio lattofosfato
e lactofosfato de calcio

540 calcium laevulinate
Or calcium laevulate. White crystalline powder, easily soluble and quite stable in solution. It is used in medicine whenever a calcium therapy is indicated.
f lévulinate de calcium
d Calciumlävulinat
i calcio levulinato
e levulinato de calcio

541 calcium pantothenate
Vitamin factor promoting cellular metabolism and protecting the skin and the liver. It is mainly used as an antacid agent capable of controlling the growth of the intestinal flora. It is further in-

dicated in convalescence and in the therapy of rickets and various allergies.
f pantothénate de calcium
d Calciumpantothenat
i calcio pantotenato
e pantotenato de calcio

542 calcium peroxide
A compound that is obtained by reaction of the sodium peroxide with the solution of a calcium salt and subsequent crystallization. A yellowish, crystalline powder used in medicine as an antacid and antiseptic agent. It is also employed for the sterilization of water.
f peroxyde de calcium
d Calciumperoxyd
i calcio perossido; perossido di calcio
e peróxido de calcio

543 calcium phosphate
Or calcium phosphate tribasic. A therapeutic agent capable of promoting recalcification. It is used as an antacid, adsorbant and antidiarrhoic agent. It may cause constipation and should be avoided in the course of a therapy based on digitalis.
f phosphate de calcium tribasique
d Tricalciumphosphat
i calcio fosfato tribasico
e fosfato de calcio tribásico

544 calcium phosphate (dibasic)
Anabolic (capable of making new living tissues from nutrient materials) and haemopoietic agent acting as a tonic and capable of improving the conditions of the nervous and muscular systems and of promoting calcification.
f phosphate de calcium dibasique
d Dicalciumphosphat
i calcio fosfato dibasico
e fosfato cálcico dibásico

545 calcium polystyrene sulphonate
Ion exchange resin used in the treatment of electrolyte abnormalities by altering absorption or excretion in the intestine.
f sulfonate de calcium polystyrène
d Calciumpolystyrolsulfonat
i solfonato di calcio polistirene
e sulfonato de calcio polistirene

546 calcium saccharosates
Various compounds which are formed when calcium hydroxide is dissolved in sugar solutions. They are used in all cases of calcium deficiency.
f saccharosates de calcium
d Calciumsaccharosaten
i saccarosati di calcio
e sacarosatos de calcio

547 calcium sodium ferriclate
Antianaemic and haemopoietic factor used in the therapy of anaemias marked by deficiency of iron in the blood.
f ferriclate de calcium sodique
d Calciumnatrionferriclat

ı calcıo sodıo ferriclato
e ferriclato de calcıo sódico

548 calcium sulphaloxate
Sulphonamide antıbacterial agent capable of actıng on pyogenic coccı, pneumococcı, menıngococcı and colıbacıllı. (See sulfadımıdıne). It is principally used for the therapy of a mıld form of ınfectıve dıarrhoea.
f sulfaloxate de calcıum
d Calcıumsulfaloxat
ı calcıo solfalossato
e sulfaloxato de calcıo

549 calcium trisodium pentetate
A chelatıng compound used for the dıagnosıs and for the treatment of poisoning caused by lead, cadmium, chromıum, iron, gold, cobalt, manganese, nickel, plutonium and other radıoactıve metals. It also fınds employment ın the therapy of thalassaemia major, haemosiderosis and haemochromatosis.
f pentétate de calcium trisodique
d Calcıumtrisodiumpentetat
ı calcıo trisodıo pentetato
e pentetato de calcıo trısódico

550 calendula
Or marıgold. Group of plants wıth composıte flower formatıons. The calendula offıcınalıs ıs used ın medıcıne for ıts stımulant properties.
f soucı
d Goldblume; Totenblume; Rıngelblume
ı calendula; fıorrancıo
e caléndula; maravılla

551 calliandra
Or pambotano. Term denoting the bark of the root of the Callıandra, a genus of tropıcal pınnate-leaved trees and shrubs wıth clustered flowers, yıeldıng an essentıal oıl, tannın, callıandreın (a glycosıde), salts of calcıum, potassıum, magnesıum etc. It has febrıfuge properties.
f callıandre; pambotano
d Pambutano
ı pambotano
e panbotano; ponbotano

552 calomel
Mercurous chlorıde. A drug used as a strong purgatıve, especıally ın the treatment of anımals. Parenterally, ıt ıs used as an antıluetıc agent. Also known as horn quıcksılver.
f calomel; chlorure de mercure
d Kalomel; Mercurochlorid
ı calomelano; mercuroso cloruro
e calomel; calomelano; cloruro mercurıoso

553 calumba
The drıed slıced root of a plant growıng ın Mozambıque. It contaıns alkaloıdal quaternary bases and non-alkaloıdal bıtter prıncıples (extremely bıtter). It ıs used as a tonıc.
f colombo

d Kalumbawurzel; Ruhrwurzel
ı colombo
e colombo

554 calumbin
The non-alkaloıdal bıtter prıncıple that ıs derived from the calumba root. It ıs the glucoside of the lactone of calumbıc acıd, a yellow amorphous acıd.
f colombine
d Kalumbın
ı colombına
e colombına

555 camazepam
A benzodiazepinic drug ındıcated for the treatment of psychic dısorders. It has a marked sedatıve actıon and may be described as a tranquıllızer without myorelaxant and cardiodepressant effects. It ıs recommended for all forms of anxıety neuroses, neurovegetative dystonias, psychosomatic ıllnesses etc.
f camazépam
d Camazepam
ı camazepam
e camazepam

556 cambogia
Or gamboge, gamboge gum. Yellowısh gum resın that ıs obtaıned from the stem of Garcınıa hamburyı and was used ın medıcıne as a hydragogue (causing a watery dıscharge from the bowels) cathartıc agent.
f gomme gutte; gutte
d Guttı; Gummıgutt
ı gambogıa; gomma gotta
e gambogıa; gutagamba

557 camphene
Unsaturated terpene exıstıng ın three ısomeric forms and present ın many essentıal oıls, e.g. turpentıne, camphor, cypress oıl etc. It may be obtaıned by catalytic ısomerızatıon of alpha-pınene or by the actıon of alkalıs on the bornyl chlorıde that ıs yıelded by pınene and hydrochlorıc acıd. It ıs used ın preparatıons for the treatment of broncho-pulmonary dısorders.
f camphène
d Kamphen
ı canfene
e canfeno

558 camphor
Generıc term for a group of dıcyclıc terpenes present ın plants. Specıfically, the colourless, crystallıne substance that ıs obtaıned from the wood of the Cinnamomum camphora found ın Japan, Taiwan and Chına. Camphor has a strong aromatıc taste and odour and an ırrıtant actıon on the skın and mucous membranes causıng local vasodılatatıon wıth an ensuıng slıght analgesia. Often used as an adjuvant ın expectorant preparatıons and, ın a solution of arachıs oıl for subcutaneous ınjectıon, ın order to pro-

voke, by local irritation, reflex stimulation, of the medullary centres.
f camphre
d Kampfer
i canfora
e alcanfor

559 camphorate
A liniment, i.e. a preparation for application to the skin either by friction or
by some dressing, formed by a mixture of
camphor and olive- or cotton-seed oil.
f camphorate
d Kampforat; Campforat
i canforato
e canforato

560 camphoric acid
Dibasic acid obtained by the oxidation
of camphor. It is used in medicine as a
laryngeal antiseptic, as well as a urinary and intestinal disinfectant.
f acide camphorique
d Kampfersäure
i acido canforico
e ácido canfórico

561 camphor liniment
Camphorated oil, i.e. a preparation consisting of 20% of camphor in arachis oil.
It is administered by injection as a reflex stimulant of the central nervous system, as well as a counter-irritant in
bronchitis and non-articular rheumatism.
f liniment de camphre
d Kampferliniment
i linimento di canfora
e linimento de alcanfor

562 camphor oil
The liquid that is obtained in the course
of the distillation of the camphor wood in
a stream of steam. It is a mixture of
terpenic hydrocarbons, safrol, eugenol
etc. It may be light or heavy according
to its purity and to the interval of
distillation. It is used as an antiseptic
agent. See also essential camphor oil and
rectified camphor oil.
f huile de camphrées
d Kampferöl
i olio di canfora
e aceite de alcanfor

563 camphosulphonic acid
Monosulphonic, optically active acid prepared by causing the camphor to react
with fuming sulphuric acid. It is used in
medicine, in the form of salt (of calcium,
sodium, piperazine, magnesium etc.), as
a cardiac and respiratory stimulant.
f acide camphosulfonique
d Kampfosulfonsäure
i acido canfosolfonico
e ácido sulfocanfórico

564 camphotamide
An analeptic, respiratory and circulatory
stimulant agent indicated for the treatment of cardiocirculatory collapse, conta-

gious diseases and intoxications.
f camphotamide
d Camphotamid
i canfotamide
e canfotamida

565 camphothymol
Oily liquid, insoluble in water and soluble in alcohol and ether, formed by a
mixture of equal parts of camphor and
thymol. It is an antiseptic and sedative
agent.
f camphothymol
d Camphothymol
i canfotimolo; timolo canfora
e canfotimolo; timolo alcanfor

566 camylofin
A parasympatholytic and anticholinergic
agent having a spasmolytic action on
smooth muscles. It is used in the treatment, and for the relief, of hepatic,
biliary and urinary colics, uterine pains,
dysmenorrhoea and various stomach disorders.
f camylofine
d Camylofin
i camilofina
e camilofina

567 candicidin
An antibiotic having an antimycotic activity particularly against Candida (a
genus of yeast-like fungi causing candidiasis). It is indicated in the treatment
of skin and vaginal candidiasis.
f candicidine
d Candicidin
i candicidina
e candicidina

568 cannabinine
The poisonous alkaloid which is found in
the essential oil of Indian hemp.
f cannabinine
d Cannabinin
i cannabinina
e canabinina

569 cannabinol
Colourless or yellow oil, smelling like
Indian hemp, capable of producing a
state of euphoria. Its tetrahydroderiva-
tive presents properties which are analogous to those of marijuana. It finds
little use in medicine.
f cannabinol
d Cannabinol
i cannabinolo
e canabinol

570 canrenone
Aldosterone-inhibiting diuretic agent. It
is chiefly used in the therapy of ascitic
and oedematous cirrhosis, chronic oedema
tous syndromes due to cardiac insufficien
cy, kidney disorders, hyperaldosteronaemia caused by hypertension and
myasthenia. In the course of treatment
with the drug, the electrolyte levels in

the plasma and azotaemia (the accumulation of nitrogenous compounds in the
blood) should be regularly checked.
f canrénone
d Canrenon
i canrenone
e canrenona

571 cantharidate
Salt of cantharic acid, a monocarboxylic
acid, isomer of cantharidine from which
it is obtained by action of the hydriodic
acid or of the chlorosulphonic acid.
f cantharidate
d Cantharidat
i cantaridato
e cantaridato

572 cantharidin
Solid crystalline substance contained
(about 1%) in several insects which have
been killed and dried with ether. It is
found in the form of crystals or lamellae
and has a burning taste. It is used in
medicine as a counter-irritant and, if
administered internally, as an aphrodisiac. Also used in the topical treatment
of verrucae and molluscum contagiosum.
f cantharidine
d Cantharidin
i cantaridina
e cantaridina

573 canthaxantine
Plant pigment found in animal tissue and
used to tan the skin without exposing it
to the sun. The drug is administered
orally and does not prevent burns.
f canthaxantine
d Canthaxantin
i cantaxantina
e cantaxantina

574 capobenic acid
A sodium salt capable of acting as thera
peutic agent in angina pectoris by promoting a moderate increase of the coronary circulation without affecting the consumption of oxygen in the myocardium.
f acide capobénique
d Capobensäure
i acido capobenico
e ácido capobénico

575 capreomycin
Antibiotic produced by Streptomyces ca-
preolus and particularly active against
Mycobacterium tuberculosis. It is especial
ly indicated when the responsible micro-
organism has grown resistant to para-
-aminosalicylic acid (PAS), cycloserine,
streptomycin and ethionamide. A cross-
-resistance with neomycin and other similar drugs may arise. Ototoxicity and
nephrotoxicity are among its possible
side effects.
f capréomycine
d Capreomycin
i capreomicina
e capreomicina

576 capronium chloride
A parasympathomimetic agent having a
good effect on the gastric plexus and
used as peripheral vasodilator in the
treatment of gastric ptosis and achlorhydria. It promotes the growth of hair
in alopecias.
f chlorure de capronium
d Kaproniumchlorid
i capronio cloruro
e cloruro de capronio

577 capsaicine
Crystalline substance constituting the bit
ter active principle of the capsicum
annuum having rubefacient and revulsive
action.
f capsaïcine
d Capsaicin
i capsaicina
e capsaicina

578 capsicin
Therapeutic agent causing reddening of
the skin and used externally in the form
of tincture, liniment or plaster. It is
generally employed to relieve rheumatic
pains, muscular stress, peripheral vascular disorders when caused by cold. It
also has an internal use as it acts on
the circulation and on the respiration receptors owing to an activity which is
very similar to that of lobeline and
veratrine.
f capsicine
d Capsizin
i capsicina
e capsicina

579 captodiame
A tranquillizer characterized by its sedative and antispasmodic properties. It is
mainly indicated in the therapeutic treatment of neurasthenia and states of anxiety connected with heart disorders. It
also finds some use in the therapy of
hyperthyroidism and of psychoses appearing in the course of the menopause.
f captodiame
d Captodiam
i captodiamo
e captodiamo

580 captopril
Agent capable of inhibiting the enzyme
involved in the formation of hormone controlling the maintenance of blood pressure. It is used in the treatment of hypertension and in heart disorders. Hypotension, proteinuria, bone marrow depression etc. may appear as side effects of
the drug.
f captoprile
d Captopril
i captoprile
e captoprilo

581 caramiphen
Therapeutic agent capable of relieving or
checking cough and characterized by a

central and peripheral mechanism. It is particularly used in the treatment of violent cough due to some upper respiratory infection. Its action is similar to that of papaverine and atropine. Also used as a bronchial spasmolytic.

f caramiphène
d Karamiphen
i caramifene
e caramifeno

582 caraway

Or caraway oil. An essential oil used as a carminative, i.e. an agent capable of expelling gas from the alimentary canal and thus relieve flatulence, griping and colic.

f essence de carvi
d Kümmelöl
i comino; olio di comino
e esencia de alcaravea

583 carbachol

Parasympathomimetic having the same action of acetylcholine, but considerably more prolonged. It is used as miotic (causing contraction of the pupil of the eye) in glaucoma and also as a tonic for intestinal and bladder muscles after a surgical operation.

f carbachol
d Carbachol
i carbacolo
e carbacolo

584 carbamazepine

Anticonvulsant agent used in the treatment of epilepsy. It is also an analgesic chiefly used for the relief of trigeminal neuralgia.

f carbamazépine
d Carabamazepin
i carbamazepina
e carbamazepina

585 carbamide

Amide of the carbamic acid (urea). A white crystalline substance, easily soluble in water and alcohol. It is excreted in the urine of mammals, birds and certain reptiles as the end-product of the protein metabolism as it is formed in the liver by amino acids. It combines with acids to form ureides. Used in medicine as a diuretic agent.

f carbamide; urée
d Carbamid; Harnstoff
i carbamide; urea
e carbamida; urea

586 carbaryl

An antiparasitic, insecticide agent having the property of inhibiting cholinesterase, the enzyme which inactivates by hydrolysis the acetylcholine produced at the endings of voluntary motor nerves and nerves of the parasympathetic division of the involuntary system by nerve impulses, and checks its accumulation.

f carbaryl

d Carbaryl
i carbarile
e carbarilo

587 carbasalate calcium

An analgesic, antirheumatic and antipyretic agent. It should not be administered in haemorrhagic diathesis.

f carbasalate calcique
d Calciumkarbasalat
i calcio carbasalato
e carbasalato cálcico

588 carbasone

Pentavalent organic arsenical compound containing 28.5% of arsenic. A chemotherapeutical agent chiefly used in amoebic dysentery for its ability to act against vegetative forms and cysts in the alimentary tract. It is generally non-toxic, although skin rashes have occasionally been noticed. It also acts against Trichomonas (a genus of flagellate Protozoa) vaginalis and intestinalis.

f carbasone
d Carbason
i carbasone
e carbasona

589 carbazochrome

A systemic haemostatic agent derived from epinephrine and characterized by a physiological mechanism of action. It is used for the prophylaxis and treatment of haemorrhagic conditions.

f carbazochrome
d Carbazochrom
i carbazocromo
e carbazocromo

590 carbenicillin

Semisynthetic penicillin quite active against Gram-negative bacteria, e.g. Escherichia coli and Pseudomonae. It finds good employment in the therapy of urinary infections as well as in systemic infections caused by susceptible organisms.

f carbénicilline
d Carbenizillin
i carbenicillina
e carbenicilina

591 carbenoxolone

Antiinflammatory agent and antacid. It is used in the treatment of gastric and duodenal ulcers and mouth ulcers, as well as of various ulcerative diseases.

f carbénoxolone
d Carbenoxolon
i carbenossolone
e carbenoxolona

592 carbimazole

Therapeutic agent capable of inhibiting the utilization of iodine for the biosynthesis of thyroid hormones. It appears capable of cancelling the action of peroxidase which brings about the oxidation of iodine. The drug is used in the treatment of primary and secondary thyrotoxicosis and of the Basedow's disease, as well as in some cases of obesity.

f carbimazol
d Carbimazol
i carbimazolo
e carbimazolo

593 carbinoxamine

Antihistamine causing desensitization by inhibiting the action of serotonin. It is used in the form of maleate for the treat ment of some acute allergies, e.g. urticaria, asthma, coryza etc.

f carbinoxamine
d Carbinoxamin
i carbinossamina
e carbinoxamina

594 carbocisteine

A mucolytic agent capable of exerting a regulating action on mucous secretions by reducing their viscosity in the respiratory tract. It also promotes expectoration by its direct lytic action on mucopolysaccharides. It further restores the physiological secretion of fluid mucus, which protects the respiratory tract, for a regenerative action on mucosal cells. It is mainly used in the treatment of chronic bronchitis, rhinitis, pharyngitis, sinusitis, otitis etc.

f carbocistéine
d Carbocistein
i carbocisteina
e carbocisteina

595 carbocromen

An elective coronary vasodilator used to augment the blood flow in those cases in which the supply of oxygenated blood through the coronary arteries is reduced. The drug can be generally used as it does not cause hypotension.

f carbocromène
d Carbocromen
i carbocromene
e carbocromeno

596 carbolic acid

Phenol. Soluble in water but preferably in alcohol, chloroform, ether or glycerin. It is obtained from coal tar and used as an antiseptic and disinfectant.

f acide carbolique; phénol
d Carbolsäure; Phenol
i acido carbolico; fenolo
e ácido carbólico; fenol

597 carbon dioxide

A gas easily dissolving in water to form carbonic acid. It is an end-product of the tissue oxidation of carbohydrates and fats. It stimulates the respiratory centres and, mixed with oxygen, is used as inhalation in cases of asphyxia, poisoning by carbon oxide, abuse of sleeping drugs and other depressants of the central nervous system, as well as to facilitate the elimination of possible inhalation of anaesthetics. It is also used to relieve persistent hiccup. In a solution (aerated water) or in the form of orally-administered bicarbonate, it promotes the secretion of gastric juice, particularly its hydrochloric acid.

f bioxyde de carbone
d Kohlendioxyd
i biossido di carbonio
e dióxido de carbono

598 carbon tetrachloride

A therapeutic agent used for its antihelminthic action against Nematoda (a phylum of round worms). It is very rarely administered orally owing to its high degree of toxicity for the liver.

f tétrachlorure de carbone
d Tetrachlorkohlenstoff
i carbonio tetracloruro
e tetracloruro de carbono

599 carboquone

Alkylating agent capable of a cytostatic action used in the therapy of the symptomatology, both subjective and objective, of lung cancer and chronic myeloid leukaemia. The use of the drug requires periodical blood counts.

f carboquone
d Carboquon
i carboquone
e carboquona

600 carbromal

A mild sedative and hypnotic drug used for the treatment of slight insomnia in cardiopathic and elderly patients. It is also indicated in cases of mild insomnia in young children, as well as to relieve anxiety and psychoses due to climacterium.

f carbromal
d Carbromal
i carbromal
e carbromal

601 carbubarb

Sedative unaccompanied by hypnotic action and used in the treatment of anxiety states, psychic instability, neurovegetative disorders and neurotic depression.

f carbubarb
d Carbubarb; Carbubarbital
i carbubarbo; carbubarbital
e carbubarbo

602 carbutamide

A hypoglycaemic agent marked by a good chemotherapeutic action and capable of stimulating insulin secretion and of activating the glycogenetic function of the liver. It is used in the treatment of fat diabetes (diabetes mellitus associated with obesity) affecting old people.

f carbutamide
d Carbutamid
i carbutamide
e carbutamida

603 carbuterol
A beta-stimulant bronchodilator acting selectively on beta-2-bronchial receptors. It is mainly used when chronic bronchitis is accompanied by bronchial spasms. The drug should not be used in the course of pregnancy.
f carbutérol
d Carbuterol
i carbuterolo
e carbuterolo

604 cardamom
The dried, not quite ripe fruit of the plant Elettaria cardamomum cultivated in Sri Lanka and South India and characterized by a long rhizome. It is used in medicine as a carminative and adsorbent agent.
f cardamome
d Kardamom
i cardamomo
e cardamomo

605 cardamom oil
An oil containing terpineol and cineole, distilled from whole fruits. Terpineol is an aromatic solvent antiseptic and cineole an irritant with antiseptic action mostly used in inhalants.
f essence de cardamome; huile volatile de cardamome
d Kardamomöl
i olio di cardamomo; essenza di cardamomo
e esencia de cardamomo

606 carfecillin
A penicillanic acid. Carbenicillin-derivative antibiotic marked by a spectrum as broad as that of ampicillin and equally effective against Pseudomonas aeruginosa and indole-positive Proteus. The acid is hydrolyzed to carbenicillin in the liver and in the bowels. It is mainly used in the treatment of infections caused by susceptible microorganisms of the urinary tract. Fever may follow the administration of the drug as a side effect.
f carfécillin
d Carfecillin
i carfecillina
e carcifilina

607 carfenazine
A tranquillizer chiefly used for the treatment of acute and chronic schizophrenia. It must be administered with great caution to patients suffering from tachycardia and liver disorders.
f carfénazine
d Carfenazin
i carfenazina
e carfenazina

608 carindacillin
Penicillanic acid. An antibiotic.
f carindacilline
d Carindacillin
i carindacillina
e carindacilina

609 carisoprodol
A relaxant of smooth muscle used to treat painful muscle spasms. It has a mild anticholinergic action and is also an antipyretic. It is also used to relieve local painful spasms, post-traumatic contractures, rheumatic inflammation and more generally myalgia.
f carisoprodol
d Carisoprodol
i carisoprodolo
e carisoprodolo

610 carmustine
Alkylating-type antineoplastic agent. It is particularly indicated in the treatment of haemolymphopathies and especially of the Hodgkin's disease in the final stages, of primitive and metastatic brain tumours, melanomas, sarcomas, tumours in the respiratory and in the intestinal tracts and in the breast.
f carmustine
d Carmustin
i carmustina
e carmustina

611 carnitine
Intracellular metabolite involved in lipid metabolism and especially in the transport of fatty acyl units across the mitochondrial membrane. It thus stimulates digestive secretions with ensuing anabolizing and antianorexic effects. It is mainly used in cases of inadequate or defective assimilation of nourishment, grave loss of weight, denutrition and overexertion.
f carnitine
d Karnitin
i carnitina
e carnitina

612 carob bean
The fruit of the Ceratonia siliqua, an evergreen tree, appearing as a brown, thick, elongated (up to 15 cm) leaf. It finds some use in medicine as the flour obtained from it is given to infants in cases of diarrhoea.
f caroubier
d Karob
i carrubo; carrubio
e algarroba; garrofero

613 carob flour
The flour that is obtained from the fruit of the Certatonia siliqua and is used in the treatment of diarrhoea affecting infants. Also known as carob gum or locust gum.
f résine de caroube
d Karobaharz
i resina di carrubo
e resina de algarroba

614 caroverine
A muscle relaxant and spasmolytic agent without atropic effect, used in the treatment of spasms occurring in the gastro-

intestinal, biliary and genitourinary tracts.
f carovérine
d Caroverin
i caroverina
e caroverina

615 carpipramine
Therapeutic agent capable of lowering the mental, intellective and emotive tone without producing obnubilation of the conscience. It is used in psychoses, schizophrenia, chronic states of delirium and as an adjuvant in detoxication treatment.
f carpipramine
d Carpipramine
i carpipramina
e carpipramina

616 carrageen
Or carragheen, Irish moss. The dried red alga Chondrus crispus, a brownish-purple substance used mainly as an emulsifying agent for cod-liver oil and in the preparation of nutrient jellies. It also finds employment in the treatment of broncho-
-pulmonary disorders.
f carraguen; mousse d'Irlande
d Carrageen; Felsenmoos; Tartschenflechte
i fuco carrageo; fuco crispo
e carragoén; liquen de Irlanda; musgo marino

617 carrot
A cultivated plant (Daucus carota of the family Umbelliferae). The aerial part of the plant, known as wild carrot herb is a diuretic and has been used for a long time for all affections of the bladder.
f carrotte
d Karotte; gelbe Rübe
i carota
e zanahoria

618 carteolol
Beta-adrenoreceptor blocking drug used as eyedrops in the treatment of glaucoma.
f cartéolol
d Carteolol
i carteololo
e carteololo

619 carticaine
Local anaesthetic agent used for regional nerve block anaesthesia and infiltration anaesthesia in surgical operations, in obstetrics and in dentistry generally together with a vasoconstrictor.
f carticaïne
d Carticain
i carticaina
e carticaina

620 caryophyllin
White ketonic substance of the terpene series, almost totally odourless, found in clove oil (see caryophyllum).
f caryophylline
d Caryophyllin
i cariofillina

e cariofilina

621 caryophyllum
Clove oil. Antispasmodic and carminative agent for internal use and an irritant, rubefacient and mild antiseptic agent, as well as analgesic for external use. It is frequently employed in dentistry for topical application into the cavity of decayed teeth or upon exposed tooth pulp.
f essence de girofle; essence d'oeillet
d Nelkenöl
i olio di garofano; essenza di garofano
e esencia de clavos de especia

622 cascara bark
A cathartic agent obtained from the bark of buckthorn tree chiefly growing in British Columbia, the Western States of the USA and Kenia. The constituents are not yet completely known but they belong to the class of hydroxymethylanthraquinones. It acts by stimulating bowel movement through the nerve plexus in the large-bowel wall. It may cause an excessive catharsis and its prolongued use may cause black pigmentation in the colon. Also known as sacred or chittam bark.
f écorce sacrée
d amerikanische Faulbaumrinde; Sagradarinde
i cascara sagrada
e cáscara sagrada

623 cascarilla bark
The bark of the branches of the Croton eluteria (cascarilla), white exteriorly and dark interiorly, characterized by a pleasant smell and a bitter taste. It is used for the treatment of gastric atony and of some types of dyspepsias.
f écorce de cascarille; écorce de chacrille; écorce éluthérienne
d Kaskarillrinde
i corteccia di cascarilla
e corteza de cascarilla

624 cascarilline
The active principle of the therapeutic properties of cascarilla. It is extremely bitter and soluble in alcohol.
f cascarilline
d Kaskarillin
i cascarillina
e cascarilina

625 casein
A protein which is present in milk in the form of caseinogen and contains phosphorus and sulphur. It is formed from caseinogen through the action of rennin in the presence of calcium. It is considered an essential protein in so far as it contains all the amino acids which are absolutely necessary for human and animal nutrition. In medicine, preparations of casein are used if the ability to digest natural protein food is impaired and also as an adjuvant in high-protein diets.

f caséine
d Casein
i caseina
e caseina

626 castor oil
A strong cathartic agent capable of releasing in the intestine by means of bile
and lipase ricinoleic acid wich stimulates
peristalsis and enteric secretion. It is
also used for external preparations in
various dermatoses owing to its soothing
action. Also known as ricinus oil.
f huile de ricin
d Rizinusöl
i olio di ricino
e aceite de ricino; aceite de castor

627 catalpa
A genus of plants growing in China,
Japan and the Americas. The most important of them, from a therapeutic stand-
point, is the Bignonia catalpa (catawaba
tree) characterized by large corded
leaves and by a capsular, pendant, ci-
gar-shaped fruit. The syrup which is extracted from the root is used in medicine
as a bronchodilator and a sedative. Also
known as trumpet tree.
f catalpa
d japanischer Trompetenbaum; Zigarrenbaum
i catalpa; bignonia
e catalpa

628 cataplasm
Or, preferably, poultice. A soft mass,
mostly of medicated clay, heated and
spread on a piece of cloth for application to inflamed areas, sores or lesions,
so as to supply a humid warmth, relieve
pain or act as a counter-irritant.
f cataplasme
d Kataplasma; Breiumschlag
i cataplasma
e cataplasma; apósito

629 catechol
Or catechin; catechinic acid. Phenolic
substance derived from phlobatannins
(members of a group of tannins which on
heating are converted into catechol) and
catechins (substance present in catechu,
q.v.). It chiefly consists of catechutan-
nic acid and the bitter substance catechin. The tannic acid contained in the
drug is released slowly when taken orally and produces an astringent effect that
has proved useful in the treatment of
diarrhoea.
f catéchine
d Catechin; Catechinsäure
i catechina
e catequina; ácido catequínico

630 catechu
Catechu nigrum. A species of acacia growing in India and Ceylon. A dense, dark
brown, astringent substance is extracted
from its wood and is known as catechu.
It is used by the natives to prepare the

betel (a stimulant masticatory) and presents astringent, expectorant and antiseptic properties that are exploited in
the treatment of some forms of diarrhoea,
burns etc. It is also possible to obtain
a glyceroalcoholic solution which is used
as a mouth and skin antiseptic.
f catechu
d Catechu
i catecù
e catecú

631 cathartic acid
Or cathartinic acid. The active principle
of the senna leaves, also known as ca-
thartine, probably a mixture of various
glycosides. It is a brown powder, very
bitter and with purgative properties.
f acide cathartique
d Cathartinsäure
i acido catartinico
e ácido catártico

632 cathine
The active principle which is obtained
from fresh or dried leaves of Catha
edulis. Chemotherapeutic agent capable
of exciting the central nervous system. It
is also used as an anorexia-producing
agent in the treatment of obesity. Prolonged use may affect cardiac functions
and cause glaucoma and prostatic hypertrophy.
f cathine
d Kathin
i catina
e catina

633 cefacetrile
Semisynthetic cephalosporin (an antibiotic)
used parenterally. It is a broad-spectrum
antibiotic acting against Gram-positive
and Gram-negative organisms. It presents
no toxicity for the kidneys and is mainly
indicated for the treatment of infections
of the respiratory tract, the genito-urinary tract, as well as of bones, articula
tions and skin.
f céfacétrile
d Cefacetril
i cefacetrile
e cefacetrilo

634 cefalexin
Semisynthetic antibiotic capable of acting
against several Gram-positive and Gram-
-negative bacteria which prove resistant
to penicillin and cephaloridine. It is
used for the treatment of infections caused by susceptible organisms.
f céfalexine
d Cefalexin
i cefalessina
e cefalexina

635 cefaloridine (or cephaloridine)
A broad-spectrum antibiotic of the cephalosporin group, capable of acting against
penicillin-resistant bacteria and used for
the treatment of respiratory, meningeal,

urinary, osteomyelitic and septicaemic infections from Gram-positive and Gram--negative microorganisms.
f céfaloridine
d Cephaloridin
i cefaloridina
e cefaloridina

636 cefamandole
A cephalosporin antibiotic for injection. This semisynthetic drug has proved resistant to a large number of beta-lactamases and is active against many Gram--positive and Gram-negative microorganisms, as well as anaerobes. It is used in the treatment of infections of the respiratory and genito-urinary tracts, of bones, joints, soft tissues, cholecystis and peritoneum. The administration of the drug must be carefully monitored in cases of renal insufficiency.
f céphamandole
d Cephamandol
i cefamandolo
e cefamandolo

637 cefazolin (or cephazolin)
A semisynthetic cephalosporine used parenterally. It is a broad-spectrum antibiotic capable of acting against Gram-positive organisms (e.g. the strain of penicillase--producing staphylococcus) and Gram-negative ones. It is a fairly potent drug against Escherichia coli, Klebsiella pneumoniae and Proteus mirabilis, and thus indicated for the treatment of infections of the respiratory, digestive and genito-urinary tracts. It is also used for the therapy of bacterial endocarditis and septicaemia caused by susceptible bacteria.
f céphazoline
d Cephazolin
i cefazolina
e cefazolina

638 cefoxitin
Cephamycin antibiotic for injection related to cephalosporins. It is resistant to beta-lactamases and is used against many Gram-positive and Gram-negative both aerobic and anaerobic germs. The drug is used for the treatment of infections of the respiratory and genito-urinary tracts, of bones, muscles, joints and skin whenever such infections are caused by susceptible organisms. Also endoabdominal and endopleural infections are treated with the drug. Possible side effects include an increase of azotaemia and reduction of creatinine clearance.
f céfoxitine
d Cefoxitin
i cefossitina
e cefoxitina

639 cellacephate
Partial mixed acetate and hydrogen phthalate ester of cellulose used in medicine as an enteric coating.
f cellacéphate
d Cellacephat
i cellacefato
e celacefato

640 cephalodroxil
A broad-spectrum antibiotic related to cephalosporin. It is active against Gram--positive and Gram-negative organisms and in particular penicillase-producing staphylococci. It is mainly used to combat infections of the respiratory, biliary and gastrointestinal tracts, as well as genitourinary disorders. There is no contra-indication in the use of the drug in children although a potential allergy to penicillin or cephalosporins must be taken into account.
f céphalodroxile
d Cephalodroxil
i cefalodrossile
e cefalodroxilo

641 cephaloglycin
An antibiotic of the cephalosporin group used in the treatment of infections of the urinary tract caused by sensitive organisms. Not to be used for patients who are hypersensitive to penicillin.
f céphaloglycine
d Cephaloglycin
i cefaloglicina
e cefaloglicina

642 cephalosporin
A term denoting antibiotic substances identified in cultures of Cephalosporium acremonium, a mycete belonging to the Sporophorinaceae family. Three of them have been identified and are now known as C, P and N cephalosporines. The most widely used in medicine is the C cephalo sporine whose active nucleus is represented by the 7-amino-cephalosporanic acid, from which various other semisynthetic cephalosporines have been obtained. Their therapeutic application removes infections caused by Gram-positive and some Gram-negative microorganisms.
f céphalosporine
d Cephalosporin
i cefalosporina
e cefalosporina

643 cephalothin
7-(2-thienylacetamido) cephalosporanic acid. A semisynthetic antibiotic obtained from cephalosporin C (see cephalosporin). It is a bactericide and bacteriostatic agent capable of acting against a large number of Gram-positive and Gram-negative microorganisms which prove resistant to penicillin. An intramuscular injection of cephalothin attains peak serum concentrations within about one hour. The drug is excreted in the urine both unaltered and as the acetyl-metabolite.
f céphalothine
d Cephalothin
i cefalotina

e cefalotina

644 cephamandole
Cephalosporin-related antibiotic characterized by a wider range of antibacterial activity when compared with previously used drugs of this group. It is particularly used against penicillase-producing staphylococci.
f céphamandole
d Cephamandol
i cefamandolo
e cefamandolo

645 cephapirin
Cephalosporin-related antibiotic capable of acting against Gram-negative and Gram-positive microorganisms, including penicillin-resistant staphylococci. It is indicated in the treatment of infections of the respiratory, genitourinary and gastro intestinal tracts, of the skin and generally solft tissues. The drug is also used in the treatment of osteomyelitis and endo carditis.
f céphapirine
d Cephapirin
i cefapirina
e cefapirina

646 cepharanthine
Therapeutic agent used for detoxication and as an antidote.
f cépharanthine
d Cepharanthin
i cefarantina
e cefarantina

647 cephradine
Semisynthetic acidoresistant cephalosporin administered orally. It is a broad-spectrum bactericidal antibiotic active against Gram-positive and Gram-negative microorganisms. The drug is used for the treatment of infections by susceptible organisms, e.g. infections of the urinary tract, soft tissue and skin.
f céphradine
d Cephradin
i cefradina
e cefradina

648 cephuroxime
An antibiotic of the cephalosporin group. It proves resistant to the ß-lactamases and is active on both Gram-positive and Gram-negative bacteria. It is indicated for the treatment of infection of the respi ratory and urinary tracts, of bone and soft tissues, as well as of various forms of septicaemia.
f céphuroxime
d Cephuroxim
i cefuroxima
e cefuroxima

649 ceruletide
Therapeutic agent causing excitation of cholecystis and biliary tract, as well as relaxation of Oddi's sphincter, stimula-tion of peristalsis in the intestine, exogenous pancreatic secretion and secretion in the gastric region. It is generally used as a diagnostic agent in the radiographic examination of the gallbladder and of the bile ducts and for the examination of the functional capacity of exocrine pancreas. Postoperative intestinal atony is also treated with the drug which, more generally, finds employment in chronic constipation. Not to be used in the presence of acute pancreatitis and grave cardiocirculatory disorders.
f céruletide
d Ceruletid
i ceruletide
e ceruletida

650 ceruse
The basic carbonate of lead used for poultices and ointments.
f blanc de plomb; céruse; carbonate basique de plomb
d Bleicarbonat; Bleiweiss
i carbonato di piombo; bianco di piombo; cerussa
e blanco de plomo; cerusa; carbonato de plomo

651 cetalkonium chloride
Quaternary ammonium disinfectant and antiseptic agent used in surgery for the disinfection of both skin and instruments.
f chlorure de cétalconium
d Cetalkoniumchlorid
i cetalconio cloruro
e cloruro de cetalconio

652 cetiedil
Peripheral vasodilator whose papaverine--like action is supplemented by an antibradykinin and antiserotonin activity. It is mildly anticholinergic, inhibits the aggregation of platelets and increases the cyclic AMP (adenosine monophosphate) level. It acts very quickly against pain without any central analgesic effect. It is mainly indicated for the treatment of senile and presenile arteriopathies, of angiopathies caused by diabetes, vasomotor disorders in the limbs, acrocianosis etc.
f cétiédil
d Cetiedil
i cetiedile
e cetiedilo

653 cetothiamine
A liposoluble derivative of thiamine used in the prophylaxis and the treatment of disorders caused by a deficiency of vitamin B_1.
f cétothiamine; dicarbétoxythiamine
d Cetotiamin
i cetotiamina
e cetotiamina

654 cetraric acid
Acid characterized by a bitter taste and found in Iceland moss in the form of

white crystals. It is little soluble in water and used for its tonic and laxative action.
f acide cétrarique
d Cetrarsäure
i acido cetrarico
e ácido cetrárico

655 cetrimonium bromide
Antiseptic and disinfectant agent which is biologically disintegrated and used in the form of powder, particularly to disinfect surgical instruments.
f bromure de cétrimonium
d Cetrimoniumbromid
i cetrimonio bromuro
e bromuro de cetrimonio

656 cetylpyridinium chloride
Antiseptic and disinfectant of the cationic group. It is used externally in ear, nose and throat disorders.
f chlorure de cétylpyridinium
d Pyridiniumchlorid
i cetilperidinio cloruro
e cloruro de cetilperidinio

657 chamomile
Or Roman chamomile, double chamomile, garden chamomile. Perennial herb cultivated for the first time in England and subsequently all over Europe. Its flowers yield an antispasmodic substance used in spastic conditions of the digestive tract and in dysmenorrhoea. Also used externally as an antiinflammatory agent. A beverage prepared with dried flower heads of chamomile is a mild sedative.
f camomille romaine
d römische Kamille; Kamille; Ackerhundskamille
i camomilla; camomilla romana
e manzanilla romana

658 chamomile oil
Blueish aromatic essential oil obtained from the flower heads of chamomile and used for various purposes in medicine.
f essence de camomille
d Kamillenöl
i essenza di camomilla
e esencia de manzanilla

659 charcoal
Common charcoal having a high absorbent power, especially when finely divided. It is used in first-aid treatment of poisoning by drugs and toxins and to combat diarrhoea. It also protects the intestinal flora if administered orally in large doses.
f charbon de bois
d Holzkohle
i carbone di legna; carbone vegetale
e carbón vegetal

660 charcoal (activated)
Charcoal being a strong adsorbant (it is called "activated" only because it meets given standards in absorbence tests), it acts, when finely divided, as a strong agent against the effects of acute poisoning by reducing the absorption of drugs or toxins. It is generally used in conjunction with other adsorbants and some disinfectants, e.g. bismuth salts. It frequently causes nausea and vomiting.
f charbon actif
d Aktivkohle
i carbone attivo
e carbón activado

661 chaulmoogra
Or kalawso. East Indian tree of the family Flacourtiaceae yielding an oil (or soft fat) from its seeds, which chiefly consists of glycerides of chaulmoogric, hydnocarpic and gorlic acid and used for the treatment of leprosy and, more generally, skin diseases.
f chaulmoogra
d Chaulmoogra; Gynocardia
i ginocardia
e ginocardia; chalmogra

662 chaulmoogra oil
See definition of chaulmoogra.
f huile de chaulmoogra
d Chaulmoograöl; Gynocardiaöl
i olio di chaulmoogra
e aceite de ginocardia; aceite de chalmogra

663 chaulmoogric acid
Unsaturated cyclic fatty acid obtained from chalmoogra and hydrocarpus oil and used in the prolonged treatment of leprosy.
f acide chaulmoogrique
d Chaulmoograsäure
i acido chaulmoogrico
e ácido chalmoógrico

664 chebule
The dried fruit of an East Indian tree yielding a substance which is used in medicine as an astringent agent. Also called myrobalan.
f badamier chébule; myrobalan
d Chebula; Myrobalan
i mirobalano
e mirobálano; polidrupa

665 cheeseflower
Or mallow. The common tall mallow from whose leaves a principle is obtained which is used for expectorant preparations.
f mauve sauvage; fausse guimauve
d Malve; Käsepappel
i malva; malva selvatica
e malva; maya silvestre

666 chelidonic acid
Cycloketonic dicarboxylic acid derived from the nucleus of pyrone (isomeric compound forming the basis of many natural products) used for spasmolytic preparations.
f acide chélidonique
d Chelidonsäure

ı acido chelidonico
e ácido quelidónico

667 chelidonine
Alkaloid of chelidonium (a genus of herbs of the family Papaveraceae) found in celandine (a diuretic and cathartic agent) and having a narcolytic and spasmolytic action that is similar to that of papaverine.
f chélidonine
d Chelidonin
ı chelidonina
e quelidonina

668 chenodeoxycholic acid
Or chenodiol. A primary biliary acid preventing the formation and promoting the dissolution of gall stones. The drug is particularly indicated for the treatment of patients subject to the risk of cholesterol stone formation.
f acide chénodéoxycholique
d Chenodeoxycholinsäure
ı acido chenodesossicolico
e ácido quenodeoxicólico

669 chenodiol
Chenodesoxycholic acid. A primary biliary acid capable of increasing the solubility of cholesterol in the bile and used for the treatment of cholesterol-related bile stones. It is further indicated for the treatment of patients suffering from lithogenous bile and thus subject to a high risk of cholesterol-stone formation.
f chénodiol
d Chenodiol
ı chenodiolo; acido chenodesossicolico
e quenodiolo

670 cherry-laurel water
A solution prepared from the fresh leaves of the cherry laurel (Prunus laurocerasus) which yield hydrocyanic acid on hydrolysis, used as carminative and a mild sedative.
f eau distillée de laurier cerise
d Kirschlorbeerwasser
ı acqua di lauro ceraso; idrolato di lauro ceraso
e agua de laurel cerezo; hidrolato de laurel cerezo

671 chervil
An aromatic annual herb native of Eastern Europe and further east in Siberia. It is used in various medicinal preparations.
f cerfeuil
d Kerbel
ı cerfoglio
e cerafolio

672 Chian turpentine
A yellow-brown semifluid oleoresin exuding from the terebinth tree, characterized by a sharp odour and taste. It is used as a raw material for synthetic camphor and for various medicinal preparations.
f térébenthine de Chypre
d Chiosterpentin
ı trementina di Chio; trementina di Cipro
e trementina de Chio

673 chicory
A perennial herb widely cultivated for its leaves and even more for its roots which yield an active principle of preparations used as bitter tonic and as a mild laxative. Also called succory.
f chicorée; intybe
d Cichorie; Zichorie
ı cicoria
e achicoria

674 chincona bark
Jesuits' bark, Peruvian bark. The dried bark of several trees of the genus Cinchona, which contain alkaloids, e.g. quinine, cinchonine, quinidine and chinchonidine, and has been used for a long time in the treatment of malaria as well as an antipyretic in other fevers and as a stomachic agent.
f écorce de quinquina
d Chinarinde; Fieberrinde
ı china rossa; china rubra; corteccia di china
e quina roja; cascarilla roja del rey

* **chinidinum   s. quinidine**

* **chininum   s. quinine**

675 chiniofon sodium
Bright-yellow powder consisting of the sodium salt of 7-iodo-8-oxyquinoline-5-sulphonic acid with sodium bicarbonate. Owing to its iodine content, it is capable of destroying amoebae in the intestinal tract in any form, i.e. either mobile or encysted. It is indicated in acute and chronic amoebiasis and administered either by mouth or as a retention enema by night.
f quiniofon sodium
d Chiniofonnatrium
ı chiniofone sodio
e sodio de quinofón

676 chionanthin
The solid resinous substance which is obtained from the root of the Chionantus virginica and constitutes the active principle of a diuretic and cholagogue drug.
f chionanthine
d Chionanthin
ı chionantina
e quionantina

677 Chionantus
Genus of trees of the family Oleaceae. The root bark of the species Chionantus virginica contains chionanthin (a glycoside) and saponin and is used as a bitter tonic, febrifuge and diuretic.
f Chionantus
d Chionantus
ı Chionantus

e Chionantus

678 chloral betaine
A compound of chloral hydrate with betaine. Hypnotic and sedative agent whose properties are similar to those of barbiturates. It is used particularly for children and old people to combat insomnia and promote sedation before an operation. The drug should not be used in cases of grave hepatopathies, cardiopathies and nephropathies.
f chloral bétaïne
d Chloralbetain
i cloralio betaina
e cloral betaina

679 chloral hydrate
Colourless crystalline substance characterized by a penetrating odour and a bitter taste. It depresses the central ner̲vous system and is thus used as a hypnotic, although its side effects, mainly nausea and stomach irritation may prove difficult to overcome. It is mainly used for children and aged people who are allergic to barbiturates. It should be avoided in cases of hepatopathies and severe cardiopathies.
f hydrate de chloral
d Chloralhydrat
i cloralio idrato
e hidrato de cloral

680 chloralodol
Hypnotic and sedative agent used in the treatment of insomnia and for the soothing of anxiety conditions and agitation in elderly patients. Not to be recommended in cases of severe nephropathies, hepatopathies and cardiopathies.
f chloralodol
d Chloralodol
i cloralodolo
e cloralodolo

681 chloraltannin
Condensation product of chloral (trichloro̲ acetaldehyde) with the tannic acid. It is̲ an amorphous greyish, hygroscopic powder used in medicine as antiseborrhoic and against the loss of hair.
f chloraltanin
d Chloraltannin
i cloraltannino
e cloraltanino

682 chlorambucil
A cytotoxic drug related to Mustine hydro̲chloride. An oncolytic agent used in neoplastic conditions of lymphoid tissues and particularly in lymphocytic leukaemia, malignant lymphomas and in the Hodgkin's granuloma which is characterized by enlargement of the lymph glands, hyperplasia of the lymphoid tissue in the spleen and liver, as well as in other organs, and anaemia. The drug has prov̲ed very effective for a large number of patients, but should be discontinued when bone marrow depression is diagnosed.
f chlorambucil
d Chlorambucil
i clorambucile
e clorambucilo

683 chloramphenicol
A broad-spectrum antibiotic especially indicated for the treatment of typhoid fever and near-fatal infections, e.g. those caused by the Salmonella typhi. Its bacteriostatic effect is obtained by inhibiting protein synthesis. Both reversible and irreversible blood disorders may represent a serious side effect. The treatment should not exceed two weeks in premature and newborn infants.
f chloramphénicol
d Chloramphenicol
i cloramfenicolo
e cloramfenicolo

684 chlorazanil
A diuretic especially used for the treatment of oedemas of various types. It is particularly effective in the therapy of mild exogenous poisoning.
f chlorazanil
d Chlorazanil
i clorazanile
e clorazanilo

685 chlorbenzoxamine
Parasympatholytic agent strongly recommended for the therapy of gastroduodenal ulcer as well as for the prophylaxis of gastrointestinal ulcerations due to histamine- and reserpine-related drugs.
f chlorbenzoxamine
d Chlorbenzoxamin
i clorbenzossamina
e clorbenzoxamina

686 chlorcyclizine
Antihistaminic drug capable of neutralizing the principal action of histamine in the body and used as an antiemetic as well as in the treatment of allergic reactions. It should be avoided in cases of severe hepatic lesions. One of its main side effects is drowsiness and nervous depression.
f chlorcyclizine
d Chlorcyclizin
i clorciclizina
e clorciclizina

687 chlordiazepoxide
7-chloro-2-methylamino-5-phenyl-3H-benzo--1,4-diazepine-4-oxide. Its hydrochloride is an antineurotic, anticonvulsant and psychotherapeutic agent. The action is carried out by depressing the central nervous system at the subcortical plane and is used for the relief of states of anxiety and tension also when associated with psychoneurotic reactions.
f chlordiazépoxyde
d Chlordiazepoxyd

i clordiazepossido
e clordiazepóxido

688 chlorexolone
A diuretic agent used in the treatment of
fluid overload and in the control of high
blood pressure. Its action is carried out
by diminishing the sodium reabsorption
in the kidney.
f chlorexolone
d Chlorexolon
i clorexolone
e clorexolono

689 chlorhexamide
Hypoglycaemic agent of the sulphonylurea
used in the initial treatment of diabetes
mellitus.
f chlorcyclamide
d Chlorcyclamid
i cloresamide
e clorhexamida

690 chlorhexidine
A broad-spectrum disinfectant and antiseptic agent used mainly to combat skin
infections caused by Gram-positive and
Gram-negative bacteria but also for general hygienic purposes and in surgery.
f chlorhexidine
d Chlorhexidin
i cloresidina
e clorhexidina

691 chloric ether
A mixture of chloroform and alcohol used
in medicine as an anaesthetic. Also
known as chloroform spirit.
f éther chlorique; chloroforme dulcifié
d Chloroformspiritus
i alcool cloroformato
e alcohol cloroformado

692 chlormadinone
A progestational (relating to the phase of
the menstrual cycle immediately preceding
the onset of menstruation in the course of
the functional activity of the corpus luteum and the secretory activity of the
endometrium) agent which is capable of
inhibiting ovulation. It is mainly indicated when there is a danger of abortion
and more generally in case of gynecological disorders caused by inadequate
luteal activity. It is also characterized
by its antigonadotropic (neutralizing the
action of a specific gonadotropin) and
antioestrogenic action.
f chlormadinone
d Chlormadinon
i clormadinone
e clormadinona

693 chlormethine
An oncolytic agent indicated in the treatment of Hodgkin's disease (the pathological enlargement of the lymph glands
accompanied by hyperplasia of the lymphoid tissue in the spleen, liver and
other organs) and similar lymphomas, e.

g. lymphosarcoma and lymphocytoma. Also
used in conjunction with radiotherapy.
f chlorméthine
d Chlormethin
i clormetina
e clormetina

694 chlormezanone
Myorelaxant with central mechanism used
in the treatment of the Parkinson's disease, as an antispasmodic (mainly in
lumbago), a tranquillizer and a mild
analgesic. It has no action at all on
smooth muscles and on the respiratory
system.
f chlormézanone
d Chlormezanon
i clormezanone
e clormezanona

695 chlorocresol
Antiseptic and disinfectant agent highly
effective against a large number of sapro-
lytic pathogenous organisms. It is more
potent than phenol but displays a greater toxicity. It acts by causing a precipitation of the protoplasmatic proteins following a penetration through the membranes due to the liposolubility.
f chlorocrésol
d Chlorocresol
i clorocresolo
e clorocresolo

696 chloroform
Colourless volatile highly toxic liquid
characterized by an ethereal odour and a
sweetish taste, which is usually obtained
by the chlorination and oxidation of
acetone or the chlorination of methane or
methyl chloride. It was used for a long
time as a general anaesthetic and also
as a carminative and anodyne agent.
f chloroforme
d Chloroform
i cloroformio
e cloroformo

697 chloroguanidine hydrochloride
Therapeutic agent used in the treatment
of malaria. It acts on the asexual forms
of the plasmodial species that are capable of attaining at least a clinical remission. The Plasmodium falciparum infection
can be radically healed with the right
dosage. The drug, beside being an effective malarial suppressive agent, renders
gametocytes non-viable.
f chlorhydrate de chloroguanidine
d Chlorguanidinhydrochlorid
i idrocloruro di cloroguanidina
e hidrocloruro de cloroguanidina

698 chloromorphine
Intermediate substance at times formed in
the course of the conversion of morphine
to apomorphine. In its pharmacological
action, it resembles diamorphine and it
may be found as a contaminant in apomorphine.

f chloromorphine
d Chlormorphin
i cloromorfina
e cloromorfina

699 chlorophenol
Mixture of the ortho- and para-isomers of monochlorphenol that is obtained through the action of chlorine on phenol. It is used for sanitary purposes. However, the term is also used to denote any derivative of phenol containing chlorine, i.e. compounds which have considerable antiseptic properties.
f chlorophénol
d Chlorphenol
i clorofenolo
e clorofenol

700 chlorophyll
The green colouring substance of plants which is essential to photosynthesis and occurs, generally in discrete bodies, only in the presence of light and only if iron is available in the living cell. It is generally extracted as a mixture containing the substance itself with other pigments, e.g. carotene and xanthophyll. Chlorophyll is used for its deodorant properties.
f chlorophylle
d Chlorophyll
i clorofilla
e clorofila

701 chlorophylline
A dicarboxylic acid that is obtained by the saponification of the chlorophyll and has the same pharmacological properties of chlorophyll.
f chlorophylline
d Chlorophyllin
i clorofillina
e clorofilina

702 chloropicrin
A shocking gas used in chemical warfare. Chemically a nitrochloroform, trichloronitromethane. It appears as a dense liquid marked by a suffocating lacrimatory vapour which also causes vomiting.
f chloropicrine
d Chloropikrin
i cloropicrina
e cloropicrina

703 chloroprocaine
Local analgesic of the ester type less toxic and far more potent than procaine. It is particularly indicated to obtain local anaesthesia by infiltration and nerve block, as well epidural and caudal anaesthesia.
f chloroprocaïne
d Chlorprocain
i cloroprocaina
e cloroprocaina

704 chloroprocaine hydrochloride
Ester-type local analgesic, very similar to procain but less toxic and more potent.

f chlorhydrate de chloroprocaïne
d Chlorprocainhydrochlorid
i idrocloruro di cloroprocaina
e hidrocloruro de cloroprocaina

705 chloropyrilene
Antihistaminic agent indicated for the therapy of allergic dermatoses, burns and insect stings.
f chloropyrilène
d Chlorpyrilen
i cloropirilene
e cloropirileno

706 chloroquin
Chemotherapeutic and chemoprophylactic agent used in all acute forms of malaria. It is capable of destroying the malarial parasite at the schizont stage of its life cycle. The drug is also used to combat hepatic amoebiasis, in rheumatoid arthritis and in collagenosis. It should not be used in the course of pregnancy, in states of anaemia, miopathies and retinopathies.
f chloroquine
d Chloroquin
i clorochina
e cloroquina

707 chlorothiazide
A diuretic agent capable of reducing the reabsorption of electrolytes in the proximal renal tubules and thus increasing the excretion of sodium, potassium and chloride ions and the concomitant excretion of water. It is capable of lowering the blood pressure and is therefore used in the treatment of oedemas and hypertension. An electrolyte imbalance accompanied by hypokaliaemia (low blood potassium content) is a possible side effect.
f chlorothiazide
d Chlorthiazid
i clorotiazide
e clorotiazida

708 chloroxilenol
Phenol-related antiseptic agent marked by a bactericide action in particular against pyogenous cocci, and used in dermatology, surgery, obstetrics etc. The drug has no effect against Gram-negative organisms.
f chloroxilénol
d Chlorxilenol
i clorossilenolo
e cloroxilenolo

709 chlorphenacemid
Therapeutic agent used in the treatment of epilepsy and of all epileptic disorders.
f chlorphénacémide
d Chlorphenacemid
i clorfenacemide
e clorfenacemida

710 chlorphenamine
Antihistaminic agent used for the relief, more specifically for the symptomatic relief of various allergic disorders, e.g.

urticaria, hay fever, contact dermatitis
etc.
f chlorphénamine
d Chlorphenamine
i clorfenamina
e clorfenamina

711 chlorphenesin

Topical antifungal agent active against
most pathogenous mycetes, as well as
staphylococci, streptococci, Escherichia
coli and Trichomonas. The drug is frequently used for the treatment of cutaneous mycosis and vaginal trichomoniasis.
A mild histolytic activity has been observed as a side effect.
f chlorphénésine
d Chlorphenesin
i clorfenesina
e clorfenesina

712 chlorphenoxamine

An anticholinergic and antihistaminic
agent, as well as a relaxant of skeletal
muscles used in the treatment of the
Parkinson's disease and of all disorders
borne of the disease, e.g. arteriosclerotic,
idiopathic and postencephalytic parkinsonism. Somnolence and vertigo are probable side effects. The drug should not
be administered in cases of prostatic
hypertrophy.
f chlorphénoxamine
d Chlorphenoxamin
i clorfenossiamina; clorfenossamina
e clorfenoxamina

713 chlorphentermine

An anorectic, sympathomimetic amine indicated for the treatment of essential
obesity associated with cardiovascular
disorders and/or diabetes. It promotes
lipidic catabolism. The drug may cause
dyspepsia.
f chlorphentermine
d Chlorphentermin
i clorfentermina
e clorfentermina

714 chlorprocainamide

Therapeutic agent used in cases of arrhythmia as it depresses the excitability
of the cardiac muscle to electrical stimulation. Atrial flutter, extrasystoles,
tachycardia, fibrillation and nodal
rhythm are also treated with chlorprocainamide.
f chlorprocaïnamide
d Chlorprocainamide
i cloroprocainamide
e clorprocainamida

715 chlorproethazine

A muscle-relaxant, tranquillizer and anti
spasmodic agent used in the treatment of
Parkinson's disease, in gynecology, ortho
pedics and rheumatology. The drug is
also indicated in the therapy of manic
schizophrenia.
f chlorproéthazine

d Chlorproethazin
i clorproetazina
e clorproetazina

716 chlorproguanil

Antimalarial drug marked by schizontici-
dal and gametocidal properties also used
for prophylactic purposes. It yields no
results when malaria is caused by
strains presenting chemoresistance.
f chlorproguanile
d Chlorproguanil
i clorproguanile
e clorproguanilo

717 chlorpromazine

Tranquillizer of the phenothiazine group
marked by sedative, psychotropic, antiemetic, antiadrenergic, peripheral anticholinergic, antihistaminic and antiserotoninic activity. The drug has proved
quite effective in some psychoses and
particularly schizophrenia. It is generally used to combat anxiety and tension,
as well as psychic disorders characterized by anxiety and agitation, senile tension, acute alcoholism and poisoning due
to barbiturates and narcotic analgesic
dependence. The drug has an adrenolytic
action. It is further used to intensify
the effect of analgesics in terminal illnesses and to control nausea, vomiting
and hiccups. An overdose may well cause
coma, convulsions, hypotension and arrhythmias.
f chlorpromazine
d Chlorpromazin
i clorpromazina
e clorpromazina

718 chlorpropamide

Antidiabetic agent of the sulphonylurea
group administered per os, mainly used
for the treatment of fat senile and hepatic diabetes and relieves the need of a
severely controlled diet.
f chlorpropamide
d Chlorpropamid
i clorpropamide
e clorpropamida

719 chlorpropandiol

A tranquillizer and a homeostatic agent
of the neurovegetative activity. It is
used to combat states of anxiety, neurovegetative dystonia, insomnia, neurosis
and, more generally, as a supporting
drug in the treatment based on neuroleptics.
f chlorpropandiole
d Chlorpropandiol
i clorpropandiolo
e clorpropandiolo

720 chlorprothixene

Chlorpromazine-related drug mainly used
in the treatment of psychoses, but also
for the detoxication of patients affected
by addition-leading drugs and alcoholism.
Also used against agitation and insomnia

in aged patients. The drug being structurally and pharmacologically related to phenothiazine, the principal side effects do not differ from those of the latter, e. g. drowsiness, sometimes jaundice, haematological and cardiovascular disorders etc.

f chlorprothixène
d Chlorprotixen
i clorprotisene
e clorprotixeno

721 chlorquinaldol

Bacteriostatic and antifungal agent used in dermatology for the therapy of cutaneous mycosis, impetigo, forunculosis and pemphigus of the newborn. Mastitis and some non-specific intestinal infections are also treated with this drug. Not to be used when peptic ulcer has been identified.

f chlorquinaldol
d Chlorquinaldol
i clorquinaldolo
e clorquinaldolo

* **chlorquine   s. chlorquinine**

722 chlorquinine

Synthetic antimalarial agent similar in its action to mepacrine (q.v.) but even less toxic. It represents the best malarial schizonticide so far known. The two salts most commonly used are the phosphate and the sulphate.

f chlorquinine
d Chlorchinin
i clorochinina
e cloroquinina

723 chlortalidone

Non-thiazide diuretic acting on the renal tubules and used for the treatment of cardiac, renal and hepatic oedema, hypertension, exogenous intoxications and poisoning caused by the retention of waste products which are normally excreted from the body. The drug is characterized by the persistance for a long time of the ensuing diuresis. Not to be administered during lactation as it blocks the secretion of milk.

f chlortalidone
d Chlortalidon
i clortalidone
e clortalidona

724 chlortetracycline

A broad-spectrum antibiotic capable of acting against Gram-positive and Gram-negative organisms. Its action is based on its capacity of inhibiting bacterial protein synthesis. A prolonged administration may cause decalcification of the teeth.

f chlortétracycline
d Chlortetracyclin
i clortetraciclina
e clortetraciclina

725 chlorthenoxazine

Analgesic, antipyretic and antirheumatic agent mainly used to relieve headache, toothache, osteoarthritis etc. It has a quick but short action.

f chlorthénoxazine
d Chlorthenoxazin
i clortenossazina
e clortenoxazina

726 chlortrianisene

Synthetic oestrogen displaying a prolonged action and used in the therapy of carcinoma of the prostata, post-partum galactorrhoea, menopausal disorders and dystrophic vaginitis.

f chlortrianisène
d Chlortrianisen
i clorotrianisene
e clorotrianiseno

727 chlorzoxazone

Relaxant of skeletal muscles and antispasmodic agent mainly used in the treatment of the Parkinson's disease and of painful contractures. The drug has also proved useful against rheumatic, arthritic, traumatic and orthopaedic anomalies. It may cause hepatic lesions and lypothymia.

f chlorzoxazone
d Chlorzoxazon
i clorzossazone
e clorzoxazona

728 cholecalciferol

Therapeutic agent used in the treatment of rickets and to promote calcification, in particular of the bones. The drug is capable of balancing vitamin D deficiency and finds therefore employment in osteoporosis, osteomalacia, psoriasis and some other dermatoses.

f cholécalciférol
d Cholecalciferol
i colecalciferolo
e colecalciferolo

729 cholesterol

The natural fatty constituent of all animal cells, i.e. a precursor of steroids. It is used topically in preparations for the protection and health of the skin.

f cholestérol
d Cholestrin; Gallenfett
i colesterolo
e colesterol

730 cholestyramine

The basic, anionic exchange resin which binds bile salts in the intestine and thus promotes their excretion in the faeces instead of allowing reabsorption from the bowels. The cholesterol level in the blood is thus reduced. It is especially used to soothe the itching due to partial biliary obstruction and in the treatment of hypercholesterolaemia. Large doses of the drug may cause steatorrhoea.

f cholestyramine
d Cholestyramin

ı colestıramına
e colestıramına

731 **cholic acid**
Trioxycholanic acid found in the bile of
man, ox, sheep and other animals. Chemı
cally, it is a steroid which, in a mixture with glycıne and taurıne, orıgınates
the glycocholic and taurocholic bılıary
acıds present ın the bıle. It is formed
ın the lıver by the cholesterol of whose
metabolısm it is the end product. It is a
whıte bıtter powder with a cholagogue
(promotıng the flow of bıle from the gallbladder) action. Also called cholalıc acıd.
f acide cholalıque; acide cholıque
d Cholsäure; Cholalsäure
ı acido colıco; acido colalıco
e ácido cólico; ácido colálico

732 **choline**
Bılıneurıne. Amıne base present ın the
blood, cerebrospınal fluıd and urıne,
often classıfıed as a member of the B com
plex of vıtamıns. It has a lıpotropıc action and is capable of dispersing fat
from the lıver. It also has a stımulant
action on ıntestınal and uterıne muscles
and heıghtens the tone of detrusor urınae.
It is maınly used ın the treatment of pre
cırrhotic fatty ınfıltration of the lıver.
f cholıne
d Cholın
ı colına
e colına

733 **choline chloride**
A lıpotropıc amıno acıd. Its action is
related to the transmethylatıon process.
The drug is ındıcated ın dısorders of the
lıver, ınıtıal cırrhosıs and ıntoxıcatıons
due to hepatotoxıc substances. It is a
mıld choleretıc and cholagogıc agent as
well and is frequently used ın conjunction with methıonıne and sorbıtol. Nausea
and occasıonally a state of excıtatıon
are possıble sıde effects.
f chlorure de cholıne
d Cholınchlorıd
ı colına cloruro
e cloruro de colına

734 **choline citrate**
A lıpotropıc agent used ın dısorders of
the lıver, ın particular when there is a
fatty ınfıltration, ınıtıal hepatic cırrhosıs, toxıc and ınfectıous hepatıtıs.
f cıtrate de cholıne
d Cholıncıtrat
ı colına cıtrato
e cıtrato de colına

735 **choline dehydrocholate**
Lıpotropıc and hydrocholeretıc agent used
ın the treatment of hepatıc dısorders,
acute and chronıc hepatıtıs, ınflammatıon
of the bılıary tract and cholecystıtıs.
f déhydrocholate de cholıne
d Cholındehydrocholat
ı colına deıdrocolato

e dehıdrocolato de colına

736 **choline phytate**
A drug that is capable of lıberatıng by
hydrolysıs ın the lıver lıpotropıc substances. e.g. cholıne, ınosıtol and the
related phosphorıc esters. It is used for
the treatment of lıver dısorders, ınfectıous hepatıtıs, bılıary cırrhosıs and
hepatıc amyloıdosıs. It is also used to
treat atherosclerosıs when thıs is due to
cholesterol, lıpoıd materıal and calca-
reous substances.
f phytate de cholıne
d Cholınphytat
ı colına fıtato
e fıtato de colına

737 **choline salicylate**
Antırheumatic and antıpyretıc agent (salı
cylıc acıd), as well as lıpotropıc agent
used ın lıver dısorders.
f salıcylate de cholıne
d Cholınsalıcylat
ı colına salıcılato
e salıcılato de colına

738 **cholinesterase**
Enzyme present ın the blood and tıssues
at the endıngs of voluntary motor nerves
and nerves of the parasympathetic dıvı-
sion of the ınvoluntary system and capable of ınactıvatıng by hydrolysıs the
acetylcholıne produced there by nerve
ımpulses and of preventıng its accumulation. The human serumal cholınesterase,
ı.e. an extremely pure concentrate of the
enzyme subjected to lyophılızatıon, is
used ın the treatment of prolonged
apnoea caused by hypersensıtıvıty to suc
cınylcholıne chlorıde or by the reduced
actıvıty of the serum enzymes. It is also
ındıcated for a supportıng therapy of
poısonıng by esters of the phosphorıc
acıd.
f cholınestérase
d Cholınesterase
ı colınesterası
e colınesterasa

739 **choline theophyllinate**
A therapeutic agent maınly used ın the
treatment of chronıc bronchıtıs. It is a
coronary and perıpheral vasodılator and
a cardıac stımulant, but above all an
expectorant, dıuretıc and spasmolytıc
drug owıng to the theophyllıneacetıc ra-
dıcal. It is also used ın heart dısorders,
dıffıculty ın breathıng derıvıng therefrom,
bronchıal asthma and kıdney trouble.
f théophyllınate de cholıne
d Cholıntheophyllınat
ı colına teofıllınato
e teofılınato de colına

740 **chromocarb**
A drug capable of protectıng the capıl-
lary vessels. A hyaluronıdase (an enzyme
whıch destroys hyaluronıc acıd and is
present ın the fıltrates of varıous bac-

terial species) inhibiting agent. It is used in the treatment of capillary fragility when diabetes or excessive hypertension has been diagnosed. Also used in the treatment of purpura petechiae, epistaxis, gingival haemorrhage etc.

f chromocarbe
d Chromocarb
i cromocarbo
e cromocarbo

741 chrysophanic acid
Acid occurring as a glucoside in rhubarb, as a rhamnoside in cascara and in chrysarobin. It is used for the treatment of skin diseases, in particular psoriasis and eczema.

f acide chrisophanique
d Crisophansäure
i acido crisofanico
e ácido crisofánico

742 chymotrypsin
An intestinal proteolytic enzyme formed by the action of trypsin on the zymogen chymotrypsinogen. It is an antiinflammatory agent used to relieve soft-tissue flogosis and oedema associated with traumatic lesions, e.g. contusions, ecchymoses, dermatoses etc.

f chymotrypsine
d Chymotrypsin
i chimotripsina
e quimotripsina

743 ciclonicate
A derivative of nicotinic acid used as a spasmolytic agent acting on the blood vessels and as a vasodilator especially active on microcirculation. It is used in the treatment of peripheral and atheromatous vascular disorders and in cerebral atherosclerosis.

f ciclonicate
d Ciclonicat
i ciclonicato
e ciclonicato

744 cimepanol
A powerful choleretic characterized by its long-lasting action. It is also a non-hypnotic and analgesic sedative. It is mainly used in the treatment of liver disorders due to malaria, alcoholism, cholecystectomy, amoebiasis, bile stones and, more generally, excess of food. It also finds uses in combating nausea, vomiting, dyspepsia, states of anxiety, constipation, neurotoxic diseases etc. It should not be used by children.

f cimépanol
d Cimepanol
i cimepanolo
e cimepanolo

745 cimetidine
Gastric antacid which selectively blocks histamine receptors mediating gastric acid secretion, both basal and induced. It also reduces the production of pepsin.

It is used in peptic ulceration and gastric hyperacidity, as well as in peptic oesophagitis, the Zollinger-Ellison syndrome (non-insulin-secreting adenoma of the islets of Langerhans associated with peptic ulceration), ulcer-related haemorrhages, erosion of the mucosa of the upper gastrointestinal tract, gastritis and duodenitis associated with gastric hypersecretion.

f cimétidine
d Cimetidin
i cimetidina
e cimetidina

746 cimicifuga racemosa
Bugbane, black cohosh. Antispastic and analgesic agent used in the treatment of spasms, headache, hypertension and tinnitus.

f actée à grappes; cimicifuga
d schwarze Schlangenwurze; traubenförmiges Schwarzkraut
i racemosa; cimicifuga
e cimicífuga; saúco

747 cinchocaine
Amide-type local anaesthetic used for surface, infiltration and spinal anaesthesia. When administered by injection it is more toxic than lidocaine and when administered locally more toxic than cocaine, but its action is more potent thus making it possible to use smaller doses. It has a considerable vasodilative action and therefore a greater risk of toxicity, so that the use of adrenaline is indicated to delay the absorption.

f cinchocaïne
d Cinchocain
i cincocaina
e cincocaina

748 cinchonidine
An alkaloid, isomer of the cinchonine, contained in the various barks of china. Colourless and tasteless crystals, soluble in alcohol and chloroform and insoluble in water. They display a tonic action and have been used as an antimalarial agent.

f cinchonidine
d Cinchonidin
i cinconidina
e cinconidina

749 cinchonine
One of the alkaloids contained in the chincona bark, a stereoisomer of the cinchonidine. It is a colourless bitter powder which was used as antimalarial agent in replacement of quinine.

f cinchonine
d Cinchonin
i cinconina
e cinconina

750 cinchophen
An uricosuric agent capable of increasing the excretion of uric acid in the urine

by inhibiting its reabsorption. The drug
is also an analgesic used for the relief
of rheumatic pains, neuralgias, sciatica
and lumbago. It should not be administer
ed to patients suffering from renal lithia
sis.
f cinchophène
d Cinchophen
i cincofene
e cincofeno

751 cinepazet
Therapeutic agent used in the treatment
of angina pectoris owing to its peripheral and coronary vasodilative action. Its
adrenergic activity reduces the effort of
the heart and the consumption of oxygen
by the myocardium. It is indicated in
coronary insufficiency and the conditions
which are common after an infarction.
f cinépazet
d Cinepazet
i cinepazeto
e cinepazeto

752 cinmetacin
Non-steroid antiinflammatory, antipyretic
and analgesic agent used in the treatment of arthritis, spondylitis and osteoarthritis. The drug must be used with
great caution as it frequently causes
nausea, anorexia, diarrhoea etc.
f cinmétacine
d Cinmetacin
i cinmetacina
e cinmetacina

753 cinnamate
A salt of cinnamic acid, q.v.
f cinnamate
d Cinnamat; zimtsaures Salz
i cinnamato
e cinamato

754 cinnamic acid
Acid occurring free and as esther in
benzoin, tolu, storax, and, with cocaine,
in coca leaves. It has been used in medi
cine to promote leucocytosis in cancer
and tuberculosis. Also called cinnamylic
acid, phenylacrylic acid.
f acide cinnamique
d Zimtsäure; Phenylacrylsäure
i acido cinnamico
e ácido cinámico

755 cinnamon
The dried inner bark of Cinnamomum
zeylanicum, which contains a volatile oil
(the cinnamon oil), characterized by a
strong aromatic odour. It is used for its
carminative action in various preparations, particularly for the treatment of
diarrhoea as the drug contains tannic
acid.
f cannelle
d Zimt
i cannella
e canela

756 cinnarizine
Piperazine-derived antihistaminic used in
the treatment of allergic disorders of the
skin and of the mucous membranes. It is
also indicated for the therapy of arteriosclerosis-related cerebral disorders.
f cinnarizine
d Cinnarizin
i cinnarizina
e cinarizina

757 cinoxacin
An antibiotic having action and use like
those of nalidix acid (an antibacterial
acid) and oxolinic acid (an antiproteus
agent). It acts against Gram-negative
germs and used in the treatment of infections of the genitourinary tract.
f cinoxacine
d Cinoxacin
i cinossacina
e cinoxacina

758 cinoxate
Agent used in dermatology; it promotes
sun-tanning and protects against sunlight
exposure and wind.
f cinoxate
d Cinoxat
i cinossato
e cinoxato

759 cisplatin
Cytotoxic platinum compound used in the
treatment of metastatic testicular and
ovarian cancer. The therapy must be
very carefully monitored as the drug
frequently affects the kidneys and causes
bone marrow suppression, ototoxicity and
some allergic reactions.
f cisplatine
d Cisplatin
i cisplatino
e cisplatino

760 citraconic acid
Acid which can be synthesized from citric
acid as an isomer of mesaconic and ita-
conic acid. It is a dicarboxylic unsaturated acid which can be considered a
maleic acid in which one hydrogen atom
of one of the =CH groups has been substituted with a methyl radical.
f acide citraconique; acide méthylmaléique
d Citraconsäure; Methylmaleinsäure
i acido citraconico; acido metilmaleico
e ácido citracónico; ácido metilmaleico

761 citric acid
A therapeutic agent which can be administered in the form of lemonade to help
digestion and in the form of sodium salt
as an anticoagulant. A mixture of citric
acid and sodium citrate is indicated for
the treatment of chronic metabolic acidosis borne of renal insufficiency.
f acide citrique
d Zitronensäure
i acido citrico
e ácido cítrico

762 citrophosphate
Compound of a citrate and a phosphate, in particular a solution of dicalcium phosphate in ammonium citrate, which is used in diseases or conditions involving a deficiency of calcium.
f citrophosphate
d Citrophosphat
i citrofosfato
e citrofosfato

763 clefamide
Amoebicide drug used in the treatment of acute and chronic intestinal amoebiasis.
f cléfamide
d Clefamid
i clefamide
e clefamido

764 clemastine
Antihistaminic and antioedematous agent indicated for the treatment of allergic conditions, e.g. itching, urticaria, etc. It also finds employment in the therapy of respiratory allergies and anomalous reactions to drugs.
f clémastine
d Clemastin
i clemastina
e clemastina

765 clemizole
Antihistaminic agent indicated in the treatment of allergic dermatoses as well as asthma, serum-related disorders, anaphylactic shock and some drug reaction.
f clémizole
d Clemizol
i clemizolo
e clemizolo

766 clenbuterol
Bronchodilator used in the treatment of bronchial asthma, chronic bronchitis and various emphysema-related disorders.
f clenbutérol
d Clenbuterol
i clenbuterolo
e clenbuterolo

767 clidinium bromide
Anticholinergic parasympatholytic agent with effects which are similar to those of atropine. It is chiefly used in the treatment of peptic ulcer.
f bromure de clidinium
d Clidiniumbromid
i clidinio bromuro
e bromuro de clidinio

768 clindamycin
Lincomycin-related semisynthetic antibiotic but better absorbed. It is indicated in the treatment of infections due to susceptible Gram-positive pathogenous microorganisms, in particular staphylococci, streptococci and pneumococci attacking the respiratory tract, the skin, ear etc.
f clindamycine
d Clindamycin
i clindamicina
e clindamicina

769 clioquinol
Antiseptic and disinfectant agent very effective against various bacteria and mycetes and chiefly used in the treatment of intestinal amoebiasis and for the protection against intestinal infections. Also indicated for vaginal trichomoniasis by topical application. The treatment, both disinfectant and antiseptic, of the intestinal tract by this drug should be limited to the amoebic and bacillary diarrhoeas, fermentative and putrefactive dyspepsias and enterocolitis due to sensitive microorganisms in order to avoid serious side effects and particularly neuropathies.
f clioquinol
d Clioquinol; Jodochlorhydroxyquin
i cliochinolo
e clioquinolo

770 clobazam
A tranquillizer. Benzodiazepine anxiolytic very similar to Diazepam (q.v.) in its effects. The drug is also used as anticonvulsant.
f clobazam
d Clobazam
i clobazam
e clobazam

771 clobenzorex
Therapeutic agent causing anorexia and thus used in the treatment of obesity in the adult. It is also indicated in the treatment of obesity from diabetes, of hypertension, arthrosis, obesity in children etc.
f clobenzorex
d Clobenzorex
i clobenzorex
e clobenzorex

772 clobetasol
Therapeutic agent having the same use and properties of clobetasone, q.v.
f clobétasol
d Clobetasol
i clobetasolo
e clobetasolo

773 clobetasone
Glucocorticoid agent for the topical treatment of inflammatory and allergic skin diseases, particularly psoriasis and eczema.
f clobétasone
d Clobetason
i clobetasone
e clobetasona

774 clobutinol
A sedative used for the relief of cough and marked by a central mechanism of action. It is indicated in the treatment of persistent cough independently of its origin. It is also an expectorant.
f clobutinol

d Clobutinol
i clobutinolo
e clobutinolo

775 clocanfamide
Non-anticholinergic agent capable of regu
lating gastric secretion and used in the
treatment of gastroduodenitis of any kind
and of gastro-duodenal ulcer.
f clocanfamide
d Clocanfamid
i clocanfamide
e clocanfamida

776 clocapramine
A neuroleptic agent used in the treatment
of schizophrenia. It may cause irritation,
anxiety and insomnia.
f clocapramine
d Clocapramin
i clocapramina
e clocapramina

777 clofazimine
Iminophenacinic dye used for its antibacterial and antiinflammatory action in the
treatment of leprosy and of the inflammations which are frequently associated
with it. The appearance of red urine,
diarrhoea and skin pigmentation are the
principal side effects.
f clofazimine
d Clofazimin
i clofazimina
e clofazimina

778 clofedanol
Antitussive agent mainly used in the
treatment of acute and/or chronic bronchitis and tracheitis. The drug sometimes
produces urticarial reactions.
f clofédanol
d Clofedanol
i clofedanolo
e clofedanolo

779 clofenamide
Saluretic, i.e. a diuretic carrying out
its action by inhibiting the reabsorption
of the sodium and thus increasing its
excretion. It is used in the treatment of
oedemas independently of their origin, as
well as anasarca, hypertension, premenstrual tension etc. In association with
other hypertensive agents, it is also used
in the treatment of disorders caused by
hypertension.
f clofénamide
d Clofenamid
i clofenamide
e clofenamida

780 clofenetamine
A muscle relaxant marked by anticholiner
gic properties and used for the treatment
of postencephalitic and arteriosclerotic
parkinsonism. It is also indicated in
neurovegetative disorders. It should not
be used before 15 years of age.
f clofénétamine

d Clofenetamin
i clofenetamina
e clofenetamina

781 clofezone
Analgesic and antiinflammatory agent
used in the treatment of rheumatic conditions, both of the phlogistic and degenerative type, as well as of periarthritis,
tendinitis, synovitis, gout, haemorrhoids,
superficial phlebitis, itching etc. Some
gastrointestinal disorder may occur as a
side effect.
f clofézone
d Clofezon
i clofezone
e clofezona

782 clofibrate
Therapeutic agent capable of reducing
blood cholesterol and fats and thus used
in patients presenting raised levels of
both. The drug is also indicated for
xanthoma tuberosum (a disease characterized by the appearance in the skin of
xanthomatous nodules) associated with
hyperlipidaemia. Side effects include
nausea, diarrhoea, dyspepsia, meteorism
and abdominal pains. The drug is capable of potentiating anticoagulants.
f clofibrate
d Clofibrat
i clofibrato
e clofibrato

783 clofibride
Therapeutic agent capable of reducing
high plasma concentrations of cholesterol
and triglycerides. It is indicated in the
treatment of atherogenous hyperlipidaemia
and atherosclerosis. It should not be
used in pregnancy and in hepatorenal
insufficiency.
f clofibride
d Clofibrid
i clofibride
e clofibrido

784 clofluonide
An antibacterial and antimycotic agent
which is used in various types of tinea
and pyodermitis. It is used topically in
the form of ointment, powder or solution.
f clofluonide
d Clofluonid
i clofluonide
e clofluonido

785 clofoctol
An antibacterial agent acting on Gram-
-positive microorganisms of the respiratory tract. It is chiefly used in paediatrics in case of acute or chronic infections of the upper respiratory tract,
bronchopneumonia and as a supporting
therapy in ear, nose and throat disorders.
f clofoctol
d Clofoctol
i clofoctolo

e clofoctolo

786 clomctacin
A pure analgesic marked by peripheral action and used to relieve acute and chronic rheumatic and traumatic pains. It is also indicated for the relief of toothache.
f clométacine
d Clometacin
i clometacina
e clometacina

787 clomethiazole
A sedative characterized by its very quick action when taken in large doses. Also a hypnotic and anticonvulsant agent. It is indicated in states of anxiety, delirium tremens and epilepsy. Also used in surgery as a preanaesthetic. Hypotension may occur as a side effect.
f clométhiazole
d Clomethiazol
i clometiazolo
e clometiazolo

788 clometocillin
An antibiotic (semisynthetic penicillin) related to penicillin G and indicated in the treatment of infections by sensitive microorganisms of the respiratory tract, of acute rheumatic conditions, sinusitis, otitis and dental infections. It is also recommended as a coadjuvant in the course of a prolonged administration of corticosteroids.
f clométocilline
d Clometocillin
i clometocillina
e clometocilina

789 clomiphene
A sex hormone indicated in cases of infertility due to failure of ovulation. Its action is exerted both on the pituitary gonadotrophic hormones and the ovary allowing ovulation. The drug is also used to correct menstrual disorders. It should not be used when ovarian cysts have been ascertained.
f clomiphène
d Clomiphen
i clomifene
e clomifeno

790 clomipramine
Antidepressant which acts by increasing the concentration of catecholamines in the brain with ensuing stimulation of the sympathetic functions. It is indicated in the treatment of endogenous depressive and manic-depressive states. Should other psychotropic drugs be administered concurrently, a strict monitoring is advisable in order to avoid serious side effects.
f clomipramine
d Clomipramin
i clomipramina
e clomipramina

791 clomocycline
A broad-spectrum antibiotic acting on all bacteria which are susceptible to tetracyclines and used in the treatment of respiratory, genitourinary and biliary infections. Its action, side effects and interactions are similar to those of tetracycline, q.v.
f clomocycline
d Clomocyclin
i clomociclina
e clomociclina

792 clonazepam
Anticonvulsant benzodiazepine marked by a greater activity than diazepam. It produces an elective inhibition of the activity of the epileptogenic foci and thus prevents the generalization of convulsions. It is used intravenously for the control of status epilepticus and orally for all epileptic clinical forms of the newborn, of the adult and for focal crises.
f clonazépam
d Clonazepam
i clonazepam
e clonazepam

793 clonidine
A vasodilator and antihypertensive agent. It reduces sympathetic activity by a central action and diminishes vascular reactivity. The drug is used in the treatment of hypertension and chronic headache. The antihypertensive effect is blocked by tricyclic antidepressants. Gastric inflammation may occur as a side effect.
f clonidine
d Clonidin
i clonidina
e clonidina

794 clonixin
Analgesic and antiphlogistic agent used in the treatment of all acute and chronic painful disorders, in particular those which are associated with inflammation.
f clonixine
d Clonixin
i clonissina
e clonixina

795 clopamide
Diuretic agent acting on the kidney and capable of increasing the urinary excretion of water, sodium and chlorine. Its activity is rather short-lived and may cause a degree of hypotension. The drug is particularly indicated in the therapy of oedemas and should be avoided in the presence of severe renal lesions.
f clopamide
d Clopamid
i clopamide
e clopamida

796 clopenthixol
A phenothiazine-related major tranquillizer chiefly used for the treatment of manic and hallucinatory states, schizo-

phrenia and advanced senile dementia. Adverse side effects include sedation and hypotension. Extrapyramidal symptoms are not very frequent.

f clopenthixol
d Clopenthixol
i clopentissolo
e clopentisolo

797 cloperastine
Antitussive agent used in all forms of cough.

f clopérastine
d Cloperastin
i cloperastina
e cloperastina

798 clopirac
Antiinflammatory, antipyretic and analgesic agent generally used in all rheumatic conditions, both inflammatory and degenerative, posttraumatic reactions, neuralgias, sciatica etc. Gastrointestinal disorders may occur as side effect.

f clopirac
d Clopirac
i clopirac
e clopirac

799 clorcotolone
A cortical steroid hormone used topically for its antiinflammatory, antiallergic, antisudoriferous and antiproliferative effects. The drug is indicated in the treatment of acute, subacute and chronic eczemas. Occlusive dressings should never be used to avoid side effects of a systemic type.

f clorcotolone
d Clorcotolon
i clorcotolone
e clorcotolona

800 clorexolone
Long-acting diuretic agent of the sulphonamide type, capable of strongly promoting natriuresis, i.e. the considerable sodium glomerular filtration that occasionally occurs with greatly increased diuresis following relief of urinary obstruction, as well as kaliuresis, i.e. the excretion of potassium in the urine. It is chiefly used against oedemas, particularly cardiac and hepatic oedemas, obesity and hypertension.

f clorexolone
d Clorexolon
i cloresolone
e cloresolona

801 clorindione
2-(4-chlorophenyl)indane-1,3-dione. Anticoagulant agent whose action inhibits the formation of prothrombin and factor VII (the factor in the blood serum which is necessary for the action of brain thromboplastin) and is used both in the prophylaxis and the long-term treatment of thrombosis, thrombophlebitis, coronary thrombosis and thromboembolism following myocardial infarction.

f clorindione
d Clorindion
i clorindione
e clorindiona

802 clorotepine
A neuroleptic agent, i.e. a tranquillizer, used in the treatment of psychosis and schizophrenia.

f clorotépine
d Clorotepin
i clorotepina
e clorotepina

803 clorprenaline
Bronchodilator (similar to ephedrine, q. v.) used in the treatment of bronchial asthma and asthmatic bronchitis, as well as of chronic pulmonary emphysema, bronchiectasia and pneumoconiosis.

f clorprénaline
d Clorprenalin
i clorprenalina
e clorprenalina

804 clortermine
Sympathomimetic amine characterized by its capacity of causing loss of appetite for food. It is therefore used in the treatment of obesity. The drug must be administered very cautiously as its side effects are considerable, e.g. tachycardia, insomnia, alterations in the libido, impotence etc.

f clortermine
d Clortermin
i clortermina
e clortermina

805 clostebol
Synthetic anabolic (promoting the process of making new living tissue from nutrient materials) agent capable of increasing the biosynthesis of protein and used in the treatment of pathological loss of weight, asthenia etc. Virilization, on a very mild scale, has been reported as a side effect.

f clostébol
d Clostebol
i clostebol
e clostebol

806 clotiapine
Dibenzothiazepine-related tranquillizer used in the treatment of psychoses, both acute and chronic. Extrapyramidal symptoms, tachycardia, disorders of eye accommodation and oedemas are among the possible side effects.

f clotiapine
d Clotiapin
i clotiapina
e clotiapina

807 clotrimazole
Broad-spectrum antifungal antibiotic including dermatophytes, yeasts, moulds etc. It has proved active against Tricho-

mona vaginalis, staphylococci and strepto cocci and is used in the treatment of dermatomicosis and, more generally, of various skin infections with Candida.
f clotrimazole
d Clotrimazol
i clotrimazolo
e clotrimazolo

808 cloxacillin
Semisynthetic penicillase-resistant penicillin acting like benzylpenicillin, i.e. penicillin G sodium whose action is based on the modification of the bacterial membrane. Cloxacillin, which can be administered on a full stomach and is rapidly absorbed, is indicated in the treatment of infections caused by pyogenic cocci, of pneumonia by diplococcus, blenorrhagia, syphilis and epidemic meningitis.
f cloxacilline
d Cloxacillin
i clossacillina
e cloxacilina

809 cloxazolam
A derivative of benzodiazepine used as anxiolytic, sedative and tranquillizing agent, as well as an anticonvulsant, in states of anxiety, in depression and neurasthenia.
f cloxazolam
d Cloxazolam
i cloxazolam
e cloxazolam

810 clozapine
Neuroleptic agent characterized by a potent sedative power and a rapid strong action against psychoses. It is indicated for the therapy of schizophrenia, both acute and chronic, of states of psychomotor excitation also in its manic forms, insomnia, depression etc. Accommodation difficulty and orthostatic hypotension may appear as side effects.
f clozapine
d Clozapin
i clozapina
e clozapina

811 clupadonic acid
The active principle of cod-liver oil and other fish-oils. It generally stimulates recovery and organic defences. It is indicated in lymphatism, rickets, adenitis, hilar adenopathies and, more generally, in the course of convalescence.
f acide clupadonique
d Clupadonsäure
i acido clupadonico
e ácido clupadónico

812 coal tar
The tar that is obtained by the destructive distillation of bituminous coal and consists of many constituents, e.g. benzene, xylenes, naphthalene, pyridine, quiniline, phenol, cresols etc. It is a keratolytic agent indicated in the treatment of eczema and psoriasis.
f goudron de houille
d Steinkohlenteer
i catrame di carbone
e alquitrán de hulla

813 cobalt gluconate
Neurotrophic (relating to the nutrition of the nervous system) agent active on the orthosympathetic and parasympathetic systems and used in the therapy of peripheral arteritis, hypotension, acrocyanosis etc.
f gluconate de cobalte
d Kobaltglukonat
i cobalto gluconato
e gluconato de cobalto

814 cobalt tetracemate
A strong antidote used against cyanide poisoning as it binds with cyanide and annuls its effects on cell metabolism.
f tétracémate de cobalte
d Kobalttetracemat
i cobalto tetracemato
e tetracemato de cobalto

815 cobamamide
A form of vitamin B_{12} which is metabolically active, i.e. a coenzyme promoting the biosynthesis of protein. It is used in the anabolic treatment of hypotrophic children to improve the value of nutrition and for their growth. Also indicated in the course of convalescence.
f cobamamide
d Cobamamid
i cobamamide
e cobamamida

816 cocaine
Methylbenzoylecgonine. Bitter crystalline alkaloid contained in the leaves of Erythroxylon coca and produced in colourless prismatic crystals or in the form of a white crystalline powder. Cocaine, or its hydrochloride, is sometimes used as a local anaesthetic especially in ophthalmology and in dentistry. It is, in fact, a very strong surface anaesthetic but is rarely used internally because of its high toxicity and the great danger of habit forming. It is also used as a mydriatic (causing the dilatation of the pupil). A prolonged use may cause corneal ulcerations.
f cocaïne
d Cocain; Kokain
i cocaina
e cocaina

817 cocarboxylase
Thiamine phosphate. A coenzyme of aneurine (vitamin B_1) capable of improving the cellular metabolism particularly in relation to the central nervous system. It is indicated in the treatment of nervous disorders and also as an analgesic. Intravenously injected high doses should be avoided as they may cause collapse.

f carboxylase
d Carboxylase
i carbossilasi
e carboxilasa

818 codeine
A mild narcotic analgesic. It is used, generally combined with acetylsalicylic acid, to relieve somatic deep pain and to suppress the cough reflex. It may thus be considered an antitussive agent. The drug is also used against diarrhoea. It should be avoided in pregnancy.
f codéine
d Codein; Kodein
i codeina
e codeina

819 cod-liver oil
Fatty oil obtained from the liver of the cod and other related fishes. It is used in medicine chiefly as a source of vitamin A and D whenever condition caused by an abnormal metabolism of calcium and phosphorus (i.e. rickets, infantile tetany, osteomalacia etc.) are diagnosed.
f huile de foie de morue
d Lebertran
i olio di fegato di merluzzo
e aceite de hígado de bacalao

820 colchicine
Alkaloid obtained from the seeds and bulbs of Colchicum autumnale and used for the relief of pain in acute gout. Also used as an antiinflammatory and analgesic agent and a mild diuretic. Side effects include vomiting, diarrhoea, nausea and abdominal pains.
f colchicine
d Colchicin
i colchicina
e colquicina

821 colestipol
An ion exchange resin capable of reducing the level of plasma cholesterol by binding with bile acids in the intestinal lumen. It is used in the treatment of hyperlipoproteinaemia (type II) and hypercholesterolaemia.
f colestipol
d Colestipol
i colestipolo
e colestipolo

822 colicroot
Any of various plants whose roots are reputed to cure colic. Also known as Indian potato or yam.
f igname indigène
d Kolikwurzel; Yams
i dioscorea selvatica
e dioscorea

823 colistin
Antibiotic indicated for renal and intestinal infections caused by Gram-negative microorganisms. Orally administered, the drug is inadequately absorbed. In case of intramuscular or intravenous injection, it is used either by itself or in association with penicillin.
f colistine
d Colistin
i colistina
e colistina

824 collagen
Insoluble fibrous protein occurring in vertebrates as the principal constituent of the fibrils of connective tissue and of the organic substance of bones. Used therapeutically, it is an enzymatic cicatrizing agent indicated in the treatment of wounds involving loss of tissue as well as in ulcers caused by chronic capillaritis.
f collagène
d Kollagen
i collageno
e colágeno

825 collodion
A viscous solution of pyroxylin in a mixture of alcohol and ether or acetone, chiefly used as a coating for wounds.
f collodion
d Kollodion; Kollodium
i collodio
e colodión

826 colophony
Rosin. Translucent brittle resin which is obtained from the oleoresin or dead wood of pine by removal of the volatile turpentine or from tall oil by removal of the fatty acid components. It contains abietic and other resin acids and is used in protective topical preparations.
f colophone
d Colophonium
i colofonia
e colofonia

827 copper acetate
The normal salt of copper forming green crystals and used topically for its astringent properties.
f acétate de cuivre
d Kupferacetat
i acetato di rame
e acetato de cobre

828 copper sulphate
The sulphate which is made by roasting copper sulfide ores and by the action of sulphuric acid on copper or copper oxides. Chiefly used as a fungicide and algicide. In medicine, it is used as an emetic in the treatment of anaemia (together with iron) and as an astringent in topical preparations. Copper poisoning may occur when it is administered in large doses.
f sulfate de cuivre
d Kupfersulfat
i solfato di rame
e sulfato de cobre

829 corticosteroids
A term denoting natural and synthetic steroids whose action is similar to that of hydrocortisone which is produced in, and extracted from, the adrenal cortex. Several of these steroids are hormones, e.g. corticosterone, cortisone, aldosterone. They are characterized by a potent antiinflammatory and salt-retaining properties. Their administration must be carefully controlled owing to their several side effects, e.g. oedema, hypertension, fracture-accompanied bone thinning, diabetes etc. Also known as corticoids.
f corticoïdes
d Kortikosteroide
i corticosteroidi
e corticoesteroides

830 corticotrophin
The pituitary hormone which controls the functions of adrenal cortex. It increases the production of adrenocortical steroids. It acts together with, and has the same uses of, other adrenal corticosteroids, in particular antiinflammatory and antirheumatic.
f corticotrophine
d Corticotrophin
i corticotrofina
e corticotrofina

831 corticotrophin-gel
Corticotrophin preparation acting very rapidly and for a prolonged time. In fact, a single injection ensures haematic levels which are therapeutically effective for up to 24 hours. Its actions are the same as those of the natural hormone and it is indicated in all diseases or disorders requiring corticosteroid therapy in patients presenting a normal function of the adrenals, e.g. collagen deficiencies, rheumatoid arthritis, dermatoses, eye affections etc.
f corticotrophine-gel
d Corticotrophin-Gel
i corticotrofina gel
e corticotrofina gel

832 corticotrophin phosphate
A long-acting corticotrophin making it possible to exploit very efficiently the therapeutic properties of the hormone. It presents the same indications of the natural corticotrophin and is indicated in all diseases and disorders requiring corticosteroid therapy in patients whose adrenals function normally, e.g. collagen diseases, acute articular rheumatism, allergies and acute inflammation affecting skin and eyes.
f phosphate de corticotrophine
d Corticotrophinphosphat
i corticotrofina fosfato
e fosfato de corticotrofina

833 corticotrophin-zinc hydroxide
Long-acting corticotrophin which is slowly absorbed and whose action, obtained by intramuscular or subcutaneous injection, produces blood levels that are therapeutically effective for at least 48 hours. It has the same action and effects of the corticotrophin-gel, q.v.
f corticotrophine-hydroxyde de zinc
d Corticotrophinzinkhydroxyd
i corticotrofina-zinco idrossido
e corticotrofina-hidróxido de zinc

834 cortisone
A substance of hormonal nature which can be extracted from the adrenal cortex and obtained by synthesis. Chemically, it corresponds to the 17-hydroxy-11-dehydrocorticosterone. It is soluble in alcohol and acetone but insoluble in water. It has a strong rotatory power. Cortisone presents a very considerable antiinflammatory and antiexudative activity; it slows down the processes of granulation and affects some metabolic processes (inducing an increase of the glycaemia, of the catabolic processes of the protides and the anabolic ones of the lipids). Owing to such properties, the drug was initially used very widely. Later, however, owing to some serious and at times dangerous side effects, it has been progressively replaced with very similar (cortisonic) products having the same therapeutic properties but less dangerous side effects.
f cortisone
d Cortison
i cortisone
e cortisona

835 cortivazol
A corticosteroid used in the treatment of asthma, allergic skin diseases, collagen diseases, hepatitis and cirrhosis. The drug also finds employment in haematology, neurology, cancerology, bacterial and viral infections etc. The activity of cyclophosphamide is reduced by the drug.
f cortivazol
d Cortivazol
i cortivazolo
e cortivazolo

836 coumazoline
An alpha-stimulant agent characterized by a topical vasoconstricting and decongesting capacity. It is used in the treatment of rhinitis, sinusitis and to cause vasoconstriction in association with local anaesthetics in the course of surgical operations in otorhinolaryngology.
f coumazoline
d Cumazolin
i cumazolina
e cumazolina

837 coumetarol
Anticoagulant capable of inhibiting the synthesis of prothrombin. It is used in the prophylaxis and in the treatment of phlebitis, embolism, angina pectoris, myocardial infarction and embolism.

f coumétarol
d Cumetarol; Coumetarol
ı cumetarolo
e cumetarolo

838 **creatinolphosphate**
Therapeutic agent indicated in the treatment of chronic myocardiopathy, in the course of convalescence following a myocardial infarction and also in the therapy of some infectious diseases.
f créatinolphosphate
d Creatinolphosphat
ı creatinolfosfato
e creatinolfosfato

839 **creosote**
Colourless or yellowish oily liquid characterized by a pungent smoky taste and containing a mixture of phenolic compounds. It is obtained by the distillation of wood tar and is especially used as an expectorant in chronic bronchitis. It is also a potent disinfectant.
f créosote
d Kreosot
ı creosoto
e creosota

840 **cresol**
A strong antiseptic used as disinfectant and preservative. It is also indicated for the relief of congestion in bronchitis, asthma and related disorders. It has a very strong histolytic action and in concentrated solution may cause local corrosion, damage to liver and kidneys and depression of the central nervous system.
f crésol
d cresol; Kresol
ı cresolo
e cresol

841 **cropropamide**
A respiratory stimulant used in the treatment of cardiovascular disorders.
f cropropamide
d Cropropamid
ı cropropamide
e cropropamida

842 **crotamiton**
Crotonyl-N-ethyl-o-toluidide. Therapeutic compound used in the treatment of scabies and for the symptomatic relief of pruritus. Its action is fairly rapid and lasts from 4 to 10 hours. Allergy to the drug occurs in infants.
f crotamiton
d Crotamiton
ı crotamitone
e crotamitona

843 **cryofluorane**
A local anaesthetic for cryanaesthesia used mainly in dentistry and stomatology, but also for surgical operations performed in a very short time and for the relief of local pains.
f cryofluorane
d Cryofluoran
ı criofluorano
e criofluorano

844 **cryptenamine**
Hypotensive agent used in the treatment of essential benign hypertension and hypertensive cardiovasculopathies. A parenteral use is recommended in acute hypertensive episodes which may occur in encephalopathies and a form of eclampsia (attack of convulsions) accompanied by essential hypertension.
f crypténamine
d Cryptenamin
ı criptenamina
e criptenamina

845 **cuprammonium dodecyl sulphonate**
Or cuprammonium lauryl sulphonate. A fungicide used in the treatment of mycosis of the skin, scalp and mucous membranes. Herpes calcinatus, pityriasis versicolour and trichophytosis are also treated by cuprammonium dodecyl sulphonate.
f laurylsulfonate de cuivre ammoniacal
d Cuprammoniumlaurylsulfonat
ı cuprammonio lauril sulfonato
e sulfonato de laurilo cuproamoniacal

846 **cyacetacide**
Therapeutic agent used in the treatment of tuberculosis and of extratubercular pulmonary diseases caused by streptomycin-resistant microorganisms. It is administered orally and may cause dyspepsia and psychic excitation.
f cyacétacide
d Ciacetazid
ı ciacetacide
e ciacetacida

847 **cyamemazine**
Phenothiazine-related neuroleptic (tranquillizing) agent used in the treatment of neurotic and depressive states caused alcoholism, of anxiety, feelings of the distress and behavioural disorders. Asthenia and sometimes hypotension may occur as side effects.
f ciamémazine
d Cyamemazin
ı ciamemazina
e ciamemazina

848 **cyanocobalamin**
A specific-action vitamin used in the prophylaxis and treatment of pernicious anaemia and some types of macrocytic anaemia. It is a cobalt-containing substance which may be produced (according to BP) by the growth of certain microorganisms or obtained from liver. Its action is that of hydroxycobalamin which, in fact, has largely replaced it. The drug has also some analgesic activity and protects the liver.
f cyanocobalamine
d Cianocobalamin

i cianocobalamina
e cianocobalamina

849 cyclacillin
Broad-spectrum penicillin-related semisynthetic antibiotic capable of acting a-
gainst Gram-positive and Gram-negative
microorganisms and used in the treatment
of infections caused by susceptible organ
isms of the respiratory, genitourinary,
biliary tracts, ear, nose and laryngeal
disorders. Its effects are similar to those
of ampicillin.
f cyclacilline; ciclacilline
d Cyclacillin
i ciclacillina
e ciclacilina

850 cyclandelate
A 3,3,5-trimethylcyclohexyl mandelate
used in the treatment of vascular disorders. It is a peripheral vasodilator char
acterized by a strong myolytic action on
the smooth-muscle cells of the vasal
walls. It is also used in the treatment of
cerebral vasculopathies, coronary and
peripheral vascular disorders and intestinal spasms and pains.
f cyclandélate
d Cyclandelat
i ciclandelato
e ciclandelato

851 cyclarbamate
Muscle relaxant and tranquillizer used in
the treatment of painful gastro-intestinal,
hepatobiliary, urinary and uterine spams
and of various muscle contractures.
f cyclarbamate
d Cyclarbamat
i ciclarbamato
e ciclarbamato

852 cyclizine
Antihistaminic agent chiefly used in the
treatment of nausea, dizziness, violent
post-operative vomiting as well as vomiting in the course of early pregnancy.
f cyclizine
d Cyclizin
i ciclizina
e ciclizina

853 cyclobarbital
A sedative and hypnotic agent characterized by a prompt action as the drug is
rapidly absorbed. It is used as a preanaesthetic and as a hypnotic after a
surgical operation.
f cyclobarbital
d Cyclobarbital
i ciclobarbitale
e ciclobarbitalo

854 cyclobenzaprine
A relaxant of skeletal muscles mainly
used for the relief of muscular spasms
and painful musculo-skeletal conditions.
Some serious side effects may occur and
the administration of the drug should be

very cautious.
f cyclobenzaprine
d Cyclobenzaprin
i ciclobenzaprina
e ciclobenzaprina

855 cyclobutirol
A choleretic agent capable of stimulating
the excretory activity of the liver and of
reducing the level of cholesterol in the
blood. It is also indicated in the treatment of arteriosclerosis. Nausea and diar
rhoea are possible side effects.
f cyclobutirol
d Cyclobutirol
i ciclobutirolo
e ciclobutirolo

856 cyclofenil
A sex hormone used in the treatment of
infertility caused by failure of ovulation.
The drug induces ovulation and stimulates the secretion of the "releasing hormones" for gonadotrophins.
f cyclofénil
d Cyclofenil
i ciclofenile
e ciclofenilo

857 cyclomethycaine
An ester-type local anaesthetic used for
the symptomatic relief of ulcers, burns,
abrasions, haemorrhoids, painful dermatoses, herpes zoster, vaginitis anal and
vulvar itching.
f cyclométhycaïne
d Cyclomethycain
i ciclometicaina
e ciclometicaina

858 cyclopentamine
A vasoconstrictor, sympathicomimetic and
decongestant agent used against congestion and oedema particularly in the treat
ment of acute, chronic and allergic rhini
tis and pharyngolaryngitis. For a quick
action it is applied locally on the nasal
mucosa. Also used in ophthalmology.
f cyclopentamine
d Cyclopentamine
i ciclopentamina
e ciclopentamina

859 cyclopenthiazide
A diuretic agent capable of increasing
the excretion of water and salts without
reducing the calcium content in the blood.
It produces hypotension with an action
that is coordinated with that of the specific hypotensive agents. The drug is
used in the treatment of oedemas and to
relieve premenstrual symptoms.
f cyclopenthiazide
d Cyclopenthiazid
i ciclopentiazide
e ciclopentiazida

860 cyclopentobarbital
Cyclopentenallylbarbituric acid. A hypnotic (barbiturate) used in the treatment of

insomnia due to emotional and/or physical stress. It is also indicated in the treatment of agrypnia (insomnia associated with intense restleness) due to neurovegetative dystonia and as a sedative.

f cyclopentobarbital
d Cyclopentobarbital
i ciclopentobarbitale
e ciclopentobarbital

861 cyclopentolate

Chemical compound whose hydrochloride is a mydriatic and cycloplegic agent used in ophthalmology both for diagnosis and therapeutic purposes. It has a quicker but shorter-lasting action than atropine. It is also used to assist optical refraction.

f cyclopentolate
d Cyclopentolat
i ciclopentolato
e ciclopentolato

862 cyclophosphamide

Cytotoxic agent administered orally or intravenously in the treatment of lymphatic leukaemia and chronic myeloid leukaemia, lymphogranulomatosis and lymphosarcoma, i.e. all malignant tumours. The drug is activated by metabolism in the liver and excreted chiefly in the urine. Alopecia, anemia and gastrointestinal disorders may occur as side effects.

f cyclophosphamide
d Cyclophosphamid
i ciclofosfamide
e ciclofosfamida

863 cyclopropane

A potent inhalation anaesthetic used to induce total anaesthesia in paediatrics and obstetrics. It may cause cardiac arrhythmia.

f cyclopropane
d Cyclopropan
i ciclopropano
e ciclopropano

864 cycloserine

An antibiotic used in tuberculosis and in the treatment of infections caused by streptomycin-, izionazid-, paraaminosalicylic acid- and pyrazinamide-resistant organisms. It is also used in the treatment of tuberculosis of the kidney and of the bladder. The drug acts by inhibiting the protein synthesis and the biosynthesis of uridin peptides of the membrane. Drowsiness and in some cases ataxia have been observed as side effects.

f cyclosérine
d Cycloserin
i cicloserina
e cicloserina

865 cyclosporin

A potent immunosuppressant antibiotic used to prevent reject phenomena following the transplant of organs or tissues. Liver and kidney disorders may occur as side effects.

f cyclosporine
d Cyclosporin
i ciclosporina
e ciclosporina

866 cyclothiazide

A chlorothiazide-derived diuretic agent which acts by inhibiting the reabsorption of water, sodium and chloride. It lowers blood pressure and is used in the treatment of oedemas and hypertension.

f cyclothiazide
d Cyclothiazid
i ciclotiazide
e ciclotiazida

867 cyclovalone

A therapeutic agent used to stimulate the activity of the liver in the treatment of exogenous and other intoxications when bile secretion is inadequate.

f cyclovalone
d Cyclovalon
i ciclovalone
e ciclovalona

868 cycrimine

Atropine-like parasympatholytic agent used in the treatment of parkinsonism. It may cause nausea and anorexia and must therefore be administered in small doses and during meals.

f cycrimine
d Cycrimin
i cicrimina
e cicrimina

869 cynarine

The active principle of the artichoke. Used as a drug, it increases the volume of the bile and the excretion of solid substances and cholesterol. It is also a diuretic and hypocholesterolaemic agent indicated in disorders of the liver and of the biliary tract.

f cynarine
d Cynarin
i cinarina
e cinarina

870 cyprodenate

Psychostimulant, anxiolytic and antidepressant agent used in the treatment of psychic depression, cerebrovascular disorders and states of pre-dementia. Also indicated in the treatment of psychosomatic diseases.

f cyprodénate
d Cyprodenat
i ciprodenato
e ciprodenato

871 cyprofloxacin

A broad-spectrum antibiotic used in the treatment of several infections and especially those affecting the respiratory and urinary tracts whenever the organisms prove resistant to penicillins and cephalosporins. Side effects can be conspicuous,

in particular some allergic reactions.
f cyprofloxacine
d Cyprofloxacin
i ciproflossacina
e ciprofloxacina

872 cyproheptadine
Therapeutic agent used to combat allergies, in particular itching dermatoses, disorders caused by some serum or drug. Increased appetite and drowsiness are the most frequent side effects.
f cyproheptadine
d Cyproheptadin
i ciproeptadina
e ciproheptadina

873 cyproterone
A steroid hormone characterized by antiandrogenic activity as well as some progestogenic activity. It is used in the treatment of prostatic carcinoma and of pathological male hypersexuality and sexual deviations; also in precocious idiopathic puberty in both sexes. The drug acts by a reversible inhibition of libido.
f cyprotérone
d Cyproteron
i ciproterone
e ciproterona

874 cysteamine
Therapeutic agent used as an antidote in severe cases of poisoning caused by paracetamol (an analgesic and antipyretic) where it probably checks liver damage by reducing the concentration of a toxic metabolite.
f cystéamine
d Cysteamin
i cisteamina
e cisteamina

875 cysteina
Sulphur-containing amino acid which easily oxidizes to cystine (the reaction is reversible). It is one of two important and related amino acids derived from propionic acid and cystine. It participates in the composition of several proteins and of the glutathione tripeptide which is widely diffused in the tissues. The

agent neutralizes the activity of collagenase and is used in the treatment of corneal ulcers and, together with antiviral ethoxuridine, of herpetic ulcers. It also functions as activator of several enzymes.
f cystéine
d Cystein
i cisteina
e cisteina

876 cystine
Sulphur-containing amino acid which participates in the constitution of several protides. It is a non-essential amino acid for man as it can be replaced by methionine. It is used for the preparation of a number of dermatological products.
f cystine
d Cystin
i cistina
e cistina

877 cytarabine
Antiviral agent used systemically in the treatment of herpes encephalitis. The drug tends to offset the action of pyrimidine by inhibiting the biosynthesis of deoxyribonucleic acid.
f cytarabine
d Cytarabin
i citarabina
e citarabina

878 cyticholine
A nucleotide coenzyme, essential factor for the biosynthesis of lecithins. The drug is used to correct alterations of the lipid metabolism in the brain due to traumas or hypoxia. It also increases the activity of the ascendant reticular substance, promotes oxygen consumption and improves cerebral circulation. It is therefore mainly used in traumatic and degenerative cerebro-vascular disorders following cranial traumas or brain surgery, as well as thrombosis, Parkinson's syndrome and cerebral arteriosclerosis.
f cyticholine
d Cyticholin
i citicolina
e citicolina

D

879 dacarbazine
An alkylating agent displaying a cytostatic action and used in the treatment
of metastatic malignant melanoma. Owing
to its several, at times very serious,
side effects, e.g. leucopenia, liver disor
ders, thrombocytopenia and, above all,
marrow suppression, the drug should be
administered under very strict supervision and careful monitoring.
f dacarbazine
d Dacarbazin
i dacarbazina
e dacarbazina

880 dactinomycin
Antibiotic which is active both on Gram-
-positive and Gram-negative bacteria but
is seldom used for the treatment of infections owing to its high degree of toxicity.
It finds employment as a cytotoxic agent
against neoplastic diseases. Bone marrow
depression may occur as a side effect.
f dactinomycine
d Dactinomycin
i dactinomicina
e dactinomicina

881 dahlia
Plant of the family Compositae. The bulbs
of its flowers contain inulin and a purple dye which is known as Dahlia B and
is related to methyl violet and, being a
mixture of methyl rosaniline chlorides, is
used as an antiseptic.
f dahlia
d Dahlie
i dalia
e dalia

882 Dalmatian insect plant
A genus of composite plants characterized
by finely divided leaves and including
the Chrisantemum cinerariae folium whose
flowers yield a substance capable of
destroying infection-carrying insects.
f chrysanthème insecticide; pyrèthre de
Dalmatie
d dalmatische Insektenblüte; Pyrethrumblüte
i crisantemo di Dalmazia; piretro di Dalmazia
e pelitre de Dalmacia

883 damiana
The dried leaf of Turnera diffusa, a
plant growing in tropical America, California and Texas, used in the past as a

tonic and an aphrodisiac.
f damiane
d Damiana
i damiana
e damiana; pastorcita

884 dammar
Oleoresin extracted from various Diptero-
carpaceae which grow in tropical Asia.
It is soluble in chloroform, turpentine
etc. and contains dammarolic acid, e-
thereal oil etc. It is used in medicine
for the preparation of poultices. Also
known as damar or gum dammar.
f résine Dammar; gomme Dammar
d Dammarharz; Felsendammar
i dammar; dammara
e resina de Dammar

885 danazol
Therapeutic agent used in endocrine disorders in case pituitary control of gonad
hormone production is required. It is a
synthetic ethisterone derivative (ethisterone is a preparation whose pharmacological action is similar to that of progesterone) displaying an inhibitory activity
on the release or the synthesis of gonado
tropins with a reversible suppression of
the pituitary-ovary and pituitary-testis
axes. It is used in the treatment of endo
metriosis, primary constitutional precocious puberty, gynecomastia and fibro-
cystic mastopathy.
f danazol
d Danazol
i danazol
e danazol

886 dandelion root
The fresh or dried root of the plant
Taraxacum officinale constituting the
taraxacum, a therapeutic agent used as
a tonic, a bitter and a mild laxative.
f racine de dent de lion; racine de pissen-
lit
d Löwenzahnwurzel; Ackerzichoriewurzel
i radice di tarassaco; radice di dente di
leone
e raíz de taraxacón; raíz de diente de león

887 dandrolene
A skeletal-muscle relaxant used in the
treatment of spasms caused by embolism,
multiple sclerosis, cerebral paralysis and
lesions affecting the vertebral column.
Jaundice may occur as a side effect.

f dandrolène
d Dandrolen
i dandrolene
e dandroleno

888 danthron
Cathartic agent acting like cascara. It is derived from anthraquinone and is capable of promoting peristalsis, particularly in the large intestine. It is used in the treatment of acute and chronic constipation as well as in intoxications due to anticholinergic drugs.
f dantrone
d danthrone
i dantrone
e dantrona

889 daphne
Spurge flax. Genus of trees and shrubs of the family Thymelaeceae, whose dried bark is used to obtain mezerium which contains a bitter glycosidal principle.
f daphné paniculé; garou
d rispenblütiger Seidelbaststrauch
i dittinella
e matapollo; torvisco

890 dapsone
Or DADPS. A sulphone drug marked by a bacteriostatic action and acting by preventing the use by bacteria of metabolites that are essential to their normal development. The drug is used in the treatment of leprosy. Tachycardia, haemolytic anaemia, allergic dermatitis, anorexia and liver damage are the principal adverse effects.
f dapsone
d Dapsone
i dapsone
e dapsona

891 daturine
A mixture of alkaloids which is found in stramonium, the leaves and flowering tops of Datura stramonium, and consists of atropine and hyoscyamine. Also known as duboisine.
f daturine
d Daturin
i daturina
e daturina

892 daunorubicin
Cytotoxic antibiotic used as an antineoplastic agent and particularly indicated in the treatment of Hodgkin's disease, acute myeloid and lymphatic leukaemia and other malignant tumours. Bone marrow depression and cardiovascular disorders may occur as side effects. Also known as daunomycin.
f daunorubicine
d Daunorubicin
i daunorubicina
e daunorubicina

893 David's root
The root of small trees of the family Rubiaceae used for their diuretic and drastic properties and also, in some South American countries, against snake's bites. Such roots contain as active principles tannic acid, a glycoside and an alkaloid.
f racine de caïnca
d Schneebeerenwurzel
i radice di cainca
e raíz de cainca

894 deanol
Dimethylaminoethanol. A chemotherapeutic drug used chiefly in the treatment of psychoneuroses.
f déanol; diméthylaminoéthanol
d Dimethylaminoäthanol
i dimetilamminoetanolo
e dimetilaminoetanol

895 deanol aceglumate
Amphetamine-related agent acting on the neuroregulatory mechanism of the reticular formation without sympathetic stimulation or anorexigenous effect. It is chiefly indicated in the treatment of intellective disorders, psychic stress, weakening of the cerebral performance of organovascular or traumatic origin and alterations and/or retardation of the psychic development in very young children.
f acéglumate de déanol
d Deanolaceglumat
i deanolo aceglumato
e aceglumato de deanolo

896 deanol acetamidebenzoate
A psychotonic agent regarded as a precursor of the acetylcholine in relation to the central nervous system. It promotes and improves the capacity to learn and to concentrate while strongly diminishing emotional hyperexcitability. It is therefore indicated in states of psychic asthenia, mild depression, difficulty in learning, temporary loss of memory and behavioural disorders in children.
f acétamidebenzoate de déanol
d Deanolacetamidbenzoat
i deanolo acetamidebenzoato
e acetamidabenzoato de deanolo

897 deanol phosphate
A non-amphetaminic psychoanaleptic agent used in the treatment of psychoasthenia and anorexia in children, as well as of psychic disorders, apathy and depression in elderly patients. It is also indicated to relieve depression in parkinsonism and cerebral affections in alcoholism.
f phosphate de déanol
d Deanolphosphat
i deanolo fosfato
e fosfato de deanolo

898 debrisoquine
Adrenergic neurone blocking agent marked by a peripheral mechanism of action. The drug is used to combat hypertension. It is rapidly absorbed and its effect lasts

for 24 hours.
f débrisoquine
d Debrisoquin
i debrisochina
e debrisoquina

899 decamethonium bromide
Decamethonium is an organic radical derived by substitution from ammonium. Its
simple salts, e.g. bromide and iodide,
are muscle relaxants which behave superficially like curare but basically differ
in their mode of action. In fact, they
block transmission at the neuromuscular
junction by depolarizing the motor-end
plate. It is used as an adjuvant of gene
ral anaesthesia, in spasms and neuro-
spastic conditions, in electroshock, contractures, as well as certain types of
hypertension.
f bromure de décaméthonium
d Dekamethoniumbromid
i decametonio bromuro
e bromuro de decametonio

900 decamethonium iodide
See definition of decamethonium bromide.
The drug is practically free from any
side effect. One of the disadvantages of
the drug is the frequent necessity to
carry out artificial respiration although
it is supposed to have a sparing effect
on the respiratory muscles.
f iodure de décaméthonium
d Dekamethoniumjodid
i decametonio ioduro
e yoduro de decametonio

901 deferoxamine
A specific chelating agent for iron used
in the treatment of pathological conditions caused by an accumulation of iron,
e.g. haemochromatosis, both primitive
and caused by cirrhosis of the liver,
blood disorders and transfusions. Also
used as an antidote to combat acute intoxication due to iron which it succeeds
in eliminating. Also called desferrioxamine.
f déféroxamine
d Deferoxamin
i deferossamina
e deferoxamina

902 dehydroacetic acid
Methylacetopyrone. Colourless and tasteless crystals, insoluble in water, obtainable by the polymerization of the ketene
or from ethyl acetoacetate. It is used as
bactericide and fungicide.
f acide déhydroacétique; méthylacétopyrone
d Dehydroessigsäure; Methylacetopyron
i acido deidroacetico; metilacetopirone
e ácido dehidroacético; metilacetopirona

903 dehydroascorbic acid
An acid obtained by the oxidation of the
ascorbic acid (vitamin C) into which it
is retransformed by reduction. Like vitamin C, the dehydroascorbic acid displays

a vitaminic antiscorbutic activity, though
considerably milder.
f acide déhydroascorbique
d Dehydroascorbinsäure
i acido deidroascorbico
e ácido dehidroascórbico

904 dehydrocholic acid
A choleretic agent, i.e. an agent capable
of promoting the excretion of bile by the
liver. It has a mild cholagogue action,
i.e. promoting the flow of bile into the
duodenum from the gallbladder and bile
ducts. It is used to treat functional disorders of the liver and other disorders
connected with intestinal digestion.
f acide déhydrocholique
d Dehydrocholsäure
i acido deidrocolico
e ácido dehidrocólico

905 dehydroemetine
Chemical compound used in the treatment
of amoebiasis.
f déhydroémétine
d Dehydroemetin
i deidroemetina
e dehidroemetina

906 dehydromorphine
Or pseudomorphine, oxydimorphine. An
alkaloid present in opium formed by the
elimination of two atoms of hydrogen from
two molecules of morphine.
f déhydromorphine; pseudomorphine
d Dehydromorphin; Pseudomorphin
i deidromorfina; pseudomorfina
e dehidromorfina; seudomorfina

907 demecarium bromide
Parasympathomimetic agent capable of
inhibiting cholinesterase. It produces
miosis (abnormal contraction of the pupil)
and blocks the mydriatic effect of atropine. It is chiefly used in the treatment
of primary and secondary glaucoma. It
may cause local inflammation as a side
effect.
f bromure de démécarium
d Demekariumbromid
i demecario bromuro
e bromuro de demecario

908 demeclocycline
A broad-spectrum antibiotic derived from
tetracycline, which has proved very effec
tive against Gram-positive bacteria,
pneumococci, gonococci, Brucella etc. It
has also been used against Haemophilus
pertussis and various large viruses. The
drug produces haematic peaks which are
higher and longer lasting than those
which follow the administration of chlortetracycline and oxytetracycline.
f déméclocycline
d Demeclocyclin
i demeclociclina
e demeclociclina

909 demegestone

A synthetic progestogen (a substance displaying an action similar to that of progesterone) used in the treatment of lutein insufficiency, e.g. menstrual disorders, pains caused by premenstrual syndrome, dysmenorrhoea, mastodynia, menopause and of endometriosis.
f démégestone
d Demegeston
i demegestone
e demegestona

910 demoxytocin
A synthetic hormone displaying oxytocic (childbirth promoting) activity and capable of stimulating the contraction of the myofibrils of the mammary glands. The drug is used in the primary and secondary uterine inadequacy to perform its normal functions, in the induction of labour and to promote uterine involution and lactation.
f démoxytocine
d Demoxytocin
i demossitocina
e demoxitocina

911 denaverine
A drug capable of arresting spasms and convulsive attacks.
f dénavérine
d Denaverin
i denaverina
e denaverina

912 deoxycortone
Or desoxycortone, desoxycorticosterone. Mineralocorticoid (hormone, e.g. aldosterone, secreted by the cortex of the suprarenal gland) used in the treatment of adrenal insufficiency associated with the Addison's disease and of hypopituitarism. Hypertension, oedemas, hypokalaemic alkalosis, cardiac insufficiency and pulmonary congestion are possible side effects in case of a prolonged use.
f désoxycorticostérone
d Desoxycorticosteron
i desossicorticosterone
e desoxicorticosterona

*   **deoxyephedrine   s. methamphetamine**

913 deoxyribonuclease
Enzyme present in yeast and in various animal and vegetable tissues and which, under suitable conditions, splits the large molecule of the deoxyribonucleic acid into smaller fragments (oligonucleotides and mononucleotides) which lose some of the characteristic reactions that are peculiar of the acid itself. The deoxyribonuclease is used for the study of the qualitative and quantitative distribution of the deoxyribonucleic acid in the cells and in the animal and vegetable tissues. It is also used as a drug to resolve clots and exudates associated with trauma and inflammation.
f désoxyribonucléase

d Desoxyribonuclease
i desossiribonucleasi
e desoxiribonucleasa

914 deptropine
Antihistaminic (chiefly antiasthmatic) agent derived from tropane. It acts as a bronchodilator and as inhibitor of serotonin. It is used against allergies and to reduce bronchial secretion. Bronchial asthma, emphysema and chronic bronchitis are also treated with this drug.
f deptropine
d Deptropin
i deptropina
e deptropina

915 dequalinium chloride
Quaternary ammonium-type antiseptic capable of acting against a very wide range of microorganisms, but not against spores. The drug is mainly used in the treatment of cutaneous and mucosal infections which are resistant to antibiotics.
f chlorure de déqualinium
d Dequaliniumchlorid
i dequalinio cloruro
e cloruro de dequalinio

916 deserpidine
Reserpine-derived major tranquillizer acting on the diencephalon and mesencephalon and used in the treatment of various mental disorders characterized by manic symptoms and delirium. Depression may occur as a side effect.
f déserpidine
d Deserpidin
i deserpidina
e deserpidina

*   **desferrioxamine   s. deferoxamine**

917 desipramine
Tricyclic amipraminum-similar antidepressant whose action is quicker than that of amipraminium. It is used in the treatment of depressive states associated with hypochondriasis and melancholia, as well as of senile dementia and psychoses which may occur in the course of the climacterium.
f désipramine
d Desipramin
i desipramina
e desipramina

918 deslanoside
Deacetyl-lantoside C. A lantoside-derivative cardiac glycoside. It presents the same properties of lantoside C but with a shorter latency period and duration. Deslanoside is particularly indicated (parenterally) in case of an emergency digitalis therapy is deemed necessary, e. g. in lung oedema, paroxysmal tachycardia, auricular fibrillation etc.
f deslanoside
d Deslanosid
i deslanoside

e deslanosida

919 desmethylamitriptyline
Antidepressant agent whose action resembles that of imipramine and amitriptyline and is sedative as well as anxiolytic. It is used in the treatment of neurotic depressive states, reactive and atypical depressions also when due to a chronic psychosis, depressive syndromes caused by a neuroleptic therapy, melancholy and psychosomatic symptoms. Better known as nortriptyline.
f nortriptyline
d Desmethylamitriptylin; Nortriptylin
i nortriptilina
e nortriptilina

920 desmopressin
A synthetic form of vasopressin used nasally for the treatment of diabetes insipidus and nocturnal enuresis. It acts as an analogue of antidiuretic hormone and is thus capable of counteracting the copious quantities of urine which are produced in both conditions. The drug is also used to combat polyuropolydipsic postoperative syndromes. Diffuse pallor and uterine cramps may occur as side effects.
f desmopressine
d Desmopressin
i desmopressina
e desmopresina

921 desomorphine
Dihydrodeoxymorphine, i.e. a modified morphine having very similar properties to morphine.
f désomorphine
d Desomorphin
i desomorfina
e desomorfina

922 desonide
Corticoid steroid characterized by a potent antiinflammatory property. It is used topically in the treatment of skin diseases showing a positive reaction to adrenal corticosteroids, e.g. various types of eczema, psoriasis, allergic dermatitis etc. A prolonged administration may cause sensibilization.
f désonide
d Desonid
i desonide
e desonido

923 desoxyephedrine
Metamphetamine. A chemical compound prepared by reducing the ephedrine or the condensation product between methylamine and benzylmethylketone. The hydrochloride forms some bitter crystals, soluble in water, which display sympathomimetic vasoconstricting properties. The drug is thus used in the treatment of mental depression, parkinsonism etc.
f désoxyéphédrine; métamphétamine
d Desoxyephedrin; Metamphetamin

i desossiefedrina
e desoxiefedrina

924 desoxymethasone
Potent synthetic corticosteroid capable of an antiinflammatory and antiallergic action which is similar to that of cortisone. It is indicated in the treatment of eczema, allergic and non-allergic dermatoses, psoriasis, itching and neurodermatitis. Prolonged administration may cause systemic reactions.
f désoxyméthasone
d Desoxymethason
i desossimetasone
e desoximetasona

925 devil's apple
Mandrake, mandragora. Herb growing in southern Europe and northern Africa formerly supposed to have aphrodisiac properties. Its root was formerly used for preparations presumed to promote conception or as a cathartic and a narcotic agent.
f mandragore; herbe aux magiciens
d Alraun; Hexenkraut; Zauberpflanze
i mandragora
e mandragora macho; berenjenilla

926 devil's shoestrings
The dried leaves and stem of the catgut (the strong and wiry root of the Tephrosia virginiana, a herb growing in eastern North America) used in the past as an anthelmintic.
f téphrosie; requienie
d Aschenwicke
i tefrosia
e tefrosia

927 dexamethasone
Corticosteroid hormone characterized by a glycocorticoid action and used in the treatment of rheumatoid arthritis, severe allergic states and conditions, traumatic ophthalmologies, Hodgkin's disease, sarcoidosis, leukaemia, multiple sclerosis etc. Gastrointestinal ulcers and haemorrhages, dyspepsia, osteoporosis and many other disorders may occur as side effects.
f dexaméthasone
d Dexamethason
i desametasone
e desametasona

928 dexamphetamine
Sympathomimetic amine capable of stimulating the central nervous system. It is twice as potent as amphetamine (for its use see amphetamine). Also indicated in some brain disorders in children, exogenous obesity and post-encephalitic parkinsonism.
f dexamphétamine
d Dexamphetamin
i desamfetamina
e desamfetamina

929 dexbrompheniramine

Antihistamine used in the treatment of allergic states and conditions, e.g. hay fever, vasomotor rhinitis, contact and atopic dermatitis as well as certain forms of bronchial asthma and conjunctivitis.

f dexbromphéniramine
d Dexbrompheniramin
i desbromfeniramina
e desbromfeniramina

930 dexchlorpheniramine
Antihistaminic agent used in the treatment of allergic states and conditions, e.g. urticaria, hay fever, allergic dermatitis, serum sickness etc.

f déxchlorphéniramine
d Dexchlorpheniramin
i desclorfeniramina; destroclorfeniramina
e desclorfeniramina

931 dexcyanidanole
Therapeutic agent capable of regulating the hepatocellular homeostasis. It produces energy, promotes regeneration and regulates the enzymatic balance of the hepatocyte. It is also capable of neutralizing endogenous and exogenous toxic substances and displays an antisteatosis action. Chiefly indicated in hepatic insufficiency syndromes.

f dexcyanidanole
d Dexcyanidanol
i descianidanolo
e descianidanolo

932 dexetimide
Therapeutic agent marked by a central and peripheral anticholinergic action and used in the treatment of parkinsonism and extrapyramidal syndrome, i.e. any syndrome ascribed to a disease of the extrapyramidal system.

f dexétimide
d Dexetimid
i dexetimide; destrobenzetimide
e dexetimida

933 dexpanthenol
An alcohol-analogue of vitamin B_3 used in the treatment of paralytic ileus and postoperative distension. Combined with choline bitartrate, it is used in the treatment of tympanitis (the swelling of the abdomen caused by gas retention in the intestine or in the cavity of the peritoneum) and of a syndrome affecting the left flexure of the colon (the splenic flexure), i.e. the acute bend at the junction of the transverse and descending section of the colon. More generally, the drug is used in cases of gastritis, cholecystitis etc.

f dexpanthénole
d Dexpanthenol
i despantenolo
e despantenolo

934 dextran
Polyglycin. A ramified-structure polysaccharide resulting from the union of residues of d-glucose produced by living microorganisms in a solution of saccharose. It appears like an amorphous gummy substance which has proved extremely important in medicine as a substitute of blood plasma in cases of haemorrhages or shock. Allergic reactions, e.g. fever, angioneurotic oedema, congestive heart failure etc. may occur as side effects. The drug is also used as a lubricant in drops for dry eyes.

f dextrane
d Dextran; Macrodex
i destrano
e dextran

935 dextranomer
Spherical beads of dextran (q.v.) used for surface application of cutaneous wounds owing to its capacity of absorbing fluid exudate by capillary action and of promoting the elimination of bacteria and tissue debris.

f dextranomère
d Dextranomer
i destranomero
e dextranomero

936 dextran sulphate
The sodium salt of sulphuric acid esters of dextran used in medicine as an anticoagulant and antiinflammatory agent. It is therefore indicated in the topical treatment of surface thrombosis and phlebitis, oedemas, haematomas and contusions.

f sulfate de dextrane
d Dextransulfat
i destrano solfato
e sulfato de dextran

937 dextriferron
Antianaemic agent used parenterally in the treatment of blood iron deficiency especially if the condition proves resistant to oral therapy. Flushes, nausea, headache etc. may occur as side effects.

f dextriferron
d Dextriferron
i dextriferron
e dextriferron

938 dextromethorphan
A cough-suppressant agent, non-analgesic. Its properties resemble those of codein and it does not inhibit expectoration. It is used in the treatment of obstinate dry cough, e.g. whooping cough.

f dextrométhorphane
d Dextromethorphan
i destrometorfano
e dextrometorfano

939 dextromoramide
Powerful analgesic and tranquillizer used to relieve severe pains, particularly in terminal diseases. It is essentially similar to morphine. Its action is rapid and lasts up to 8 hours. Also used as a narcotic. It should not be administered together with barbiturates. Respiratory

depression may occur as a side effect.
f dextromoramide
d Dextromoramid
i destromoramide
e dextromoramide

940 dextropropoxyphene
A mild narcotic analgesic used to relieve weak and transient pains, generally as a coadjuvant for paracetamol whenever this drug proves inadequate. Its effects are similar to those of codeine, but the side effects, e.g. nausea, vomiting, constipation, are milder. Overdosage may cause coma with depressed respiration.
f dextropropoxyphène
d Dextropropoxyphen
i destropropossifene
e dextropropoxifeno

941 dextrose
Or glucose, grape sugar. The principal hexose which occurs in fruit, plants, honey and in the blood and tissue of animals. Chemically, a carbohydrate used in medicine either orally or intravenously as a source of calories. It is rapidly absorbed in the gastrointestinal tract and metabolized by energy-producing pathways or stored in the liver as glycogen. Apart from being a very important source of energy for all living organisms, it is used in the treatment of hypoglycaemia caused by large doses of insulin.
f dextrose
d Dextrose
i destrosio
e dextrosa

942 dextrose phosphate
Intermediate in carbohydrate metabolism, capable of promoting the phosphorylation of glucose and its penetration into the cells. It is chiefly used as a source of energy and also in the treatment of hepatic disorders.
f phosphate de dextrose
d Dextrosephosphat
i destrosio fosfato
e fosfato de dextrosa

943 dextrotartaric acid
The ordinary tartaric acid, a constituent of grape juice. When administered in the form of a long drink (tartaric lemonade) it aids digestion and gastric motility.
f acide dextrotartarique
d Dextroweinsteinsäure; Rechtsweinsäure
i acido destrotartarico
e ácido dextrotartárico

944 dextrothyroxine
An anti-hyperlipidaemic agent capable of reducing plasma cholesterol concentration. It is mainly used in the treatment of arteriosclerosis, coronary sclerosis, diabetes and hypothyroidism. Tachycardia and angina may occur as side effects in patients suffering from ischaemic heart

diseases.
f dextrothyroxine
d Dextrothyroxin
i destrotirossina
e dextrotiroxina

945 diamorphine
Narcotic analgesic similar to morphine and characterized by a rapid action. It is used for the relief of dry obstinate cough and above all to relieve pain in terminal diseases and heart infarction. Its side effects are milder than those of morphine, but the euphorant action of the drug may easily bring about addiction.
f diamorphine
d Diamorphin
i diamorfina; diacetilmorfina
e diamorfina

946 diastase
Amylase; any of a group of enzymes found in animal and plant tissues and capable of breaking down starch and glycogen to yield dextrins and maltose. Its richest source is the pancreas. It is used in medicine in the treatment of amylaceous and fermentative dyspepsias, gastritis, infantile diarrhoea and as an adjuvant in diets chiefly consisting of carbohydrates.
f diastase
d Diastase
i diastasi
e diastasa

947 diazepam
A minor benzodiazepine tranquillizer, anxiolytic and sleep-inducing agent with anticonvulsant properties. It has a meso-diencephalic site of action (i.e. acting on the limbic system). It is mainly indicated for the relief of anxiety and as a hypnotic, as well as in reduction of muscle tone in spasticity, withdrawal symptoms of acute alcoholism, phobic psychoses etc. Given intravenously it is an anticonvulsant for status epilepticus. Fatigue, drowsiness, sometimes ataxia, blurred vision and intestinal disorders are among the principal side effects.
f diazépam
d Diazepam
i diazepam
e diazepam

948 diazoxide
Hyperglycaemic agent used to reduce blood pressure in severe hypertension and to raise blood sugar level in hypoglycaemia by blocking insulin secretion by betacells of pancreas. It is therefore indicated, in addition to hypertension, in the treatment of idiopathic hypoglycaemia in infancy, functional insular neoplasms, hypoglycaemia-causing tumours, glucogen thesaurismosis etc. Severe crises of hypoglycaemia may be treated intravenously with the drug. It has no effect on coronaric and cerebral flow. Hypertricosis

(abnormal hair growth), oedemas, nausea and vomiting, as well as diabetes and hypotension may occur as side effects.

f diazoxyde
d Diazoxyd
i diazossido
e diazóxido

949 dibekacin

Aminoglycoside antibiotic capable of acting against Gram-positive and Gram-negative microorganisms. It is a semisynthetic agent derived from kanamycin B. Its side effects include possible damage to the ear vestibule.

f dibékacine
d Dibekacin
i dibecacina
e dibecacina

950 dibenzepin

Tricyclic antidepressant agent derived from benzodiazepine and indicated in the treatment of depression and depression-inducing anxiety. Sleep disorders and anticholinergic-activity reactions are the most common side effects.

f dibenzépine
d Dibenzepin
i dibenzepina
e dibenzepina

951 dibenzyl-ß-chloroethylamine

Chemical compound, related to nitrogen-mustard compounds, capable of inhibiting the stimulatory actions of adrenaline and noradrenaline thus reversing the rise in blood pressure provoked by adrenaline. It thus acts against injected and circulating adrenaline and, when administered in large doses, against adrenergic nerve stimulation without blocking the cardiac effects of adrenaline. Stimulation of the central nervous system and orthostatic hypotension may occur as side effects. It is sometimes used as a diagnostic in phaeochromocytoma (a chromaffinome in the adrenal medulla) and as hypotensive agent. Also used to treat frostbite and the Reynaud's disease.

f dibenzyl-ß-chloroéthylamine
d Dibenzyl-ß-chloroethylamin
i dibenzil-ß-cloroetilamina
e dibenzil-ß-cloroetilamina

952 dibromopropamidine

An antiseptic and disinfectant agent used in the treatment of conjunctivitis and blepharitis.

f dibromopropamidine
d Dibromopropamidin
i dibromopropamidina
e dibromopropamidina

953 dibronityrosine

Therapeutic agent capable of removing iodine from the thyroid by substituting it with bromine. The replacement reduces the functional activity of the gland and displays the sedative action that is char-

acteristic of bromine. It is mainly used in the treatment of mild neurogenic hyperthyroidism and hyperthyroidism-associated neuroses.

f dibronityrosine
d Dibronityrosine
i dibronitirosina
e dibronitirosina

954 dichloralphenazone

Sedative and hypnotic agent having an action which is similar to that of chloral but less irritant for the gastric mucosa. It is indicated in anxiety-associated insomnia, hyperexcitability and moderate pains. It may also be used as a sedative in daytime to alleviate nervous tension. The drug is a combination of chloral hydrate and phenazole and is reconverted to parent compounds by the liver. Skin rashes, headache and blood disorders may occur as side effects.

f dichloralphénazone
d Dichloralphenazon
i dicloralfenazone
e dicloralfenazona

955 dichlorisone

Adrenal corticosteroid used in the treatment of phlogistic and allergic dermatoses, e.g. contact dermatitis and infantile eczema, itching, haemorrhoids etc.

f dichlorisone
d Dichlorison
i diclorisone
e diclorisona

956 dichlorohydroxymethylquinoline

Oxine-related synthetic bacteriostatic agent used as ointment in the treatment of bacterial and mycotic cutaneous infections.

f dichlorohydroxyméthylquinoline
d Dichlorohydroxymethylquinoline
i dicloroidrossimetilchinolina
e diclorohidroximetilquinolina

957 dichlorophen

Anthelmintic agent particularly active against tapeworms. High doses display a taenicide action. It is also a purgative. Nausea, vomiting and intestinal colic may occur as side effects.

f dichlorophène
d Dichlorophen
i diclorofene
e diclorofeno

958 dichlorophenarsine

A base whose hydrochloride prepared in an alkaline buffer solution is used intravenously in the treatment of syphilis.

f dichlorophénarsine
d Dichlorophenarsine
i diclorofenarsina
e diclorofenarsina

959 dichlorphenamide

A diuretic agent capable of inhibiting carbonic anhydrase activity in the distal

renal tubules. It promotes the excretion
of sodium chloride, potassium and bicarbonate ions with urine. It is used as an
antioedemic agent in heart diseases and
in chronic pulmonary heart. Also used in
the treatment of respiratory failure caused by chronic bronchitis and in the prophylaxis and treatment of glaucoma.
f dichlorfénamide
d Dichlorphenamid
i diclorfenamide
e diclorfenamida

960 diclofenac
A non-steroid anti-inflammatory, analgesic and antipyretic agent used in the
treatment of various forms of rheumatism,
rheumatoid arthritis, inflammatory and
degenerative disorders (e.g. osteoarthritis), ankylosing spondylitis, inflammations and oedemas which are not of a
rheumatic or post-traumatic origin.
f diclofénac
d Diclofenac
i diclofenac
e diclofenac

961 dicloxacillin
A semisynthetic penicillin capable of re-
sisting inactivation by gastric acidity
and penicillase-resistant. It acts against
Gram-positive organisms, in particular
Diplococcus pneumoniae, Streptococcus pyo
genes and Staphylococcus aureus, and is
used in the treatment of susceptible bacterial infections affecting nose, ear and
larynx, as well as infections of the respi
ratory and genito-urinary tracts, of
joints, bones and soft tissues. More specifically, dicloxacillin is used to treat
all those infections that are caused by
penicillase-producing staphylococci hardly
sensitive to penicillin G. Some allergic
reactions may occur as side effect.
f dicloxacilline
d Dicloxacillin
i diclossacillina
e dicloxacilina

962 dicophane
Strong insecticide used in the form of
powder or of lotion against fleas and
lice. Absorption must be avoided as the
drug is very toxic.
f dicophane
d Dicophan
i dicofano
e dicofano

963 dicoumarol
Anticoagulant capable of reducing the
prothrombin plasmatic level. It is used
in the prophylaxis and the treatment of
post-operative and post-traumatic thrombo
phlebitis, pulmonary embolism and the
thrombosis-related occlusion of blood vessels.
f dicoumarol
d Dicumarol
i dicumarolo

e dicumarolo

* **dicyclomine   s. dicycloverine**

964 dicycloverine
Spasmolytic agent acting on smooth-muscle
fibres and nerve fibres. The drug is
used in the treatment of ulcerative and
spastic colitis and of spastic constipation,
as well as of primary dysmenorrhoea.
f dicyclovérine
d Dicycloverin
i dicicloverina
e dicicloverina

965 dienoestrol
Synthetic oestrogenous hormone indicated
for the treatment of atrophic and senile
vaginitis as well as vulvovaginitis, and
to relieve menopausal symptoms. Also
used in the treatment of prostatic carcinoma. It may cause the stimulation of
the mammary gland.
f diénestrol
d Dienestrol
i dienestrolo
e dienestrolo

966 diethazine
A therapeutic agent displaying both a
sympatholytic and parasympathomimetic
action, used in the treatment of postence
phalitic and senile Parkinson's syndrome
as well as to combat adverse effects of
neuroleptics and organic phosphorus intoxications.
f diéthazine
d Diethazin
i dietazina
e dietazina

967 diethylamine salicylate
Salicylic derivative for external use. It
is highly absorbable and displays an
antirheumatic and antiinflammatory action.
It is mainly used in the topical treatment of articular and muscular rheumatism and arthritis.
f salicylate de diéthylamine
d Diethylaminesalicylat
i dietilamina salicilato
e salicilato de dietilamina

968 diethylcarbamazine
Synthetic therapeutic agent having a specific action in several filarial infections.
It is particularly indicated in bancrof-
tian filariasis and in loiasis (a helmin-
thic infection) owing to its capacity of
destroying the microfilaria almost immediately and thus preventing the spread-
ing of the infection. It is further capable of sterilizing the adult worm and of
killing it thereby avoiding the relapse of
microfilaraemia. Gastrointestinal disorders
accompanied by pains and sometimes
ataxia may occur as side effects.
f diéthylcarbamazine
d Diethylcarbamazin
i dietilcarbamazina

e dietilcarbamazina

969 diethylstilbestrol
Synthetic compound produced as colourless crystals or as a crystalline powder and displaying oestrogenic properties. Administration by injection is preferable to oral administration as the latter may easily cause nausea and vomiting. It is chiefly used in the treatment of menopausal syndrome, natural hormone deficiency, and of certain forms of malignant disease (e.g. prostatic carcinoma). It is also used to inhibit lactation. Oedemas, anorexia and occasionally skin diseases may occur as side effects.
f diéthylstilbestrol
d Diethylstilbestrol
i dietilstilbestrol
e dietilstilbestrol

970 difemerine
A spasmolytic anticholinergic agent with a topical anaesthetic action used in the treatment of spastic conditions of the digestive, biliary and genitourinary tracts. Mydriasis (enlargement of the pupil) may occur as a side effect.
f difémérine
d Difemerin
i difemerina
e difemerina

971 difenamizole
Analgesic, antiphlogistic and antipyretic agent, as well as a relaxant of the skeletal muscles. It is mainly used in the treatment of cervical-rib syndrome (syndrome produced by the compression of the lower cord of the brachial plexus by a cervical rib), neuralgias, headache, lumbago, herpes zoster, postoperative genito-urinary disorders etc.
f difénamizole
d Difenamizol
i difenamizolo
e difenamizolo

972 difenpiramide
Alpha-piridilphenylacetamide. A non-steroid antiinflammatory, analgesic and anti pyretic agent used for the treatment of phlogistic and degenerative rheumatism, osteoarthrosis, sciatica, lumbago, bursitis and many other inflammatory diseases. It is very easily tolerated.
f difenpiramide
d Difenpiramid
i difenpiramide
e difenpiramida

973 difetarsone
Antiparasitic agent capable of stopping the growth of protozoa or destroying them. The drug is indicated in the treatment of evolutive and chronic intestinal amoebiasis, inflammation of the intestinal mucosa caused by Trichomonas and oxyuria, i.e. the infestation with a nematoid worm of the genus Enterobius previously known as Oxyuris. In addition to intestinal disorders, the drug may cause arsenic poisoning.
f difétarsone
d Difetarsone
i difetarsone
e difetarsona

974 diflucortolone
Corticosteroid for topical use indicated in inflammatory skin diseases, e.g. contact dermatitis, eczema, psoriasis, lichen, both planus and verrucose, insect bites, burns etc.
f diflucortolone
d Diflucortolon
i diflucortolone
e diflucortolona

975 diflunisal
Acetylsalicylic acid-related analgesic marked by a fairly long duration and by its total lack of effect on platelet function. It is chiefly indicated for the soothing of pain in osteo-arthritic conditions, in traumas, toot-ache etc. Dyspepsia, nausea and other gastric disorders may occur as side effects.
f diflunisal
d Diflunisal
i diflunisal
e diflunisal

976 difluprednate
Antiinflammatory, antiallergic and antipruritic corticosteroid for topical use. It is indicated in the treatment of inflammatory, allergic and pruriginous dermatoses, eczematous dermatoses and eczema, as well as psoriasis, lichen and related disorders. Depigmentation and cutaneous atrophy may occur as side effects.
f difluprednate
d Difluprednat
i difluprednato
e difluprednato

977 diftalone
A non-steroid antiinflammatory agent used chiefly in the treatment of rheumatoid arthritis. Also indicated in the treatment of lumbago, sciatica, thrombophlebitis, as well as in ophthalmologic, ear, nose and throat and skin disorders. A temporary leukopenia and agranulocytosis may occur as side effects.
f diftalone
d Diftalon
i diftalone
e diftalona

978 digitalis
Crude foxglove extract used in the past as a cathartic and emetic drug and presently as the principal agent used in cardiology owing to its capacity of slowing down and regularize the cardiac rhythm, of increasing the energy of the systoles and completing the function of the diastoles and consequently improving the pres

sure and increase diuresis. Its therapeutic properties are due to the particular
glycosides which can be extracted from
its leaves, the most important of which
is the digitoxin or digitalin. Also known
as foxglove, purple foxglove.
f digitale; digitale pourprée; grande digitale
d Digitalis purpurea; roter Fingerhut
i digitale
e digital; dedalera roja

979 digitoxin
Glycoside ($C_{41}H_{34}O_{12}$) of the leaves of
Digitalis purpurea, which is considered
its principal active constituent. It stimulates the action of the heart, prolongs
diastole, strengthens systole and promotes
diuresis. It is therefore used in the
treatment of cardiac disorders, particular
ly heart failure, atrial flutter and supra
ventricular tachycardia. It may cause a
severe form of brachycardia owing to its
cumulative effects.
f digitoxine
d Digitoxin
i digitossina
e digitoxina

980 digoxin
A foxglove derivative capable of increasing the contraction energy of the heart
and thus slowing down its rate. It is an
excellent and quick-acting cardiotonic
agent chiefly used in heart failure and
in rhythm disorders. Various gastrointestinal disorders may occur in case of over
dosage.
f digoxine
d Digoxin
i digossina
e digoxina

981 dihexyverine
Antispastic and parasympatholytic agent
acting as a mild local anaesthetic and
used in the treatment of biliary and urinary spasms as well as of spastic conditions occurring after childbirth. Its prin
cipal side effect is tachycardia.
f dihexyvérine
d Dihexyverin
i diesiverina
e dihexiverina

982 dihydralazine
Antihypertensive agent marked by a central and peripheral mechanism of action
and a capacity of acting on the central
nervous system. It is mainly indicated
in the treatment of systemic and especial
ly renal hypertension and of eclamptic
(eclampsia is an attack of convulsion
associated with pregnancy) states. Nausea, anorexia and a decrease of the volume of urine may occur as side effects.
f dihydralazine
d Dihydralazin
i didralazina; diidralazina
e didralacina; dihidralacina

983 dihydrocodeine
Analgesic and mild narcotic agent very
similar to codein but more potent in its
analgesic action. It is also used as an
antitussive drug in the treatment of dry
cough due to inflammation or infection of
the upper respiratory tract and of tuberculosis. It tends to cause nervous depres
sion and, if administered by injection,
constipation.
f dihydrocodéine
d Dihydrocodein
i diidrocodeina
e dihidrocodeina

984 dihydroergocornine
Alkaloid constituent of ergotoxine. Adrena
lytic and sympatholytic agent chiefly
used for lowering the blood pressure.
f dihydroergocornine
d Dihydroergocornin
i diidroergocornina
e dihidroergocornina

985 dihydroergocristine
A vasodilator. The drug is marked by an
alpha-adrenolytic action on the arteries,
a tonic action on the veins and as a con
sequence an increase of the peripheral
tissular circulation. It is used in the
treatment of cerebral and peripheral vascular disorders, e.g. hypertension, headache, cervical osteochondritis and toxaemia occurring in the course of pregnancy.
f dihydroergocristine
d Dihydroergocristin
i diidroergocristina
e dihidroergocristina

986 dihydroergotamine
Adrenosympatholytic agent capable of
stimulating the smooth muscles of the
blood vessels, in particular the venous
ones. It is indicated in the treatment of
syndromes marked by orthostatic circular
disorders and obstinate migraine.
f dihydroergotamine
d Dihydroergotamin
i diidroergotammina
e dihidroergotamina

987 dihydroergotoxine
An adrenergic blocking agent capable of
activating the cerebral metabolism and
inhibiting the platelet aggregation. It is
therefore indicated in the treatment of
cerebral arteriosclerosis and in those cerebral disorders which affect the elderly.
It also finds useful employment in hypertension, cranial traumas and peripheral
vascular disorders. Nasal congestion may
occur as a side effect.
f dihydroergotoxine
d Dihydroergotoxin
i diidroergotossina
e dihidroergotoxina

988 dihydroquinidine
Quinidine-related cardiac depressant,

more active and less toxic than the parent drug. The drug is used in the treatment of arrhythmias not necessarily associated with cardiac insufficiency as well as complete arrhythmias, auricular flutter, paroxystic tachycardia, frequent extrasystoles etc. Ventricular tachycardia and myocardial depression may occur as side effects.

f dihydroquinidine
d Dihydroquinidin
i diidrochinidina; idrochinidina
e quinidina; dihidroquinidina

989 dihydrostreptomycin
Streptomycin-related broad-spectrum antibiotic, but less toxic than streptomycin itself. It is used in the treatment of tuberculosis and of infections due to strains of Gram-positive and Gram-negative organisms, e.g. Escherichia coli. The drug is also used to combat enteric infections. A degree of deafness and some allergic reactions have been observed as side effects.

f dihydrostreptomycine
d Dihydrostreptomycin
i diidrostreptomicina
e dihidrostreptomicina

990 dihydroxystearic acid
The acid which is found as a glyceride in castor oil, obtained from the seeds of the Ricinus communis.

f acide dihydroxystéarique
d Dihydroxystearinsäure
i acido diidrossistearico
e ácido dihidroxiesteárico

991 dihydroxytachysterol
A drug closely related to Vitamin D and characterized by similar action. It is used in the treatment of osteomalacia when this has proved resistant to Vitamin D. The drug is also used in the treatment of osteodystrophy caused by chronic renal failure and of hypoparathyroidism. It must be administered with great caution in the presence of hypercalcaemia.

f dihydroxytachystérol
d Dihydroxytachysterol
i diidrossitachisterolo
e dihidroxitaquisterolo

992 diiodohydroxyquinoline
Antimicrobial and antiseptic agent, hardly absorbable and acting on the intestinal germs which cause non-specific colitis and enterocolitis accompanied by diarrhoea. It is also used in the prophylaxis of amoebic dysentery and trichomoniasis.

f diiodohydroxyquinoléine
d Dijodohydroxyquinolein
i diiodoidrossichinolina
e diyodohidroxiquinoleina

993 diiodothyrosine
Synthetic iodized amino acid resembling the natural substance found in the thyroid. It is characterized by its capacity of moderating the action of thyrosin and antagonizing the effects of thyroid hormone overproduction. It is therefore used in the treatment of the Basedow's disease, in cases of hyperthyroidism and physiological disorders caused by thyroid dysfunction.

f diiodothyrosine
d Dijodothyrosin
i diiodotirosina
e diyodotirosina

994 diisopromine
Choleretic and cholagogue agent displaying an antispasmodic action especially on the extrahepatic biliary tract. It is used in the treatment of dyspepsia and in some hepatic affections, e.g. initial cirrhosis of the biliary tract and biliary lithiasis.

f diisopromine
d Diisopromin
i diisopromina
e diisopromina

995 dilazep
A vasodilator acting particularly on the coronary vessels and used in the treatment of insufficiency of the latter and of angina pectoris, as well as for the prevention and therapy of myocardial infarction and therapy of senile heart.

f dilazep
d Dilazep
i dilazep
e dilazep

996 diloxanide
Antiamoebic drug used in the treatment of intestinal and hepatic amoebiasis.

f diloxanide
d Diloxanid
i dilossanide
e diloxanida

997 diltiazem
Therapeutic agent used in the treatment and in the prevention of angina pectoris. The drug inhibits the entry of calcium into the heart muscle and helps in avoiding overstressing during physical exercise. It is also used to relieve the anginose pains which are caused by myocardial infarction.

f diltiazem
d Diltiazem
i diltiazem
e diltiazem

998 dimazole
Benzothiazole-related antimicrobial agent effective as a fungicide especially against dermatophytes and Candida albicans. It is used as therapeutic agent in dermatomycosis.

f dimazol
d Dimazol
i dimazolo
e dimazolo

999 dimefline

A cardiovascular and respiratory analep
tic agent chiefly used in the treatment
of intoxications due to hypnotic drugs
and neurodepressant poisons. Being a-
bove all a respiratory stimulant, it is
sometimes used in surgery in cases of
anaesthesia-related incidents.

f diméfline
d Dimefline
i dimeflina
e dimeflina

1000 dimelazine

Cardiocircular analeptic agent used in
the treatment of intoxications caused by
hypnotics or neurotropic agents and to
stimulate respiration in patients suffering from preanaesthetic respiratory depression or apnoea.

f dimélazine
d Dimelazin
i dimelazina
e dimelacina

1001 dimemorfan

Central-action, non-narcotic antitussive
agent used in the treatment of cough associated with acute and chronic respiratory diseases, e.g. pneumonia, bronchitis, silicosis, tracheitis etc.

f dimémorfane
d Dimemorfan
i dimemorfano
e dimemorfano

1002 dimenhydrinate

Antihistaminic agent displaying a generic antiallergic action and used for a
palliative treatment of allergic conditions. It is also used to combat vertigo
and has proved useful to relieve symptoms in the Menière's syndrome. It may
cause somnolence.

f dimenhydrinate
d Dimenhydrinat
i dimenidrinato
e dimenhidrinato

1003 dimercaprol

A compound which was originally used
as an antidote for arsenic-containing
vesicants as damage to the skin may be
prevented by the application either before or within two hours after contamina
tion. The drug is also a good antidote
in poisoning by antimony, bismuth, gold,
chromium and nickel, yet useless in lead
poisoning. It has proved very effective
in the treatment of haemorrhagic encepha
litis or exfoliative dermatitis caused by
organic arsenicals used in the treatment
of syphilis. Allergies, hypertension, tachycardia and convulsions may occur as
side effects.

f dimercaprol
d Dimercaprol
i dimercaprolo
e dimercaprolo

1004 dimetacrine

Tricyclic antidepressant agent used in
the treatment of depressive states born
of both periodic and cyclical manic-depressive psychoses, as well as in mixed
psychoses, organic, involutional neurotic
and reactive depressions. The drug is
also used to combat states of depression
due to oligophrenia and psychopathies.
Tachycardia, urinary retention and sleep
disorders may occur as side effects.

f dimétacrine
d Dimetacrin
i dimetacrina
e dimetacrina

1005 dimethindene

Antihistamine of the pyridine group. It
is chiefly used as an antipruritic agent.

f diméthindène
d Dimethinden
i dimetindene
e dimetindeno

1006 dimethisterone

Orally administered active progestational
agent having no androgenic, oestrogenic
or anabolic activity. It is used in the
treatment of amenorrhoea and the threat
of abortion. It is also indicated to relieve premenstrual tension.

f diméthistérone
d Dimethisteron
i dimetisterone
e dimetisterona

1007 dimethoxanate

Phenothiazine-related cough suppressant
agent used in the treatment of acute and
chronic dry cough of any origin but
especially in laryngitis, tracheitis, tracheobronchitis etc. Allergic dermatitis
may occur as side effect.

f diméthoxanate
d Dimethoxanat
i dimetossanato
e dimetoxanato

1008 dimethylpiperazine

An anthelminthic agent particularly active in oxyuriasis (the infestation with
a nematode worm) and ascariasis (the
infestation of the gastro-intestinal canal
with the parasite Ascaris lumbricoides).

f diméthylpipérazine
d Dimethylpiperazin
i dimetilpiperazina
e dimetilpiperacina

1009 dimethylsulphoxide

Antiphlogistic and analgesic agent used
topically for the treatment of traumas
and of acute neuralgias. It is also used
as a solvent for various pharmaceutical
preparations. Skin flaking and eczema
may appear as side effects.

f diméthylsulfoxyde
d Dimethylsulfoxyd
i dimetilsulfossido
e dimetilsulfóxido

1010 dimethylthiambutene
Analgesic and antipyretic agent. It is the dimethyl analogue of diethylthiambutene, an analgesic for subcutaneous administration to small animals.
f diméthylthiambutène
d Dimethylthiambuten
i dimetiltiambutene
e dimetiltiambuteno

1011 dimetofrine
A sympathomimetic agent marked by a hypertensive, cardiokinetic and bradycardiac action. It should not be used in hypertensive patients. Intravenous administration may cause an acute sensation of cold.
f dimétofrine
d Dimetofrin
i dimetofrina
e dimetofrina

1012 dimoxyline
A vasodilator, in particular a coronary and pulmonary vasodilator whose action resembles that of papaverine, chiefly used in the treatment of heart diseases, pulmonary embolism and peripheral vasospasms. Inflammation of the gastrointestinal tract may occur as a side effect.
f dimoxyline
d Dimoxylin
i dimossilina
e dimoxilina

1013 dinoprost
Prostaglandin F_{2a} with the same effects of dinoprostone.
f dinoprost
d Dinoprost
i dinoprost
e dinoprost

1014 dinoprostone
Prostaglandin E_2 marked by oxytocic (parturition-hastening) action and used for the induction of labour, therapeutic abortion, intrauterine foetal death and hydatidiform mole.
f dinoprostone
d Dinoproston
i dinoprostone
e dinoprostona

1015 dioctyl sodium sulphoccinate
Anionic surface active agent. It diminishes the surface tension of the faecal mass making it possible for water to be imbibed and thus soften the faecal matter. It is mainly used as a cathartic agent, combined with phenolphthalein or anthraquinones. It is also used to soften hardened cerumen.
f dioctyl sulfoccinate de sodium
d Natriumdioctylsulfoccinat
i sodio diottil sulfoccinato
e sulfoccinato de sodio dioctílico

1016 diodone
Iodine-containing contrast medium used in urography and in angiography.
f diodone
d Diodon
i diodone
e diodona

1017 diosmin
Therapeutic agent capable of protecting the walls of blood and lymphatic vessels and mainly used in the treatment of oedemas of various origin, as well as of venous disorders in the course of pregnancy, functional and fibromatous haemorrhagies, haemorrhoids, capillary fragility and various other vasculopathies.
f diosmine
d Diosmin
i diosmina
e diosmina

1018 diphemanil methylsulphate
Anticholinergically acting parasympatholytic agent having effects which resemble those of atropine and used for the treatment of peptic ulcer, gastritis, hyperacidity and pyloric spasms. Urine retention may occur as side effect.
f méthylsulfate de diphémanil
d Diphemanilmethylsulfat
i difemanilio metilsolfato
e sulfato de difemanilo

1019 diphenadione
An anticoagulant agent characterized by a hypoprothrombinaemic action probably due to an active metabolite and used in the prophylaxis and treatment of embolism and thrombophlebitis. It is also used in the treatment of serious heart diseases. Hepatopathies and nephropathies may occur as side effects.
f diphénadione
d Diphenadion
i difenadione
e difenadiona

1020 diphenoxilate
Synthetic therapeutic agent capable of reducing intestinal motility and used against diarrhoea, both acute and chronic, caused by food poisoning and other causes. It is related to morphine and may produce drowsiness, nausea, vomiting and tachycardia as side effects.
f diphénoxilate
d Diphenoxilat
i difenossilato
e difenoxilato

1021 diphenylamine
Aniline-derived colourless compound whose solution in sulphuric acid is used as an indicator in the volumetric analysis of ferrous salts.
f diphénylamine
d Diphenilamin
i difenilammina
e difenilamina

1022 diphenylhydantoin
Or dithizone. A moderately hydroscopic, white and odourless powder capable of absorbing carbon dioxide when exposed to air. It is characterized by a strong anticonvulsant action on the motor cortex and a hardly noticeable hypnotic effect. Owing to these properties, the drug is widely used in the treatment of epilepsy, i.e. in major and psychomotor attacks, as it has no effect at all in minor epilepsy. Gastrointestinal disorders, hyperplasia of the gums and several neurological symptoms may occur as side effects. Also ataxia and mental disorders have been observed. Also known as phenytoin.
f diphénylhydantoïne; phénitoïne
d Diphenylhydantoin; Phenytoin
i difenilidantoina; fenitoina
e difenilhidantoina; fenitoina

1023 diphenylhydramine
Widely used antihistaminic drug characterized by a rapid action and a short duration. It is also used in association with ephedrine in the therapy of bronchial asthma.
f diphénylhydramine
d Diphenylhydramin
i difenilidrammina
e difenilhidramina

1024 diphenylpyraline
Antihistaminic agent used in the treatment of various allergic diseases, e.g. rhinitis, urticaria etc. Gastric disorders and drowsiness may occur as side effects.
f diphénylpyraline
d Diphenylpyralin
i difenilpiralina
e difenilpiralina

1025 dipipanone
Practically non-hypnotic analgesic agent marked by an action which is very similar to that of atropine and morphine, yet without causing respiratory depression. It is widely used to combat pains borne of surgical operations, cancer, neuritis etc. It may cause depression.
f dipipanone
d Dipipanon
i dipipanone
e dipipanona

1026 dipiproverine
Parasympatholytic agent characterized by an anticholinergic mode of action and used as antispasmodic in gastroenterology, gynecology and urology.
f dipiprovérine
d Dipiproverin
i dipiproverina
e dipiproverina

1027 dipivefrin
A therapeutic agent which is metabolized to adrenaline after absorption and is used as eye drops for chronic open-angle glaucoma where it penetrates through the cornea more readily than adrenaline. A stinging sensation in the eyes may occur as side effect.
f dipivefrin
d Dipivefrin
i dipivefrin
e dipivefrin

1028 dipotassium clorazepate
A psychodepressant of the diazepinic series used in the symptomatic treatment of states of anxiety, feelings of acute distress, menopausal and senile states of fear and depression. It has no hypnotic action and does not lead to habituation.
f clorazépate dipotassique
d Dikaliumclorazepat
i clorazepato dipotassico
e clorazepato dipotásico

1029 diprophylline
A dihydroxypropyltheophylline used in medicine for the treatment of angina pectoris and bronchospasms. The drug is a diuretic agent and can also act as a muscle relaxant and antiasthmatic, but not as an analeptic. Gastric pains, allergic dermatoses and cardiovascular disorders, particularly in children, have been observed as side effects. Also known as dyphylline.
f diprophylline
d Diprophyllin
i diprofillina
e diprofilina

1030 dipyridamole
A coronary vasodilator displaying a papaverine-like mode of action. It is a vasal smooth-muscle relaxant and owing to the improved circulation it represents the best trophic influence on the myocardial tissue. It is therefore used in the treatment of coronary stenosis, acute or chronic coronary insufficiency, senile heart and, more generally, the pathological conditions which follow a myocardial infarction.
f dipyridamole
d Dipyridamol
i dipiridamolo
e dipiridamolo

1031 dipyrocetyl
Pyrocatechin-related antirheumatic agent used in various forms of acute rheumatic diseases.
f dipyrocétyl
d Dipyrocetyl
i dipirocetile
e dipirocetilo

1032 distigmine bromide
Parasympathomimetic anticholinesterase agent used in the treatment of intestinal atony occurring after a surgical operation, meteorism, myasthenia, urinary retention and partial paralysis caused by poliomyelitis or traumas.

f bromure de distigmine
d Distigminbromid
i distigmina bromuro
e bromuro de distigmina

1033 disulfiram
Tetraethylthiuram disulphide, capable of
blocking alcohol metabolism at the stage
of acetaldehyde. The drug is harmless
and causes no unpleasant symptoms unless alcohol is subsequently ingested,
in which case violent nausea, vomiting
and collapse may occur. In extreme
cases, death has occurred from the treat
ment; hence the necessity of the strictest
medical supervision in the administration
of the drug.
f disulfirame
d Disulfiram
i disulfiram
e disulfiram

1034 disulphamide
A diuretic agent acting on the distal
renal tubules and inhibiting carbonic
anhydrase. It is chiefly used in the
treatment of oedemas caused by water
retention. A condition of hypokalaemia
may occur as a side effect.
f disulfamide
d Disulfamid
i disulfamide
e disulfamida

1035 ditazole
Antiphlogistic and analgesic agent capable of inhibiting platelet aggregation
and used for the prevention and the
therapy of atherosclerotic and thrombotic
affections of the vascular system, valve
prosthesis (mechanical substitute for a
diseased cardiac valve), organ trans-
plants and retinopathies due to diabetes.
More generally, the drug is used to
treat vascular inflammatory conditions.
f ditazole
d Ditazol
i ditazolo
e ditazolo

1036 dithiazine iodide
Anthelmintic characterized by a paralyzing action and particularly effective
against worms of the genus Oxyuris.
f iodure de dithiazine
d Dithiazinjodid
i ditiazina ioduro
e yoduro de ditiacina

1037 dithranol
Antiparasitic agent used topically in the
treatment of psoriasis and other chronic
skin affections. It appears to be acting
by reducing the rate of skin cell formation. Hypersensitivity to the drug should
be ascertained before administration by
applying it to a small area of skin.
f dithranol
d Dithranol
i ditranolo

e ditranolo

1038 dixanthogen
An antiparasitic agent used in medicine
for the treatment of some dermatoses,
particularly if accompanied by itching.
f dixanthogène
d Dixanthogen
i dixantogene
e dixantogeno

1039 dixyrazina
A phenothiazine neuroleptic agent used
in the treatment of anxiety, states of
confusion, neurovegetative diseases and
as a preanaesthetic and antiemetic.
f dixyrazine
d Dixyrazin
i dissirazina
e dixiracina

1040 dobutamine
A sympathomimetic agent marked by a
strong inotropic effect on the heart. It
stimulates cardiac beta-adrenoreceptors
directly bringing about an increase in
the cardiac rate. It is used in the form
of an infusion in the treatment of shock.
It does not provoke constriction of peripheral blood vessels and an increase in
blood pressure. Cardiac arrhythmia may
occur as a side effect.
f dobutamine
d Dobutamin
i dobutamina
e dobutamina

1041 dofamium chloride
Antiseptic agent used to combat bacterial
and candida-related infections of the
mouth and the throat.
f chlorure de dofamium
d Dofamiumchlorid
i dofamio cloruro
e cloruro de dofamio

1042 domiodol
Therapeutic agent capable of reducing
the denseness and viscosity of tracheo-
bronchial secretions and of promoting
their elimination. It is therefore used
in affections of the upper respiratory
tract, bronchitis and bronchiolitis, asthma, emphysema, tracheitis, rhinitis etc.
f domiodol
d Domiodol
i domiodolo
e domiodolo

1043 domiphen bromide
Antiseptic agent used, generally in the
form of tablets, for oral infections and
sore throat. It has a mild histolytic (the
spontaneous dissolution of living organic
tissues) action.
f bromure de domiphène
d Domiphenbromid
i domifenio bromuro
e bromuro de domifeno

1044 domperidone
Antiemetic agent with dopamine antagonist action, used to control vomiting and
nausea caused by cancer chemotherapeutic treatment. Cardiac dysrhythmias and
involuntary movements may occur as side
effects.
f dompéridone
d Domperidon
i domperidone
e domperidona

1045 dopamine
Sympathomimetic agent capable of acting
as neurotransmitter. It is secreted by
neurons sited in the ventral hypothalamus and crossing over along axons to
the medial eminence from which it releases into the hypophyseal portal system factors having an inhibitory or releasing effect. It is therefore indicated
in the treatment of shock and of the
symptoms of chronic manganese poisoning
which can easily be mistaken with those
of parkinsonism. The administration of
the drug must be carefully monitored as
side effects are very numerous and may
be serious.
f dopamine; hydroxytyramine
d Dopamin; Hydroxytyramin
i dopamina; idrossitirammina
e dopamina; hidroxitiramina

1046 dosulepin
Tricyclic antidepressant having a mildly
tranquillizing effect and generally used
in the treatment of states of anxiety and
depression. A blurred vision, nausea,
sometimes vomiting and fatigue may occur as side effects.
f dosulépine
d Dosulepin
i dosulepina
e dosulepina

1047 doxapram
A stimulant of the central nervous system used to stimulate respiration. It is
therefore used in the treatment of respiratory depression of varying type and
also in the treatment of poisoning by
neurotropic drugs.
f doxapram
d Doxapram
i doxapram
e doxapram

1048 doxepin
Tricyclic antidepressant with mildly tran
quillizing properties. It is used in the
treatment of depressive neuroses and,
more generally, of depressive states accompanied by anxiety and agitation.
Drowsiness is the more common side effect of the drug.
f doxépine
d Doxepin
i dossepina
e doxepina

1049 doxorubicin
Cytotoxic antibiotic used in neoplastic
diseases. It probably acts affecting preformed ribonucleic acid. It is metabolized internally and eliminated with urine.
Haemoblastosis and some sarcomas are
treated with the drug. Bone marrow depression, leukopenia, stomatitis, cardio-
toxicity and gastrointestinal disorders
may occur as side effects.
f doxorubicine
d Doxorubicin
i dossorubicina
e doxorubicina

1050 doxycycline
Tetracycline-related cytotoxic antibiotic
used in the treatment of neoplastic diseases, as well as septicaemic or localized infections caused by both Gram-positive and Gram-negative bacteria. It differs from other tetracyclines as it is not
excreted by the kidneys, and thus used
when renal impairment appears as a
complicating factor. Gastrointestinal disorders and nausea may occur as adverse
effects.
f doxycycline
d Doxycyclin
i dossiciclina
e doxiciclina

1051 doxylamine
Ethanolamine-related antihistaminic drug
marked by a short duration and a rapid
excretion. It is used in the treatment
of allergic and anaphylactic disorders,
e.g. rhinitis, spastic bronchitis, asthmatic bronchitis and conjunctivitis. Agra
nulocytosis may occur as a side effect.
f doxylamin; doxylamine
d Doxylamin
i dossilamina
e doxilamina

1052 dried stomach extract
Preparation obtained from the mucosa of
the stomach of healthy animals, in particular pigs. It is an opotherapeutic
agent capable of balancing a deficiency
or outright lack of gastric digestive
enzymes with pepsin that is physiologically adequate, lab ferment and the
relative coenzymes. It is therefore indicated in the treatment of gastric achylia,
chronic or acute gastric catarrh, dyspepsia and other stomach disorders.
f extrait sec d'estomac
d Magentrockenauszug; Stomachi Extractum
i estratto secco di stomaco; stomaco estrat
to
e extracto seco del estómago

1053 droperidol
Butyrophenone-related tranquillizer and
sedative agent mainly used in the treatment of schizophrenia, anxiety and manic states. The drug may be used as a
preanaesthetic in conjunction with some
analgesics. Respiratory depression,

apnoea and muscular rigidity may be observed as adverse effects.

f dropéridol
d Droperidol
i droperidolo
e droperidolo

1054 droprenilamine
Therapeutic agent capable of promoting and increasing myocardial tonicity by diminishing the coronary vascular resistance without affecting the coronary arteries. It displays an antiadrenergic action and consequently combats arrhythmias. It is chiefly used in coronary insufficiency, both organic and functional, and in ischaemic myocardiopathies.

f droprénilamine
d Droprenilamin
i droprenilamina
e droprenilamina

1055 dropropizine
A drug capable of relieving cough as it is intrinsically a bronchodilator and expectorant agent. It is also used when bronchial spasms and respiratory insufficiency are diagnosed.

f dropropizine
d Dropropizin
i dropropizina
e dropropicina

1056 drostanolone
Androstane-derived anabolic steroid used in the treatment of metastatic mammary tumours, in particularly after mastectomy. Excess fluid and water retention may occur as side effects.

f drostanolone
d Drostanolon; Drostanol
i drostanolo
e drostanolo

1057 drotaverine
Spasmolytic and vasodilator as well as beta-blocking agent on cardiac receptors used to combat gastrointestinal and pyloric spasms, cholelithiasis (the presence of gall-stones in the gall-bladder or the bile duct), tenesmus (the pain which is associated with the excretion of faeces or urine), spastic constipation etc.

f drotavérine
d Drotaverin
i drotaverina
e drotaverina

1058 drotebanol
Narcotic and antitussive agent. It is chiefly used in the treatment of various types of cough. Nausea, vomiting, vertigo, either constipation or diarrhoea may occur as side effects.

f drotébanol
d Drotebanol
i drotebanolo

e drotebanolo

1059 D-xylose
Or alpha-D-xylopyranose. Glucose-resembling sugar easily absorbed from the normal small intestine. It is obtained by the hydrolization of straw or maize cobs. It has a low rate of metabolism and is therefore excreted almost 30% unchanged in the urine. It is used for the diagnosis of malabsorption from the gastrointestinal tract. Some abdominal discomfort and diarrhoea may occur as adverse effects.

f D-xylose
d D-xylose
i xilosio (D)
e xilosio (D)

1060 dyclonine
A topical anaesthetic agent particularly used in the treatment of mucous or cutaneous diseases, e.g. eczema, dermatosis, erythemas, independently of the presence of an inflammatory process. Its action is both immediate and prolonged. It is also used in various affections of the eyes.

f dyclonine
d Dyclonin
i diclonina
e diclonina

1061 dydrogesterone
An anabolic steroid mainly used in the treatment of premenstrual disorders, dysmenorrhoea, irregular intermenstrual bleeding, endometriosis, any condition caused by progesterone insufficiency, i.e. threatened and habitual abortion. The drug does not inhibit ovulation and has no contraceptive effect.

f dydrogestérone
d Dydrogesteron
i didrogesterone
e didrogesterona

1062 dyflos
Di-isopropyl phosphorofluoridate, a long-acting anticholinesterase used in the treatment of glaucoma.

f dyflos
d Dyflos
i dyflos
e diflos

1063 dysopyramide
Antiarrhythmic and cardiosedative agent used in the treatment of cardiac arrhythmias, extrasystoles and paroxysmal tachycardia. Sight and gastrointestinal disorders may occur as side effects.

f dysopyramide
d Dysopyramid
i disopiramide
e disopiramida

E

1064 **earth almond**
Chufa, ground almond, sedge root. Plant growing in Mediterranean countries, whose tubers, characterized by a bitter and resinous taste and a pleasant smell, are used as a tonic and a diuretic.
f amande de terre; souchet comestible
d Zypergraswurzel; Erdmandel
i cipero commestibile; babbagigi
e chufa

1065 **eau de Javelle**
Or Javelle water. A solution of sodium hypochlorite used as a disinfectant.
f eau de Javel
d Eau de Javelle; Natriumhypochloritlösung
i acqua di Javel
e agua de Javel

1066 **ecbolic**
Or oxytocic. Relating to any drug or agent capable of causing contraction of the uterus and thus promoting parturition or abortion. The principal ecbolics are ergot, extract of pituitary and quinine, which display a direct action on the uterine muscle, in particular in the course of childbirth.
f oxytocique
d geburtsauslösend; geburtsbeschleunigend; abortauslösend
i ossitossico
e ecbólico; ocitócico

1067 **ecgonine**
Tropine carboxylic acid. A crystalline substance obtained from the crude alkaloids of coca leaves or by the hydrolysis of cocaine. The ordinary form is laevorotatory and its methyl-benzoyl derivative is cocaine.
f ecgonine
d Ecgonin
i ecgonina
e ecgonina

1068 **echinacea root**
The dried root, or rhizome, of either Echinacea pallida or Echinacea angustifolia which used to be employed in medicine for the treatment of ulcers and boils.
f racine de rudbeckie
d Echinacearinde; Sonnenhutwurzel
i radice di echinacea
e raíz de equinácea

1069 **econazole**
Antifungal agent capable of acting against most pathogenic fungi and Gram-positive bacteria. The drug is used in practically all forms of dermatomycosis, in various infections caused by Gram-positive organisms, in onychomycosis, vulvovaginal mycosis etc.
f éconazole
d Econazol
i econazolo
e econazolo

1070 **ecothiopate iodide**
A potent and long-acting anticholinesterase used in the treatment of both simple and secondary glaucoma and some forms of strabismus. Iritis and vision disorders may occur as adverse effects.
f iodure d'écothiopate
d Echothiopatjodid
i ecotiopato ioduro
e yoduro de ecotiopato

1071 **ectylurea**
Sedative and tranquillizing agent acting on the cortex and the subcortex. It is rapidly absorbed and its action lasts from three to four hours. Generally used to sooth tension and hyperemotivity.
f ectylurée
d Ektylurea
i ectilurea
e ectilurea

1072 **edestin**
Vegetable protein of the globulin group, contained in the seeds of hemp, cotton and castor oil. A white crystalline substance soluble in water and in diluted acids. It contains all essential amino acids and is capable of sustaining growth in a diet which is free from other proteins. It is used for the analytical determination of pepsin. Also known as hemp-seed globulin.
f édestine
d Hanfsamenglobulin; Edestin
i edestina
e edestina

1073 **edetic acid**
Ethylenediamine-N,N,N',N-tetra-acetic acid, a powerful sequestering agent for metallic salts os sodium salts. It is used in medicine for the treatment of lead poisoning.

f acide édétique
d Edetinsäure
i acido edetico
e ácido edético

1074 edrophonium
An anticholinesterase agent. Its chloride is an antidote against d-tubocurarine (alkaloid used in surgery to induce muscle relaxation of the non-depolarizing type without deep anaesthesia) and is sometimes used intravenously in the treatment of myasthenia gravis. An increased salivation and occasionally arrhythmia have been observed as side effects.
f édrophonium
d Edrophonium
i edrofonio
e edrofonio

1075 effervescent powder
Mixture of substances which, once brought together in a solution, give rise to a rapid production of gas. Effervescent powders are frequently used in medicine as therapeutic substances are added to the bicarbonate and to the acid (citric or tartaric acid) representing the basis of the preparations.
f poudre aérophore; poudre gazifère
d Brausepulver
i polvere effervescente
e polvo efervescente

1076 efloxate
Coronary vasodilator capable of increasing the rate of circulation without affecting the cardiac and respiratory frequency rate. It is used in the treatment of acute and chronic coronary insufficiency and of angina pectoris.
f efloxate
d Efloxat
i eflossato
e efloxato

* **egg albumin   s. ovalbumin**

1077 Egyptian henbane leaves
The leaves of the Hyoscyamus muticus used as a source of hyoscyamine, an alkaloid occurring in dextrorotatory and laevorotatory forms. The racemic form is atropine.
f feuilles de jusquiame d'Egypte
d ägyptische Bilsenkrautblätter
i foglie di giusquiamo d'Egitto
e hojas de beleño de Egipto

1078 elastic collodion
Or flexible collodion. Solution of gun-cotton, colophony and castor oil in a mixture of alcohol and ether used to produce an unschrinking skin-protective film.
f collodion élastique
d elastisches Kollodium
i collodio elastico; collodio flessibile
e colodión elástico; colodión flexible

1079 elder flowers
The dried corollas and stamens of the flowers of Sambucus nigra with a proportion of buds, pedicels and ovaries. The volatile oil derived from them is a fragrant astringent used in medicine for the preparation of eye and skin lotions.
f fleurs de sureau
d Aalhornblüten; Fliederblüten
i fiori di sambuco
e flores de saúco

1080 elecampane
The dried rhizome and roots of Inula helenium whose leaves are used as an adulterant of digitalis. Also called scab wort.
f aunée; inule; oeil de cheval
d Alant; Brustalant; Helenenkraut
i elenio; enula campana; erbella
e alia; astabaca; énula campana

1081 elemi
An oleoresin obtained by the incision of the trunk of several plants of the Burseraceae family. It is characterized by an aromatic smell (resembling that of fennel) and a pungent taste. It is used in the preparation of various pharmaceutical products, mainly ointment and blister plasters.
f élémi
d Elemi; Amyrin
i elemi
e elemi

1082 elemicin
A component of the elecampane oil, a colourless liquid which, by ebullition with alcoholic potash is transformed into the propenylic isomer isoelemycin.
f élémicine
d Elemicin
i elemicina
e elemicino

1083 eleosaccharum
Any essential oil to which a given quantity of sugar has been added. Also known as oleosaccharum.
f oléosaccharure
d Ölzucker
i oliosaccarato
e oleosacaruro

1084 eleostearic acid
Crystalline unsaturated fatty acid existing in two stereoisomeric forms, the alpha acid occurring as the glycerol ester (especially in tung oil) and the beta acid obtained by irradiation from the alpha acid. It is found as a glyceride in the Japanese and Chinese wood oil.
f acide élaïostéarique
d Eläostearinsäure
i acido eleostearico
e ácido eleomargárico

1085 elixir

A liquid containing active therapeutic drugs, to which a syrup, glycerin or alcohol has been added to form a mixture which keeps off the unpleasant taste.

f élixir
d Elixir
i elixir
e elixir

1086 elm bark

Or slippery elm bark. The dried bark of the Ulmus fulva, a small American tree, which contains mucilage and some tannin and is used internally as a demulcent (a substance capable of protecting mucous membranes and soothing irritation) or, if mixed with hot water, as a poultice. The same bark, cut into strips, has been used by abortionists as tent-like insertions in the neck of the uterus.

f écorce d'orme pyramidal
d Elmenrinde; Rüstenrinde
i corteccia d'olmo
e corteza de olmo

1087 eluate

The ultimate result of the process of separating the constituents of powdered substances by washing with a given solvent.

f éluat
d Eluat
i eluato
e eluato

1088 elution

The process of separating the constituents of powdered substances by using a selective solvent. Should the eluates be subsequently reabsorbed in suitable media, the separation may be completed. The technique is frequently used in the production of pharmaceuticals.

f élution
d Elution; Eluieren
i eluizione
e elución

1089 embelic acid

Or embelin. An acid obtained from the berries of Embelia ribes. Yellow-orange crystals having anthelmintic properties, in particular against tapeworm.

f acide embélique
d Embeliasäure
i acido embelico
e ácido embélico

1090 embramine

Antihistaminic drug used in the treatment of some allergic dermatoses, e.g. urticaria, allergic rhinitis and angioneurotic oedema. Drowsiness and sometimes vertigo are possible side effects, in particular if alcohol is ingested after administration.

f embramine
d Embramin
i embramina
e embramina

1091 embryonis extractum

A preparation obtained from the embryonic tissues found in cows that have been slaughtered while pregnant. An opotherapeutic (relating to treatment of a disease by extracts of ductless glands) agent capable of activating general metabolism. It is used as a eutrophic, tonic and restorative in the treatment of disorders due to organic disorders and senescence, as well as for immature infants or seriously deficient development of the newborn.

f extrait embryonal
d Embryoextract
i embrione estratto
e extracto embrional

1092 emepronium

A parasympatholytic and antispasmodic agent, in particular for the urinary bladder. It is therefore used to reduce tone in the urinary bladder in cases when this is responsible for pains and urinary frequency or incontinence. Its toxic effects are similar to those of atropine, e.g. constipation, accommodation disorders and dryness of the passage from the mouth to the pharynx. Its bromide is preferred.

f émépronium; bromure d'émépronium
d Emepronium; Emeproniumbromid
i emepronio; emepronio bromuro
e emepronio; bromuro de emepronio

1093 emetine

Alkaloid (methylcephaeline) contained (0.5 to 2%) in various species of ipecacuanha. An amorphous, yellowish, bitter-tasting and easily alterable powder, soluble in water and alcohol and used in medicine as an emetic. It is chiefly used, either as the hydrochloride or as a complex iodide with bismuth, in the treatment of amoebiasis and schistosomiasis.

f émétine
d Emetin
i emetina
e emetina

1094 emetonium iodide

Anticholinergic agent characterized by an action which is both spasmolytic and capable of inhibiting gastric secretion. It is thus used in the treatment of gastric and duodenal peptic ulcer, biliary dyskinesia (the functional spasm which affects the sphincter of Oddi in the terminal section of the bile duct), pylorospasm and spastic colitis. Constipation and accommodation disorders are frequent side effects.

f iodure d'émétonium
d Emetoniumjodid;
i emetonio ioduro

e yoduro de emetonio

1095 emmenagogue
Relating to any agent capable of promoting menstruation, e.g. oestrogen hormones or aloe which is capable of congesting the pelvic organs.
f emménagogue
d Menstruationsmittel
i emmenagogo
e emenagogo

1096 emollient
Any substance capable of protecting the mucous membranes from which it subtracts water when they are inflamed. Emollients are mostly oils, oils with a paraffin basis, as well as insoluble substances, e.g. French chalk.
f émollient
d erweichendes Mittel
i emolliente
e emoliente

1097 emulsifier
Surface-active agent capable of promoting the formation and stabilization of an emulsion. Also called emulgator or emulsifying agent.
f émulsifiant
d Emulgator; Emulgierungsmittel
i emulsionante
e emulgador; emulgente

1098 emulsion
Intimate mixture of two incompletely miscible liquids, e.g. oil and water, in which one of the liquids is dispersed in the other in the form of fine droplets. Emulsions of liquid or solid substances in an aqueous liquid with an emulsifier are frequently used in the preparation of medicinal drugs to eliminate some bad taste.
f émulsion
d Emulsion
i emulsione
e emulsión

1099 emylcamate
A tranquillizer and a muscle relaxant characterized by its capacity to act without impairing attention and reflex reactions. It is used in the treatment of states of anxiety and exaggerated emotivity, restlessness, irritability, neuro-vegetative disorders, insomnia etc. Headache and often a pronounced dryness of the mouth are possible side effects.
f émylcamate
d Emylcamat
i emilcamato
e emilcamato

1100 endrisone
Topical antiinflammatory steroid for ophthalmic use characterized by the total absence of the adverse effects for the eyes which so frequently occur with the use of adrenocortical steroids. The drug is recommended for the treatment of inflammatory and allergic diseases affecting the eyes and their adnexa, exaggerated reactions to miotics (pupil-contracting agents) and all primary and secondary phlogistic processes of the anterior segment of the eye.
f endrisone
d Endrison
i endrisone
e endrisona

1101 enflurane
Nonflammable rapidly-acting inhalation anaesthetic used in surgery for the induction and the maintenance of general anaesthesia. It depresses without any delay the pharyngeal and laryngeal reflexes and does not stimulate salivary or bronchial secretions. It has an adequate myorelaxant action during the maintenance phase. Localized spasms have been observed following profound anaesthesia.
f enflurane
d Enfluran
i enflurano
e enflurano

1102 English liverwort
A lichen having demulcent properties and used in the past in the treatment of liver disorders.
f lichen peltigère
d Lebermoos; Hundsschildflechte
i lichene canino
e liquen canino

1103 enoxolone
Antiphlogistic and antipruritic agent used in dermatology for the treatment of acute and subacute cutaneous inflammations, itch, eczemas, erythemas, psoriasis and some infectious dermatoses in association with antibacterial agents. The drug is also used in the treatment of haemorrhoids, anusitis and proctitis. It is applied topically in the form of ointment.
f énoxolone
d Enoxolon
i enoxolone
e enoxolone

1104 enviomycin
Antibiotic drug acting on Gram-positive and Gram-negative microorganisms and in particular on tuberculous ones.
f enviomycine
d Enviomycin
i enviomicina
e enviomicina

1105 ephedra
The dried stem of Ephedra sinica (a Chinese herb known as ma-huang) and of Ephedra equisetina or Ephedra gerardiana containing the alkaloids ephedrine and pseudoephedrine, acting as vasoconstrictor and cardiac stimulant agents.
f éphédra

d Ephedra; Mahuang; Meerträubchen
i efedra; uva marina
e efedra; uva de mar; belcho de mar

1106 ephedrine
White crystalline alkaloid found in various plants of the genus Ephedra. Of the various stereoisomeric forms, the laevorotatory one is the truly active form. It can be prepared synthetically by the electrolytic reduction of the phenylmethyl dichetone in the presence of phenylamine. Ephedrine displays considerable therapeutic properties. It has a milder but longer-lasting action than adrenaline and can, thus differing from the latter, be administered orally. As a hydrochloride, sulphate etc. it is used in the treatment of cardiovascular diseases, bronchial asthma and as a stimulant of the nervous system.
f éphédrine
d Ephedrin
i efedrina
e efedrina

1107 ephedrine hydrochloride
The hydrochloride of the alkaloid contained in the Ephedra sinica and other species of Ephedra, presently mostly prepared synthetically. A sympathomimetic agent analogous to epinephrine, q.v., but better tolerated and more easily metabolized than the latter. For its uses, see definition of ephedrine.
f hydrochlorure d'éphédrine
d Ephedrinhydrochlorid
i efedrina cloridrato
e hidrocloruro de efedrina

1108 epicillin
Semisynthetic penicillin, a broad-spectrum antibiotic acting on Gram-positive and Gram-negative bacteria. It is used in the treatment of infections of the ear, throat and nose as well as infections interesting the bronchopulmonary, genito-urinary, digestive and hepatobiliary systems. The drug has also proved useful in the treatment of osteoarticular infections and acute articular rheumatism. It is equally useful as an antibiotic protective agent in relation to various viral diseases, prolonged cortisone or antimitotic therapies and surgical operations requiring a long time.
f épicilline
d Epicillin
i epicillina
e epicilina

1109 epiestriol
A natural female hormone characterized by a trophic vulvo-vaginal action and used in the treatment of postmenopausal degenerative alterations in the vaginal mucous membrane (atrophic vaginitis), inflammation of the vagina and of the cervix uteri, vaginitis due to infection or to parasites, some forms of dysmenor-

rhoea and, more generally, in various disorders occurring in the course of the menopause.
f épiestriol
d Epiestriol
i epiestriolo
e epiestriolo

1110 epimestrol
Oestrogenous hormone capable of stimulating ovulation and used to treat infertility caused by anovular cycles (the recurrent interruption of the process of ovulation), secondary amenorrhoea and oligomenorrhoea. Nausea, vomiting and headache accompanied by flushes occur as adverse effects.
f épimestrol
d Epimestrol
i epimestrolo
e epimestrolo

1111 epinephrine
Adrenal-medulla hormone used as simpathomimetic. The drug strengthens and accelerates the cardiac activity, produces hyperglycaemia, mydriasis, intensification of metabolism and in some instances hypertension. It is used as a vasoconstrictor, e.g. in local anaesthesia, and to combat asthma. Cutaneous and mesenteric ischaemia have been observed as side effects.
f épinéphrine
d Epinephrin
i epinefrina
e epinefrina

1112 epirizole
Antiphlogistic, central and peripheral analgesic agent used in the treatment of lumbago, cervicobrachial (relating to neck and arm) syndromes, arthropaties, neuralgias, inflammations of the respiratory tract, cystitis, adnexitis (inflamed condition of the accessory organs of the uterus), perineal lacerations and postoperative and post-traumatic pains. Anorexia, nausea and vomiting may occur as side effects.
f épirizole
d Epirizol
i epirizolo
e epirizolo

1113 epithiazide
A chemotherapeutical agent used in the treatment of hypertension. Also known as epithizide.
f épithiazide
d Epitizid
i epitizide
e epitizida

1114 epitiostanol
An antiestrogenous drug used in the palliative treatment of mastocarcinoma (breast cancer) in an advanced stage. Its virilising effect is not as frequent as it occurs with a testosterone treat-

ment.
f épitiostanol
d Epitiostanol
i epitiostanolo
e epitiostanolo

1115 epomediol
A choleretic agent which does not stimulate contractions of the gall bladder. It
promotes a more copious flow of bile and
a greater excretion of biliary constituents, e.g. bilirubin, biliary salts,
cholesterol etc. It is generally used in
the treatment of disorders which are
caused by inadequate production of bile,
hepatopathies, obdurate constipation,
halitosis, postprandial somnolence, chole
cystitis and disorders which may occur
following a cholecystectomy.
f épomédiol
d Epomediol
i epomediol
e epomediol

1116 epoprostenol
A prostaglandin which is produced endo-
genously and is characterized by potent
vasodilator properties. It is administered intravenously to inhibit platelet aggregation. It preserves platelet function
in the course of cardiac bypass procedures and charcoal haemoperfusion. In
renal dialysis, it may be used as an
alternative to heparin to prevent coagulation.
f époprosténol
d Epoprostenol
i epoprostenolo
e epoprostenolo

1117 eprazinone
A mucolytic and expectorant agent used
against cough. The drug is a bronchial
antispasmodic and a respiratory analeptic administered in chronic bronchitis
and asthma. It may be administered oral
ly and rectally.
f éprazinone
d Eprazinon
i eprazinone
e eprazinona

1118 eprozinol
Bronchospasmolytic, antiasthmatic and
antidyspnoeic agent as well as a sedative without depressive action on the
respiratory centre. It is used in the
treatment of dyspnoea associated with
asthma, chronic respiratory disorders
affecting bronchi and lungs, emphysema,
the bronchiolitis and bronchoalveolitis
affecting infants and, more generally to
prevent asthmatic crises.
f éprozimol
d Eprozimol
i eprozimolo
e eprozimolo

* **ergocalciferol   s. calciferol**

1119 ergoclavine
The equimolecular mixture of two less-
-important alkaloids of ergot, the ergosine, whose action is similar to that of
ergotoxine, q.v., and ergosinine which
forms ergosine upon boiling with alcoholic potash.
f ergoclavine
d Ergoclavin
i ergoclavina
e ergoclavina

1120 ergocornine
An alkaloid in which the lysergic acid,
the dimethylpyruvic acid, the proline
and valine are tied together by amidic
bonds. It has the property of reducing
the blood pressure.
f ergocornine
d Ergocornin
i ergocornina
e ergocornina

1121 ergocristine
Ergot alkaloid formed by lysergic acid,
dimethylpyruvic acid, proline and phenylalanine. Colourless crystals insoluble
in water.
f ergocristine
d Ergocristin
i ergocristina
e ergocristina

1122 ergocryptine
Ergot alkaloid formed by lysergic acid,
dimethylpiruvic acid, proline and leucine.
f ergocryptine
d Ergocryptin
i ergocriptina
e ergocriptina

1123 ergometrine
Ergot alkaloid corresponding to the oxy-
propylamide of the lysergic acid. Also
called ergobasine or ergonovine. Colourless crystals weakly soluble in water
and soluble in alcohol with blue fluorescence. It is not used as such in medicine, but in the form of its salts, i.e.
tartrate, hydrochloride and maleate,
which are capable of exerting a tonic
and a uterine haemostatic action.
f ergométrine
d Ergometrin
i ergometrina
e ergometrina

1124 ergosine
Ergot alkaloid formed by lysergic and
pyruvic acid, proline and leucine. Its
action is similar to that of ergotoxine.
It is obtained by treating with acids
the ergosinine.
f ergosine
d Ergosin
i ergosina
e ergosina

1125 ergosinine

Ergot alkaloid isomeric with ergosine which it forms on boiling with alcoholic potash.

f ergosinine
d Ergosinin
i ergosinina
e ergosinina

1126 ergosterol

Or ergosterine. Phytosterol which is found in plants, yeast, ergot and some higher fungi. Ergosterol is converted on irradiation into vitamin D_2 or calciferol. It is used in the treatment of rachitic conditions as it is the basis for the preparation of calciferol.

f ergostérol; ergostérine
d Ergosterin; Ergosterol
i ergosterolo
e ergosterol; ergosterina

1127 ergot

The dried sclerotium of a fungus which attacks the ovary of rye, breaking down the rye proteins. Ergot contains several active principles, the most important being the alkaloids ergotoxine, ergotamine and ergometrine. The latter is the alkaloid which is principally responsible for the characteristic action of ergot on the pregnant uterus, in which it increases the strength of the contractions and the tone of the muscle. Fluid extracts of ergot, injected intravenously, increase the blood pressure owing to the contraction of the arterioles. The drug has also a haemostatic importance particularly as a prophylactic against post--partum haemorrhage.

f ergot de seigle
d Mutterkorn
i segale cornuta
e centeno corniculato

1128 ergotamine

Ergometrine-resembling ergot derivative characterized by a vasoconstricting and alpha-adrenoreceptor blocking activity. It is chiefly used in the treatment of headaches, especially when associated with sensory disturbances, and postpartum atony.

f ergotamine
e Ergotamin
i ergotamina
e ergotamina

1129 ergotine

The extract of ergot. Solid preparation of ergot which is obtained by precipitating the aqueous extract with alcohol and evaporating the filtrate. It is characterized by a strong action on the smooth muscular fibres, particularly those of the uterus of which they provoke the typical contractions. It is owing to this property that ergotine has been used in the treatment of metrorrhagia (the irregular intermenstrual bleeding that is superimposed on the normal cycle) and as an

abortifacient. However, the latter use is extremely dangerous as the provoked contractions do not resemble the physiological ones and fail to undergo the phase of relaxation.

f ergotine
d Ergotin
i ergotina
e ergotina

1130 ergotinine

The first alkaloid isolated from ergot. It is a white crystalline powder and, in fact, a mixture of the alkaloids ergocorninine, ergocryptinine and ergocristinine. Pharmacologically, it is an inactive compound.

f ergotinine
d Ergotinin
i ergotinina
e ergotinina

1131 ergotoxine

Ergot-derived active alkaloid characterized by its important action on smooth muscle, in particular that of the uterus. Marked contractions are generally caused by small doses, whereas large doses may cause spasms. The gravid uterus is more sensitive to the drug but here again large doses cause constriction of the smaller blood vessels and bradycardia, so that the capillary endothelium is damaged and a condition of gangrene may result. The drug inhibits the vasoconstrictor response to injected adrenaline and a fall of blood pressure occurs with the latter.

f ergotoxine
d Ergotoxin
i ergotossina
e ergotoxina

1132 eriodictyol

Or eriodictin. Vitamin P (citrin). A colouring matter which is present in lemon juice, orange peel and paprika. It is regarded as an essential dietetic component which is capable of strengthening the capillary walls and preventing haemorrhages caused by scurvy.

f ériodictyol
d Eriodictyol
i eriodictiolo
e eriodictiol

1133 eritryl tetranitrate

Coronary vasodilator capable of strengthening the myocardial metabolism and used in the prophylaxis and treatment of angina pectoris. It is sometimes administered in association with digitaloid agents, anticoagulants, antidepressants and tranquillizers.

f tétranitrate d'éritryle
d Eritryltetranitrat
i eritrolo tetranitrato; eritriltetranitrato
e tetranitrato de eritrilo

1134 erucic acid

Unsaturated acid of the acrylic series, isomeric with brassidic and cetoleic acids and contained as glyceride in the cod-liver oil.
f acide érucique
d Erukasäure
i acido erucico
e ácido erúcico

1135 erythromycin
Bacteriostatic and bactericidal antibiotic particularly active against Staphilococcus aureus, group-A streptococci, pneumo cocci, Haemophilus influenzae, Brucella etc. It is only mildly effective against Escherichia coli and Proteus vulgaris and fails entirely to act against viruses and mycetes. It supposedly acts by inhibiting the biosynthesis of proteins. Gastrointestinal disorders and allergic cutaneous reactions may occur as side effects.
f érythromycine
d Erythromycin
i eritromicina
e eritromicina

1136 erythrosine
Red substance formed from tyrosine with nitric acid, a dyestuff used as an indicator of bacterial plaques on the teeth.
f érythrosine
d Erythrosine
i eritrosina
e eritrosina

1137 eseridine
An alkaloid contained in the Calabar beans, colourless crystals soluble in alcohol and in ether, used in veterinary medicine as a laxative.
f éséridine
d Eseridine
i eseridina
e eseridina

1138 eseridine salicylate
A cholinesterase inhibitor capable of stimulating gastrointestinal motility and secretion. The drug is indicated in the treatment of gastritis caused by hypochlorhydria, hypostenia-related dyspepsia, functional cholepathia and, more generally, digestive disorders and consti pation.
f salicylate d'éséridine
d Eseridinsalicylat
i eseridina salicilato
e salicilato de eseridina

1139 eserine
Physostigmine. Alkaloid which is obtained from the Calabar beans and is used in medicine for its capacity to inhibit the action of cholinesterase, i.e. the enzyme that destroys the acetylcholine liberated at parasympathetic nerve-endings. It also potentiates the muscarinic effects of acetylcholine and increases the cholinergic response at neuromuscular

junctions in skeletal muscle. In association with posterior pituitary extract, eserine has been used to restore intestinal tone after a surgical operation.
f ésérine; physostigmine
d Eserin; Physostigmin
i eserina; fisostigmina
e eserina; fisostigmina

1140 essence
Or essential oil. Any of a wide class of volatile odoriferous oils of vegetable origin capable of giving the plants a particular odour and sometimes a characteristic property. Essential oils are obtained from various parts of the plants, e.g. leaves, flowers, bark etc., by expression, extraction or steam distillation, and are frequently used for pharmaceutical preparations.
f essence; huile essentielle; huile volatile
d Essenz; aromatisches Öl
i essenza; olio essenziale
e esencia; aceite esencial

1141 essential camphor oil
Or rectified camphor oil. A colourless or yellowish liquid with the odour of camphor and formed by terpenes, safroles, acetaldehyde, camphor, terpineol, eugenol etc. It is used in medicine as a counter-irritant and rubefacient to be applied to rheumatic and inflamed joints. It is also used as a parasiticide. Also known as white camphor oil.
f huile essentielle de camphrée
d aromatisches Kampferöl
i essenza di canfora; olio essenziale di canfora
e esencia de alcanfor

* estradioestradiol s. oestradiol

1142 estramustine
Chemical combination of nor-iprite and oestradiol displaying cytotoxic activity in neoplastic diseases by acting on the cancerous cells with hardly noticeable oestrogenic effect, androgenous and anabolizing activity. It is mainly used in the treatment of cancer of the prostata and any endocrine-related tumours which prove resistant to oestrogens and other cytotoxic agents. Leukopenia, thrombocytopenia and perineal pains may occur as adverse effects.
f estramustine
d Estramustin
i estramustina
e estramustina

1143 etafenone
Vasodilator and tonic agent used to combat angina pectoris and particularly indicated when pathological conditions of the coronary arteries are accompanied by myocardial hypoxaemia.
f étafénone
d Ätafenon
i etafenone

e etafenona

1144 etamiphyllin
A cardiac and respiratory analeptic drug used in the treatment of heart and lung diseases of an infectious and toxic nature, asthmatic bronchitis, cardiac insufficiency and asphyxia.
f étamphylline
d Ätamphyllin
i etamifillina
e etamifilina

1145 etamocycline
Broad-spectrum, tetracycline-derivative antibiotic mainly used in the treatment of infections affecting the bronchopulmonary, genitourinary, hepatobiliary and ocular organs. It is also used in the treatment of other Gram-positive and Gram-negative infections of ear, nose and throat, cutaneous diseases, as well as an adjuvant in treatments involving corticosteroids.
f étamocycline
d Ätamocyclin
i etamociclina
e etamociclina

1146 ethacridine
Acridine-derived antiseptic which is active on several Gram-positive and Gram--negative organisms and used externally in the treatment of cutaneous diseases. It also finds employment internally in the treatment and prophylaxis of acute and chronic gastroenteritis, diarrhoeas and constipation, in particular spastic constipation.
f éthacridine
d   Äthacridin
i etacridina
e etacridina

1147 ethacrynic acid
A diuretic agent acting on the ascending limb of Henle and on the cells of the proximal and distal renal tubules. It is mainly used in the treatment of oedemas, in particular of cardiac, pulmonary and renal origin, and of ascites due to cirrhosis. Its action, if the drug is injected, is almost immediate (ab. 15 seconds) and lasts for 6 to 8 hours.
f acide étacrinique
d Äthacrynsäure
i acido etacrinico
e ácido etacrínico

1148 ethadione
An anticonvulsant agent used in the treatment of petit-mal epilepsy, i.e. the form of epilepsy which is characterized by sudden loss of conscience without any falling or incontinence and lasting not more than a few seconds. Photophobia, agranulocytosis, muscular weakness, alopecia and retinal or optical-nerve diseases may occur as side effects.
f éthadione

d Äthadion
i etadione
e etadiona

1149 ethamivan
Respiratory analeptic agent characterized by an action and effects which are similar to those of nikethamide, q.v. It is mainly used in the treatment of acute or chronic respiratory disorders, intoxications caused by barbiturates, comatose conditions accompanied by cyanosis and, though rarely, some heart disorders. The action is prompt but of short duration. Dyspnoea and excitation may occur as side effects.
f éthamivan
d Äthamivan
i etamivan
e etamivan

1150 ethamsylate
Therapeutic agent capable of combating capillary fragility. It is mainly used to prevent haemorrhages and ruptures especially in patients suffering from hypertension, arteriosclerosis, diabetes and obesity. It is also used in the treatment of disorders affecting the venous circulation, of ecchymoses and as an antidote of coagulants.
f éthamsylate
d Äthamsylat
i etamsilato; etanisilato
e etamsilato

1151 ethanediamine
Ammonia-smelling liquid used in organic synthesis and pharmaceutically in a 30% solution as a solvent for theophylline. Its dihydrochloride decomposes in the body like ammonium chloride and has been used to acidify the urine in the form of ketain-coated tablets.
f éthanediamine
d Äthanediamine
i etanediamina
e etanediamina

1152 ethanolamine
An amino acid soluble in water, esterified with oleic acid. It is used in medicine as a sclerosing agent for varicose veins.
f éthanolamine
d Äthanolamine
i etanolamina
e etanolamina

1153 ethanolamine acetylleucinate
An antivertigo drug used in the treatment of the Menière's syndrome and also of horizontal vertigo (vertigo experienced while lying down and probably caused by labyrinthine disorders), post--traumatic, postoperative and toxic vertigo. The drug is also used to combat vertigo caused by drugs, atheromatosis, hypertension and acute cephalalgia.
f acétyleleucinate d'éthanolamine

d Äthanolaminacetylleucinat
i etanolamina di acetilleucinato
e acetilleucinato de etanolamina

1154 ethanolamine oleate
Chemical compound prepared in sterile solution with benzyl alcohol and used as a sclerosing agent in the treatment of varicose veins.
f oléate de éthanolamine
d Äthanolaminoleat
i etanolamina oleato
e oleato de etanolamina

1155 ethaverine
A spasmolytic agent acting like atropine and used in the treatment of spastic conditions of the biliary and digestive tracts, peptic ulcer, coronary spasms, spastic cephalalgia and dysmenorrhoea. It may cause glaucoma.
f éthavérine
d Äthaverin
i etaverina
e etaverina

1156 ethchlorvynol
A sedative and hypnotic agent characterized by a rapid but short-lasting action on the nervous central system. It is used in the treatment of insomnia. Hypotension, nausea, vomiting and a sense of depression may occur as side effects.
f éthchlorvynol
d Äthchlorvynol
i etclorvinolo; etilclorovinolo
e etilclorvinolo

1157 ethenzamide
Antirheumatic agent whose action resembles that of the salicylates. It is also an antipyretic agent. Gastrointestinal disorders may occur as adverse effects.
f éthenzamide
d Äthenzamid
i etenzamide
e etenzamida

1158 ether
Ethyl ether. An inhalation anaesthetic agent less toxic than chloroform yet bound to cause some initial excitement. It is a colourless, volatile, highly inflammable liquid characterized by a peculiar odour and a sweetish taste. Also known as sulphuric ether.
f éther; éther sulfurique
d Äther; Äthyläther; Schwefeläther
i etere; etere etilico; etere solforico
e éter; éter etílico; éter sulfúrico

1159 ethinamate
An open-chain ureide. Hypnotic agent which is not related to barbiturates and acts on the cortex and the subcortex inducing a sleep which is very similar to the normal physiological one. It is therefore used in the treatment of mild insomnia of any origin.
f éthinamate

d Äthinamat
i etinamato
e etinamato

1160 ethinylandrostenediol
Synthetic derivative of androstenediol characterized by sex-hormone activity. The drug can be administered orally because the ethinyl radical protects the molecule in the stomach and in the intestine.
f éthinylandrosténediol
d Äthinylandrostenediol
i etinilandrostenediolo
e etinilandrostenediolo

1161 ethinyloestradiol
Synthetic female sex hormone, very active when orally administered, indicated in ovarian insufficiency, menopausal symptoms, metrorrhagia, prostatic or mammary carcinomata, senile vaginitis, some forms of osteoporosis, pruritus vulvae etc. The drug also acts as a contraceptive when taken in association with a progesterone derivative.
f éthinylestradiol
d Äthinylöstradiol
i etinilestradiolo
e etinilestradiolo

1162 ethionamide
A bacteriostatic agent used in the treatment of tuberculosis, both pulmonary and extrapulmonary, e.g. renal, intestinal or cerebral tuberculosis. The drug, whose action resembles that of amino salicylate, is generally used in association with other antituberculous agents. Gastrointestinal disorders and dizziness may occur further to a prolonged treatment.
f éthionamide
d Äthionamid
i etionamide
e etionamide

1163 ethisterone
Ethinyltestosterone, anhydrohydroxyprogesterone, pregneninolone. Synthetic agent which is not found in animal bodies and is characterized by a therapeutic action similar to that of progesterone, though with less strength. It differs from the latter as it is roughly equally active when injected or administered orally.
f éthistérone
d Äthisteron; Äthinyltestosterone
i etisterone
e etisterona; etiniltestosterona

1164 ethoglucid
A cytotoxic agent particularly indicated in the treatment of metastatic tumours and for a local or regional therapy. Its action is retarded and lasts for a long time. Some histolytic action and hypotension have been observed as side effects.
f éthoglucide

d Äthogluzid
i etoglucide
e etoglucido

1165 **ethoheptazine**
Analgesic agent used to soothe mild to moderate pains. Its action resembles that of pethidine. It is also used to relieve headache caused by painful nervous conditions. Vertigo and nausea, sometimes accompanied by epigastric pains, may occur as side effects.
f éthoheptazine
d Äthoheptazin
i etoeptazina
e etoeptacina

1166 **ethophylline**
Theophylline derivative capable of dilating the coronary arteries and stimulating the cardiac muscle. It acts on the vasal musculature like the xanthine derivatives. The drug also acts as diuretic and improves the glomerular infiltration thus inhibiting the reabsorption of water and electrolytes. It may cause convulsions in children as a side effect.
f éthophylline
d Äthophyllin
i etofillina
e etofilina

1167 **ethophyllin nicotinate**
Antiischaemic agent acting centrally and peripherally. It dilates the cerebral coronary and peripheral blood vessels and is marked by an action which sets in slowly but lasts for a long time. It has a positively inotropic action and thus increases the cardiac output. It is a hypocholesterolaemic and fibrinolytic agent, as well as a bronchodilator. The drug is mainly indicated in the treatment of cerebrovascular disorders, inadequate blood supply to the heart and all peripheral vascular diseases.
f nicotinate d'éthophylline
d Äthophyllinnikotinat
i etofillina nicotinato
e nicotinato de etofilina

1168 **ethopropazine**
Autonomic blocking agent with an associated spasmolytic action. The drug is used in extrapyramidal parkinsonism, hypertension, tremors and all those nervous syndromes which originate in the corpus striatum. Also known as profenamine, isothiazine or phenopropazine.
f éthopropazine; isothiazine
d Ethopropazin; Prophenamin
i etopropazina; fenopropazina
e etopropacina

1169 **ethosuximide**
Alpha-ethyl-alpha-methylsuccinimide. An anticonvulsant and antiepileptic agent used in the therapy of the primary and secondary forms of the petit mal. It may be used at any age. Should the petit mal occur together with major epilepsy attacks, the drug may be used in association with other anticonvulsants, e.g. phenobarbital, primidone etc. Gastric, nervous and psychic disorders may appear as side effects.
f éthosuximide
d Äthosuximid
i etosuccimide
e etosuximida

1170 **ethotoin**
Hydantoin-related anticonvulsant with central mechanism of action and long-lasting effect. It is used in the treatment of major epilepsy in association with various other drugs. Dizziness, general atony, occasionally tremors and nervousness have been observed as side effects.
f éthotoïne
d Äthotoin
i etotoina
e etotoina

1171 **ethoxazene**
Local anaesthetic agent acting on the urinary mucosa and used in the treatment of the pains which usually accompany urethritis, cystitis and pyelitis. The drug causes an orange or red colouring of the urine and should not be administered in the presence of severe hepatopathies.
f éthoxazène
d Äthoxazen
i etossazene
e etoxazeno

1172 **ethoxazorutoside**
Water-soluble rutin derivative showing effects which are very similar to those of vitamin P and mainly used in the treatment of capillary fragility and as a cardiac tonic. It is also a diuretic, antiphlogistic and hypotensive agent. Also employed in the therapy of capillaries of the brain, eyes and coronaries, which show signs of fragility.
f éthoxazorutoside
d Äthoxazorutosid
i etossazorutoside
e etoxazorutosida

1173 **ethoxzolamide**
Diuretic sulphamidic agent inhibiting the carbon anhydrase and used to reduce intraocular pressure by diminishing the secretion of aqueous humour. It is thus indicated in the treatment of acute and secondary glaucoma, hypertensive uveitis and, more generally, to reduce an abnormal tension of the ocular globe. It should not be used in pregnancy and in the presence of hepatic or renal insufficiency. Allergic reactions and anorexia have been observed as side effects.
f éthoxzolamide; étozolamide
d Äthoxzolamide
i etossizolamide

e etozolamida

* **ethyl aminobenzoate   s. benzocaine**

1174 ethylamphetamine
Hunger-inhibiting drug used in the treatment of obesity. A sympathomimetic agent
administered orally.
f éthylamphétamine
d Äthylamphetamine
i etilanfetamina; etamfetamina
e etamfetamina; etilanfetamina

1175 ethyl biscoumacetate
A coumarin-series anticoagulant capable
of inhibiting vitamin K in the liver by
reducing the concentration of prothrombin in the blood and depressing the
synthesis of the factor VII, IX and X
by the liver. The drug is indicated in
the prophylaxis and therapy of thrombosis, thrombophlebitis, embolism, as
well as arteriosclerosis and infarction
in association with other drugs.
f biscoumacétate de éthyle
d Äthylbiscoumazetat
i etilbiscumacetato
e biscumacetato de etilo

1176 ethyl chloride
General inhalation anaesthetic and local
anaesthetic agent. It acts in the same
way as methyl chloride owing to its
swift evaporation and consequent cooling.
As narcosis lasts for only a short time,
the drug is only useful in surgical operations which can be carried out in
minutes. Some circulatory and respiratory depression may occur as a side effect.
f chlorure d'éthyle
d Äthylchlorid
i etile cloruro; cloroetano
e cloruro de etilo

1177 ethyl dehydrocholate
Or edogestrone. A progestational steroid.
Choleretic and cholagogue (promoting
the flow of bile) agent very similar to
hydrocholic acid in its action, although
this is carried out more gradually and
with longer-lasting effects. The drug is
used to stimulate the secretion of bile
without increasing the content of bile
solids.
f déhydrocholate de éthyle
d Äthyldehydrocholat
i etile deidrocolato
e dehidrocolato de etilo

1178 ethylefrine
Ethylphenylephrine, ethylnorepinephrine.
A cardiovascular analeptic (arousal-causing agent) similar to synephrine both
for potency and absorption capacity but
marked by a much longer action. It increases the rate of circulation without
altering the cardiac output and has a
general tonic effect on the vascular
system. It is indicated in various states

of circulatory collapse and shock, orthostatic disorders, hypotension and in the
course of anaesthesia and prolonged surgical operations.
f éthyléphrine
d Äthylnorepinephrine
i etilefrina
e etilefrina

1179 ethylfelmine
Therapeutic agent capable of combating
hypotension and stimulating the central
nervous system owing to its direct action
on the vasomotor centres. It is chiefly
used in the treatment of essential hypotension, orthostatic and secondary hypotension, as well as hypotension caused
by cranial traumas, concussion, brain
tumours and, more generally, intoxications of various nature.
f éthylfelmine
d Äthylfelmin
i etilfelmina
e etilfelmina

1180 ethyl glutamate
Therapeutic agent used to protect the
liver and to balance the metabolism of
nitrogen. It is used in the treatment of
hyperammoniaemia related to liver disorders and accompanied by neurotoxic
reactions, hepatic coma and those acute
complications which may be the result of
alcoholism, stress and excessive physical
and mental work.
f glutamate d'éthyle
d Äthylglutamat
i etile glutamato
e glutamato de etilo

1181 ethylhydrocupreine
Quinine-derivative used in the past for
the treatment of pneumonia and now
replaced by far more effective sulphonamide drugs. Its hydrochloride may be
used in solutions and topically for eye
affections. It cannot be used internally
as it is too toxic.
f éthylhydrocupréine
d Äthylhydrocuprein
i etilidrocupreina
e etilhidrocupreina

1182 ethylhydroxyethylcellulose
A purgative which acts by increasing
the volume of the faeces and stimulating
intestinal peristalsis.
f éthylhydroxyéthylcellulose
d Äthylhydroxyäthylcellulose
i etulosio; cellulosa etilidrossietiletere
e etulosio; celulosa etilhidroxietiléter

1183 ethylmorphine
Homologue of morphine which is obtained
by ethylation. Its hydrochloride resembles codeine in its action and is used
both for coughs and as a sedative and
a narcotic. Though considerably weaker
than heroin, the drug is much safer.
f éthylmorphine

d Äthylmorphin
i etilmorfina
e etilmorfina

1184 ethylmorphine hydrochloride
Analgesic and bechic agent capable of depressing the cough centre in various forms of irritation of the respiratory tracts, in tuberculosis and pleurisy. The analgesic effect is brought about by the transformation of the drug into morphine. It is also used in the treatment of neuralgias, painful colics and pain-caus ing eye infections. Also known as code-thyline.
f chlorhydrate de codéthyline; chlorure d'éthylmorphine
d Äthylmorphinchlorhydrat
i etilmorfina cloridrato
e cloridrato de etilmorfina

1185 ethyl nitrite
Compound which, dissolved in alcohol, has been used in the treatment of angina pectoris, though with indifferent effects, and also as a diaphoretic (perspiration-inducing). Also known as nitrous ether.
f nitrite d'éthyle; éther nitreux; éther éthylnitreux
d Äthylnitrit; Salpetrigsäureäthylester
i nitrito di etile; etere nitroso
e nitrito de etilo

1186 ethylnoradrenaline hydrochloride
Sympatheticomimetic amine indicated as a bronchial antispasmodic. The drug is closely related to adrenaline.
f hydrochlorure d'éthylnoradrénaline
d Äthylnoradrenalin-hydrochlorid
i etilnoradrenalina idrocloruro
e hidrocloruro de etilnoradrenalina

1187 ethyloestrenol
Anabolic steroid, derived from androstenolone. It promotes protein biosynthesis and supports the reformation of bone matrix. The drug is chiefly used to balance the weakness and depression which frequently accompany convalescence, as well as in cases of pronounced thinness.
f éthylestrénol
d Äthylöstrenol
i etilestrenolo
e etilestrenolo

1188 ethyl orthoformate
A bronchial antispasmodic agent used in the treatment of cough and characterized by a hypnotic and anaesthetic properties. The drug is indicated for the relief of persistent cough independently of its origin and for pre- and post-operative therapy.
f orthoformiate d'éthyle
d Äthylorthoformiat
i etile ortoformiato
e ortoformiato de etilo

1189 ethylphenacemide

Anticonvulsant agent mainly used in the treatment of epilepsy (both petit mal and major epilepsy) and of lesions affecting the temporal lobe. Gastrointestinal disorders may occur as side effects.
f éthylphénacémide
d Äthylphenacemid
i etilfenacemide
e etilfenacemida

1190 ethynodiol
Derivative of the synthetic progestogen 19-nor-testosterone, which is used as an oral contraceptive with oestradiol. It does not cause virilization, nor does it promote anabolism. It is also used for the treatment of luteinic insufficiency, menopausal symptoms and dysmenorrhoea. Cardiovascular disorders, malignant hypertension and erythema nodosum may occur as adverse effects.
f éthynodiol; étynodiol
d Äthynodiol
i etinodiolo
e etinodiolo

1191 etidocaine
Local anaesthetic agent used for infiltra tion anaesthesia, as well as for peridural, caudal, peripheral nerve block anaesthesia. The drug should not be used for children and great caution is required with patients suffering from liver or kidney lesions.
f étidocaïne
d Ätidocain
i etidocaina
e etidocaina

1192 etiroxate
Therapeutic agent capable of reducing hyperlipidaemia and cholesterolaemia. It is therefore used in the treatment of primary hyperlipoproteinaemia, i.e. the condition in which there is an excess of lipids in the blood, of type II and IIB.
f étiroxate
d Ätiroxat
i etirossato
e etiroxato

1193 etodroxizine
A tranquillizer characterized by a sleep-inducing effect and used in sleep disor ders which are caused by a state of anxiety. Drowsiness may occur as side effect.
f étodroxizine
d Ätodroxizin
i etodrossizina
e etodroxicina

1194 etofibrate
Anti-hyperlipidaemic, hypocholesterolaemic and antiatherosclerotic agent used in the treatment of hyperlipidaemic conditions, coronary, cerebral and vascular diseases, as well as retinopathies. The drug should be used with great caution during pregnancy and in the presence

of renal and liver disorders.
f étofibrate
d Ãtofibrat
i etofibrato
e etofibrato

1195 etoloxamine
An antihistaminic and antiallergic drug used in the treatment of various allergies.
f étoloxamine
d Ãtoloxamin
i etolossamina
e etoloxamina

1196 etolozine
Diuretic agent capable of inhibiting the reabsorption of the sodium thus increasing its elimination, as well as a hypertensive. It is used in the treatment of oedemas, in particular oedemas of cardiac, hepatic and renal origin. The drug also finds employment in the treatment of essential hypertension. Some allergic reactions have been observed as side effects.
f étolozine
d Ãtolozin
i etolozina
e etolozina

1197 etomidate
Fairly strong anaesthetic injected intramuscularly for the induction of total anaesthesia. The injection itself is rather painful and may cause hypotension.
f étomidate
d Ãtomidat
i etomidato
e etomidato

1198 etoperidone
A psychotropic drug used in the treatment of states of anxiety and depression, psychosomatic disorders, senile involution and presenile dementia. The drug acts by normalizing the biochemical transmitting bases of the emotive processes. Alcoholic drinks should be avoided during the therapy with the drug.
f étopéridone
d Ãtoperidon
i etoperidone
e etoperidona

1199 etoxuridine
Antiviral agent used in the treatment of most forms of herpetic keratitis. The effects of the drug should be carefully monitored in the course of pregnancy or in patients suffering from metaherpetic keratitis.
f étoxuridine
d Ãtoxuridin
i etossuridina
e etoxuridina

1200 etretinate
A synthetic derivative of retinoic acid used in the treatment of severe difficult-to-heal psoriasis and other grave disorders of skin growth. Among its side effects, teratogenic actions, exfoliation of the skin, hair loss and functional liver disorders have been observed. It should not be used in pregnancy.
f étrétinate
d Ãtretinat
i etretinato
e etretinato

1201 etybenzatropine
Parasympatolytic and atropine-like anticholinergic agent with a central mechanism of action. It is indicated in extra pyramidal syndromes and in the Parkinson's disease also when caused by medicinal drugs. Glaucoma and hypertrophy of the prostata may occur as adverse effects.
f étybenzatropine
d Ãtybenzatropin
i etibenzatropina
e etibenzatropina

1202 etymemazine
A major tranquillizer which may act as hypnotic if administered in large doses. It resembles phenothiazine agents in its action when the latter do not include the chloro group. It is indicated in the therapy of insomnia and of states of anxiety. It should be avoided during pregnancy and administered with caution to children.
f étymémazine
d Ãtymemazin
i etimemazina
e etimemazina

1203 eucalyptol
Ether ($C_{10} H_{18} O$) which can be derived from the cis isomer of the terpene hydrate by the elimination of a molecule of water. It is contained in the oil of various species of eucalyptus and in the oil of cajeput. An oily, colourless liquid, smelling like camphor and with a pungent taste. It is used in the symptomatic treatment of infections affecting the upper respiratory tract either in tablets or pastilles. It can also be inhaled with steam or as a spray. Combined with zinc oxide it is occasionally employed in temporary dental fillings. Also known as cineole.
f eucalyptol
d Eucalyptol
i eucaliptolo
e eucaliptol

1204 eucalyptus oil
Oil obtained by rectification of the oil that is extracted from the leaves of various species of eucalyptus. A colourless or pale-yellow liquid characterized by a camphor-like odour and a pungent taste. It contains eucalyptol (cineole) and D-pinene together with other terpenes. It is used medicinally as an anti

septic, a deodorant and to combat catarrhal colds. It is also employed to relieve asthma and bronchitis through inhalations. It may be administered internally in capsules or as an emulsion in the treatment of catarrhal inflammation of mucous membranes, in particular of the respiratory tract and the bladder.
f huile d'eucalyptus
d Eucaliptusöl
i olio di eucalipto
e aceite de eucalipto

1205 eucatropine
A parasympatholytic mydriatic agent acting without altering accommodation and intraocular tension. Used especially for ophthalmologic examination. Its effects generally last for two to six hours. The drug should not be employed in the presence of glaucoma.
f eucatropine
d Eucatropin
i eucatropina
e eucatropina

1206 eugenic acid
Or eugenitic acid. Constituent of the oil of cloves generally used as an antiseptic and to reduce sensitivity in dentistry. It is an allylic derivative of guaiacol.
f acide eugénique; eugénol
d Eugensäure; Eugenol
i acido eugenico; eugenolo
e ácido eugénico; eugenol

1207 eupeptic
Relating to any drug or substance capable of aiding the process of digestion.
f eupeptique
d eupeptisch
i eupeptico
e eupéptico

1208 excipient
Any binding agent which make it possible for powdered drugs to be made into pills or tablets. It is generally an inert substance constituting a vehicle, e.g. simple sugar and mucilage of acacia, gum arabic, lanolin, starch etc. Excipients should have no therapeutic action of their own. Also known as vehicle.
f excipient; véhicule
d Träger; Arzneiträger; Vehikel
i eccipiente; veicolo
e excipiente; vehículo

1209 excitant
Any agent capable of promoting or arousing physiological and/or nervous and mental activity. Also any agent capable of increasing organic activity.
f excitant
d Anregungsmittel; Reizmittel
i eccitante
e excitante

1210 exiproben
A total choleretic agent capable of promoting an increase of the biliary flow which is three times the basic value with a proportional physiological elimination of all the components of the bile itself. The choleresis thus obtained sets in within thirty minutes and lasts three to four hours. The drug is indicated in diseases of the biliary tract, gall stones, cholangitis, dyspepsia caused by inadequate bile flow, hyperbilirubinaemia and hypercholesterolaemia that are due to hepato-biliary disorders.
f exiprobène
d Exiproben
i esiprobene
e exiprobeno

1211 expectorant
Any agent tending to promote expectoration, i.e. the discharge of mucus and bronchial secretion. Expectorants act by increasing the secretion itself, thus rendering it less viscous and easier to discharge. The action may be reflex, central or peripheral.
f expectorant
d Expektorans; Abhustemittel
i espettorante
e expectorante

1212 extract
Any pharmaceutical preparation of a vegetable or animal substance which contains the active principle but no cellular material. Extracts may be dry, soft and liquid and most of them are only used orally as they inevitably contain extraneous material. Extracts which are free from such material, e.g. from protein, and can be administered parenterally, are called injections.
f extrait
d Extrakt; Auszug
i estratto
e extracto

F

1213 fabiana
Pichi herb. A shrub (Fabiana imbricata of the Solanaceae family) growing in Peru, whose flowering twigs are used in the form of extract or decoction as a diuretic and disinfectant of the urinary tract owing to the active principles contained therein.
f fabiane; pichi
d Fabiane; Pichikraut
i fabiana; pichi
e fabiana; pichi

1214 factor VIII
Blood clotting factor which is defective in haemophilia and in the von Willebrand's disease (a form of non-thrombocytopenic purpura). The drug is used intravenously to check uncontrollable bleeding.
f facteur VIII
d Faktor VIII
i fattore VIII
e factor VIII

1215 factor IX
Antihaemophilic human globuline B. A biological antihaemorrhagic agent which makes it possible to obtain in a comparatively short time an adequate increase of the coagulation factors II, VII, IX and X. It is indicated for the treatment of haemophilia B and in all haemorrhages caused by hereditary or acquired alterations of the clotting process related to inadequacy of factors II, IX and X. In the acquired forms it is advisable to use the agent together with vitamin K. A possible risk of serum hepatitis must be taken into consideration.
f facteur IX
d Faktor IX
i fattore IX
e factor IX

1216 factor antihaemorrhagicus
Non-saponifiable fraction of liver of mammals containing antihaemorrhagic active principles. It is indicated in the treatment and above all prevention of haemorrhage-causing uterine dysfunctions, e.g. menorrhagia and metrorrhagia. It does not affect the normal menstrual flux.
f facteur antihémorragique
d Antiblutungsfaktor
i fattore antiemorragico

e factor antihemorrágico

1217 factor coagulans sanguinis
Blood-clotting factor. Coagulating principle extracted from the blood of mammals. Physiological haemostatic agent capable of accelerating the natural process of coagulation without affecting the vasal lumen. The drug is indicated in the treatment of gastro-intestinal, vesical and post-partum haemorrhages and in surgical inaccessible haemorrhages.
f facteur de coagulation sanguine
d Blutkoagulationsfaktor
i fattore coagulante del sangue
e factor de coagulación del sangre

1218 fairy flax
Or purging flax. European annual herb having white or yellow-white flowers followed by seeds which are a cathartic and diuretic therapeutic substance.
f lin cathartique; lin purgatif
d Purgierlein; Berglein
i lino purgativo; lanaiola
e lino purgante; cantilagua

1219 false pellitory of Spain
Or masterwort. Tall perennial herb with a ramified rhizome, containing an active principle which is used to aid digestion. It grows in Europe.
f imperatoire des montagnes
d Rossfenchel; echter Haarstrang
i imperatoria; erba rena
e imperatoria romana

1220 false sarsaparilla
A North America perennial herb with long-stalked ternate leaves and greenish flowers, whose aromatic roots are used as a substitute for sarsaparilla. The dried roots of the latter are used as an olfactory agent in the form of extract or infusion.
f salsepareille indienne
d falsche Sarsaparille; indische Sarsaparille
i salsapariglia falsa; salsapariglia d'India
e zarzaparilla de las Indias

1221 famotidine
A gastric histamine receptor blocking agent used in the treatment of peptic ulcers, gastric hyperacidity, oesophageal reflux and prophylaxis of gastro-intes-

tinal bleeding. Tiredness, dizziness, constipation and anorexia have been observed as side effects.

f famotidine
d Famotidin
i famotidina
e famotidina

1222 fatty acid
Any of the saturated or unsaturated monocarboxylic acids occurring naturally in the form of glycerides in fats and fatty oils. They are chiefly used for the production of a large number of derivatives.

f acide gras
d Fettsäure
i acido grasso
e ácido graso

1223 fatty acid glyceride
Any ester which is formed by glycerol with a fatty acid.

f glycéride d'acide gras
d Fettsäureglycerinester
i gliceride di acidi grassi
e glicerido de los ácidos grasos

1224 fatty alcohol
Any alcohol of the aliphatic series.

f alcool gras
d Fettalkohol
i alcool grasso
e alcohol graso

1225 fazadinium bromide
Neuromuscular blocking agent, non depolarizing. It is chiefly used to promote relaxation prior to a surgical operation and to facilitate the endotracheal intubation. Inflammatory reactions at the site of the injection have been observed as side effects, as well as some episodes of tachycardia.

f bromure de fazadinium
d Fazadiniumbromid
i fazadinio bromuro
e bromuro de fazadinio

1226 featherfew
Or feverfew, wild chrysanthemum. A bitter herb characterized by febrifuge properties.

f chrysanthème matricaire; herbe vierge; grande camomille
d Bocksblumenkraut; Goldfederich; Mutterkraut
i erba amara; amareggiola; matricale
e amargaza; matricaria; arrugas

1227 febarbamate
A tranquillizer which combines the action of the barbituric acid with that of meprobamate (a tranquillizing substance) and is used to treat anxiety and tremors. Gastrointestinal disorders may occur as side effects.

f fébarbamate
d Febarbamat
i febarbamato
e febarbamato

1228 feclobuzone
Phenylbutazone-derived antirheumatic agent displaying analgesic, antipyretic and antiphlogistic action. Generally used in the treatment of inflammatory diseases in the presence of infections and contusions, rheumatic disorders, arthritis, neuritis, gout, superficial thrombophlebitis etc. Gastric inflammation may occur as a side effect.

f féclobuzone
d Feclobuzon
i feclobuzone
e feclobuzona

1229 fedrilate
Antitussive drug indicated in the treatment of chronic bronchitis, asthma and emphysema. The drug should not be used for babies or very young children. Gastric disorders may occur as adverse effects.

f fédrilate
d Fedrilat
i fedrilato
e fedrilato

1230 felon herb
European perennial herb which was used for the treatment of the felon or whitlow, a suppurative infection of a finger, almost always of the terminal phalanx. Also known as mouse ear or mugwort.

f épervière; oreille de souris; piloselle
d Nagelkraut; kleines Habichtskraut
i pelosella; ieracio
e pilosela; hieracio

1231 felypressin
A vasoconstrictor which is sometimes used with local analgesic drugs instead of adrenaline. It is a polypeptide that is less likely to cause cardiac arrhythmias than sympathomimetic vasoconstrictors and does not interfere with antidepressant drugs.

f félypressine
d Felypressin
i felipressina
e felipresina

1232 fenadiazole
Non-barbituric hypnotic agent. It is used to combat insomnia and particularly psychogenic insomnia. A dependance is possible. Nausea and headache may occur as adverse effects.

f fénadiazole
d Fenadiazol
i fenadiazolo
e fenadiazolo

1233 fenalamide
Myotropic spasmolytic agent characterized by its direct action on the smooth musculature. It strengthens the analgesic action of aminophenazone and noramidopyrine sodium metansulfonate and is

used in the treatment of all spastic
states of the smooth muscles, in urology,
obstetrics, ginecology and, more general_
ly, in surgery. A prolonged rectal administration or parenteral injections of
the drug may cause hypotension.
f fénalamide
d Fenalamide
i fenalamide
e fenalamida

1234 fenalcomine
A dilator of the coronary arteries and
capillaries, used in the treatment of
angina pectoris and the conditions which
usually follow a myocardial infarction.
f fénalcomine
d Fenalcomin
i fenalcomina
e fenalcomina

1235 fenamisal
Phenyl aminosalicylate. Chemotherapeutic
agent capable of arresting the growth
of the tubercle bacillus and used in the
treatment of tuberculosis. Nausea, vomiting, diarrhoea, skin rashes, hepatic
necrosis and other adverse effects may
occur.
f fénamisal; aminosalicylate de phényle
d Fenamisal; Phenylaminosalicylat
i fenamisal; fenil aminosalicilato
e fenamisal; aminosalicilato de fenilo

* **fenasal   s. niclosamide**

1236 fenbufen
Non-steroid, analgesic and antiphlogistic
agent which is rapidly absorbed and
whose action is fairly long-lasting. It
is used to relieve pains, inflammation
and any other disorder of a rheumatic
nature, as well as osteoarthrosis and
rheumatoid arthritis. The principal coun_
terindication is hypersensitivity to acetylsalicylic acid. Gastrointestinal disorders, pyrosis, nausea and sometimes
cramps may occur as side effects.
f fenbufène
d Fenbufen
i fenbufene
e fenbufeno

1237 fencamfamin
A psychostimulant agent used in the
treatment of physical and psychic stress,
depression, neurovegetative dystonias
and all those near-pathological conditions which accompany convalescence. It
must be used very cautiously in patients
suffering from angina pectoris.
f fencamfamine
d Fencamfamin
i fencamfamina
e fencamfamina

1238 fencarbamide
An antispasmodic agent mainly used to
cause the dilatation of the uterine neck
and to balance the hypertonic conditions

of the genital organs. The action is simi_
lar to that of papaverine.
f fencarbamide
d Fencarbamid
i fencarbamide
e fencarbamida

1239 fencibutirol
Non-cholagogue choleretic capable of increasing the secretory activity of the
liver. It has both a hydrocholeretic
(increasing the volume of the bile without stimulating the secretion of its solid
constituents) effect and an effect stimulating the production of the specific
components of the bile by the hepatic
cell. It is used in pathological conditions of the biliary tract, i.e. cholecystitis, bile stones, acute and chronic
biliary insufficiency and disorders which
follow a surgical operation on the organ.
f fencibutirol
d Fencibutirol
i fencibutirolo
e fencibutirolo

1240 fendiline
Coronary vasodilator used in the treatment of coronary insufficiency and consequent angina pectoris, sclerosis of the
coronary arteries, as well as prophylactic and therapeutic treatment of myocardial infarction. Allergic reactions have
occasionally been observed as side effects.
f fendiline
d Fendilin
i fendilina
e fendilina

1241 fenethylline
Centrally acting cerebral stimulant indicated in the treatment of states of
asthenia, psychic disorders, physical
and mental stress and states of depression. It is also used against difficulty
to concentrate. It may cause insomnia
if taken before going to bed.
f fénéthylline
d Fenethyllin
i fenetillina
e fenetilina

1242 fenfluramine
Therapeutic agent used against obesity
and characterized by a central and
peripheral effect. It does not act as a
psychotonic and causes no hypertension.
It is also used to combat both psychogenic and diabetes-related obesity. An
increase of the body temperature and
drowsiness may occur as side effects. It
should not be administered in pregnancy.
f fenfluramine
d Fenfluramin
i fenfluramina
e fenfluramina

1243 fenipentol
A choleretic agent with intense and pro-

longed hydrocholeretic action accompanied by an increased secretion of the bilirubin and indicated in case it is advisable to increase the flow of bile, e.g. cholelithiasis (gall stones in the gall bladder), cholecistopathies and disorders of the liver. It should not be administered if there is some occlusion in the common bile duct.

f fénipentol
d Fenipentol
i fenipentolo
e fenipentolo

1244 fennel
The dried fruit of Foeniculum vulgare growing in some European countries and in Russia. It contains a volatile oil of a particular aromatic taste and odour, specifically liminene, pinene, fenchone and anethole and is used therapeutically as a carminative.

f fenouil
d Fenchel
i finocchio
e hinojo

1245 fennel-leaf water dropwort
Water fennel. A European poisonous herb characterized by fibrous roots and yielding a principle which is used in some therapeutic preparations.

f fenouil aquatique; ciguë aquatique
d Wasserfenchel; Wasserkörbel
i finocchio d'acqua; fellandrio
e hinojo de agua; felandrio

1246 fennel oil
Colourless or yellowish fennel-smelling liquid with a sweetish taste. It is obtained by distilling in a steam stream the fruits of the fennel. It contains camphene, aldehyde and anisic acid. It is used in medicine for its expectorant and stomachic properties.

f essence de fenouil
d Fenchelöl
i essenza di finocchio
e esencia de hinojo

1247 fennel water
Saturated solution of fennel oil in distilled water used as a stimulant and a carminative agent.

f eau de fenouil
d Fenchelwasser
i acqua di finocchio
e agua de hinojo

1248 fenofibrate
Anti-lipidaemic, hypocholesterolaemic, anti-atherosclerotic agent used in the treatment of simple hypercholesterolaemia, both plain and accompanied by xanthomatosis, hypertriglyceridaemias, mixed hyperlipidaemias, angina pectoris, cerebrovascular insufficiency, peripheral arteriopathies, diabetic retinopathy and, more generally, arterial hypertension. Gastrointestinal disorders are sometimes occurring as side effects.

f fénofibrate
d Fenofibrat
i fenofibrato
e fenofibrato

1249 fenoprofen
Anti-inflammatory and analgesic agent, non-steroid, used in the systematic treatment of chronic rheumatic polyarthritis, arthrosis deformans, spondylosis (the non inflammatory and degenerative disease of the spine), spondylarthrosis, tendinitis etc. Gastroenteric disorders and skin rashes may occur as side effects. The drug should not be used in pregnancy.

f fénoprofène
d Fenoprofen
i fenoprofene
e fenoprofeno

1250 fenoterol
Bronchoselective beta-adrenoreceptor agonist marked by an extremely limited action on the cardio-circulatory system. It is used in bronchial asthma and also, by means of intravenous infusion, to inhibit premature labour. Tachycardia and arrhythmias may occur as adverse effects. The drug should be avoided in the presence of hyperthyroidism.

f fénotérol
d Fenoterol
i fenoterolo
e fenoterolo

1251 fenoxedil
Peripheral and cerebral vasodilator capable of acting on the smooth vascular musculature so as to attain a gradual relaxation allowing an improvement of the peripheral and cerebral blood flow. The drug further promotes the oxygen and glucose utilization by the cells by checking platelet aggregation on the onset. It is mainly used in the treatment of peripheral and cerebral vascular insufficiency and in the prophylaxis of the symptoms which are caused by ischaemia affecting the cerebral region.

f fénoxédil
d Fenoxedil
i fenoxedil
e fenoxedil

1252 fenozolone
A stimulant of the central nervous system used in the therapeutic treatment of mental stress, especially when accompanied by loss of memory, difficulty in concentration and difficulty of adaptation in children. It should not be administered in pregnancy and in the presence of psychosis.

f fénozolone
d Fenozolon
i fenozolone
e fenozolona

1253 fenpentadiol
A tranquillizer and a sedative as well as a psychotropic agent used in the treatment of neuropsychic disorders, restlessness and neurovegetative diseases. Owing to its stimulant properties, the drug also finds employment in behavioural anomalies and mild senile dementia. Feeling of tiredness and a disturbed digestion may occur as side effects.
f fenpentadiol
d Fenpentadiol
i fenpentadiolo
e fenpentadiolo

1254 fenpipramide
2,2-diphenyl-4-piperidinobutyramide. A chemical compound used as a spasmolytic agent.
f fenpipramide
d Fenpipramid
i fenpipramide
e fenpipramida

1255 fenprometamine
Phenylpropylmethylamine. Colourless liquid not easily soluble in water. A sympathicomimetic amine with vasoconstricting properties, also capable of relieving nasal congestion.
f phénylpropylamine
d Phenylpropylamin
i fenprometammina
e fenprometamina

1256 fenproporex
Lipolytic agent used to treat obesity by strongly reducing the appetite for food, when the condition is caused by wrong diet or anxiety. The drug may be used in the presence of diabetes, arthrosis, respiratory insufficiency and cardiovascular disorders.
f fenproporex
d Fenproporex
i fenproporex
e fenproporex

1257 fenpyramine
A myotropic spasmolytic agent and an analgesic. Its action is quick and long-lasting. It is used in the treatment of all painful and spastic states of the smooth musculature of hollow organs, of the uterus and, generally, in surgery. The therapy should be carefully monitored in the first few months of pregnancy.
f fenpyramine
d Fenpyramin
i fenpiramina
e fenpiramina

1258 fenquizone
Sulfonamide diuretic acting on the proximal renal tubules, the ascending ramus of the Henle's hoop and the proximal tract of the distal convoluted tubule. The diuretic increment occurs gradually but lasts for an appreciable length of time. It promotes sodium loss, whereas

the loss of potassium is moderate. The drug is used for the treatment of disorders accompanied by the retention of salt and water and as an adjuvant in the therapy of arterial hypertension.
f fenquizone
d Fenquizon
i fenchisone
e fenquisona

1259 fenspiride
Anti-inflammatory and anti-exudative agent acting selectively on the respiratory apparatus. Its bronchodilator and antidyspnoeic activity makes it very useful in the treatment of primary and secondary asthma, chronic respiratory insufficiency, acute and chronic bronchitis, pulmonary emphysema, bronchiectasis (dilatation of the bronchi secondary to structural alterations in the bronchial walls) accompanied by spastic episodes, laryngitis etc.
f fenspiride
d Fenspirid
i fenspiride
e fenspirida

1260 fentanyl
Analgesic and general anaesthetic agent whose action is prompter and more potent than that of morphine, but of short duration, i.e. approximately 30 minutes. It is used as an analgesic marked by a very rapid action. Respiratory depression may occur as a side effect.
f fentanyl
d Fentanyl
i fentanil
e fentanil

1261 fentiazac
Non-steroid antiphlogistic, analgesic and antipyretic agent used both in internal medicine and surgery whenever an analgesic and antipyretic action is required together with an antiphlogistic action. The drug may adversely affect patients suffering from some gastrointestinal disorder.
f fentiazac
d Fentiazac
i fentiazac
e fentiazac

1262 fenticlor
A therapeutic antimycotic agent used in the treatment of dermatomycosis and other infectious dermatoses caused by bacteria. Inflammatory reactions may occur as side effect.
f fenticlor
d Fenticlor
i fenticloro
e fenticloro

1263 fentonium bromide
Neurotropic spasmolytic agent with a selective synaptic action on the parasympathetic system in the region of the

intramural plexus and without atropine-
-related effects or orthosympathetic ganglion-blocking effects. The drug is used
in the treatment of pathological disorders of the motor and secretory activity
of the gastroenteric and urinary apparatus, bronchospastic states, the habitual
vomiting of the suckling and prior to
radiologic examinations.
f bromure de fentonium
d Fentoniumbromid
i fentonio bromuro
e bromuro de fentonio

1264 fenugreek oil
The essence which is obtained from a le-
guminous annual Asian herb, used in
pharmaceutical preparations, in particular veterinary medicines.
f huile de fenugrec
d Bockshornsamenöl
i olio di fieno greco
e aceite de alholva

1265 feprazone
Pyrazole-derivative antiphlogistic agent
acting like an adrenal corticoid, i.e.
by protecting cellular membranes and
inhibiting the activation of bradykinin
(polypeptide produced by the action of
enzymes on protein). It has an elective
concentration in the seat of an inflammation and a strong effect against
oedemas. It is used in the treatment of
painful and non-painful phlogistic states
of a non-infectious nature and as a
coadjuvant in the chemotherapy of inflam
mations caused by bacteria and viruses.
The retention of salt and water, as well
as pyrosis may occur as side effects.
f féprazone
d Feprazon
i feprazone
e feprazona

1266 fermentation amyl alcohol
Fusel oil. Bad-smelling acrid oily liquid
obtained in small quantities as a bypro-
duct in alcoholic fermentation, e.g. of
potatoes, grain or molasses, and consist
ing of a mixture mainly of alcohols, e.
g. isobutyl, active amyl, isopentyl etc.
Also known as fusel oil and potato spirit.
f alcohol amylique brut; huile de fusel
d roher Amylalkohol; Fuselöl
i olio di fusel; alcool amilico greggio
e aceite de fusel; alcohol amílico en bruto

1267 ferric acetate
The normal acetate of iron which is used
in medicine as a tonic.
f acétate ferrique
d Ferriacetat
i acetato ferrico
e acetato férrico

1268 ferric ammonium citrate
Complex salt containing varying amounts
of iron, either as red crystals or as

brownish-yellow powder, or as green
crystals of powder. Both varieties are
used in medicine for the treatment of
anaemia due to iron deficiency.
f citrate ferrico-ammoniacal
d Eisenammoniumcitrat
i citrato di ferro ammoniacale
e citrato de hierro amoniacal; citrato férrico amoniacal

1269 ferric ammonium sulphate
Ammonioferric alum, a crystalline compound marked by a violet colour and
used, like ferric chloride, as an astringent and styptic agent.
f alun ferrique ammoniacal
d Ferriammoniumsulfat; Ammoniumeisenalaun
i solfato ferrico d'ammonio
e sulfato férrico-amónico

1270 ferric chloride
Deliquescent salt obtained in anhydrous
form as a dark crystals which form seve
ral crystalline hydrates. It is used in
medicine in a water solution or tincture
as an astringent or styptic agent.
f chlorure ferrique
d Eisenchlorid; Ferrichlorid
i cloruro ferrico
e cloruro férrico

1271 ferric glycerophosphate
Iron glycerophosphate, a therapeutic pre
paration consisting of ferric hydroxide
and glycerophosphoric acid and used for
its tonic properties.
f glycérophosphate ferrique
d Eisenglycerophosphat
i glicerofosfato ferrico
e glicerofosfato férrico

1272 ferric hydroxide
Hydrated form of iron oxide, a reddish-
-brown gelatinous precipitate which,
mixed with a suspension of magnesium
oxide, is used as an antidote for arsenical poisoning.
f hydroxyde ferrique
d Ferrihydroxyd; Eisenhydroxyd
i idrossido ferrico
e hidróxido férrico

1273 ferric valerate
Or ferric valerianate. A salt, insoluble
in water, which has been used in the
past in the treatment of anaemia.
f valérianate ferrique
d Eisenvalerianat
i valerianato ferrico
e valerianato férrico

1274 ferricyanide
A salt of ferrocyanic acid used as an
antidote in the treatment of acute,
subacute and chronic intoxication caused
by thallium. The agent is also used to
reduce the biological half-life of caesium
radioactive isotopes in patients who
have been exposed to them.
f ferrocyanure

d Ferrocyanid
i ferrocianuro
e ferrocianuro

1275 ferritin
Iron-protein complex (it contains iron
in its trivalent form) which plays a
very important part in the transport,
absorption and storage of iron. It is
particularly indicated in the treatment
of idiopathic hypochromic anaemia of
pregnancy, hypochromic anaemia of infancy and puberty, anaemia caused by
intoxications and anaemia caused by deficiency.
f ferritine
d Ferritin
i ferritina
e ferritina

1276 ferrocarbonate
Therapeutic agent indicated in the treat
ment of anaemia due to deficiency of
iron in the blood, as well as anaemia
occurring during prolonged convalescence.
Also used in the prophylaxis of infections.
f ferrocarbonate
d Ferrokarbonat
i ferrocarbonato
e ferrocarbonato

1277 ferrocholinate
Therapeutic agent used in the treatment
of anaemia caused by iron deficiency
and of primary and secondary anaemic
syndrome. Intestinal disorders may occur
as a side effect.
f ferrocholinate
d Ferrocholinat
i ferrocolinato
e ferrocolinato

1278 ferroglycinate
Therapeutic agent used in the treatment
of nutritional anaemia, as well as anaemia caused by some infectious disease,
hypermenorrhoea, pregnancy, defective
absorption, acute and chronic haemorrhages.
f ferroglycinate
d Ferroglycinat
i ferroglicinato
e ferroglicinato

1279 ferromaltose
Ferric preparation used for intramuscular iron therapy for its slow and gradual release of trivalent iron. It is
recommended for all hypochromic iron
deficiency anaemias caused by nutritional inadequacy, pregnancies, haemorrhagies and various chronic diseases.
The drug should not be administered in
the presence of haemochromatosis and
hyperferric anaemias.
f ferromaltose
d Ferromaltose
i ferromaltosio
e ferromaltosa

1280 ferropolysaccharate
Antianaemic agent used as a source of
easily absorbable iron in the treatment
of anaemias caused by iron deficiency.
Gastrointestinal disorders may occur as
side effects.
f ferropolysaccharate
d Ferropolysaccharat
i ferropolisaccarato
e ferropolisacarato

1281 ferrous aspartate
A ferrous salt usually indicated in case
of iron deficiency and hypoferric anaemia independently of its nature. It is
administered orally and very easily
absorbed. The drug is not widely used
as it causes gastrointestinal inflammation accompanied by vomiting and diarrhoea.
f aspartate ferreux
d Eisenaspartat
i aspartato ferroso
e ferroaspartato

1282 ferrous gluconate
Haematinic agent indicated for the treatment of secondary iron deficiency anaemia as it increases the haemoglobin rate
and improves cellular respiration. It is
also useful to combat certain forms of
cephalalgia and some dermatoses. A
prolonged administration may cause
haemosiderosis.
f gluconate ferreux
d Ferroglukonat
i gluconato ferroso; ferro gluconato
e gluconato ferroso

1283 ferrous sulphate
Therapeutic agent with an antianaemic
action mainly used in the treatment of
hypochromic and hypoferric anaemias.
The drug is also used in the treatment
of some dermatoses and nervous disorders.
f sulfate ferreux
d Ferrosulfat
i solfato ferroso
e sulfato ferroso

1284 fibracillin
Broad-spectrum, semisynthetic penicillin
active on Gram-positive and Gram-negative bacteria. It is particularly used in
the treatment of infections of the upper
and lower respiratory tract, tonsillitis,
sinusitis and otitis.
f fibracilline
d Fibracillin
i fibracillina
e fibracilina

1285 fibrin
Insoluble protein formed from the soluble
protein of blood plasma fibrinogen by
the action of the enzyme thrombin. As
a therapeutic agent, fibrin acts as a
topical haemostatic drug. Once left in
situ it is completely absorbed without

causing any inflammation. It is often employed in neurosurgery, hepatic, pulmonary and cardiac surgery and in surgical operations on ear, nose, throat and teeth.
f fibrine
d Fibrin
i fibrina
e fibrina

1286 fibrinogenum humanum cryodessicatum
Dried human fibrinogen. Lyophilic preparation completely sterile and formed by the soluble fraction of the human plasma, which, by adding thrombin, is transformed into fibrin. It is thus a haemostatic agent capable of promoting blood coagulation. It is used to treat thrombocytopenic and fibrinogenopenic haemorrhagic syndromes, various forms of haemophilia etc. Hepatotoxicity may set in as a side effect of the treatment.
f fibrinogène humain lyophilisé
d Fibrinogenum humanum cryodessicatum
i fibrinogeno umano liofilizzato
e fibrinógeno humano liofilizado

1287 fibrinolysin (human)
A proteolytic enzyme capable of dissolving the blood clots through the proteolysis (the process of hydrolysis of proteins to soluble degradation products) of the fibrin network. It is used in the treatment of phlebothrombosis, thrombophlebitis, pulmonary embolism and more generally arterial thrombosis with the exception of cerebral and coronary thrombosis. Some febrile and allergic reactions have been observed as side effects.
f fibrinolysine (humaine)
d Fibrinolysin (humanes)
i fibrinolisina (umana)
e fibrinolisina (humana)

1288 filicin
The mixture of ether-soluble acidic substances which are found in the male fern and whose prolonged use can cause poisoning.
f filicine; acide filicique
d Filixin; Filixsäure
i filicina
e ácido filícico; filicina

1289 fipexide
A stimulant used in the treatment of physical, psychic, mental and neurotic asthenia, especially when the latter is caused by depression in elderly patients. Intermittent and recurrent loss of memory in alcoholics is also treated with fipexide. Anorexia and some insomnia may occur as side effects.
f fipexide
d Fipexid
i fipesside
e fipexida

* **fivefinger root**   s. ginseng

1290 flavianic acid
Yellow crystalline acid mainly used in precipitating arginine, tyrosine or organic bases. It finds employment in medicine as an antiseptic.
f acide flavianique
d Flaviansäure
i acido flavianico
e ácido flaviánico

1291 flavone
A phenylchromone derivative. Specifically, a general term for various polyhydroxy and methoxy derivatives of this chemical compound which are available in the form of glycosides and have the property of strongly reducing the fragility of the capillary blood vessels. The action is more prolonged than that of khellin (visammin, $C_{14}H_{12}O_5$), q.v. and is thus exploited in the treatment of vascular spasms associated with coronary disorders, visceral and ureteral spasms and acute and chronic bronchial asthma.
f flavone
d Flavon
i flavone
e flavona

1292 flavoxate
Specific antispasmodic agent used mainly in bladder disorders and particularly in the symptomatic treatment of primary dysuria (the condition whereby the passing of urine is difficult and/or painful), as well as dysuria caused by cystitis and urethritis. The drug also finds employment in the treatment of the painful conditions which frequently follow prostatectomy. Headache and dryness of the mouth may occur as adverse effects.
f flavoxate
d Flavoxat
i flavossato
e flavoxato

1293 fleawort seed
The seed of the Plantago psyllium which swell and become gelatinous when moist. A lotion obtained from the seeds has a mild laxative effect.
f graine de plantain; psyllium
d Flohsamen; Heusamen
i seme di conizza
e semilla de zaragatona

1294 flecainide
Therapeutic agent used in the treatment of arrhythmias, in particular when the irregular heart rhythms reach a level at which life is threatened. It may cause hypotension and some depression of the central nervous system.
f flécaïnide
d Flecainid
i flecainide
e flecainida

1295 Fleming's tincture of aconite
A strong aconite tincture produced for

external use and employed in the treatment of neuralgia.
f teinture d'aconit forte
d starke Akonittinktur
i tintura di aconito forte
e tintura de acónito de Fleming

1296 flopropione
Spasmolytic agent acting on the smooth musculature and used for the relief of painful spasms caused by intestinal and biliary disorders. It is a cholagogue (promoting the flow of bile into the duodenum) and capable of regulating intestinal motility.
f flopropione
d Flopropion
i flopropione
e flopropiona

1297 florantyrone
Hydrocholeretic agent used in the treatment of mild hepatic insufficiency and as a prophylactic in gallbladder dysfunction and biliary dyskinesia. It is also used to accelerate convalescence after an operation on the biliary tract.
f florantyrone
d Florantyron
i florantirone
e florantirona

1298 floredil
Therapeutic agent capable of activating and stabilizing mitochondrial enzymes. The drug is used in the treatment of ischaemic coronaropathies.
f florédil
d Floredil
i floredile
e floredilo

1299 floverine
Papaverine-related spasmolytic agent with specific action on the intestinal smooth muscles and on the sphincter of Oddi. It is chiefly used for the treatment of cephalalgia caused by bad digestion, biliary dyscinesia, cholecistitis with or without bile stones, disorders due to cholecistectomy and inflammation of the sphincter of Oddi (odditis).
f flovérine
d Floverin
i floverina
e floverina

1300 floxuridine
Therapeutic agent (5-fluorouracil-derivative) used for the palliative treatment of a cancer which cannot be operated on and/or be treated with any other antineoplastic drug. Side effects may be very numerous, e.g. leukopenia, thrombo cytopenia, gastrointestinal cramps, etc.
f floxuridine
d Floxuridin
i flossuridina
e floxuridina

1301 fluanisone
A tranquillizer chiefly used in the treatment of acute and chronic psychoneuroses characterized by states of anxiety and insomnia. It also finds employment in surgery when used in association with analgesic. Hypotension and tachycardia are possible side effects.
f fluanisone
d Fluanisone
i fluanisone
e fluanisona

1302 fluazacort
Term indicating an antiphlogistic corticosteroid for topical use. It has practically no systemic effects and is indicated in the treatment of cutaneous diseases that respond to a local therapy, e.g. eczemas, inflammatory and allergic dermatitis, psoriasis, intertrigo, neurodermatitis itching etc. The drug should be used very cautiously during pregnancy.
f fluazacort
d Fluazacort
i fluazacort
e fluazacort

1303 flucloxacillin
Antibiotic. A penicillase-resistance agent active against most Gram-positive and Gram-negative microorganisms.
f flucloxacilline
d Flucloxacillin
i fluclossacillina
e flucloxacilina

1304 flucytosine
Antifungal agent particularly active as an antimetabolite on Candida albicans, Aspergillus fumigatus etc., as well as on the agents of the chromoblastomycosis. It is therefore used in the treatment of systemic infections caused by sensitive fungi, e.g. septicaemic, diffuse and chronic candidiasis, granulomatous cryptococcosis etc. Vomiting, diarrhoea, nausea, leukopenia and thrombocytopenia are possible side effects.
f flucytosine
d Flucytosin
i flucitosina
e flucitosina

1305 fludrocortisone
Synthetic corticosteroid hormone characterized by a very potent mineralocorticoid action and an appreciable glucocorticoid effect. As a topical preparation, it is used in the treatment of inflammatory and allergic cutaneous diseases, e. g. contact dermatitis, seborrhoeic (with hypersecretion of sebum) dermatitis, neurodermatitis, eczema, psoriasis, intertrigo (erythema of mucocutaneous surfaces), etc. Hypopigmentation, some burning and inflammation as well as topical dryness of skin may occur as side effects.
f fludrocortisone

d Fludrocortison
i fludrocortisone
e fludrocortisona

1306 fludroxycortide
Flurandrenolone acetonide. Steroid drug having an action which is similar to that of topical corticosteroids and used in the treatment of inflammatory and allergic cutaneous diseases, e.g. seborrhoeic dermatitis, psoriasis, intertrigo etc. Burning, dryness, hypopigmentation and local atrophy of the skin may occur as side effcts.
f flurandrénolone (acétonide); fludroxycortide
d Fludroxycortid; Flurandrenolon
i flurandrenolide; fludrossicortide
e fludroxicortida

1307 flufenamic acid
Antiphlogistic, antirheumatic and analgesic agent used in the treatment of chronic polyarthritis, in particular of painful cervical and lumbar arthritis. The drug is not a corticosteroid and should not be used in pregnancy, gastritis and in presence of gastric ulcer. It is not indicated for children.
f acide fluFénamique
d Flufenaminsäure
i acido flufenamico
e ácido flufenámico

1308 flugestone
Or fluorogestone. Progestational steroid.
f flugestone
d Flugeston
i flugestone
e flugestona

1309 flumedroxone
Fluororate derivative of progesterone indicated in the therapy of cephalalgia and premenstrual nervousness. It is also used for the prophylaxis of the same disorders. Some pains in the breast and hepatic dysfunction may occur as side effects.
f flumédroxone
d Flumedroxon
i flumedrossone
e flumedroxona

1310 flumequine
Antiseptic of the urinary tract mainly indicated in the treatment of cystitis, pyelonephritis and, more generally, infections of the same tract due to urinary lithiasis or a surgical operation. Nausea, intensified peristalsis and intestinal disorders in general are possible side effects.
f fluméquine
d Flumequin
i flumequina
e flumequina

1311 flumethasone
A topical corticosteroid used for the treatment of inflammatory and allergic cutaneous diseases, particularly in eczemas and psoriasis. Occlusive dressings should be avoided in case of topical application owing to the danger of systemic toxicity. Hypopigmentation, dryness and local skin atrophy are possible side effects.
f fluméthasone
d Flumethason
i flumetasone
e flumetasona

1312 flunarizine
Peripheral vasodilator indicated in the treatment of peripheral vasculopathies, e.g. intermittent claudication, varicose ulcers, paresthaesias, vertigo, Menière's syndrome, nausea, nystagmus and cinesia (the feeling of nausea which is caused by motion, e.g. seasickness).
f flunarizine
d Flunarizin
i flunarizina
e flunarizina

1313 flunisolide
Potent synthetic corticosteroid with antiphlogistic and antiallergic activity. It is mainly used in the treatment of allergic rhinitis by nasal spray. Nasal irritation and dripping and reddening of the throat are frequent side effects. The drug should not be used in the presence of nasal infections due to fungi or bacteria.
f flunisolide
d Flunisolid
i flunisolide
e flunisolido

1314 flunitrazepam
Fluorurate derivative of the benzodiazepine having hypnotic and sedative action. As an anxiolytic and muscle-relaxant agent, it is used in all forms of insomnia. Even a small dosis (1–2 mg) of the drug alters appreciably the EEG up to 18 hours after ingestion, which obviously requires particular precaution in its administration.
f flunitrazépam
d Flunitrazepam
i flunitrazepam
e flunitrazepam

1315 fluochlorone acetonide
Glucocorticoid indicated for the topical treatment of some skin diseases, e.g. dermatosis, psoriasis, seborrhoic eczemas etc. Hypopigmentation, dryness and local skin atrophy may occur as side effects.
f acétonide de fluochlorone
d Fluochloronacetonid
i fluoclorone acetonide
e acetonida de fluoclorona

1316 fluocinolone acetonide
A topical corticosteroid used in the treatment of inflammatory and allergic derma

toses, contact dermatitis, seborrhoeic dermatitis, intertrigo etc. The drug must be used cautiously in pregnancy. Occlusive dressings must be avoided owing to the danger of systemic toxicity.
f acétonide de fluocinolone
d Fluocinolonacetonid
i fluocinolone acetonide
e acetonida de fluocinolona

1317 fluocinonide
Topical corticosteroid characterized by an antiinflammatory action and used in the therapy of dermatoses, in particular eczema and psoriasis. The drug is highly liposoluble and has a low surface tension, thus increasing the possibility of cellular penetration. Hypersensitivity reactions may occur following a prolonged administration.
f fluocinonide
d Fluocinonid
i fluocinonide
e fluocinonida

1318 fluocortin
Corticosteroid agent marked by antiphlogistic properties and used topically in the treatment of eczema (vulgar, degenerative, seborrhoeic and nummular), dermatitis affecting the newborn, neurodermatitis, psoriasis, chronic lupus erythematosus, 1st degree burns and insect bites. Occlusive bandages should be avoided in the application whenever mycosis and folliculitis have been diagnosed.
f fluocortine
d Fluocortin
i fluocortina
e fluocortina

1319 fluocortolone
A topical corticosteroid characterized by a chiefly glucocorticoid action and a negligeable catabolizing action on proteins, as well as hardly any effect on the mineral balance of the organism. It is used orally for the treatment of rheumatic and allergic diseases and, more generally, in all pathological conditions of internal organs, in dermatology and ophthalmology requiring a systemic therapy based on corticosteroid agents. Owing to its numerous adverse effects, the drug should be used cautiously and the treatment with it carefully monitored.
f fluocortolone
d Fluocortolon
i fluocortolone
e fluocortolona

* **fluopromazine   s. triflupromazine**

1320 fluorescein
Eosin-related orange-coloured dye obtained from phthalic anhydride and resorcinol. It is used in medicine as a pH indicator and in ophthalmology.
f fluorescéine

d Fluorescein
i fluoresceina
e fluoresceina

1321 fluoresone
A tranquillizer, analgesic and sedative agent used in the treatment of spastic conditions of the gastrointestinal tract, of peptic ulcers and, more generally, neurovegetative disorders.
f fluorésone
d Fluoreson
i fluoresone
e fluoresona

1322 fluoroacetic acid
Fluorine-containing organic acid distinguished into mono- bi- and tri- fluoroacetic acid according to the number of fluorine atoms present in it. The monofluoroacetic acid is a highly toxic liquid for all animals and in the form of sodium salt or methyl ester one of the most potent poisons known to man as it is capable of blocking one of the most important biochemical process, the tricarboxylic acids cycle.
f acide fluoroacétique
d Fluoressigsäure
i acido fluoroacetico
e ácido fluoroacético

1323 fluorohydrocortisone
Synthetic cortisonic agent whose action has proved more potent than that of cortisone itself.
f fluorohydrocortisone
d Fluorhydrocortison
i fluoroidrocortisone
e fluorohidrocortisona

1324 fluoromethalone
Potent synthetic corticosteroid to be used topically in the treatment of inflammatory and allergic dermatitis. The drug also finds employment in non-infectious ocular inflammations, e.g. acute and chronic iritis, scleritis, keratitis and severe conjunctivitis. A burning sensation may occur as side effect and affect both skin and eyes.
f fluorométhalone
d Fluoromethalon
i fluorometalone
e fluorometalona

1325 fluorouracil
Antineoplastic agent which operates by blocking the biosynthesis of the nucleic acids. It is thus used as a palliative in the treatment of solid mammary tumours as well as tumours of the colon or the rectum. As the drug is bound to cause anaemia, it should not be administered to patients suffering from it. Leukopenia, thrombocytopenia and sometimes intestinal lesions have occurred as adverse effects.
f fluorouracile
d Fluorouracil

i fluorouracile
e fluorouracilo

1326 fluoxymesterone
Androgenic and anabolic steroid used in the male to treat impotence, hypogonadism and prostatic hypertrophy, and in the female to treat mammary tumours and diseased conditions of the mammary glands. In addition to water retention, the drug may cause hirsutism in women.
f fluoxymestérone
d Fluoxymesteron
i fluossimesterone
e fluoximesterona

1327 flupenthixol
A tranquillizer agent with an antidepressant and anxiolytic action but little sedative effect. It is used in the treatment of depressive and anxiety states, as well as affective disorders. Insomnia, hypotension and extrapyramidal disturbances may occur as side effects.
f flupenthixol
d Flupenthixol
i flupentissolo
e flupentixolo

1328 fluperolone
Glucocorticoid agent indicated for the treatment of inflammatory and allergic skin diseases, e.g. contact dermatitis, psoriasis, eczema etc. Hypopigmentation, dryness and topical atrophy of the skin may occur as side effects.
f flupérolone
d Fluperolon
i fluperolone
e fluperolona

1329 fluphenazine
A phenothiazine-related major tranquillizer capable of acting on the subcortical area, i.e. the reticular substance. The drug is indicated in the treatment of various psychoses as an anxiolytic, especially in the manic forms. It is also characterized by a marked antiemetic action which makes it very effective and useful when it is necessary to combat and check nausea and vomit after a surgical operation.
f fluphénazine
d Fluphenazin
i flufenazina
e flufenacina

1330 fluprednisolone
Corticosteroid characterized by a glucocorticoid action and indicated in the therapy of rheumatism, phlogistic and allergic skin diseases, collagenosis, traumatic ophthalmopathies, both inflammatory and allergic, acute and chronic leukaemia, adrenocortical insufficiency, Hodgkin's disease, multiple sclerosis etc. Dyspepsia, sometimes gastrointestinal ulcers and haemorrhages, acne, hirsutism and various other side effects may occur in the course of the treatment.
f fluprednisolone
d Fluprednisolon
i fluprednisolone
e fluprednisolona

1331 fluprednylidene
Or fluprednidene. A topical corticosteroid used for the treatment of inflammatory and allergic skin diseases, e.g. contact dermatitis, eczema, psoriasis, intertrigo etc. Hypopigmentation and atrophy of the skin may occur as side effects.
f fluprednidène
d Fluprednyliden
i fluprednidene
e fluprednideno

1332 flurazepam
Benzodiazepine characterized by a specific hypnotic action and capable of reducing the latency period (the time between the application of a stimulus and the resulting effect) and normalizing the required sleep duration. There is no reaction whatsoever on awakening. The drug is indicated in all forms of insomnia, but its administration must be carefully monitored as side effects are multiple and may be serious.
f flurazépam
d Flurazepam
i flurazepam
e flurazepam

1333 flurbiprofen
Antiphlogistic, analgesic and antipyretic agent capable of inhibiting the synthesis of prostaglandin. It is mainly used in the treatment of osteoarthrosis, rheumatism and ankylosing spondylitis (poker spine). Dyspepsia and pyrosis are probable side effects.
f flurbiprofène
d Flurbiprofen
i flurbiprofene
e flurbiprofeno

1334 flurothyl
Convulsion-producing drug indicated in the treatment of grave depression and of some forms of schizophrenia. Amnesia, arrhythmia and sometimes vertebral fractures have been observed as side effects.
f flurothyle
d Flurothyl
i flurotile
e flurotilo

1335 fluroxene
Trifluoroethyl vinyl ether. Volatile anaesthetic ether with a rapid but short-lasting action. The drug may cause some tendency to infections and should not be used in patients suffering from severe cardiopathies.
f fluroxène
d Fluroxen
i flurossene
e fluroxeno

1336 fluspirilene
Strong tranquillizer indicated in the treatment of acute and chronic psychoses of a schizophrenic nature. Involuntary movements, extrapyramidal dysfunctions and low blood pressure may occur as side effects.
f fluspirilène
d Fluspirilen
i fluspirilene
e fluspirileno

1337 flybane
Any of several plants which are used for the destruction of houseflies.
f conise; herbe aux mouches
d Mückenkraut; Dürrwurz
i conizza; baccherina
e coniza

1338 folescutol
A capillary-protecting agent used in the therapy of vascular disorders in the course of pregnancy, menopause, premenstrual syndromes, the effects of oral contraceptives, oedemas, haemorrhoids, retinopathies and nephropathies. More generally, it is indicated for the treatment of capillary fragility.
f folescutol
d Folescutol
i folescutolo
e folescutolo

1339 folic acid
Vitaminic factor which is effective in the therapy of some anaemias of an alimentary origin and, though partially, of pernicious anaemia. The acid was first isolated from the spinach leaves, but later identified in many green plants, in certain fungi, in yeasts, microorganisms and animal tissues. The acid is presently produced synthetically and used for the manufacture of several anti-anaemic preparations.
f acide folique
d Folsäure; Blattsäure
i acido folico
e ácido fólico

1340 folinic acid
A chemical compound; colourless crystals not easily soluble in water, which can be obtained by the catalytic reduction of the folic acid. Its therapeutic properties are analogous to those of the folic acid.
f acide folinique
d Folinsäure
i acido folinico
e ácido folínico

1341 fominoben
Respiratory analeptic agent indicated in the treatment of chronic cough, bronchitis, emphysema, right atrio-ventricular valve insufficiency, difficulty in breathing due to bronchopneumopathic disorders and hypoxaemia occurring in the course of a surgical operation or in childbirth.
f fominobène
d Fominoben
i fominobene
e fominobeno

1342 formaldehyde
The simplest of the aldehydes; a colourless gas, very irritating, which is obtained by passing the vapour of methyl alcohol over a heated copper catalyst. Formalin is the aqueous solution formed in this way. Formaldehyde has the property of making proteins insoluble and consequently immune from bacterial decomposition. Being in itself far too toxic and corrosive, formaldehyde is used in the form of condensation products, mostly with carbohydrates which hydrolyze very slowly in the body. As a therapeutic drug, it is particularly useful in the treatment of throat infections.
f formaldéhyde; méthanol; méthylaldéhyde
d Formaldehyd; Oxymethylen
i formaldeide; aldeide formica
e formaldehido; aldehido fórmico

1343 formalin
The formaldehyde aqueous solution which is obtained by passing the vapour of methyl alcohol and air over a heated copper catalyst. It is too toxic and too corrosive to be used as an antiseptic or a disinfectant and therefore employed in the form of condensation products. It is chiefly used for the conservation of anatomical parts or organic material. Also known as formaldehyde solution or formol.
f formol
d Formalin; Formaldehydlösung
i formalina
e formalina

1344 formamide
The amide of formic acid, which is closely related to urea. It is the only amide that is not solid at normal temperature and, as a slightly viscous liquid, it boils at 20°C. It is used as a solvent.
f formamide
d Formamid; Ameisensäureamid
i formammide
e formamida; ámida fórmica

1345 formebolone
Anabolic steroid with myotrophic action, capable of stimulating protein phosphorus, calcium anabolism and oxidoreduction processes. The drug is chiefly used in asthenia, grave loss of weight, conditions of debility in the course of convalescence, osteoporosis and delayed fusing of fractures. It is also indicated for the promotion of growth in underdeveloped children, increased catabolism and cachexia, i.e. the very grave state of general ill health marked by anaemia, wasting and muscular weakness.

f formébolone
d Formebolon
ı formebolone
e formebolona

1346 formic acid
An acid whose name is derived from the
Formica rufa (red ant) in which it was
identified in the blood of animals in the
17th century. It is now prepared by the
oxidation of methyl alcohol or formaldehyde and by partial decomposition of
oxalic acid. It acts as a strong irritant
when injected through the sting of ants,
bees and nettles and is used in the
treatment of influenza and rheumatism,
as well as to reduce muscular tremor.
f acide formique
d Ameisensäure
ı acido formico
e ácido fórmico

1347 formocortal
Fluximesterone. Antiphlogistic steroid
with an action that is a 100 times more
potent than that of hydrocortisone. The
drug is metabolized and eliminated even
if it has been absorbed and is generally
used for the treatment of inflammatory
and allergic skin diseases.
f formocortal
d Formocortal
ı formocortal
e formocortal

1348 formocresol
A mixture of tricresol and formalin (in
equal parts) which is inserted in the
root canal of a tooth in order to combat
the infection which is caused by necrosis of the pulp.
f formocrésol
d Formocresol
ı formocresol
e formocresol

1349 formosulphathiazole
Or formosulphacetamide, formosulphaben-
zamide; a substance which is hardly
soluble and thus badly and inadequately
absorbed by the intestines. It breaks
down very slowly in them to formaldehyde and sulphathiazole and exerts a
very potent antibacterial action. It is
thus indicated in the treatment of enteritis and enterocolitis due to bacterial
flora. It has also been used in the
treatment of cholera but without a truly
positive benefit.
f formosulfathiazole
d Formosulfathiazol
ı formosulfatiazolo
e formosulfatiazolo

1350 formyl
Monovalent radical (HCO-) which can be
obtained from formaldehyde or formic
acid and is present in many compounds,
e.g. the formyl fluoride (HCOF), a colourless liquid that is dangerous as it libe-

rates hydrofluoric acid and carbon oxide
on decomposition. It is used in organic
syntheses and considered as chemical
aggressive agent.
f formyle
d Formyl
ı formile
e formilo

1351 foscellulose
An ion-exchange (the phenomenon whereby labile positive or negative ions in
solution may replace ions of similar
charge in materials with which the solution comes into contact) agent capable
of exerting an inhibiting action on calcium excretion and assimilation. It is
chiefly used in the prophylaxis and the
therapy of calcium salts urolithiasis in
case of idiopathic hypercalciuria, sarcoidosis, intoxication caused by vitamin
D and in idiopathic hypercalcaemia affecting children. A prolonged administration may cause a negative calcium balance.
f foscellulose
d Foszellstoff
ı foscellulosa
e foscelulosa

1352 fosfestrol
A synthetic phosphorated oestrogen used
in the treatment of cancer, in particular
of prostatic cancer. The acid phospha-
tases in the cancerous cells of the prostata promote the liberation of active
diethyldihydroxystilbene. Gynaecomastia
and testicle hypoplasia may occur as
side effects.
f fosfestrol
d Fosfestrol
ı fosfestrolo
e fosfestrolo

1353 Fowler's solution
Or potassium arsenite solution. Solution
of arsenic trioxide in potassium hydroxide, neutralized with hydrochloric acid.
It is used whenever administration of
arsenic is indicated.
f solution de Fowler; soluté d'arsénite de
potasse
d Fowlersche Lösung; Kaliumarsenitlösung
ı liquore arsenicale del Fowler; arsenito
di potassio
e liquor arsenical de Fowler; solución de
arsenito potásico

1354 framycetin
Antibiotic derived from neomycin and
used topically against staphylococci
(Gram-positive cocci). It is used in the
treatment of skin infections and, if
taken orally, of pulmonary pus-exuding
or producing conditions, acute arthritis,
fistulas (unnatural communications between an organ and the body surface or
between an organ and another) caused
by acute or chronic osteitis and otor-
rhoea. The drug is also indicated for

gastro-enteritis and bowel sterilization.
f framycétine
d Framycetin
i framicetina
e framicetina

1355 frangula
A shrub growing practically all over Europe. When fresh it displays emetic properties, thus making it necessary to let it mature for at least one year before using it for therapeutic purposes. The frangula contains anthraquinone derivatives and has a cascara-like cathartic action. It is generally administered in the form of a liquid extract.
f frangule
d Faulbaum (rinde)
i frangola
e frángula

1356 franguline
A rhamnoside derived from a trihydroxy-methylanthraquinone present in the seeds, bark and roots of various species of the genus Rhamnus. Orange-colours, water--insoluble crystals which are easily hydrolyzed with acids yielding emodin and rhamnose. It has a purgative action. Also known as rhamnoxanthine.
f franguline; cascarine; rhamnoxanthine
d Frangulin; Cascarin
i frangulina; cascarina
e frangulina; ramnoxantina

1357 fringe tree
Poison tree, snowdrop tree. Small tree or shrub growing in the southern region of North America, whose bark used to be dried and used as a diuretic.
f arbre à franges; arbre à neige
d Schneeflockenbaum
i chionanto
e quionanto; árbol de nieve

1358 fructose
Fruit sugar, laevulose. Ketohexose isomer with D-glucose and found in nature in sweet fruits, in honey and in the form of inulin, as well as in dahlia tubers and in the Helianthus tuberosus. It is also obtained by the inversion of cane sugar. It is used therapeutically in the treatment of diabetes. Fructose is also secreted by the seminal vesicle and is present in semen.
f fructose; lévulose; sucre de fruit
d Fruktose; Fruchtzucker
i fruttosio; levulosio
e fructosa; azúcar de frutas

1359 fuchsine
Bright green water-soluble crystals used as a stain for Microbacterium tuberculosis. It is a mixture of pararosaniline and rosaniline hydrochloride and used as a potent germicide and in the treatment of varicose ulcers, burns, carbuncles and impetigo. The term itself includes a group of dyes, e.g. rosani-

line, employed for bacteriological and histological staining.
f fuchsine
d Fuchsin
i fucsina
e fucsina

1360 Fuller's earth
A good adsorbant agent used in poisoning caused by the weedkiller paraquat which it binds effectively. It may be administered orally and also, in particular cases, directly into the stomach by means of a naso-gastric tube. It is generally given with magnesium sulfate to provoke complete evacuation in the attempt to empty the bowels of the paraquat.
f terre à foulon; terre de Fuller
d Fullererde
i terra per folloni; terra di Fuller
e tierra de Fuller

1361 fumagillin
Antibiotic chemical compound produced by cultures of Aspergillus fumigatus, which has been used for the treatment of amoebiasis owing to its amoebicide property.
f fumagilline
d Fumagillin
i fumagillina
e fumagilina

1362 fumarate
The ester of the fumaric acid. Sodium fumarate is used in medicine for its laxative action.
f fumarate
d Fumarat
i fumarato
e fumarato

1363 fumaric acid
Ethylene dicarboxylic acid found in combination with cetraric acid in the lichen acid from Iceland moss and in Fumaria officinalis, isomeric form of maleic acid. An intermediate in the Krebs cycle.
f acide fumarique
d Fumarsäure
i acido fumarico
e ácido fumárico

1364 fumarine
Alkaloid present in opium and some other plants, especially Dicentra spectabilis. It is a compound of the isoquinoline group and finds employment owing to its anaesthetic properties. The drug causes convulsions of an epileptic character.
f fumarine; protopine
d Fumarin; Protopin
i fumarina
e fumarina

1365 fumigacin
An antibiotic, probably the lactone of the helvolic acid (an antibiotic secreted

by a parasitic mould) capable of acting against staphylococci. The drug has proved too toxic to be administered parenterally.
f fumigacine
d Fumigacin
i fumigacina
e fumigacina

1366 fumigatin
Therapeutic substance obtained from Aspergillus fumigatus having bactericidal properties. Chemically, hydroxymethoxy toluquinone.
f fumigatine
d Fumigatin
i fumigatina
e fumigatina

1367 fungicide agent
Any agent or substance capable of destroying fungi and their spores or of inhibiting their growth.
f fongicide
d pilztötendes Mittel; Pilzmittel
i fungicida
e fungicida

1368 furazabol
Steroid anabolic agent capable of lowering the level of cholesterol and triglycerides in the blood. It is indicated in the treatment of arteriosclerosis, hypercholesteraemia, disorders of the protein metabolism, osteoporosis, malnutrition, convalescence etc. Sodium retention and obstructive jaundice (a type of jaundice in which the excreted bile is reabsorbed into the blood through blood capillaries and lymphatics owing to obstruction in the bile duct) may occur as adverse effects. Menstrual disorders are also possible.
f furazabol
d Furazabol
i furazabolo
e furazabolo

1369 furazolidone
Antimicrobial agent particularly active against intestinal Gram-negative bacteria. It is poorly absorbed and mainly used in the treatment of gastro-enteritis and diarrhoea. Allergic reactions, nausea, vomiting rashes and haemolysis in certain patients may occur as side effects.
f furazolidone
d Furazolidon
i furazolidone
e furazolidona

1370 furethidine
Chemical compound used as a narcotic analgesic.
f furéthidine
d Furethidin
i furetidina
e furetidina

1371 furfenorex
Anorexia-inducing agent used in the treatment of simple obesity in the adult, as well as constitutional, post-partum and postmenopausal obesity. It is also indicated in the treatment of hypertension, some forms of arthrosis and in cases in which obesity is complicated by diabetes.
f furfénorex
d Furfenorex
i furfenorex
e furfenorex

1372 furoate
The salt or ester of the furoic acid.
f furoate
d Furoat
i furoato
e furoato

1373 furoic acid
Or pyromucic acid. Either of two crystalline monocarboxylic acids derived from furan (the liquid obtained from wood oils of certain pines) and chiefly used as a preservative.
f acide pyromucique
d Brenzschleimsäure
i acido furoico
e ácido furoico

1374 furonazide
Therapeutic agent whose toxicity is less potent than isoniazid, used in the treatment of tuberculosis, both pulmonary and extrapulmonary. Also indicated for the prevention of the disease. Allergic reactions and mild gastro-intestinal disorders may occur as side effects.
f furonazid
d Furonazid
i furonazide
e furonazida

1375 furosemide
A diuretic agent acting on the proximal and distal renal tubules and whose effect may be due to aldosterone antagonism but not to carbon anhydrase. The drug is indicated in the therapy of acute pulmonary oedema, cardiac and cerebral oedema, cirrhosis and an overdosage of digitalis-related therapeutic agents.
f furosémide; frusémide
d Furosemid
i furosemide; frusemide
e furosemida

1376 fursulthiamine
Thiamine-related liposoluble agent mainly used in the therapy of neuralgias, rheumatic pains, toxic and postoperative neuritis and, generally, myalgias. Anaphylactic reaction in case of parenteral administration may occur as a side effect.
f fursulthiamine
d Fursulthiamine

i fursultiamina
e fursultiamina

1377 fusafungine
Aerosol-administered antibiotic chiefly
for the treatment of infections of the upper respiratory tract caused by Gram-po
sitive and Gram-negative microorganisms.
The drug has also proved useful in
cases of inflammation. More generally,
it is used in acute and chronic bronchitis, influenza, sinusitis, laryngitis,
tonsillitis etc.
f fusafungine
d Fusafungin
i fusafungina

e fusafungina

1378 fusidic acid
An antibiotic obtained from cultures of
Fusidinium coccineum and active against
Gram-positive bacteria, especially the
pyogenic Staphylococcus resistant to oth
er antibiotics. It is indicated in the
therapy of septicaemic infections, osteomyelitis, abscesses of various types and
infected wounds. Dyspepsia has been
observed as a side effect.
f acide fusidique
d Fusidinsäure
i acido fusidico
e ácido fusídico

G

1379 galactose
Hexose ($C_6H_{12}O_6$), glucose, talose, mannose stereoisomer, known in two optically active forms, "d" and "l". The d-galactose is obtained by hydrolysis and occurs in the trisaccharide raffinose, in the cerebrosides of brain and nerve tissue and in the galactans of seaweeds and lichens. It is a white dextrorotatory substance, i.e. colourless exagonal crystals, less sweet than glucose and less water-soluble. It forms mucidic acid on fermentation.
f galactose
d Galactose
i galattosio
e galactosa

1380 galacturonic acid
Uronic acid deriving from galactose and existing in two forms, alpha and beta. It occurs naturally, combined with sugar, in mucilages and pectins of which it is a constituent.
f acide galacturonique
d Galacturonsäure
i acido galatturonico
e ácido galacturónico

1381 galangale
Or galingale. Pungent aromatic rhizome produced in Eastern Asia by plants which are related to the true gingers and were once used for therapeutic purposes.
f souchet long
d Zyperwurz; wilder Galgant
i cipero lungo; galanga
e juncia

1382 galanga oil
Essential oil obtained by the distillation of the rhizomes of Alpinia officinarum. It is a yellowish-green liquid characterized by an aromatic smell and a taste of camphor, soluble in alcohol and ether and principally constituted by cineole (oxide or ether of the menthene group of terpenes), pinene (terpene hydrocarbon) and eugenol (phenolic compound). It is used as an aromatic, stomachic and stimulant agent.
f essence de galanga
d Galgantöl
i olio di galanga
e aceite de galanga

1383 galanga root
The pungent aromatic rhizome that is related to the true ginger and used to produce the galanga oil.
f racine de galanga
d Galgantwurzel
i rizoma di galanga
e raíz de galanga

1384 galantamine
Or galanthamine, nivaline. A chemical compound extracted from Galathus woronowii. Colourless crystals not easily soluble in water and used, albeit very rarely, for the treatment of neuritis. Intrinsically, an anticholinesterase and analeptic agent.
f galantamine; nivaline
d Galantamin; Nivalin
i galantamina
e galantamina

1385 galbanum
Oleo gum resin obtained by means of incisions of the plants of the genus Ferula. It has a pale-yellow or reddish colour, a fennel-like balsamic odour and a bitter aromatic taste. It is used in medicine for the preparation of revulsive plaster, as well as a stimulant in bronchitis.
f galbanum
d Galbanumsaft; Mutterharz
i galbano
e gálbano

1386 galbanum oil
A yellow liquid characterized by a penetrating odour and containing pinene (terpene hydrocarbon) and cadinene (a dicyclic sesquiterpene containing a reduced naphthalene skeleton). It is used as a stimulant in bronchitis.
f essence de galbanum
d Galbanumöl
i olio essenziale di galbano
e esencia de gálbano

1387 galega
Small genus of tall perennial Eurasian herbs of the Leguminosae family. They are characterized by compound leaves and racemose flowers. The extract obtained from the plant, in particular the Galega officinalis, is used as a sudoriparous and galactagogue agent. Its alkaloid, the galegine, q.v., has proved

useful in the treatment of diabetes.
f galéga; rue des chèvres
d Geissraute; Bockskraut; Geissklee
i galega; avanese; lavanese; ruta capraia
e ruda cabruna

1388 galegine
A guanidin-derivative contained in the
leaves of the Galega officinalis. White,
hygroscopic water- and alcohol-soluble
crystals marked by a hypoglycaemic
capacity and therefore indicated in the
therapy of diabetes, particularly in
insulin-resistant cases.
f galégine
d Galegin
i galegina
e galegina

1389 galenic
Or galenical. Relating to pharmaceutical
preparations which contain one or more
organic ingredients having a mostly
complex and variable composition and
not always easy to analyze. Such preparations are generally obtained through
physical and mechanical operations
rather than through a chemical reaction,
e.g. creams, decoctions, syrups, tinc-
tures etc. More generally, a galenical
preparation is prepared on the strength
of a detailed prescription by the chemist
under his control and responsibility.
f galénique
d galenisch
i galenico
e galénico

1390 galingale
Or greater galangal, Indian galgant.
The rhizome of the Alpinia galanga obtained in cylindrical irregular chunks,
of a brown-reddish colour and an aromatic pungent taste. The odour is rather
pleasant. It is used in medicine as a
carminative agent.
f galanga de l'Inde; grand galanga
d grosser Galgant
i galanga maggiore
e galanga mayor

1391 gallamine triethiodide
Synthetic curare derivative used as a
relaxant of skeletal muscles. Though
less effective than tubocurarine, it is
appreciably less toxic in particular
with regard to respiration and blood
pressure. It is frequently used as a
coadjuvant in narcosis, in convulsive
shock therapy, tetanus, poisoning caused
by strychnin and paraplegia.
f triéthiodure de gallamine
d Gallamintriethiodid
i gallamina triiodoetilato
e triiodoetilato de galamina

1392 gallate
Salt of the gallic acid. Also an ester of
the gallic acid. Gallate is also a salt
of the gallium hydrate which, owing to

its amphoterous character with strong
bases, behaves like an acid.
f gallate
d Gallat
i gallato
e galato

1393 gallic acid
White crystalline acid found in several
plants both in the free form and combined in tannins from which it is obtained
by the action of moulds or alkalis. It
is used in medicine as an antiseptic in
the treatment of cutaneous diseases, as
well as a disinfectant.
f acide gallique
d Gallussäure
i acido gallico
e ácido gálico

1394 gallnut
Or nutgall. Term denoting the excres-
cences on the twigs of the oak caused
by the stimulation of the tissues by the
deposition of the eggs of the gall wasp
(a hymenopterous gallfly). Pharmaceutical preparations based on galls are
used for external application as astrin-
gents.
f galle de chêne; noix de galle
d Gallapfel; Eichapfel
i galla; noce di galla
e nuez de agalla

1395 gamma benzene hexachloride
Potent and very rapid insecticide, larvi-
cide and acaricide used in the form of
a dust, spray or smoke. Also used in
the treatment of acari infestation (topically), lice infestation and, generally,
skin parasites. Some allergic reaction
may occur as side effect.
f hexachlorure de gamma benzène
d Gammabenzenhexachlorid
i esaclorocicloesano
e hexaclorociclohexano

1396 ganglioside
Each of a group of glycolipides intimately related to the cerebrosides and
mainly present in the ganglion cells of
the nervous system and in increased
amounts in certain lipoidoses. Gangliosides can be extracted from bovine cerebral cortex and act as neurotrophic
agents as they stimulate the anatomical
and functional recovery of the nervous
fibres which may have been damaged
by toxic, traumatic or degenerative
causes. Gangliosides thus find employment in the treatment of acute or chronic nervous disorders of a central and
peripheral origin, which show some
alteration of the nerve excitability and/
or transmission.
f ganglioside
d Gangliosid
i ganglioside
e gangliosido

1397 garlic
Allium. The bulb of the garlic plant, which contains a glucoside together with a volatile oil derived from its hydrolysis and whose chief constituent is allyl sulphide. It is characterized by a penetrating strong odour and is used in medicine as a diuretic, diaphoretic and expectorant agent.
f ail
d Knoblauch
i aglio
e ajo

1398 garlic oil
The oil which is distilled from Allium sativum and contains allyl sulphide. It is used therapeutically as a stimulant of bronchial secretion in bronchitis.
f essence d'ail
d Knoblauchöl
i essenza di aglio
e esencia de ajo

1399 gefarnate
A mixture of stereoisomers of 3,7-dimethylocta-2,6-dienyl-5,9,13-trimethyltetradeca-4,8,12 trienoate. The drug, which is capable of protecting the gastroduodenal mucosa by a physiological mechanism, is used in the treatment of peptic ulcers, gastritis and enterocolitis. Cutaneous disorders of an allergic nature have been observed as side effects.
f géfarnate
d Gefarnat
i gefarnato
e gefarnato

1400 gelatin
The glutinous material which is obtained from animal tissues by prolonged boiling. Specifically, a transparent colloidal protein, either colourless or pale-yellowish, hard and brittle when dry but easily dissolving in hot water and forming a jelly on cooling. It finds various uses in medicine.
f gélatine
d Gelatine
i gelatina
e gelatina

1401 gelatination
The chemical process whereby the mass of a sol solidifies into a gelatinous state without separation of liquid.
f gélatinisation
d Gelatinierung
i gelatinizzazione
e gelatinización

1402 gelsemine
Crystalline alkaloid derived from the root of the yellow jasmine. It is used in medicine for its depressant effect on the central nervous system and on the heart, whereby it causes a considerable fall in the blood pressure. The gelsemine tincture, or tincture of gelsemium,

is used in the treatment of trigeminal neuralgia and headache.
f gelsémine
d Gelsemin
i gelsemina
e gelsemina

1403 gelseminine
Alkaloid isolated from the rhizomes of gelsemium sempervirens; white crystals, insoluble in water, soluble in alcohol and capable of an antineuralgic and antispasmodic action.
f gelséminine
d Gelseminin
i gelseminina
e gelseminina

1404 gemeprost
Synthetic prostaglandin capable of acting on the uterus to promote the delivery of the foetus. The drug is also used to prepare the uterus for the surgically performed termination of the pregnancy.
f géméprost
d Gemeprost
i gemeprost
e gemeprost

1405 gemfibrozil
Therapeutic agent capable of reducing the concentration of lipids in the blood, i.e. triglycerides and cholesterol. Should hyperlipidaemia be diagnosed, the drug requires a particular diet. Abdominal pains, diarrhoea, cephalalgia, a varying degree of impotence, pains in the limbs and sometimes blurred vision may occur as side effects.
f gemfibrozil
d Gemfibrozil
i gemfibrozil
e gemfibrozil

1406 gentamicin
Antibiotic produced by Micromonospora purpurea, whose action is effective against Gram-positive and Gram-negative bacteria, e.g. Escherichia coli, Staphylococcus aureus, Pseudomonas aeruginosa etc. Oral administration is not advisable. The drug is generally used in conjunction with ampicillin and kanamycin. Some allergic disorders may occur as side effects.
f gentamicine
d Gentamicin
i gentamicina
e gentamicina

1407 gentiamarin
Glycoside, either a $C_{16}H_{22}O_{11}$ or $C_{16}H_{20}O_{10}$, which is found in the roots of gentian, q.v.
f gentiamarine
d Gentiamarin
i genziamarina
e genciamarina

1408 gentian
The dried and fermented rhizome of the
Gentiana lutea, of the Gentianaceae
family, containing bitter glycosides,
sugars, enzymes and gentisic acid, q.v.
It finds use in medicine as a bitter
tonic chiefly to stimulate appetite and
gastric secretions.
f gentiane
d Enzian
i genziana
e genciana

1409 gentian root
The root of the Gentiana lutea which
contains bitter glycosides, some enzymes
and gentisic acid. It stimulates the
salivary and gastric secretions and
strengthens the peristaltic action of the
gastroenteric tract. It is thus capable
of combating anorexia and displaying a
tonic activity in dyspeptic conditions
and in convalescence. Its use as a febri
fuge is due to its association with chincona.
f racine de gentiane
d Enzianwurzel
i radice di genziana
e raíz de genciana

1410 gentisate
The salt of the gentisic acid. Sodium
gentisate is used in medicine in the
treatment of acute articular rheumatism.
f gentisate
d Gentisat
i gentisato
e gentisato

1411 gentiseine
Xanthone trioxyderivative; a crystalline,
needle-shaped substance which is obtained by treating gentisine with hydroiodic
acid.
f gentiséine
d Gentisein
i gentiseina
e gentiseina

1412 gentisic acid
Dioxybenzoic acid also known as genti-
sinic or hydroquinone-benzoic acid.
Colourless, water- and alcohol-soluble
crystals which are obtained, together
with fluoroglycine by fusion of the gentisein or by treating hydroquinone with
potassium bicarbonate in the presence
of potassium sulphide. It is used in
medicine for its analgesic and diaphoretic (perspiration-inducing) properties.
f acide gentisique
d Gentisinsäure
i acido gentisinico; acido gentisico
e ácido gentisínico; ácido gentísico

1413 gentisine
Methyl ether of the gentiseine; a yellow
pigment isolated from the root of gentian
obtained from yellow, needle-shaped
crystals which are soluble in water and

alcohol.
f gentisine
d Gentisin
i gentisina
e gentisina

1414 geraniol
Primary aliphatic unsaturated alcohol
present in several essential oils either
free or in the form of ester. It is an
oily, colourless liquid generally used
as an insect bait.
f géraniol
d Geraniol
i geraniolo
e geraniol

1415 geranium oil
An essential oil obtained by distillation
in a stream of steam of the fresh leaves
of several species of Pelargonium. An
oily, colourless or greenish liquid used
in medicine as an antiseptic.
f essence de géranium; essence de pélar-
gonium
d Geraniumöl
i olio di geranio; essenza di geranio
e esencia de geranio

1416 germander
A shrubby European perennial plant
characterized by bright rose or red-
-purple flowers which are used in medicine as stimulating and astringent
agents. Better known as wall germander.
f germandrée officinale; chasse-fièvre;
sauge amère
d echter Gamander; Edelgamander; Bathen
gel
i camedrio; germandria; querciola
e camedrio; germandrina

1417 gestodene
Sex hormone capable of acting as an
oral contraceptive if combined with an
oestrogenic agent. Its action is similar
to that of progesterone.
f gestodène
d Gestoden
i gestodene
e gestodeno

1418 gestonorone caproate
Long-acting progestational hormone
displaying an antigonadotropic activity
and having an inhibitory effect on the
pituitary-hypothalamic system. It is
used in the conservative (capable of
restoring function and preserving
health) treatment of prostatic adenoma
and endometrial carcinoma. A temporary
weakening of sexual potency has been
observed in male patients as a side
effect. Other adverse effects include a
possible and mild alteration of the blood
chemistry particularly in patients suffer
ing from liver diseases.
f caproate de gestonorone
d Gestonoroncaproat
i gestonorone caproato

e caproato de gestonorone

* **gestronol hexanoate** s. **gestonorone caproate**

1419 ghatti gum
Or indian gum. Gum obtained from the dhawa and related trees. It is used as a substitute for gum arabic in some therapeutic preparations.
g gomme ghatti
d indisches Gummi; Dhaura-Gummi
i gomma ghatti; gomma indiana
e goma ghatti

1420 ginger
The rhizome of Zingiber officinale, a reed-like plant that is cultivated in several tropical countries. The ginger used for pharmacological preparations is known as unbleached Jamaican ginger whose rhizome is properly cleaned to remove the corky surface and dried in the sun. It contains a pungent active principle, the gingerol which is used in medicine as a carminative and stimulant agent.
f gingembre
d Ingwer
i zenzero
e jengibre

1421 ginger-grass oil
Essential oil obtained from ginger grass, resembling palmarosa oil and characterized by an odour which is like that of the common ginger. It is used in medicine as a stimulant.
f essence de gingergrass
d Gingergrasöl
i essenza di gingergrass
e esencia de gingergras

1422 ginger oil
The essential oil that is obtained from ginger and contains phellandrene (a monocyclic terpene) and zingiberene (a monocyclic sesquiterpene), isomeric with bisabolene, the sesquiterpene found in lemon oil and pine-needle oil.
f essence de gingembre
d Ingweröl
i essenza di zenzero; olio di zenzero
e esencia de jengibre

1423 ginseng
The Panax ginseng, a perennial herb growing in North-West Asia, to whose roots a great deal of beneficial attributes and particularly tonic and aphrodisiac properties have been ascribed. In fact, some oestrogenous substances are found in the drug together with a saponine and other compounds. It finds some use in Europe and in America as a general and sexual tonic. It also appears that the drug offers some therapeutic advantages by increasing the resistance to disease, accelerating the convalescence periods and stimulating

the activity of vital centres and of the autonomic nervous system.
f ginseng
d Ginseng
i ginseng
e ginseng

1424 gitalin
Mixture of digitalis glucosides that is used in congestive heart failure, auricular flutter and rapid auricular fibrillation. More specifically, the glucoside which is found in the leaves and the seeds of the purple foxglove (Digitalis purpurea).
f gitaline
d Gitalin
i gitalina
e gitalina

1425 gitaloxine
Digitalis purpurea-derived cardiotonic glycoside used in the treatment of all forms of cardiac insufficiency which demand a digitalis therapy. The drug must be administered with caution as its side effects may be serious (anorexia, diarrhoea, arrhythmias etc.).
f gitaloxine
d Gitaloxin
i gitalossina
e gitaloxina

1426 gitoformate
Digitalis-related cardiotonic agent used in the treatment of cardiac insufficiency requiring digitalin-based therapy. It is particularly effective in strongly reducing the fibrillation and the auricular flutter, as well as paroxystic tachycardia. Nausea, vomit and anorexia may occur as side effects.
f gitoformate
d Gitoformat
i gitoformato
e gitoformato

1427 gitogenin
The active part of the gitonin, q.v. from which it is obtained by heating in the presence of hydrochloric acid $(C_{50}H_{44}O_4)$.
f gitogénine
d Gitogenin
i gitogenina
e gitogenina

1428 gitonin
A saponine contained in the seeds and leaves of purple foxglove (Digitalis purpurea). Amorphous powder, little soluble in water, capable of yielding by hydrolysis gitogenin, galactose and xilose.
f gitonine
d Gitonin
i gitonina
e gitonina

1429 gitoxigenin
The aglycone which is formed on com-

plete acid hydrolysis of the gitoxin, a physiologically active cardiac glycoside, in which it is present combined with digitoxose, a desoxymethylpentose capable of exerting an appreciable influence on the cardiotoxic action of the aglucones.
f gitoxigénine
d Gitoxigenin
i gitossigenina
e gitoxigenina

1430 gitoxin
Glycoside isolated from the leaves of digitalis and consisting of gitoxigenin and three molecules of digitoxose. It is physiologically active and capable of strengthening the cardiac output and improving the peripheral circulation by stimulating the vagus.
f gitoxine
d Gitoxin
i gitossina
e gitoxina

1431 glacial acetic acid
Acetic acid with usually less than one percent of water and produced as a pungent caustic hygroscopic liquid which crystallizes rapidly and is used as a good solvent for several medicinal preparations.
f acide acétique glacial
d Eisessig; kristallisierte Essigsäure
i acido acetico glaciale
e ácido acético glacial

1432 glafenine
Analgesic agent especially used to relieve acute or chronic pains, e.g. cancer, acute neuralgias, migraine. Also used in gynecology and, generally, in surgery. Nausea, oliguria and anuria may occur as adverse side effects.
f glafénine
d Glafenin
i glafenina
e glafenina

1433 Glauber salt
Sodium sulphate. Saline hydragogue purgative, hardly or very little absorbed, marked by a rapid action and capable of extracting water from the tissues of the intestinal wall. It is used in the treatment of chronic constipation caused by inadequate peristalsis.
f sel de Glauber; sulfate de sodium
d Glaubersalz; Natriumsulfat
i sale di Glauber; solfato di sodio
e sal de Glauber; sulfato sódico

1434 glaziovine
Psychotonic agent mainly used as an anxiolytic as it is capable of checking psychic tension through a mechanism of psychic regulation without sedative effects. It is indicated in anxiety neuroses associated to depression.
f glaziovine; glaziovinum

d Glaziovinum
i glaziovina
e glaziovina

1435 gleditsia
Or gleditschia, thorny acacia. A genus of thorny trees of the Leguminosae family, with pinnate or bipinnate leaves and small greenish spikes of flowers. The bark of the root and the seeds of the gleditsia (Sophora tormentosa) were used in the past to combat cholera.
f arbre à miel; carouge à miel
d dreidornige Gleditschie; amerikanischer Bohnenbaum
i sofora
e sófora; acacia de tresespina

1436 glibenclamide
Oral antidiabetic drug which is obtained by synthesis and is indicated in the treatment of mild forms of diabetes both of pancreatic and hepatic origin. It supports the action of insulin. A prolonged therapy with the drug, which is somewhat habit-forming, should be accurately monitored. Nausea, vomiting, dizziness, headache and jaundice may occur as side effects.
f glibenclamide
d Glibenclamid
i glibenclamide
e glibenclamido

1437 glibornuride
Antidiabetic agent of the sulphonylurea group characterized by a hypoglycaemic activity due to the release of endogenous insulin from the beta cells of the pancreas. The drug is used for the treatment of diabetes mellitus of the adult.
f glibornuride
d Glibornurid
i glibornuride
e glibornurido

1438 gliclazide
Hypoglycaemic agent of the sulphonylurea series capable of inhibiting platelet aggregation and thus reduce the cardiovascular complications of diabetes mellitus independently of the presence of obesity.
f gliclazide
d Gliclazid
i gliclazide
e gliclazido

1439 glipizide
Oral antidiabetic drug capable of stimulating the secretion of insulin by the beta-cells of the pancreas. The absorption is rapid and the action fairly prolonged as the drug has a biological half-life of 2.5/4 hours. It is particularly indicated in the treatment of diabetes mellitus in the adult.
f glipizide
d Glipizid
i glipizide

e glipizido

1440 gliquidone
Orally administered antidiabetic drug which is rapidly metabolized by the liver and excreted in the faeces. The effect is thus of short duration, but the drug is indicated whenever there appears to be a danger of hypoglycaemia.
f gliquidone
d Gliquidon
i gliquidone
e gliquidona

1441 glisoxepide
Hypoglycaemic agent of the sulphanylurea series indicated in the therapy of diabetes mellitus in the adult. Episodes of hypoglycaemia, though very rarely, may occur as side effects, as well as intestinal disorders, allergic reactions, epigastric pains and jaundice.
f glisoxépide
d Glisoxepid
i glisossepide
e glisoxepido

1442 globulinum humanum antihaemophilicum B
A preparation (lyophilized) of human plasma protein constituted by a concentrate of the clotting factors II, IX and X with a content of factor IX which is at least 60 times greater than that of fresh plasma. It is a biological antihaemorrhagic agent and acts by increasing the coagulation factors IX and X. It is therefore indicated in the treatment of haemophilia B and in haemorrhages caused by hereditary or acquired blood alterations which are due to the lack of factors II, VII, IX and X. The substance must be used with great caution in patients suffering from allergic diathesis. The common term for the drug is antihaemophilic human globulin B.
f globuline antihémophilique humaine B
d menschliches Antihämophilieglobulin B
i globulina umana antiemofilica B
e globulina humana antihemofílica B

1443 glucagon
Hormonal substance elaborated by the pancreas and displaying an action which is opposite to that of insulin, i.e. capable of bringing about a mobilization of the glycogen of the liver and a consequent increase of the glycaemia. In fact, it promotes the activation of the phosphorylasis with an action which is similar to that of adrenaline but without causing an increase of the blood pressure. It is a peptide hormone.
f glucagon
d Glukagon
i glucagone
e glucagón

1444 glucamethacin
Or glucomethacin. Non-steroid antiinflam-

matory, antipyretic, antirheumatic agent used in the treatment of rheumatoid arthritis, gout, various diseases of the osteo-articular apparatus and primary or secondary neoplastic osteopathies.
f glucométacine
d Glucomethacin
i glucometacina
e glucometacina

1445 glucitol
Hexahydric alcohol formed by reduction of glucose (or sorbose) and chiefly used as a humectant agent and softener. Also known as sorbitol.
f glucitol; sorbitol; glucohexitol; sorbite
d Sorbit
i sorbite; glucoesitolo
e sorbitol; glucohexitol

1446 glucocorticoid
Each steroid hormone containing two oxidized carbon atoms and capable of affecting carbohydrate metabolism. Cortisone, corticosterone, hydroxycorticosterone etc. belong to the glucocorticoids.
f glucocorticoïde
d Glukokortikoid
i glucocorticoide
e glucocorticoido

1447 glucofranguline
A purgative agent acting on the colon. It is indicated in all forms of constipation, particularly in the elderly.
f glucofranguline
d Glucofrangulin
i glucofrangulina
e glucofrangulina

1448 gluconate
The salt of the gluconic acid. Particularly important are the calcium and magnesium gluconates which derive from the d-gluconic acid and are widely used in medicine.
f gluconate
d Gluconat
i gluconato
e gluconato

1449 gluconic acid
Characteristic saccharinic acid which is obtained by the oxidation of glucose or cane sugar. It occurs in the oxidative changes caused in glucose by some strains of bacteria and by some moulds. A white water-soluble powder mainly used for the preparation of gluconates and as a sequestering agent.
f acide gluconique
d Gluconsäure
i acido gluconico
e ácido glucónico

1450 glucosulphamide
Antibiotic capable of acting against Gram-positive cocci and principally used in the treatment of bronchopulmonary infections. The absorption of the drug

is very rapid and peak plasma level easily attained. Gastrointestinal disorders may occur as adverse effects.
f glucosulfamide
d Glucosulfamid
i glucosolfamide
e glucosulfamida

1451 glucurolactone
A therapeutic agent capable of promoting the synthesis of glucuronic acid in the liver and therefore of recovering from the effects of poisoning. It is indicated in the treatment of endogenous and exogenous poisoning, e.g. alcohol intoxication.
f glucurolactone; glucuronolactone
d Glucurolakton
i glucurolattone; glucuronolattone
e glucurolactona

1452 glucuronic acid
Uronic acid which can be derived from glucose by oxidation of the primary alcoholic group. It exists in two optically active forms, d. and l. Its d-isomer is a syrupy, alcohol-soluble liquid that is obtained by reduction of the dilactone of the d-saccharic acid. The acid, which is present in human urine together with phenols and alcohol, plays an important part in the processes of disintoxication of the organism owing to its capacity of binding in ethereal combination, through its aldehydic group, with some toxic substances that prove difficult to oxidize and may be both endogenous, e.g. phenolic bodies deriving from intestinal putrefaction, and exogenous, e.g. camphor, morphine and chloral hydrate, rendering them harmless or at least less harmful.
f acide glucuronique
d Glukuronsäure
i acido glucuronico
e ácido glucurónico

1453 glutamic acid
Crystalline non essential dicarboxylic amino acid existing in two optically isomeric forms and occurring generally as the dextrorotatory t-form both free and combined in glutamine and several proteins in plants and animals. It is used in medicine owing to its beneficial effect on crises of petit mal and some forms of retarded mental development in children. Also called glutaminic acid.
f acide glutamique
d Glutaminsäure
i acido glutammico
e ácido glutámico; ácido glutamínico

1454 glutamin
Amide of the glutaminic acid. Needle-shaped, colourless crystals poorly soluble in water and insoluble in alcohol. It is a neutral amino acid copious in proteins and occurring free in many animal tissues. It is the only amino

acid capable to pass rapidly from the blood to the cerebral tissue where it can be metabolized. In addition to its biochemical importance, the acid is used in medicine as a coadjuvant in the therapy of alcoholism, retarded mental development and epileptic syndromes.
f glutamine
d Glutamin
i glutammina
e glutamina

1455 glutaral
Antibacterial and antimycotic agent mainly used for the disinfection of surgical and medical instruments. It is also indicated (the glutaraldehyde) for the treatment of warts.
f glutaral
d Glutaral
i glutaral; glutaraldeide
e glutaral; glutaraldehido

1456 glutaric acid
Dicarboxylic aliphatic acid, also known as pyrotartaric acid. It occurs naturally in the fat of the sheep wool and in the juice of unripe beets. It is used in organic synthesis. The amino derivative which is obtained by substituting a hydrogen atom of one of the $-CH_2-$ groups is the glutamic acid.
f acide glutarique
d Glutarsäure; Brenzweinsäure
i acido glutarico
e ácido glutárico; ácido pirotartárico

1457 glutathione
A tripeptide, naturally occurring, which is present in most living tissues and has a physiologically important role apparently depending on the very reactive sulphydryl group (-SH) which is present in the molecule, whose easy oxidation to the corresponding disulphide (-S-S-) makes possible a participation in oxidation-reduction systems. The drug is indicated in the treatment of acute and chronic hepatitis, cirrhosis, intoxication by drugs, ammonia, heavy metals and gas. It is also used to combat complications borne of some antineoplastic therapy.
f glutathione
d Glutathion
i glutatione
e glutationa

1458 gluten
Tenacious though elastic protein substance which is present (about 10%) in wheat flour. The extract of gluten, being very rich in nitrogen, can provide a high protein intake in small bulk. Also known as vegetable albumen.
f gluten; fibrine végétale
d Gluten; Kleber; Weizenkleber
i glutine; fibrina vegetale
e gluten; fibrina vegetal

1459 glutethimide
Rapid-action hypnotic of moderate duration. As it is slowly excreted in the
bile and urine, it is administered in
small doses which have a mild sedative
action. It does not depress respiration
or reduce blood pressure.
f glutéthimide
d Glutethimid
i glutetimide
e glutetimida

1460 glybuthiazol
Approved name for a chemical compound
capable of displaying a hypoglycaemic
action.
f glybuthiazole
d Glybuthiazol
i glibutiazolo
e glibutiazol

1461 glybuzole
Thiadiazole-related hypoglycaemic agent
used in the treatment of diabetes in the
adult.
f glybuzole
d Glybuzol
i glibuzolo
e glibuzolo

1462 glyceride
Any of a wide class of compound which
are esters of glycerol, particularly fatty
acids, occurring naturally as fats or
fatty oils or are made synthetically.
They are classed as mono-, di- and triglycerides according to the number of
hydroxyl groups of glycerol esterified.
f glycéride
d Glycerid
i gliceride
e glicerido

1463 glycerol
Or glycerin. Sweet, syrupy hygroscopic
trihydroxy alcohol which occurs combined
as glycerides and is formed by the
alcoholic fermentation of sugars or as
a by-product in the manufacture of
soaps. Used as therapeutic agent, it is
a mild laxative that is administered
orally. It is also used as a sweetening
agent in the preparation of some drugs.
f glycérine
d Glycerin
i glicerina
e glicerina

1464 glycerophosphate
Intermediate in triglyceride synthesis
formed by reduction of trihydroacetone
phosphate. The glycerophosphates exist
in two isomer forms, alpha and beta,
respectively derived from the alpha and
beta glycerophosphoric acids. The salts
used in pharmacology are a mixture of
the two isomers and are used (in particular the alkaline ones) as tonic in the
treatment of anaemia and nervous disorders, as well as of some forms of chronic rheumatism.
f glycérophosphate
d Glycerophosphat
i glicerofosfato
e glicerofosfato

1465 glycerophosphoric acid
Dibasic acid usually obtained by the
esterification of a glycerin hydroxyl
with a molecule of phosphoric acid. A
syrupy, colourless, tasteless, water-solu
ble liquid which is used in the preparation of glycerophosphates.
f acide glycérophosphorique
d Glycerinphosphorsäure
i acido glicerofosforico
e ácido glicerofosfórico

1466 glycerylguethol
A mucolytic and expectorant agent used
as an antiseptic in bronchopulmonary
disorders and in the treatment of rhinitis, tracheobronchitis, acute and chronic
bronchitis.
f glycérylguéthol
d Glycerylguethol
i gliguetolo
e glicerilguetolo

1467 glyceryl trinitrate
Vasodilator used for the symptomatic or
prophylactic treatment of angina pectoris.
It may be administered sublingually in
order to obtain a very rapid absorption
or applied as gel to the skin (a trans-
cutaneous therapy) for a sustained and
continuous absorption in prophylaxis.
The drug is also used in case of cardiac failure in the course of hypotensive
surgery or cardiac surgery to prevent
infarction. Allergic reactions have been
observed following application to the
skin.
f trinitrate de glycéryle
d Glyceryltrinitrat
i glicerile trinitrato
e trinitrato de glicerilo

1468 glycobiarsol
Amoebicide agent used in the treatment
of intestinal amoebiasis, particularly in
the chronic stage of the disease. It is
also used in the topical treatment of the
type of vaginitis which is caused by the
parasitic flagellate protozoa Trichomonas
vaginalis. In the latter case, vaginal
inflammation may occur as a side effect.
f glycobiarsol
d Glycobiarsol
i glicobiarsolo
e glicobiarsolo

1469 glycocholic acid
Substance contained in the bile resulting
from the conjugation of cholic acid with
glycine. It occurs mostly in the form of
sodium salt, the principal constituent of
the salts of the human bile.
f acide glycocholique
d Glycocholsäure

ı acıdo glıcocolıco
e ácido glicocólico

1470 glycolic acid
Translucent crystallıne compound occur-
rıng in nature in unrıpe grapes and ın
sugar beets. It ıs also obtaıned by
hydrolysis of chloroacetic acıd or as an
ıntermedıate ın the manufacture of ethylene glycol. It ıs used, among other
things, for the preparatıon of esters,
e.g. glycolates.
f acıde glycolıque
d Glycolsäure
ı acıdo glıcolıco
e ácıdo glıcólıco

1471 glyconiazide
Antıtuberculous agent whose actıvıty ıs
similar to that of ısonıazıd, q.v., and
at least four tımes as potent. It is used
ın the treatment of tuberculosıs, general_
ly admınıstered together with other antıtuberculous therapeutic substances.
Neurovegetatıve dısorders, ınsomnıa and
eıghtened body temperature may occur
as sıde effects.
f gluconıazıde
d Glyconıazıd
ı gluconıazıde
e gluconıazıda

1472 glycopyramide
Hypoglycaemic agent of the sulphonylurea group used ın the treatment of
diabetes in the adult. Nausea, vomıtıng,
vertıgo, headache, some allergıc reactıons accompanıed by fever and develop-
mental dısorder of the blood have been
observed as adverse effects.
f glycopyramıde
d Glycopyramid
ı glıcopıramıde
e glıcopırámıda

1473 glycopyrronium bromide
Antıcholınergıc and antıspasmolytıc agent
mainly used for symptomatic treatment
of peptıc ulcer, but also of cholecystıtıs
and spastıc colıtıs. Epısodes of tachy-
cardıa have been observed as sıde effect.
f bromure de glycopyrronıum
d Glycopyrronıumbromid
ı glıcopırronıo bromuro
e bromuro de glicopırronıo

1474 glycyclamide
Synthetıc hypoglycaemıc agent of the
sulphonylurea group to be admınıstered
orally. The drug ıs capable of stımulatıng the secretion of ınsulın and ınhıbıt-
ıng the degradatıon of the latter by the
lıver. The result ıs a stronger action
of the ınsulın. It ıs used ın the treatment of dıabetes mellıtus ın the adult
and of senıle bılıary diabetes. An antı-
dıuretıc action has been occasıonally
observed as a sıde effect.
f glycyclamıde
d Glycyclamid

ı glıcıclamıde
e glıcıclamıda

1475 glycyrrhetinic acid
Or glycyrrhetin. Pentacyclıc terpene
whose structure is similar to that of the
steroıds. It ıs obtaıned by hydrolysıs
of the glycyrrhızıc acıd. An amorphous,
bıtter-tasting substance dısplayıng phy-
sıologıcal actıons which are sımılar to
those of the deoxycortıcosterone.
f acıde glycyrrhétınıque
d Glycyrrhetınsäure; Glycyrrhetin
ı acıdo glıcırretıco; glıcırretına
e ácıdo glıcírrıco; glıcırrıcına

1476 glymidine
Oral antıdıabetic capable of stımulatıng
the secretion of endogenous insulin. The
drug is rapidly and totally excreted and
used in the treatment of dıabetes mellıtus wıthout acıdosıs or ketosıs. Allergıc
cutaneous disorders have been observed
as sıde effects.
f glymıdıne
d Glymıdın
ı glımıdına
e glımıdına

1477 glyoxylic acid
Dıhydroxyacetic acid which ıs responsible for the typıcal purple reaction ın
the Hopkıns-Cole test for tryptophan or
for proteıns contaınıng tryptophan.
f acıde glyoxylıque
d Glyoxylsäure
ı acıdo glıossılıco
e ácıdo glıoxílıco

1478 glypentide
Hypoglycaemic, non-halogenated sulphonylurea capable of exertıng a dırect
stimulation of the beta cells of the pancreas. It ıs chıefly used ın the treatment of diabetes mellıtus ın adults.
f glypentıde
d Glypentıd
ı glıpentıde
e glıpentıdo

* **glyphernasin   s. tryparsamide**

1479 goldenseal
Perennıal American herb havıng a thick
knotted yellow rootstock and large
leaves. The dried rhızome and roots of
the herb are used as an alternatıve and
tonıc agent. Also called hydrastıs.
f hydrastıs; sceau d'or
d Hydrastıs; Goldsıegel
ı ıdraste
e hidrastıs

1480 gold salt
Antıphlogıstıc agent whose mechanısm of
action ıs stıll unknown. It ıs seemıngly
specıfıc for rheumatoid arthrıtıs. It ıs
admınıstered orally or by ıntra-muscular
ınjectıon. Toxıc reactıons, dıarrhoea,
nausea, dermatıtıs and stomatıtıs normal_

ly occur as side effects. Bone marrow depression, nephritis and hepatitis may be caused by the drug.

f sel d'or
d Goldsalz
i sale d'oro
e sal de oro

1481 gomenol
Volatile oil obtained by the distillation of the fresh leaves of Melalauca leucadendro and used for the preparation of a balm in broncho-pulmonary affections. Also called oleogomenol.

f goménol
d Gomenol
i gomenolo; olio essenziale di niauli
e gomenol

1482 gonadorelin
Hormone produced in the hypothalamus and capable of stimulating the synthesis and secretion of gonadotropins and especially luteinizing hormone (LH) from the hypophysis. It is used as pulsatile subcutaneous or intravenous injection for the treatment of amenorrhoea and infertility due to ovarian hormone deficiency. Also useful as dynamic test in case the functionally of the hypothalamus/hypophysis gives rise to suspicion. An increase of plasmatic LH makes it possible to diagnose a deficiency of the hypothalamus. In the opposite case, an alteration or impairment of the hypophysis can be diagnosed.

f gonadoréline
d Gonadorelin
i gonadorelina
e gonadorelina

1483 gonadotrophin
Any substance which is capable of regulating the activity of the gonads. All these substances are found in extracts of the pituitary glands, in the blood and in the urine of pregnant women or animals and are respectively known as pituitary gonadotrophin, serum gonadotrophin and chorionic gonadotrophin. The pituitary gonadotrophins that are known are a) the follicle-stimulating hormone which is capable of regulating the development of the Graafian follicle in the female and spermatogenesis in the male and b) the luteinizing hormone, a peptide hormone secreted by the anterior pituitary, which controls the luteinization in the female and the production of testosterone in the male. The hormone is used in infertility and in delayed puberty.

f gonadotrophine
d Gonadotrophin
i gonadotrofina
e gonadotrofina

1484 goserelin
Hormone, analogue of gonadotrophin-releasing hormone and used for depot administration, i.e. the administration of the drug in a form which is only slowly absorbable, in abdominal wall for the treatment of cancer of the prostate glands. Goserelin acts by strongly reducing the production of male sex hormones. Loss of libido, hypermastia and an intensification of bone pain may occur as side effects.

f goséréline
d Goserelin
i goserelina
e goserelina

1485 gramicidin
Broad-spectrum antibiotic acting as bacteriostatic agent and effective against Gram-positive bacteria, e.g. pneumococci, staphylococci etc. The drug is also used for local application to skin burns and wounds, as well as for infections of the nose and mouth. Some histolytic action has been observed as side effect.

f gramicidine
d Gramicidin
i gramicidina
e gramicidina

1486 grindelia
Grindelia herb, gumweed. A genus of plants with 25 species of perennial herb with yellow flowers which contain an essential oil displaying a general sedative action on the nervous centres.

f grindélia
d Grindelie; Grindelienkraut
i grindelia
e grindelia

1487 grisein
Antibiotic produced by the Streptomyces griseus, fairly active against Gram-positive and Gram-negative germs and against some fungi. A red, water-soluble powder and contains some iron.

f griséine
d Grisein
i griseina
e griseina

1488 griseofulvin
An antibiotic effective against fungi and particularly against several species of Microsporum, Epidermophyton and Trichophyton by inhibiting the biosynthesis of the nucleic acids. It is also used in the treatment of gout. Nausea, epigastric pains, diarrhoea, vertigo and sometimes angioneurotic oedema may occur as adverse effects.

f griséofulvine
d Griseofulvin
i griseofulvina
e griseofulvina

1489 groundsel
Herb of the genus Senecio vulgaris which is sometimes used as an emmenagogue, i.e. an agent capable of promoting or assisting menstruation. Also known as

ragwort or simson.
f petit séneçon; séneçon des oiseaux
d Greiskraut; Baldgreiskraut
i senecione; erba calderina; verzellina
e senecio; hierba cana

1490 guaiacol
Cathecol-monomethyl ether. Either a colourless or pale-yellow liquid or a crystalline substance characterized by a strong odour and a pungent taste. It can be obtained by the fractional distillation of creosote or produced synthetically from cathecol. Chemically, it is the main constituent of creosote and is related to phenol. Owing to its disinfectant and deodorant properties, it is occasionally indicated in the therapy of bronchiectasis.
f guaïacol
d Guajakol; Guajacol
i guaiacolo
e guayacol

1491 guaiapate
Very potent antitussive agent acting by a central mechanism, i.e. on the cough reflex. Respiration is not affected and the drug is therefore very useful in the treatment of cough that is associated with infections of the upper respiratory tract.
f guaiapate
d Guaiapat
i guaiapato
e guaiapato

1492 guaiaphenesin
A good substitute of guaiacol characterized by a high degree of solubility and palatability. It is an expectorant and antitussive agent as well as a pulmonary antiseptic and thus useful in the treatment of both acute and chronic respiratory affections, e.g. bronchitis, tracheitis, obstinate cough etc., and pulmonary disorders which may follow a surgical operation. If administered in high doses, the drug may act as skeletal-muscle relaxant.
f guaiaphénésine
d Giaiaphenesin
i guaiafenesina
e guaiafenesina

1493 guaiazulen
Antiphlogistic agent characterized by its capacity of reconverting into epithelium mucous membranes, in particular the gastric one. The action is both spasmolytic and antiallergic, so that it can be used in many forms of gastric disorders, especially when there is little tolerance for other drugs. It can also be used for the topical treatment of stomatitis.
f guaiazulène
d Guaiazulen
i guaiazulene
e guaiazuleno

1494 guamecycline
Tetracycline-derived synthetic antibiotic mainly used in the treatment of chronic pulmonary affections, bronchial asthma, chronic bronchitis, spastic bronchitis etc. Mild gastrointestinal disorders may occur as side effects.
f guamécycline
d Guamecyclin
i guameciclina
e guameciclina

1495 guanazodine
Adrenergic neurone blocking drug used in the treatment of hypertension. Orthostatic hypotension, diarrhoea, a weakening of libido and some allergic reactions may occur as adverse effects.
f guanazodine
d Guanazodin
i guanazodina
e guanazodina

1496 guanethidine
Chemical compound used in the treatment of hypertension, eclamptic and preenclamptic toxaemia, paroxysmal tachycardia and glaucoma. Orthostatic hypotension, diarrhoea and urine retention may occur as side effects.
f guanéthidine
d Guanethidin
i guanetidina
e guanetidina

1497 guanochlor
Antihypertensive adrenergic neurone blocking agent. The drug also inhibits dopamine-betaoxydases and depletes the amines of the central nervous system. It is used in the treatment of primary and renal hypertension, i.e. a hypertension accompanied by renal artery abnormality. Hypotension, vertigo, diarrhoea, depression etc. may occur as side effects.
f guanochlor
d Guanochlor
i guanocloro
e guanocloro

1498 guanoxan
Hypotensive agent characterized by its ability to decrease both the systolic and diastolic blood pressure. It is an adrenergic blocking drug used in the treatment of mild essential hypertension also when the latter is of a nephrogenetic origin.
f guanoxan
d Guanoxan
i guanoxano
e guanoxano

1499 guar gum
A binding agent used for the preparation of tablets and for the thickening of food. It acts by absorbing moisture from the intestine and causing a feeling of satiety by the bulk it forms. It is

therefore used in the treatment of obesity and of diabetes as it helps to stabilize blood glucose levels. A feeling of excessive fullness, meteorism and diarrhoea may occur as side effects.
f guar
d Guar-Gummi
i gomma guar
e goma guar

* **gum dragon** s. tragacanth

1500 **guttapercha**
Or gutta percha. The dried and purified latex of various plants. It is a rubber-
-like substance from which solutions are obtained that are used in chloroform as substitute for collodion in the preparation of plastic skins. The same substance is also used as a temporary filling in dentistry.
f guttaperche
d Guttapercha
i guttaperca
e gutapercha

1501 hachimycin
Antimycotic, antibacterial and antiprotozoal agent capable of checking the
growth or destroying anaerobic bacteria.
It finds frequent use in the treatment
of trichomoniasis and candidiasis.
f hachimycine
d Hachimycin
i achimicina
e haquimicina

1502 haemafacient
Or haematopoietic. Relating to haemo-
poiesis, i.e. the process of production
of the blood cells, both in the uterine
and extrauterine phase of life. Thus,
any agent or substance capable of improving the quality of the blood and of
increasing its quantity.
f hémopoïetique
d hämatobildend
i emopoietico
e hemopoyético

1503 haemagglutinin
A generic term denoting an antibody
determining the agglutination of the red
corpuscles. It may occur naturally or
be induced by the injection of erythrocytes from one species into a member of
a different species, in which case the
cells act as an antigen capable of stimulating the production of the antibody.
f hémagglutinine
d Hämagglutinin
i emoagglutinina
e hemaglutinina

1504 haematin
Or hydroxyhaemin. Amorphous substance
of a dark blue colour derived from the
oxidation of haem (the ferroprotopor-
phyrin which forms the non-protein part
of the molecules of haemoglobin, myo-
globin and cytochrome C). Its principal
feature is the conversion of iron from
the ferrous to the ferric state.
f hématine
d Hämatin
i ematina
e hematina

1505 haematoporphyrin
Iron-free compound obtained by treating
haemoglobin or haematin with strong
acids. A dark-red powder, insoluble in
water but easily soluble in alkalis,

alcohol and concentrated sulphuric acid,
it is used in medicine as a nerve tonic
for it stimulates the sympathetic system.
It also finds employment in the treatment of depressive states, organic weakness and convalescence.
f hématoporphyrine
d Hämatoporphyrin
i ematoporfirina
e hematoporfirina

1506 haematoxylin
The colourless to yellowish crystalline
phenolic compound that is found in
logwood and mainly used as a biological
stain owing to its prompt oxidation to
haematein, the crystalline phenolic
quinonoid compound which constitutes the
essential dye in logwood extracts.
f hématoxyline
d Hämatoxylin
i ematossilina
e hematoxilina

1507 haemin
Dark-brown crystalline porphyrin derivative which is formed in old blood
clots. It is insoluble in water and soluble in acids or alkalis. It can be obtained by the action of glacial acetic
acid and sodium chloride on oxyhaemo-
globin. The crystals are used in medicine for the identification of blood.
f hémine
d Hämin
i emina
e hemina

1508 haemoglobin
The iron-containing protein pigment in
the red blood cells functioning primarily
in the transport of oxygen from the
lungs to the tissues of the body.
f hémoglobine
d Hämoglobin
i emoglobina
e hemoglobina

1509 haemostatic
Any agent which is capable of arresting
haemorrhage, in particular any drug
capable of shortening the clotting time
of blood.
f hémostatique
d Blutstillungsmittel; Hämostatikum
i emostatico
e hemostático

1510 halazone
Chlorine antiseptic resembling chloramine and containing approximately 25% of available chlorine. It is used for the sterilization of water and in the topical treatment of vaginal infection, leukorrhoea and trichomoniasis.
f halazone
d Halazon
i alazone
e halazona

1511 halcinonide
Glucocorticosteroid mainly used in the treatment of phlogistic and allergic cutaneous diseases, e.g. contact dermatitis, seborrhoea, eczema, psoriasis etc. Dryness, hypopigmentation and local atrophy of the skin may occur as side effects.
f halcinonide
d Halcinonid
i alcinonide
e halcinonida

1512 halibut liver oil
Brownish fatty oil extracted from the liver of the halibut and used therapeutically as a source of vitamin A.
f huile de foie de flétan
d Heilbuttleberöl
i olio di fegato di passera di mare
e aceite de hígado de hipogloco

1513 hallucinogen
Any substance, e.g. a drug or toxin, capable of producing hallucination, i.e. the subjective experience of a perception when there is no corresponding stimulus explaining it.
f hallucinogène
d Halluzinogen
i allucinogeno
e halucinatorio

1514 haloperidol
The first of the butyrophenone series of the major tranquillizers. The product is a sedative psychotropic agent and tranquillizer as well as a long-acting antiemetic. It has no hypnotic action. Experiments with rats and ewes show that the compound acts by depressing the prolactin-inhibiting factors. See chlorpromazine for possible side effects.
f halopéridol
d Haloperidol
i aloperidolo
e haloperidolo

1515 haloprogin
Antimycotic agent for topical application. It is very effective against several fungi and is used in the treatment of tinea pedis, tinea cruris, tinea manuum and tinea versicolor. Burning, local inflammation, the appearance of blisters and in some cases maceration may occur as adverse effects. The drug should be used with great caution in the course of pregnancy.
f haloprogine
d Haloprogin
i aloprogin
e haloprogin

1516 halopyramine
Antihistamine acting like chlorpromazine and used as an antiemetic and in the treatment of allergic reactions.
f halopyramine
d Halopyramin
i alopiramina
e halopiramina

1517 halothane
The most widely used volatile anaesthetic agent. Some preanaesthetic is generally intravenously injected to combat the excitation caused by the drug. Halothane is particularly indicated in asthmatic patients and is frequently administered jointly with some curare preparation or nitrogen peroxide. Hypotension and bradycardia may occur as side effects.
f halothane
d Halothan
i alotano
e halotano

1518 hamamelis
Commonly known as witch hazel. Genus of shrubs or small trees whose leaves and bark display medicinal properties. Owing to the considerable content of tannic substances, hamamelis has an astringent and haemostatic action, as well as a vasoconstricting one which seems to be particularly effective on the venous walls. Such properties are exploited for the preparation of therapeutic substances used in the treatment of varicose veins, haemorrhoids etc.
f hamamélis
d Hamamelis; Hexenhasel
i amamelide
e hamamelis

1519 hamamelis water
A fluid extract obtained from the bark of hamamelis, used pharmaceutically for preparations used in the therapy of venous complaints.
f eau d'hamamélis
d Hamamelisrindenwasser
i acqua di amamelide
e agua de hamamelis

1520 harmaline
Crystalline alkaloid obtained from the seeds and roots of Peganum harmala, which is oxidizable into harmine (a drug used in the treatment of paralysis agitans). It is characterized by anthelmintic properties and has been tentatively used to combat malaria and some nervous disorders. It is also a coronary dilator.
f harmaline
d Harmalin

i armalina
e harmalina

1521 hawthorn
Flowering shrub or small tree of the genus Crataegus characterized by white or pink flowers used in medicine as a sedative.
f aubépine
d Weissdorn; Hagedorn
i biancospino
e espino blanco; majuelo

1522 hedaquinium chloride
Antibacterial disinfectant agent and fungicide principally used in the treatment of cutaneous mycotic infections, e.g. epidermophytosis, trichophytosis etc. The preparation is applied topically.
f chlorure d'hédaquinium
d Hedaquiniumchlorid
i edaquinio cloruro
e cloruro de hedaquinio

1523 hedge mustard herb
Annual herb with pale yellow flowers which was used in medicine as a diuretic and an expectorant.
f herbe aux chantres
d wilder Senf; Raukenkraut
i erba cornacchia; erba di irione
e erisimo; hierba de los cantores

1524 heliotropin
Piperonal. White crystalline solid substance obtained by the oxidation of safrol, the main constituent of the oil of sassafras, and sometimes used in pharmaceutical preparations.
f héliotropine; pipéronal
d Heliotropin; Piperonal
i eliotropina; piperanolo
e heliotropina; piperonal

1525 helixin
Antitussive and bronchial antispasmodic agent with mucolytic and bronchodilating action. It is indicated in the treatment of acute and chronic bronchitis, emphysema, silicosis and disorders affecting the respiratory tract.
f hélixine
d Helixin
i elixina
e helixina

1526 hemlock
Coniin-alkaloid containing plant growing in Europe. It is a poisonous herb with finely cut leaves and small white flowers, whose extract has been used in medicine as a respiratory sedative.
f ciguë
d Schierling
i cicuta
e cicuta

1527 hemlock spruce
Pinus canadensis, hemlock bark, i.e. a bark that is obtained from the Tsuga canadensis, a North American tree, and is used in medicine as an astringent agent.
f sapin du Canada
d Hemlocktanne
i abete del Canada
e abeto del Canadá

1528 hemp
Tall herb cultivated in Asia from which hashish, a narcotic drug which is chewed, smoked or drunk for its intoxicating dream-like effects, is derived. Also called marihuana.
f chanvre; chanvre indien; haschisch
d indischer Hanf; Haschischkraut
i canapa indiana; hascish
e cáñamo; marihuana

1529 henbane
The dried flower top and leaves of Hyoscyamus, an annual or perennial genus of herb growing in uncultivated land in Europe, Western Asia and North Africa. It is a poisonous plant owing to the presence in it of alkaloids, e.g. hyoscyamine, hyoscine and atropine.
f hannebane; jusquiame
d Hyoszyamus; schwarzes Bilsenkraut
i giusquiamo
e beleño

1530 heparin
High-molecular weight polysaccharide formed by molecules of glucuronic acid alternated with acetylglucosamine with two or three residues of sulphuric acid which give the molecule a considerable acidity and negative charge. It is the most important physiological anticoagulating agent; its antithromboplastinic action interferes in the passage prothrombin-thrombin and is probably related to its high acidity. It is inactive if taken orally and must therefore be injected. Its sodium salt, prepared from mammalian lung or liver is also used intravenously or by intravenous drip. Its calcium salt may be used subcutaneously without the danger of causing haematomas in order to prevent thromboembolism. It is also employed in the treatment of lipid metabolism and athero sclerosis.
f héparine
d Heparin
i eparina
e heparina

1531 hepronicate
Peripheral vasodilator and antihyperlipidaemic agent used in the treatment of peripheral disorders of blood vessels, arteriosclerosis-related circulatory insufficiency, cyanosis of the extremities caused by a degree of coldness which would not affect a healthy individual, strong headache and Bürger's disease.
f hépronicate
d Hepronicat

i epronicato
e hepronicato

1532 heptabarb
A short-action barbiturate acting on the subcortex and used to obtain a mild sedative condition or, in higher doses, as a hypnotic. In the latter case the action lasts up to 10 hours.
f heptabarb
d Heptabarb
i eptabarbitale
e heptabarbital

1533 heptaminol
Stimulating agent mainly used in cardio-vascular disorders, e.g. stenosis of the coronary arteries. It is synergic with cardiotonic drugs and tends to antagonize the action of neurolepsis-producing agents. It also acts as a stimulant in states of great fatigue.
f heptaminol
d Heptaminol
i eptaminolo
e heptaminolo

1534 hesperetin
The flavone which occurs as a glucoside in hesperidin, q.v. Trihydroxyderivative of the methoxyflavone.
f hespérétine
d Hesperetin
i esperetina
e hesperetina

1535 hesperidin
Rhamnoglucoside of the hesperetin occurring in the extract of bitter oranges. A bitter powder weakly soluble in water. It is one of the components of citrin, i.e. the vitamin of the capillary permeability.
f hespéridine
d Hesperedin
i esperidina
e hesperidina

1536 hetacillin
Penicillanic-acid derivative. Antibiotic acting on Gram-positive and Gram-negative microorganisms but without any effect on penicillin-resistant staphylococci. The drug is principally used in the treatment of infections affecting the respiratory, gastrointestinal, cutaneous and urinary systems. Some allergic skin reactions may occur as side effect.
f hétacilline
d Hetacillin
i etacillina
e hetacilina

1537 hexachlorobenzene
Benzol derivative. Needle-shaped colourless crystals used in organic syntheses and as a fungicide.
f hexachlorobenzène
d Hexachlorobenzol
i esaclorobenzene

e hexaclorobenzeno

1538 hexachlorocyclohexane
Generic term denoting a group of nine isomeric compounds, the most important of which is the gamma isomer (galla benzene hexachloride) which is a potent insecticide, stronger and more rapid in its action than DDT. It is used as a spray and, in solution with alcohol, against lice. As an emulsion it is employed against scabies.
f hexachlorocyclohexane
d Hexachlorcyclohexan
i esaclorocicloesano
e hexaclorociclohexano

1539 hexachlorophane
A topical antiseptic widely used in soaps, creams and lotions. It acts against Gram-positive and Gram-negative microorganisms and is used in the treatment of acne, impetigo and various other cutaneous disorders.
f hexachlorophane
d Hexachlorophan
i esaclorofano
e hexaclorofano

1540 hexachlorophene
Hexachloroderivative of the diphenylmethane. Colourless powder insoluble in water and soluble on weakly alkaline solutions. It is an antiseptic agent particularly active against Gram-positive microorganisms.
f hexachlorophène
d Hexachlorophen
i esaclorofene
e hexaclorofeno

1541 hexacyprone
A choleretic agent, i.e. capable of stimulating the activity of the liver and used whenever bile secretion is inadequate as it promotes intestinal digestion and the absorption of fats.
f hexacyprone
d Hexacypron
i esaciprone
e hexaciprona

1542 hexafluorenium bromide
A muscle relaxant agent and inhibitor of the plasmatic cholinesterasis. It is mainly used in association with succinyl choline, as it is capable of inhibiting its enzymatic hydrolysis, thus prolonging its action, in the course of anaesthesia so as to intensify the neuromuscular block. Respiratory depression, brachycardia, alterations in blood pressure etc. may occur as side effects. Also known as hexafluronium bromide.
f bromure d'hexafluronium
d Hexafluroniumbromid
i esafluronio bromuro
e bromuro de hexafluronio

1543 hexamethonium

Organic radical derived by substitution from ammonium, i.e. the hexamethylene member of a series of polymethylene bis (trimethylammonium radicals which are called methonium bases). The salts, e.g. the bromide, block the transmission of the ganglionic synapse and reduce the sympathetic tone, causing vasodilation and a fall in blood pressure. They are therefore used in medicine for the treatment of hypertension and sometimes in surgery to reduce the level of bleeding. The drug also inhibits gastrointestinal motility. Orthostatic hypotension as well as postural hypotension are possible side effects, together with paralysis of accommodation, urine retention and sometimes impotence.

f hexaméthonium
d Hexamethonium
i esametonio
e hexametonio

1544 hexamidine
Broad-spectrum bacteriostatic and bactericide agent effective against sulphonamide- and antibiotic-resistant pathogenic microorganisms. It is also used as a topical antiseptic in the treatment of conjunctivitis, blepharoconjunctivitis, otitis media, rhinitis and sinusitis.

f hexamidine
d Hexamidin
i esamidina
e hexamidina

1545 hexamine
Antiseptic agent used in the treatment of urinary infections. Such treatment requires that the urine is rendered acid by the administration of ammonium chloride which liberates formaldehyde from the drug. Haematuria and painful micturition have been observed as side effects.

f hexamine
d Hexamin
i esamina
e hexamina

1546 hexapropymate
Propinylcyclohexanol carbamate. Hypnotic and sedative agent acting on the cortical and subcortical planes of the central nervous system and used in the treatment of insomnia caused by emotions if a short period of sleep is advisable. The drug has also been used in the therapy of neurovegetative disorders.

f hexapropymate
d Hexapropymat
i esapropimato
e hexapropimato

1547 hexcarbacholine bromide
Carbolonium bromide. A choline-derived myorelaxant and antispasmodic agent used in surgery to obtain the relaxation of skeletal muscles and in orthopaedics. Its action is generally prolonged. Respiratory depression and hypotension are possible side effects.

f bromure d'hexcarbacholine
d Hexcarbacholinbromid
i escarbacolina bromuro
e bromuro de escarbacolina

1548 hexetidine
Antiseptic and trichomonicidal agent principally used in the treatment of infections of the vagina and of the cervix. The drug is also employed for its capacity of promoting heaemostasis and cicatrization.

f hexétidine
d Hexetidin
i esetidina; esedina
e hexetidina

1549 hexobarbital
Hexobarbitone. Barbiturate hypnotic marked by a short or very short duration of action. It is generally used in surgery as a preanaesthetic and also to treat insomnia and convulsions.

f hexobarbital
d Hexobarbital
i esobarbitale
e hexobarbital

1550 hexobendine
Therapeutic agent whose principal action is the dilatation of arterioles and capillaries, in particular those concerned with coronary circulation. It is therefore employed in the treatment of angina pectoris especially when it is associated with coronary and/or aortic sclerosis.

f hexobendine
d Hexobendin
i esobendina
e hexobendina

1551 hexocyclium methylsulphate
Atropine-resembling cholinergic blocking agent mainly used as an antispasmodic and analgesic agent in the treatment of peptic and duodenal ulcers, biliary dyskinesia and gastric hypersecretion. Accommodation disorders have been observed as side effects.

f méthylsulfate d'hexocyclium
d Hexocycliummethylsulfat
i esocilio metilsolfato
e metilsulfato de hexocilio

1552 hexoestrol
Or hexanoestrol. Synthetic oestrogen indicated in the treatment of menopausal symptoms and as an antimitotic in carcinoma of the prostate and of the breast. The drugs may block post-partum lactation.

f hexestrol
d Hexöstrol
i esestrolo
e hexestrolo

1553 hexoprenaline
Beta-stimulant bronchodilator which acts without affecting the cardiovascular sys-

tem unless administered in excessive doses. It is principally used in the treatment of bronchial asthma, spasmodic and emphysematous bronchitis.

f hexoprénaline
d Hexoprenalin
i esoprenalina
e hexoprenalina

1554 hexylcaine
Ester-type local anaesthetic used for surface and infiltration purposes and for spinal anaesthesia. Its effects are similar to those of lignocaine, q.v.

f hexylcaïne
d Hexylcain
i esilcaina
e hexilcaina

1555 hexylresorcinol
Anthelmintic agent with paralysing mechanism of action. It is particularly active against Ascarides, Ancylostoma, Trichocaephalus, Oxyuris and some taeniae. Skin disorders may occur as side effect. Not to be used when peptic ulcers have been diagnosed or in the course of pregnancy.

f hexylrésorcinol
d Hexylresorcinol
i esilresorcinolo; esilresorcina
e hexilresorcinol; hexilresorcina

1556 hippuric acid
Benzoyl glycine, benzamino acetic acid found in the urine of herbivorous animals and to a much lesser degree of man and representing a detoxication product. It occurs in white crystals, soluble in hot water, and used in medicine for the treatment of gout and rheumatism.

f acide hippurique
d Hippursäure
i acido ippurico
e ácido hipúrico

1557 histamine
Crystalline base found in ergot and some other plants and usually combined in animal tissues. It is formed by histidine by decarboxylation and can be released in a free form by several substances, e.g. curare and morphine. It is responsible for the dilation and increased permeability of blood vessels and, with larger doses, haemoconcentration, which play an important part in allergic reactions.

f histamine
d Histamin
i istamina
e histamina

1558 histamine phosphate
Histamine acid phosphate. Vasodilator with action particularly effective on capillaries and arterioles. It is further characterized by its capacity to produce the contraction of the involuntary muscles of the uterus, the guts and bronchi, and to stimulate gastric secretions. The drug is used in the treatment of peripheral vascular disorders, as a diagnostic agent in the test of gastric functions and as a desensitizing agent in pathological conditions caused by hypersensitivity to histamine.

f phosphate d'histamine
d Histaminphosphat
i istamina bifosfato
e bifosfato de histamina

1559 histapiperidine
Quickly-acting antihistaminic and antispasmodic agent, capable of inhibiting any feeling of nausea caused by motion. The drug is used in the treatment of various allergies, e.g. pollinosis and bronchial asthma, in some spastic painful states in the course of surgery and as a preanaesthetic. It also finds employment in some cutaneous diseases, particularly when accompanied by itching.

f histapipéridine
d Histapiperidin
i istapiperidina
e histapiperidina

1560 histapyrrodine
Antihistaminic agent used in the treatment of cutaneous allergic diseases, e.g. urticaria, respiratory diseases, e.g. rhinitis, coryza, asthma, and angioneurotic oedema. The drug is also indicated in the treatment of premenstrual tension.

f histapyrrodine
d Histapyrrodin
i istapirrodina
e histapirrodina

1561 histidine
Crystalline basic amino acid occurring in most proteins but not considered as essential in the human diet. It is present in urine in the course of pregnancy. Chemically it is intimately related to histamine and probably represents a precursor of the latter. Histidine is synthesized by microorganisms and by plants and is formed by decomposition of most proteins. Its soluble hydrochloride has been used in the treatment of peptic ulcer but without significant results.

f histidine
d Histidin
i istidina
e histidina

1562 histidylurea
Biological substance capable of promoting the healing of wounds and sores by the formation of scar tissue. The drug provides those substances which are required for the cellular proliferation and is therefore used therapeutically for ulcerations, decubitus sores,

rhagades, burns etc.
f histidylurée
d Histidylurea
i istidilurea
e histidilurea

1563 hollow root
The dried roots of the North American plants Dicentra canadensis and Dicentra cucullaria containing respectively the alkaloids protopine and corydine, and cryptopine together with two other alkaloids. It is used in medicine in the form of decoction as a tonic and as a diuretic.
f racine de corydale
d Lerchenspornwurzel
i radice di coridale
e rizoma de aristoloquia cava

1564 hollyhock
The common hollyhock whose flowers have apparently the same properties of marshmallow whose root yields a mucilaginous solution used in the symptomatic treatment of coughs and inflammation of the buccal membrane.
f mauve arborée; passerose
d Pappelrose
i malvarosa; malvone
e malva arbórea; malva rósea

1565 homatropine
The tropine ester of mandelic acid. Crystalline colourless, odourless, bitter solid substance, soluble in alcohol and less easily in water. It is used in medicine in the form of hydrobromide, sulphate and hydrochloride. Homatropine has the same pharmaceutical properties of atropine but displays lesser secondary actions and a shorter-lasting mydriatic effect. It is therefore widely used in ophthalmology whenever it is desirable to have a mydriasis of short duration.
f homatropine
d Homatropin; Tropinmandelat
i omatropina; tropinmandelato
e homatropina

* homatropine bromide   s. homatropine hydrobromide

1566 homatropine hydrobromide
Soluble salt of homatropine having approximately the same actions as those of atropine, though less powerful and of a shorter duration. It is especially used as a mydriatic and cycloplegic agent and particularly indicated for ophthalmoscopic examinations and in refraction as it displays less tendency to increase intraocular pressure.
f hydrobromure d'homatropine
d Homatropinhydrobromid
i omatropina bromidruro
e bromhidruro de homatropina

1567 homatropine methylbromide

Homatropine based compound used internally as a secretion inhibitor and as an antispasmodic agent. Though acting like atropine, it is less toxic and is mainly used in the treatment of disorders of the gastro-intestinal tract.
f méthylbromure d'homatropine
d Homatropinmethylbromid
i omatropina metilbromuro
e metilbromuro de homatropina

1568 homocysteine
Amino acid occurring as intermediate product in the transformation of the methionine into cysteine. The sodium salt of the laevorotatory form is sometimes used as a surrogate of the sodium glutammate.
f homocystéine
d Homocystein
i omocisteina
e homocisteina

1569 homogentisic acid
Crystalline, colourless, easily soluble in water, alcohol and ether solid substance occurring in some plants. The acid is a physiological intermediate product of the metabolism of the phenylalanine and of the tyrosine. It is rapidly metabolized in the human organism. In alkaptonuria, it appears unchanged in the urine; the blackening is due to the formation of melanins.
f acide homogentisique
d Homogentisinsäure
i acido omogentisinico
e ácido homogentísico

1570 homophenazine
Phenothiazine derivative used in the treatment of neuro- and psycho-vegetative disorders, particularly when they occur in menopause and in old age. Also used as a coadjuvant in psychotherapy and in functional disorders.
f homophénazine
d Homophenazin
i omofenazina
e homofenacina

1571 hop oil
The volatile oil which is extracted from the fruits of the common hops and used as a sedative.
f essence de houblon
d Hopfenöl
i essenza di luppolo
e esencia de lúpulo

1572 hordenine
Chemical compound extracted from the barley seeds. A colourless, tasteless, water-soluble solid substance capable of forming well crystallized salts with the acids. The most frequently used of the latter is the sulphate which is used in medicine in the treatment of intestinal infections and as heart tonic.
f hordénine
d Hordenin

i ordenina
e hordenina

1573 humic acid
One of the organic acids derived from the carbonization processes of vegetable organisms and widely diffused in soil and humus. They subsequently appear in natural waters.
f acide humique
d Humussäure; Huminsäure
i acido umico
e ácido húmico

1574 hyaluronic acid
Viscous mucopolysaccharide occurring mainly in connective tissues or their derivatives, e.g. skin, cartilage and vitreous humour. It is produced by acetylglucosamine units combined with glucuronic acid.
f acide hyaluronique
d Hyaluronsäure
i acido ialuronico
e ácido hialurónico

1575 hyaluronidase
An enzyme that is capable of promoting the dispersal and the absorption of injected solutions, both subcutaneous and intramuscular. The drug also hastens the resorption of blood and fluid in body cavities. Collaterally, it acts as an antiphlogistic, antioedemic and mildly anaesthetic agent.
f hyaluronidase
d Hyaluronidase
i ialuronidasi
e hialuronidasa

1576 hydnocarpic acid
The acid which is found, together with chaulmoogric acid, in the hydnocarpus oil extracted from the seeds of Hydnocarpus (wightiana). It is used in medicine, either subcutaneously and intramuscularly, in the treatment of leprosy.
f acide hydnocarpique
d Hydnocarpussäure
i acido idnocarpico
e ácido hidnocárpico

1577 hydracrylic acid
Ethylenelactic acid. Alcohol acid, isomeric with ethylene lactic acid, forming acrylic acid on heating.
f acide hydracrylique
d Hydracrylsäure
i acido idracrilico
e ácido hidracrílico

1578 hydralazine
Vasodilator antihypertensive agent capable of reducing a pathologically high blood pressure and of increasing the flow of blood through the kidneys. The action is rather slow but of long duration. Tachycardia, marrow depression, acute rheumatoid syndrome and systemic lupus erythematosus syndrome may occur as side effects.
f hydralazine
d Hydralazin
i idralazina
e hidralacina

1579 hydrargaphen
Potent disinfectant active on Gram-positive and Gram-negative microorganisms. It is used in the treatment of ear, nose and throat disorders and in dermatology for the therapy of trichophytosis, onychomycosis and trichomoniasis.
f hydrargaphène
d Hydrargaphen
i idrargafene
e hidrargafeno

1580 hydrargyrum chloride (yellow)
Orally administered drug owing to its purgative action. It is particularly used in veterinary practice. Also used as an antiluetic in the form of an oil solution or suspension. Also known as mercuric oxide (yellow).
f chlorure d'hydrargyre
d Hydrargyrumchlorid
i mercurio ossido giallo
e cloruro de hidrargiro

1581 hydrastine
Alkaloid related to narcotine and obtained from the golden seal. It depresses the heart rate and raises the blood pressure. The solution of its hydrochloride is injected hypodermically for the purpose.
f hydrastine
d Hydrastin
i idrastina
e hidrastina

1582 hydrastinine
Alkaloid produced by the oxidation of hydrastine. Its hydrochloride is used in medicine as a vasoconstrictor and an antihaemorrhagic agent. It is capable of contracting the uterus and is applied to the eye to dilate the pupil.
f hydrastinine
d Hydrastinin
i idrastinina
e hidrastinina

1583 hydrastis rhizome
North American plant whose main active constituents are two alkaloids, hydrastine and berberine. The drug and the two alkaloids cause uterine contractions and are used in menorrhagia. It is also frequently used as a bitter tonic in the form of tincture or extract and, applied topically, for the relief of inflammation affecting mucous membranes. Vasoconstriction and convulsions may occur as adverse effects.
f hydraste
d Hydrast
i idraste
e hidrasta

1584 hydrazine
Colourless liquid which solidifies at 2°C.
and boils at 113.5°C. It is used in organic chemistry as a reducing agent and
in the synthesis of heterocyclic compounds. Hydrazines are frequently used
in the manufacture of therapeutic drugs.
f hydrazine
d Hydrazin
i idrazina
e hidracina

1585 hydrazoic acid
Or hydronitric acid. Chemical compound
forming salts known as hydrozoates,
hydronitrides or azides. The acid is a
very potent protoplasmic poison.
f acide azothydrique; acide hydrazoïque
d Stichstoffwasserstoffsäure; Azoimid
i acido azotidrico
e ácido hidronítrico

1586 hydriodic acid
Hydrogen iodide. Colourless, water-soluble acid, used as a reducing agent and
reagent. Its solution and syrup are
used in the treatment of tertiary syphilis, arthritis, aneurysm and exophthal-
mic goitre.
f acide iodhydrique
d Jodwasserstoffsäure
i acido iodidrico
e ácido yodhídrico

1587 hydrobromic acid
Hydrogen bromide. A colourless penetrating gas used as a reagent in organic
synthesis. A 10% solution of the acid is
used in medicine in the treatment of disorders affecting the central nervous
system as it soothes cerebral excitement,
controls epileptic fits and, more generally, displays a sedative action.
f acide bromhydrique
d Bromwasserstoffsäure
i acido bromidrico
e ácido bromhídrico

1588 hydrochloric acid
Hydrogen chloride. Colourless gas forming a concentrated solution in water and
used as a caustic and corrosive agent.
A more diluted solution, i.e. about 10%,
is used in the treatment of achlorhydria,
i.e. the absence of free hydrochloric
acid in the gastric juices and in hypochlorhydria, i.e. inadequate amount of
free hydrochloric acid in the gastric
juices.
f acide chlorhydrique
d Chlorwasserstoffsäure
i acido cloridrico
e ácido clorhídrico

1589 hydrochlorothiazide
Thiazide-related diuretic agent capable
of increasing the urinary excretion of
sodium, chloride and, in smaller quantities, potassium. It is generally used
in the treatment of obesity, oedemas of

various origins and in chronic hypotension. Some allergic reactions and thrombocytopenia, sometimes also in the fetus,
have been observed as side effects.
f hydrochlorothiazide
d Hydrochlorothiazid
i idroclorotiazide
e hidroclorotiazida

1590 hydrocodone
Antitussive agent marked by a central
mechanism of action. Its analgesic and
strongly sedative action is similar to
that of codeine. The drug does not appear to promote expectoration. Its
prompt effects last for a fairly long
time.
f hydrocodone
d Hydrocodon
i idrocodone
e hidrocodona

1591 hydrocortisone
Chemical compound, 17-hydroxycortico-
sterone or 17-oxycorticosterone. It differs from cortisone owing to the presence
in position 11 of an oxydryle in place
of an atom of oxygen. It can be obtained from the adrenal cortex, but it is
more frequently produced by synthesis
(from biliary salts or vegetable steroids). It has the same pharmacological
properties of cortisone and is used in
the form of derivatives, e.g. acetate,
butylacetate, succinate etc.
f hydrocortisone
d Hydrocortison
i idrocortisone
e hidrocortisona

1592 hydrocyanic acid
Prussic acid. Toxic compound occurring
in several (cyanogenetic) glycosides.
In its pure state, it is a mobile, colourless, bitter-almond-smelling liquid. It
is prepared by treating a cyanide with
sulphuric acid or by the decomposition
of the formamide. It is extremely toxic
for man and other animals as it inhibits
the vital activity of the protoplasms.
Its action is very rapid: it first excites
and subsequently rapidly paralyses the
bulbar respiratory centre. If inhaled in
the form of gas, death may well be
instantaneous. It is used therapeutically
in very small doses to treat spasmodic
and whooping cough.
f acide cyanhydrique; acide prussique
d Cyanwasserstoffsäure; Blausäure
i acido cianidrico; acido prussico
e ácido cianhídrico; ácido prúsico

1593 hydroflumethiazide
Diuretic agent whose mode of action
resembles that of chlorothiazide and
hydrochlorothiazide, but is considerably
more potent. The drug is used as an
antioedemic especially in cardiopathies
and as an antihypertensive agent. Some
allergic reactions and anaemia have been

observed as side effects.
f hydroflumethiazide
d Hydroflumethiazid
i idroflumetiazide
e hidroflumetiazida

1594 hydrofluoric acid
Hydrogen fluoride. Poisonous, water-solu
ble gas which, abundantly diluted with
air, has found some use as an inhalant
in pulmonary tuberculosis.
f acide fluorhydrique
d Fluorwasserstoffsäure; Flussäure
i acido fluoridrico
e ácido fluorhídrico

1595 hydrogen peroxide
Colourless compound prepared in aqueous
solutions in various ways, e.g. by the
electrolysis of sulphuric acid, and which
can be concentrated generally by distillation. It is usually employed in a dilut
ed form as a disinfectant and antiseptic
agent.
f eau oxygénée; peroxyde d'hydrogène
d Wasserstoffsuperoxid; Hydrogenperoxyd
i acqua ossigenata; perossido di idrogeno
e agua oxigenada; peróxido de hidrógeno

1596 hydrogen sulphide
Hydrosulphuric acid, sulphuretted hydrogen. Colourless gas characterized by a
smell resembling that of a foul egg. It
is used to precipitate metallic sulphides.
f acide sulfhydrique
d Schwefelwasserstoff
i acido solfidrico; idrogeno solforato
e ácido sulfhídrico; hidrógeno sulfurado

1597 hydromorphone
Analgesic agent qualitatively acting like
morphine yet with greater intensity. It
is used for the relief of any acute or
chronic pain. Respiratory depression and
constipation have been observed as side
effects.
f hydromorphone
d Hydromorphon
i idromorfone
e hidromorfona

1598 hydroquinone
Therapeutic agent used in the treatment
of cutaneous diseases. It reduces pigmen
tation by acting on the thyrosine-thy-
roxinase-melanin system. Some depigmentation and cutaneous irritation may occur as side effects.
f hydroquinone
d Hydrochinon
i idrochinone; idrochinolo
e hidroquinona

1599 hydrotalcite
Antacid used in the treatment of gastric
hyperacidity, peptic ulcers, gastritis,
dyspepsia and heartburn associated with
oesophageal reflux and hiatal hernia.
Bouts of diarrhoea may set in as side
effect.

f hydrotalcite
d Hydrotalcit
i idrotalcite
e hidrotalcita

1600 hydroxocobalamin
Vitamin B_{12a}. It is an important member
of a group of similar compounds capable
of promoting erythropoiesis and represent
ing the anti-pernicious-anaemia principle
of purified liver extracts (British Pharmaceutical Codex' definition, 1968). The
drug is used in the treatment of subacute combined degeneration of the spinal
cord. Chemically, a cobalamine containing an oxydryle bound to the cobalt
atom.
f hydroxocobalamine
d Hydroxycobalamin; Hydroxocobalamin
i idrossicobalammina
e hidroxocobalamina

1601 hydroxyalanine
Serine. Amino acid which occurs in protein hydrolysates.
f hydroxyalanine; sérine
d Hydroxyalanin; Serin
i idrossialanina; serina
e hidroxialanina; serina

1602 hydroxyamphetamine
Sympathicomimetic agent principally used
in the treatment of acute and chronic
rhinitis. Ventricular arrhythmias, hypertension, vertigo, tremors etc. may occur
as adverse effects.
f hydroxyamphétamine
d Hydroxyamphetamin
i idrossianfetamina
e hidroxianfetamina

1603 hydroxyapatite
A calcium salt used therapeutically as
a source of calcium and phosphorus in
osteoporosis, osteomalacia and rickets.
f hydroxyapatite
d Hydroxyapatit
i idrossiapatite
e hidroxiapatita

1604 hydroxybutyloxide
A true choleretic acting directly on the
hepatic cells to promote the production
of bile whose flux is increased by
about 150% although the ratio of the
biliary components remains unchanged.
The choleresis thus produced lasts between six and eight hours without displaying a cholecystokinetic action. A
spasmolytic musculotropic action on the
biliary tract is observable. The drug is
used in hepatic insufficiency, cholecisti
tis, biliary dyskinesia, cholelithiasis
etc.
f hydroxybutyloxyd
d Hydroxybutyloxyd
i idrossibutilossido
e hidroxibutilóxido

1605 hydroxycarbamide

Antineoplastic agent which is rapidly absorbed and excreted by the kidneys. The drug acts by interfering with the synthesis of the deoxyribonucleic acid and is chiefly used in the therapy of chronic myeloid leukaemia.
f hydroxycarbamide
d Hydroxycarbamid
i idrossicarbamide
e hidroxicarbamida

1606 hydroxychloroquine
Chloroquine-derived antirheumatic agent mainly used in the treatment of rheumatoid arthritis and lupus erithematosus. It also find employment in antimalarial therapy. A temporary anaemia and leuko penia, pigmentary alterations, dizziness, headache etc. may occur as side effects.
f hydroxychloroquine; oxychloroquine
d Hydroxychlorochin
i idrossiclorochina
e hidroxicloroquina

1607 hydroxydione sodium succinate
Intravenously injected steroid, without hormone-like action, used as a general anaesthetic. Particularly indicated for the narcosis of diabetic patients. It also finds employment in obstetrics and to check delirium tremens.
f hydroxydione succinate de sodium
d Hydroxydionnatriumsuccinat
i idrossidione sodio succinato
e hidroxidiona sodio succinato

1608 hydroxyethylamine
Base occurring in cephalin and closely related to choline. It is used in medicine as a sclerotizing agent for varicose veins.
f hydroxyéthylamine
d Hydroxyethylamine
i idrossietilammina
e hidroxietilamina

1609 hydroxylamine
Chemical compound which can be derived from ammonium by the substitution of a hydrogen atom with an oxydryl group; a solid crystalline substance, colourless and deliquescent, easily soluble in alcohol. It is used as a reducing and oxidizing agent.
f hydroxylamine
d Hydroxylamin
i idrossilammina
e hidroxilamina

1610 hydroxymalonic acid
Or tartronic acid. Propanol diacid, i.e. an acid obtained by the oxidation of glycerol or by the reduction of meso-xalic acid.
f acide hydroxymalonique; acide tartro-nique
d Hydroxymalonsäure; Tartronsäure
i acido idrossimalonico; acido tartronico
e ácido hidroximalónico; ácido tartrónico

1611 hydroxypethidine
Pethidine derivative marked by the same analgesic activity yet with apparently better results when used as a general anaesthetic by intravenous injection.
f hydroxypéthidine
d Hydroxypethidin
i idrossipetidina
e hidroxipetidina

1612 hydroxyprocaine
The hydrochloride of diethylamino-ethanol-4-aminosalicylate. Local anaesthetic with an action resembling that of procaine but quicker and longer-lasting.
f hydroxyprocaïne
d Hydroxyprocain
i idrossiprocaina
e hidroxiprocaina

1613 hydroxyprogesterone
A progestational steroid acting on the uterus to prepare the endometrium to receive the fertilized ovum. It has the same use as progesterone, q.v.
f hydroxyprogestérone
d Hydroxyprogesteron
i idrossiprogesterone
e hidroxiprogesterona

1614 hydroxyprogesterone caproate
Therapeutic agent indicated in case of habitual or threatened abortion, of dismenorrhoea, amenorrhoea and fibrous mastitis. The drug is also used in advanced cancer in the uterus. Allergic reactions, asthma, sometimes epileptic crises and jaundice may occur as side effects.
f caproate d'hydroxyprogestérone
d Hydroxyprogesteroncaproat
i idrossiprogesterone caproato
e caproato de hidroxiprogesterona

1615 hydroxyproline
Hydroxypyrrolidine carboxylic acid, a glucogenic amino acid which is used therapeutically in the treatment of disorders of the metabolism.
f hydroxyproline
d Hydroxyprolin
i idrossiprolina
e hidroxiprolina

1616 hydroxystilbamidine
Serotonin. Substance which is released from blood platelets by an antigen-antibody reaction. It has antifungal and antiprotozoal properties and has proved effective against Gram-positive microorganisms. Mainly used in the treatment of North American blastomycosis, visceral leishmaniosis and grave infections caused by Candida albicans. The drug has also found employment in the palliative treatment of myelomatosis. Dizziness, tachycardia, hypotension, vomiting, excessive salivation, sometimes incontinence and oedema of the face and/or eye lids may occur as adverse effects.

f hydroxystilbamidine
d Hydroxystilbamidin
i idrossistilbamidina
e hidroxistilbamidina

1617 hydroxyzine
Central nervous system depressant indicated for the relief of tension and anxiety in grave emotional disturbances. In the treatment of psychoses, it has proved less effective than chlorpromazine. Headache, excessive drowsiness and some times convulsions may occur as side effects. Overdosage may cause coma.
f hydroxyzine
d Hydroxyzin
i idrossizina
e hidroxicina

*** hyosciamine s. atropine**

1618 hyoscine hydrobromide
Parasympatholytic agent characterized by central and peripheral actions which resemble those of atropine but for the fact that the drug produces central depression and hypnosis instead of stimulation, thus tending to slow the heart. It is widely used in surgery when the hypnotic effect suggests it instead of atropine and as an antiemetic for air- or sea-sickness.
f hydrobromure d'hyoscine
d Hyoszinhydrobromid
i ioscina idrobromuro
e hidrobromuro de hioscina

*** hyoscine methobromide s. hyoscine hydrobromide**

1619 hypnagogue
Sleep-inducing agent. Hypnotic. Relating to any drug capable of promoting the process of inducing sleep.
f hypnagogue; hypnotique d'induction; endormisseur
d Einshlafmittel; Hypnagogum
i ipnagogo; induttore del sonno
e hipnagogo; adormecedor

1620 hypobromite
Any salt of hypobromous acid, q.v. The sodium hypobromite is used to estimate the content of urea in the urine.
f hypobromite
d Hypobromit
i ipobromito
e hipobromita

1621 hypobromous acid
Unstable compound whose salts and in particular sodium hypobromite are used to liberate nitrogen in the measurement of urea in the urine. It is a very weak acid with oxidizing properties.
f acide hypobromeux
d unterbromige Säure
i acido ipobromoso
e ácido hipobromoso

1622 hypochlorite
Any salt of hypochlorous acid wich is used in medicine as disinfectant and antiseptic.
f hypochlorite
d Hypochlorit
i ipoclorito
e hipoclorito

1623 hypochlorous acid
Oxyacid corresponding to the monovalent chlorine and only known in its aqueous solution or in the form of salts (hypochlorites). Owing to the action of light and heat, the aqueous solutions of the acids mostly decompose into hydrochloric acid and oxygen and in a much smaller proportion into chlorine and chloric acid. The acid itself is used as a bleaching agent and its hypochlorites as disinfectants and antiseptics in surgery.
f acide hypochloreux
d unterchlorige Säure
i acido ipocloroso
e ácido hipocloroso

1624 hypoiodous acid
Oxyacid of the monovalent iodine unknown in its free state and only known in solution. It is a very weak acid and therefore its salts are strongly hydrolyzed. Solutions are very unstable and their odour resembles that of iodoform. Chemically, they function as oxidizing agents.
f acide hypo-iodeux
d unterjodige Säure
i acido ipoiodoso
e ácido hipoyodoso

1625 hyponitrous acid
Oxyacid obtained in the form of its salts by oxidation of hydroxylamine or by reduction of nitrites. It is useful in organic synthesis, both as an oxidizing and a reducing agent.
f acide hyponitreux
d Untersalpetrigsäure
i acido iponitroso
e ácido hiponitroso

1626 hypophosphite
Salt of the hypophosphorous acid. The calcium, iron, sodium and potassium hypophosphites are used in medicine as tonics in the treatment of anaemia and neurasthenia.
f hypophosphite
d Hypophosphit
i ipofosfito
e hipofosfito

1627 hypophosphoric acid
Acid of pentavalent phosphorus which gives rise to the hypophosphates. It is an unstable tetrabasic acid only obtained in the form of its salts.
f acide hypophosphorique
d Unterphosphorsäure
i acido ipofosforico

e ácido hipofosfórico

1628 hypophosphorous acid
Oxyacid of the phosphorus forming the hypophosphites which are used in the treatment of anaemia and neurasthenia owing to their tonic properties.
f acide hypophosphoreux
d Unterphosphorigsäure
i acido ipofosforoso
e ácido hipofosforoso

1629 hypromellose
Indigestible plant residue resembling methylcellulose but only used as a lubri cating agent in eye drops.
f hypromellose
d Hypromellose
i ipromellosi
e hipromelloso

1630 hyssop
Hissopus officinalis from whose leaves an oil is extracted (hyssop oil) by distillation in a stream of steam. It is a colourless, alcohol- or ether-soluble liquid used for the preparation of solutions or syrups having eupeptic, expectorant and carminative properties.
f hysope
d Hyssop; Ysop
i issopo
e hisopo

I

1631 iboga
The Tabernanthe iboga and related species growing in West Africa and from
whose roots, bark and leaves a crystalline alkaloid is obtained which the
natives use as a tonic and aphrodisiac
agent. The alkaloid itself is called
ibogaine.
f iboga
d Iboga
i iboga
e iboga

1632 ibrotamide
Tranquillizer, sedative and mildly hypnotic agent indicated in the treatment
of hyperemotivity, anxiety, irritability
and, both therapeutically and prophylac
tically, of all those symptoms which are
associated with menstruation and menopause. The drug is also used to cure
neurovegetative disorders.
f ibrotamide
d Ibrotamid
i ibrotamide
e ibrotamida

1633 ibuprofen
Non-steroid antiphlogistic agent eight to
32 times more effective than acetylsalicylic acid. As an analgesic and an anti
pyretic, it is indicated in the treatment
of sciatica, lumbago etc. The inflammation of the affected parts is due to its
capacity to inhibit prostaglandin synthesis. Gastrointestinal disorders may occur
as side effects.
f ibuprofène
d Ibuprofen
i ibuprofen
e ibuprofeno

1634 ibuproxam
A non-steroid antiinflammatory, analgesic and antipyretic agent, approximately
30 to 35 times as potent as acetylsalicylic acid. The drug is used in the
treatment of rheumatic and inflammatory
conditions, both acute and chronic, occurring in the osteoarticular system. It
is also indicated in osteoarthrosis, particularly dorsal, cervical and lumbar
arthrosis, lumbago, sciatica, radiculo-
neuritis (acute febrile polyneuritis) and
rheumatoid arthritis. Headache and some
epigastric pains may occur as side effects.

f ibuproxam
d Ibuproxam
i ibuproxam
e ibuproxam

1635 ichthyol
Or, preferably, ichthammol. Mixture of
the ammonium salts of the sulphonic
acids which is obtained from the oily
substance derived by the destructive
distillation of the fossilized remains of
fish and, more generally, marine animals. It is a brownish-black viscous
liquid used in medicine as an antiseptic
in cutaneous infections and as a vasoconstrictor. It appears that its properties are due to organic sulphur compounds.
f ichtyol
d Ichthyol
i ittiolo
e ictiol

1636 idoxuridine
Antiviral agent used in the topical treat
ment of herpes infections, in particular
herpetic keratitis. Its antiviral mechanism is based on the substitution of uridine to thymidine in the DNA structure
of the virus. Also used in ophthalmology.
f idoxuridine
d Idoxuridin
i idossuridina
e idoxuridina

1637 ifenprodil
Cerebral vasodilator having an alpha-
-blocking sympatholytic action and used
in the treatment of cerebrovascular insuf
ficiency and diseases, functional disorders caused by arterial hypertension,
vertigo and vasculopathies affecting the
retina.
f ifenprodil
d Ifenprodil
i ifenprodil
e ifenprodil

1638 Ignatius bean
Or Saint Ignatius bean. The greenish,
yellowish seed of a Philippine woody
vine acting and used like nux vomica,
a poisonous seed which contains several
alkaloids, mainly strychnine and brucine.
f fève de St. Ignace; fève isagurique
d Ignatiusbohne; Ignazbohne
i fava di S. Ignazio

e haba de San Ignacio; haba de Isagur

1639 imipramine
Antidepressant agent capable of blocking
the neural re-uptake of noradrenaline,
dopamine and 5-hydroxytryptamine. The
hydrochloride of the base is a white
odourless crystalline powder characterized by a pungent taste that is followed
by a sensation of numbness. Anticholinergic actions, e.g. dryness in the
mouth, a blurred vision and precipitation of glaucoma, constipation and retention of urine, as well as cardiac arrhythmias may occur as adverse effects.
f imipramine
d Imipramin
i imipramina
e imipramina

1640 imipraminoxide
The principal metabolite of imipramine,
q.v. a tricyclic antidepressant used in
the treatment of psychic depression, ner
vous and senile depression. Vision disorders, constipation, tachycardia and
hypotension may occur as adverse effects.
f imipraminoxyde
d Imipraminoxyd
i imipraminossido
e imipraminóxido

1641 immunoglobulin G
Concentrate of antibodies obtained from
the human plasma and used to attain
short-term immunity against some viral
infection, e.g. hepatitis.
f immunoglobuline G
d G-Immunoglobulin
i immunoglobulina G
e inmunoglobulina G

1642 immunoglobulins
Proteins characterized by a known antibody activity and some other proteins,
e.g. the gamma proteins, related to them
by chemical structure and thus antigenic
specificity. There are six distinct
classes of immunoglobulins so far known
in man. Analogous proteins are found in
animals. Immunoglobulins represent the
expression of the humoral immunity. The
term is also generically used to denote
any type of antibody. Immunoglobulins
are found in the blood plasma and other
fluids and tissues of normal and immu-
nologically stimulated individuals.
f immunoglobulines
d Immunoglobuline
i immunoglobuline
e inmunoglobulinas

**1643 immunoglobulinum equinum antilympho-
cyticum**
Concentrate of prevailingly IgG immunoglobulins isolated from the serum of
horses immunized with splenic cells of
human lymphocytes and standardized in
lymphocytotoxic units. It displays a
depressive action on the lymphocytes in

the blood, as well as on lymphoid organs and on antibody biosynthesis. The
drug is used in medicine to prevent the
reject phenomena which may occur in
transplantations and in the treatment of
autoimmune diseases. Sensitization to
proteins may occur as adverse effect.
f immunoglobulinum equinum antilympho-
cyticum
d Immunoglobulinum equinum antilympho-
cyticum
i immunoglobulinum equinum antilympho-
cyticum
e immunoglobulinum equinum antilympho-
cyticum

1644 immunoglobulinum equinum normale
Immunosuppressor used to avoid reject
phenomena in organic and tissular homo-
transplantation. It is also indicated in
the therapy of some autoimmune diseases.
Systemic reactions with fever, thrombophlebitis at the site of the intravenous
injection, tachycardia and hypotension
may occur as side effects.
f immunoglobulinum equinum normale
d Immunoglobulinum equinum normale
i immunoglobulinum equinum normale
e immunoglobulinum equinum normale

1645 immunoglobulinum equinum ophthalmicum
Equine ophthalmic immunoglobulin, i.e.
each of the immunoglobulins isolated
from the serum of horses treated with
extracts of eye and retina. A cicatrizant
agent used in the treatment of traumatic
or infectious lesions of the cornea and
the conjunctiva, of retinitis, choroiditis
myopica, retinitis and in cases of detachment of the retina. It is also indicated in the weakening of the sight caus
ed by a variety of aetiological causes.
f immunoglobulinum equinum ophthalmicum
d Immunoglobulinum equinum ophthalmicum
i immunoglobulinum equinum ophthalmicum
e immunoglobulinum equinum ophthalmicum

1646 immunoglobulinum equinum tropycum
Immunoglobulins isolated from the serum
of horses treated with embryonal extracts of pigs. A cicatrizant agent also
capable of an analgesic action and used
in the treatment of severe burns, non-
-exudative dermatosis and some types of
ulcers.
f immunoglobulinum equinum tropycum
d Immunoglobulinum equinum tropycum
i immunoglobulinum equinum tropycum
e immunoglobulinum equinum tropycum

**1647 immunoglobulinum humanum antimorbil-
licum**
Sterile preparation of immunoglobulin G
containing antibodies that are specific
against the morbillous virus. It is used
for the passive immunization against
measles and is indicated in the prophylaxis of individuals exposed to contagion,
especially unweaned babies and very
small children, as well as debilitated

patients and patients recently vaccinated with live attenuated vaccine. Hardly any side effects have been observed.

f immunoglobulinum humanum antimorbil-licum

d Immunoglobulinum humanum antimorbil-licum

i immunoglobulinum humanum antimorbil-licum

e immunoglobulinum humanum antimorbil-licum

1648 immunoglobulinum humanum antiparoti-ticum

Sterile preparation of immunoglobulin C containing specific antibodies against the parotitic virus. It is used for the passive immunization against parotitis and indicated in the prophylaxis of indi viduals who are exposed to contagion. It is also used in the treatment of the disease itself and in order to prevent complications. The immunity generally lasts up to four weeks.

f immunoglobulinum humanum antiparoti-ticum

d Immunoglobulinum humanu antiparoti-ticum

i immunoglobulinum humanum antiparoti-ticum

e immunoglobulinum humanum antiparoti-ticum

1649 immunoglobulinum humanum antipertus-sicum

Sterile preparation of immunoglobulin G containing specific antibodies against the Bordella-pertussis-produced toxin. It is used for the passive immunization against pertussis and indicated for the prophylaxis of all those who may be exposed to contagion, in particularly babies and very young children. It also finds employment in the treatment of paroxysmal and intense catarrhal stages of the disease. The immunity generally lasts for up to four weeks.

f immunoglobulinum humanum antipertus-sicum

d Immunoglobulinum humanum antipertus-sicum

i immunoglobulinum humanum antipertus-sicum

e immunoglobulinum humanum antipertus-sicum

1650 immunoglobulinum humanum antirabicum

Sterile preparation of immunoglobulin G containing specific antibodies against the rabies virus, and thus used in the passive prophylaxis of rabies in asso-ciation with vaccinal therapy.

f immunoglobulinum humanum antirabicum

d Immunoglobulinum humanum antirabicum

i immunoglobulinum humanum antirabicum

e immunoglobulinum humanum antirabicum

1651 immunoglobulinum humanum anti-Rho D

Homologous specific antibody for the passive immunization against Rho (D)

factor in Rho(D), D-negative women fol-lowing Rh incompatible pregnancies. It acts by neutralizing the Rho(D) factor of the fetal erythrocytes in the maternal circulation during pregnancy or at the moment of birth, thus preventing the formation of (active) antibodies by the mother, which are the cause of the haemolytic disease of the newborn in ensuing Rh-incompatible pregnancies.

f immunoglobulinum humanum anti-Rho D

d Immunoglobulinum humanum anti-Rho D

i immunoglobulinum humanum anti-Rho D

e immunoglobulinum humanum anti-Rho D

1652 immunoglobulinum humanum antirubeo-licum

Sterile preparation of immunoglobulin G containing specific antibodies against the rubella virus. It is used for the passive immunization against rubella and used for the prophylaxis of non-im-munized individuals which may be expos ed to contagion. The immunity lasts for up to four weeks.

f immunoglobulinum humanum antirubeo-licum

d Immunoglobulinum humanum antirubeo-licum

i immunoglobulinum humanum antirubeo-licum

e immunoglobulinum humanum antirubeo-licum

1653 immunoglobulinum humanum antivacci-nicum

Sterile preparation of immunoglobulin G containing antibodies that are specific against the vaccine virus. It is used for the passive immunization against smallpox and indicated in the prophy-laxis and treatment of complications borne of smallpox vaccination, in parti-cular postvaccinal encephalytis. Also indicated in passive smallpox prophy-laxis in individuals for whom vaccina-tion may prove dangerous as well as in instances of accidental inoculation of vaccinal virus.

f immunoglobulinum humanum antivacci-nicum

d Immunoglobulinum humanum antivacci-nicum

i immunoglobulinum humanum antivacci-nicum

e immunoglobulinum humanum antivacci-nicum

1654 immunoglobulinum humanum normale

Filtration-sterilized preparation contain-ing the near totality of the immunoglo-bulins G of the human plasma together with small quantities of other plasma proteins or almost all immunoglobulins G of placental origin together with small quantities of other human proteins. Nor-mal human immunoglobulins contain the antibodies of normal adults and may be fluid or lyophilized. It must be injected only intramuscularly. It is used in the

prophylaxis and in the treatment of
measles, rubella, poliomyelitis, infectious hepatitis, pertussis and parotitis.
The immunity generally lasts from 4 to
6 weeks. It is also indicated against
hypogammaglobulinaemia.
f immunoglobulinum humanum normale
d Immunoglobulinum humanum normale
i immunoglobulinum humanum normale
e immunoglobulinum humanum normale

1655 immunoserum antianthracicum
Specific immunotherapeutic agent for the
prophylaxis and the treatment of anthrax in man, both in the most frequent
clinical form, i.e. the cutaneous one,
and in the internal intestinal or pulmonary one. It is generally administered
in association with antibiotics. The glo-
bulin-containing serum is obtained from
horses which have been hyperimmunized
against the anthrax bacillus.
f sérum antianthrax
d Antikarbunkelserum
i siero anticarbonchio
e suero antiántrax

1656 immunoserum antibotulinum
Antibotulinic serum used in the treatment
of poisoning caused by the ingestion of
food contaminated by Clostridium botu-
linum.
f sérum antibotulinique
d Antibotulinserum
i siero antibotulinico
e suero antibotulínico

1657 immunoserum anticlostridium mixtum
Polyvalent immunoserum against gaseous
gangrene in grave wounds which are
marked by laceration or devitalization
and contaminated by infectious material.
The drug is also indicated, both in man
and animals, for the treatment of the
actual disease especially in association
with human gamma globulins.
f immunoserum anticlostridium mixtum
d Immunoserum anticlostridium mixtum
i immunoserum anticlostridium mixtum
e immunoserum anticlostridium mixtum

1658 immunoserum antidiphthericum
Diphtheria antitoxin. Used for the prophylaxis and the treatment of diphtheria.
f sérum antidiphtérique
d Anti-Diphtherie-Serum; Diphtherieantitoxin
i siero antidifterico
e suero antidiftérico

1659 immunoserum antilatrodecticum
Immunotherapeutic agent of equine origin
endowed with the specific capacity of
neutralizing the poison of Latrodectus
mactans (the black widow), a genus of
spiders whose bite may prove fatal.
f immunoserum antilatrodecticum
d Immunoserum antilatrodecticum
i immunoserum antilatrodecticum
e immunoserum antilatrodecticum

1660 immunoserum antirabicum
Preparation obtained by purification of
a native immunotherapeutic substance
containing the antiviral globulins endow
ed with the specific capacity of neutralizing the rabies virus. It is used both
for prophylaxis and therapy.
f immunoserum antirabicum
d Immunoserum antirabicum
i immunoserum antirabicum
e immunoserum antirabicum

1661 immunoserum antitetanicum
Immunotherapeutic agent for human and
veterinary use, obtained by purification
of a native immunoserum containing the
antitoxic globulins endowed with the
specific capacity of neutralizing the
toxins elaborated by the Clostridium
tetany. Some allergic reactions may occur as side effects.
f immunoserum antitetanicum
d Immunoserum antitetanicum
i immunoserum antitetanicum
e immunoserum antitetanicum

1662 immunoserum antivenenosum
Immunotherapeutic agent for human and
animal use, obtained by purification of
a native serum containing the antitoxic
globulins endowed with the specific
capacity of neutralizing poisoning by
snakes, in particular vipers.
f immunoserum antivenenosum
d Immunoserum antivenenosum
i immunoserum antivenenosum
e immunoserum antivenenosum

1663 imolamine
Coronary vasodilator used in the treatment of angina pectoris. Also indicated
for the relaxation of vascular spasm
associated with myocardial infarction.
The drug should not be used in the
course of pregnancy.
f imolamine
d Imolamin
i imolamina
e imolamina

1664 inactivated lactobacilli
The vaccine from bacteria which is
found in the vagina of women affected
by trichomonal infection and which is
capable of provoking the immune response to infections including trichomoniasis and thus helps to prevent further
infections.
f lactobacilles inactivés
d inaktivierte Laktobacilli
i lattobacilli inattivati
e lactobacilos inactivados

1665 indapamide
Frusemide-derivative marked by a non-
-thiazidic diuretic action and indicated
in the therapy of essential arterial hypertension particularly when accompanied
by cardiac, cerebral, renal and ocular
complications. Orthostatic hypotension

has been observed as a side effect.
f indapamide
d Indapamid
i indapamide
e indapamida

1666 Indian hemp extract
North American plant yielding a substance which was used by natives to soothe inflammation of the joints caused by rheumatism.
f extrait de chanvre indien
d Indischhanfextract
i estratto di canapa indiana
e extracto de cáñamo indiano

1667 indicarmin
A diagnostic agent which is used when testing the functional capacity of the kidneys, i.e. in chromocystoscopy.
f indicarmin
d Indicarmin
i indigo carmine
e indaco carmín

1668 indigo
Blue vat dye originally obtained from plants by hydrolysis of the indican and oxidation by air of the resulting indoxyl and at present synthesized. The natural product contains indican, a glucoside of indoxyl from which the colouring matter is obtained by oxidation.
f indigotier
d Indigo
i indaco
e añil tinctorio

1669 indigo carmine
The soluble blue dye which is the sodium salt of indigosulphonic acid and is used as a biological dye (see indigosulphonic acid).
f carmin d'indigo
d Indigokarmin
i carminio d'indaco
e carmín de índigo

1670 indigosulphonic acid
Soluble form of indigo which is obtained by treating it with fuming sulphuric acid. The sodium salt, also called indigo carmine, is injected intravenously in the cytoscopic test for the function of the kidney.
f acide indigosulfonique
d Sulfoindigosäure
i acido indigosulfonico
e ácido indigosulfónico

1671 indigotin
Or indigo blue. The principal colouring matter of the natural indigo which is synthesized as a blue crystalline powder characterized by a coppery lustre generally by oxidation of synthetic indoxyl with air in the presence of alkali. The substance is insoluble and can be reduced to indigo white (a compound).
f indigotine; bleu indigo

d Indigoblau; Indigotin
i indigotina
e indigotina; azul indigo

* **indoledione s. isatin**

1672 indometacin
Antiphlogistic, analgesic and antipyretic agent characterized by an action which is more potent than that of hydrocortisone, phenylbutazone and the salicylates. The drug, whose effects generally last for a long time, is used in the treatment of chronic polyarthritis, osteoarthritis, spondylitis. Gastrointestinal disorders and ulcerations, nausea, vomiting, diarrhoea etc. may occur as adverse effects.
f indométacine
d Indometacin
i indometacina
e indometacina

1673 indoprofen
Non-steroid antiinflammatory antianalgesic agent used in various rheumatic syndromes, i.e. both articular and extra-articular rheumatic pains. The drug should be administered with great caution in the presence of gastritis, gastroduodenal ulcers, ulcerous colitis, as well as in pregnancy and in the course of lactation.
f indoprofène
d Indoprofen
i indoprofene
e indoprofeno

1674 indoramin
Alpha-adrenoreceptor blocking agent used in the treatment of hypertension and peripheral vascular disorders. Also used in the prophylaxis of migraine. Sedation may develop as a side effect.
f indoramine
d Indoramin
i indoramina
e indoramina

* **inflatine s. lobeline**

1675 infusion
Any solution which is injected intravenously for therapeutic purposes. The term also denotes a dilute solution which contains the water-soluble extract of a vegetable drug and is obtained by the maceration of the drug itself with usually boiling water and the straining of the liquid without pressing the residues. Drugs which contain easily soluble active principles can be subjected to infusion with cold water.
f infusion
d Infus
i infuso
e infusión

1676 injection
The act of injecting a drug or a sub-

stance into the body. Also, the solution or the suspension of a drug which is administered under or through the skin or mucous membrane by means of a syringe.

f injection; solution pour injection
d Einspritzung;   Einspritzungsflüssigkeit; Injection
i iniezione; liquido medicinale da iniettare
e inyección; líquido por inyección

1677 inosine
The nucleoside that is produced by the elimination of phosphoric acid from inosinic acid. It is a cell consituent easily capable of penetrating the cell membrane and of promoting the energetic ATP (adenosine triphosphate) and 2,3 diphosphoric acid, essential for cell metabolisms and in particular myocardial metabolism. The drug is therefore indicated in the treatment of acute and chronic myocarditis, senile heart, myocardial infarction and conduction disorders.

f inosine
d Inosin
i inosina
e inosina

1678 inositol
Vitamin B complex factor capable of regulating lipid and cholesterol metabolism and strengthening capillary walls. The drug is used as an adjuvant in the treatment of hepatopathies and arteriosclerosis, as well as of alterations of lipid metabolism and the condition of capillary walls. It also finds employment in some allergies.

f inositole
d Inositol
i inositolo
e inositolo

*** insect flower   s. pyrethrum**

1679 insect powder
Any chemical preparation used to exterminate insects or to check their infestation. In particular, the powder obtained from the pyrethrum flowers.

f poudre insecticide; poudre de pyrèthre
d Insektenpulver
i polvere insetticida; polvere di piretro
e polvo de pelitre; insecticida

1680 insulin injection
Sterile solution of insulin marked by the capacity of promoting the oxidation of glucose, of reducing the sugar level in the blood and strengthening the formation of glycogen. The drug is used in the treatment of diabetes mellitus and in diabetic coma.

f solute injectable d'insuline
d Insulininjektion
i insulina iniettabile
e insulina inyectable

1681 insulin injection (biphasic)
Sterile suspension of bovine insulin crystals in a neutral solution of swine insulin, i.e. a combination of slow-acting insulin with normal insulin. Owing to its biphasic effect, a rapid one followed by a prolonged one, the drug is used in the continuous treatment of mild and moderate forms of diabetes. A resistance to insulin may develop as a side effect, as well as some allergies and hypoglycaemic crises.

f solute injectable d'insuline biphasique
d Biphaseninsulininjektion
i insulina iniettabile bifasica
e insulina inyectable bifásica

1682 insulin injection (neutral)
Sterile solution of buffer insulin marked by a prompt hypoglycaemic effect of a comparatively short duration. The drug has proved very useful in the treatment of severe diabetes and diabetic coma. Insulin resistance and hypoglycaemia may occur as side effects.

f solute neutre injectable d'insuline
d Neutralinsulininjektion
i insulina iniettabile neutra
e insulina inyectable neutra

1683 insulin zinc suspension
Sterile buffer solution of the amorphous form of insulin with addition of zinc chloride. It is used for the treatment of various severe forms of diabetes. Its mechanism of action is similar to that of insulin but lasts for a longer time.

f suspension tampon d'insuline zinc
d Insulinzink-Pufferlösung
i insulina zinco sospensione
e suspensión de insulina zinco

1684 interferon
Proteic substance capable of interfering with the development of the virus by inhibiting its multiplication within the cell. It is thus regarded as an antiviral substance produced by the cells themselves after the penetration of the virus.

f interféron
d Interferon
i interferone
e interferón

1685 interferon alpha-2a
An antiviral agent deriving from bacteria by recombination techniques (the formation of new combinations of genes in fertilization and the formation of new combinations of linked genes in new heritable characters or new combinations of such characters). It is used to check the growth of AIDS-associated tumours, Kaposi's sarcoma. It is in some cases preferred to conventional cytotoxic agents as it is less likely to bring about a further elimination of immune system. Accurate monitoring is advisable. Some effects on the central nervous system and the cardiovascular system have been

observed in the course of the treatment.
f interféron alpha-2a
d Alpha-2a-Interferon
i interferone alfa-2a
e interferón alfa-2a

1686 interferon alpha-2b
Interferon alpha-2a-similar agent but only used for the treatment of "hairy cell" leukaemia. Experiments are being carried out for use in other forms of cancer.
f interféron alpha-2b
d Alpha-2b-Interferon
i interferone alfa-2b
e interferón alfa-2b

1687 interferon alpha NI
Interferon having the same properties of the alpha-2b.
f interféron alpha-NI
d Alpha-NI-Interferon
i interferone alfa-NI
e interferón alfa-NI

1688 intermedin
Chromatophorotrophic hormone. Hormone elaborated by the intermediate lobe of the hypophysis and capable of regulating the distribution of the cutaneous pigment of some inferior invertebrates, which is contained in particular cells (the melanophores, xanthophores etc.). In man, the function of intermedin which is also found in other parts of the hypophysis and in the diencephalon, appears to be identical with that of the melanocyte-stimulating hormone.
f intermédine
d Intermedin
i intermedina
e intermedina

1689 insulin
Antidiabetic hormone secreted by the islets of Langerhans of the pancreas as a response to the hyperglycaemia which follows each meal. Insulin causes an increase of the velocity of introduction of glycogen into the cells, the glycogen--genesis and the lipidic synthesis in the liver and in the adipose tissue. In diabetes mellitus, insulin functions as a regulator of the various metabolic processes.
f insuline
d Insulin
i insulina
e insulina

1690 inulin
Polysaccharide occurring as a reserve substance in some vegetables. It is a linear polymer constituted by the union of various molecules of d-fructose in the furanic form. A white, colourless powder, completely tasteless, little soluble in water in which, particularly at high temperature, it forms colloidal solutions. Thus differing from amide and glycogen,

it provides dyeing action with iodine. It is used in bacteriology as a culture medium, in the preparation of particular food items for diabetics and as a source of fructose. In medicine, it is used for the study and tests of renal function owing to its property to cross the glomerules of the kidney without being reabsorbed by the tubules.
f inuline
d Inulin
i inulina
e inulina

* **invertose   s. invert sugar**

1691 invert sugar
Mixture of D-glucose and D-fructose occurring naturally in fruit and honey. Also produced from a solution of cane sugar by hydrolysis.
f sucre inverti
d Invertzucker
i zucchero invertito
e azúcar invertido

1692 iocarmic acid
A radio-opaque substance.
f acide iocarmique
d Jokarminsäure
i acido iocarmico
e ácido yocármico

1693 iodamine
Radioopaque contrast medium used for the radiographic and radioscopic examination of the urinary tract, the biliary tract and the blood vessels. Allergic reactions and vomiting may occur as side effects.
f iodamine
d Jodamin
i iodamina
e yodamina

1694 iodate
The salt of iodic acid. Iodates are a generally colourless crystalline substance and are obtained by oxidation of the corresponding iodides. Neutral and acid salts are known. The acid salt of potassium is used in alkalimetry. Some of them display topical antiseptic properties.
f iodate
d Jodat; jodsaures Salz
i iodato
e yodato

1695 iodecoilium
Sterile complex of iodine with acylcolaminoformylmethylpyridinium, an antiseptic and disinfectant agent used in the prophylaxis and in the treatment of superficial infections caused by iodine--susceptible organisms. Inflammations and topical allergic reactions have been observed as side effects.
f iodécoile
d Jodekoil

i iodecoilio
e yodecoilio

1696 iodic acid
A colourless water-soluble crystalline
substance, the most stable of the iodic
oxyacids and is formed by the oxidation of iodine. It acts as an astringent
and is widely used in analytical chemistry.
f acide iodique
d Jodsäure
i acido iodico
e ácido yódico

1697 iodide
Salt or ester of hydriodic acid. Iodides
are the most important compounds of
iodine. The iodine contained in the
iodides is easily displaced by chlorine
and bromine with the formation of the
corresponding chlorides and bromides.
All iodides and especially the potassium
ones, represent a widely used form of
therapeutic administration of iodine.
f iodure
d Jodid
i ioduro
e yoduro

1698 iodinated glycerol
An expectorant agent mainly used in the
treatment of bronchitis and cough characterized by a hard and obstinate mucus.
Iodopropylene glycerol.
f iodoglycérol
d Jodoglycerol
i iodoglicerolo
e yodoglicerolo

1699 iodine
Chemical element belonging to the halogen group and found in nature only
combined with several metals. It is
usually obtained as a black, shining
crystalline substance that is readily
volatile forming a vapour characterized
by a pungent odour. Commercially,
iodine is obtained from seaweed and
from the deposits of sodium iodate found
with the nitrate in Chile, but is also
found in seawater, rocks, soils and
underground brines. It is soluble in
several solutions of iodides and in alcohol and used externally as a disinfectant and sterilizer. Deficiency of iodine
in the diet is often the cause of goitre
and, though indirectly, of cretinism. It
is frequently used in expectorants and
in the treatment of thyrotoxicosis.
f iode
d Jod
i iodio
e yodio

1700 iodine ointment
Ointment which is prepared by dissolving iodine in water with potassium
iodide and mixing in a simple ointment
base. It is generally used in medicine

as a counter-irritant in the treatment
of chilblains, fibrositis and swollen
glands.
f pommade iodée
d Jodsalbe
i pomata iodata
e pomada de yodo

1701 iodine trichloride
A yellowish crystalline substance characterized by a very penetrating odour,
which is easily soluble in water and
in organic solvent. It is used in the
form of weak solution as an antiseptic.
f trichlorure d'iode
d Jodtrichlorid
i iodio tricloruro
e tricloruro de yodio

1702 iodized oil
Viscous oily liquid smelling like garlic
and obtained by treating a fatty vegetable oil with iodine or hydriodic acid.
It is used as a contrast medium in
X-ray.
f huile iodée
d Jodöl
i olio allo iodio
e aceite yodado

1703 iodo-acetic acid
Crystalline acid obtained by reaction of
chloroacetic acid and a metallic iodide.
It is frequently used in biochemical
research owing to its inhibiting effect
in several enzymes.
f acide iodo-acétique
d Jodessigsäure
i acido iodoacetico
e ácido yodoacético

1704 iodobehenate
The salt of iodobehenic acid, a solid
mono-iodo derivative of behenic acid
(the crystalline fatty acid found in the
form of esters, especially in the fats
and oils that are obtained from seeds
and in some waxes) made by reaction of
erucic acid and hydriodic acid.
f iodobéhénate
d Jodbehenat
i iodobeenato
e yodobehenato

1705 iodocasein
Iodized casein (the protein occurring in
milk in the form of caseinogen and containing phosphorus and sulphur) containing 5.7% iodine. Its properties are similar to those of the inorganic preparations but it is better tolerated. It is
mainly used as a purifying agent and
in order to attain loss of weight.
f iodocaséine
d Jodkasein
i iodocaseina
e yodocaseína

1706 iododesoxycytidine
Antiviral agent used in the treatment of

herpetic keratitis, simplex and zoster herpes and follicular conjunctivitis. The drug, in the form of ointment, must be applied on the ocular sac several times in 24 hours.

f iododésoxycytidine
d Jododeoxycytidin
i iodoxicitidina
e iodoxicitidina

1707 iodoform
Tri-iodomethane. Yellow crystalline substance characterized by a penetrating odour and very unpleasant taste. It is easily soluble in fixed and volatile oils and used as an antiseptic in the form of a paste, particularly as wound dressing. Owing to its toxicity on absorption, the drug has now been superseded.

f iodoforme; méthane triiodé
d Jodoform; Trijodmethan
i iodoformio; metano triiodato
e yodoformo; triyoduro de metano

1708 iodogelamine
Organic compound of iodine. Weakly oxidative agents make it possible for the drug to release iodine and examethylentetramine which strengthens the purifying and stimulant action of iodine. It is mainly used in the treatment of arteriosclerosis, blood-vessel disorders occurring in old age and degenerative arthropathies. Skin rashes and occasionally conjunctivitis may occur as side effects.

f iodogélamine
d Jodogelamin
i iodogelammina
e yodogelamina

1709 iodopeptone
Iodine combined in organic form with peptone (the product of partial hydrolysis of proteins). It contains 18% of metallic iodine. It is easily and promptly assimilated and used in the treatment of arteriosclerosis, bronchial asthma, emphysema, lymphoadenitis, adenopathies and rheumatism. Also used against obesity.

f iodopeptone
d Jodopepton
i iodopeptone
e yodopeptona

1710 iodophthalein
Tetraiodophenolphthalein. A yellow disodium salt formed by the iodination of phenolphthalein in a solution of sodium hydroxide and used as a contrast medium in radiographies of the gall bladder and of the bile duct (cholecysto graphy) as it is excreted in the bile and opaque to x-rays. Nausea, vomiting, diarrhoea, vertigo and hypotension are possible adverse effects.

f iodophtaléine
d Jodophthalein
i iodoftaleina

e yodoftaleina

1711 iodopovidon
Organic compound of iodine, water-soluble and capable of releasing in a slow process the inorganic iodine in contact with skin and mucous membranes. As a disinfectant, it is less irritant than iodine and is thus used as a topical antiseptic and a non-toxic preparation for the treatment of oral infections.

f iodopovidone
d Jodopovidon
i iodopovidone
e yodopovidona

1712 iodopsin
The photosensitive substance that is contained in the cones of the retina, similar to rhodopsin and porphyropsin. It is formed from vitamin A and is very important in daylight vision.

f iodopsine
d Jodopsin
i iodopsina
e yodopsina

1713 iodopyrin
Colourless crystalline substance containing about 40% of iodine and used in medicine as a source of iodide ions. Antipyrin iodide.

f iodopyrine
d Jodopyrin
i iodopirina
e yodopirina

1714 iodothiouracil
Thyroid inhibitor used to combat hyperthyroidism occurring on a moderate scale. In addition to iodine effects, it displays the thiouracil-caused effects. Also used as a preoperative treatment in thyroidectomy.

f iodothiuracil
d Jodothiuracil
i iodotiuracile
e yodotiouracilo

1715 iodothyreoglobulin
Therapeutic agent marked by a pharmaco dynamic action which is characteristic of the thyroid gland. It stimulates the basal metabolism by increasing the cellular oxidative processes and is indicated in the treatment of hypothyroidism, mixoedema, retarded development and hypothyroidism-related obesity. Insomnia and tachycardia have been observed as side effects in hypersensitive patients.

f iodothyréoglobuline
d Jodothyreoglobulin
i iodotireoglobulina
e yodotireoglobulina

1716 iodoxyquinolinesulphonic acid
Pale yellowish powder readily dissolving in water with effervescence to yield a solution marked by a deep-orange colour. It is used as an amoebicide, generally

in the form of tablets but also in solution.
f acide iodoxyquinolinesulfonique
d Jodoxyquinolinsulfonsäure
i acido iodossiquinolinsolfonico
e ácido yodoxiquinolinsulfónico

1717 iofendylate
Radio-opaque contrast medium used in myelography, cholangiography. Also used in radioscopy to examine abscesses and fistulae.
f iofendylate
d Jofendylat
i iofendilato
e iofendilato

1718 ion exchange
The phenomenon whereby labile ions in solution replace ions of similar charge in those materials with which the solution comes into contact. The process has been often employed for the softening of water, when calcium and magnesium ions are replaced by sodium ions. Such property is inherent in several synthetic resins which are used therapeutically so as to reduce the sodium ions in the body in case of oedema or in order to substitute potassium ions with sodium ions in anuria and also to eliminate the calcium ions from blood to prevent clotting during storage in blood banks.
f échange ionique
d Ionenaustausch
i scambio ionico
e intercambio iónico

1719 ion exchange resin
Insoluble material of high molecular weight containing either acidic groups for the exchange of cations or basic groups for the exchange of anions. It is used in medicine in order to reduce the sodium content of the body or acidity in the stomach, as well as in ion exclusion and in all ion exchange processes.
f résine échangeuse d'ions
d Ionenaustauscherharz
i resina per lo scambio ionico
e resina intercambiadora de iones

1720 ionone
Violet-smelling terpene found in two isomeric forms. The ß-isomer is an intermediate in the synthesis of the vitamin A.
f ionone
d Jonon; Ionon
i ionone
e yonona

1721 iothalamic acid
Contrast medium used in diagnostic radiology in the form of its meglumine (organic base used for the preparation of salts of iodinated organic acids) and sodium salts.
f acide iothalamique

d Iothalaminsäure; Jodtalmin
i acido iotalamico
e ácido yotalámico

1722 ipecac
Ipecacuanha. Small South American and Indian plant displaying two kinds of roots having respectively a trophic and reserve function. Its dried rhizome and roots are no longer used in medicine but represent a source of emetine, i.e. an expectorant agent and also, if administered in an adequate dosis, an emetic and diaphoretic.
f ipéca; ipécacuana
d Ipecacuanha; Brechwurzel
i ipecacuana
e ipecacuana

1723 ipomoea resin
The crude resin which is obtained from Ipomoea, a genus of tropical and subtropical herbs and shrubs whose roots or tubers have cathartic properties. The ipomoea resin is a complex mixture mainly of glycosides of jalapinolic acid and derivatives and is marked by quick-acting and drastic purgative action.
f résine de jalap d'orizaba
d Orizabaharz
i resina di orizaba
e resina de escamonea mexicana

1724 ipratropium
Anticholinergic agent, mainly used in its bromide form, for its bronchodilator action in the treatment and prophylaxis of bronchitis. Also indicated for the treatment of obstructive bronchitis in the course of, and after surgical operation, of mild and moderate asthma, as well as chronic and infective asthma.
f ipratropium; bromure d'ipratropium
d Ipratropium; Ipratropiumbromid
i ipratropio; ipratropio bromuro
e ipratropio; bromuro de ipratropio

1725 iprindole
Tricyclic antidepressant agent mainly used in the treatment of the depressive phase of endogenous manic-depressive psychosis. It resembles imipramine both in its uses and effects.
f iprindole
d Iprindol
i iprindolo
e iprindolo

1726 iproclozide
Analeptic agent capable of inhibiting the monoamine oxidase and used as an antidepressant mainly in the prophylaxis of angina pectoris and the treatment of anorexia and serious loss of weight. The action of drug is rather slow but long-lasting. Not to be prescribed for patients suffering from psychic disorders.
f iproclozide
d Iproclozid
i iproclozide

e iproclozida

1727 iproheptine
Antihistaminic agent used in the treatment of acute or chronic rhinitis.
f iproheptine
d Iproheptin
i iproeptina
e iproeptina

1728 iproniazid
Term denoting the phosphate of the isopropylic derivative of the hydrazide of the isonicotinic acid. It was used in the past in psychiatry as an antidepressant. It is now employed as a monoamine oxidase inhibitor.
f iproniazide
d Iproniazid
i iproniazide
e iproniazida

1729 iron dextran injection
Parenteral injection administered for the treatment of iron-deficiency anaemia. In case of intravenous infusion, anaphylactic reactions may occur as side effects.
f injection de fer-dextrane
d Eisendextraninjektion
i iniezione di ferro-destrano
e inyección de ferrodextran

1730 iron sorbitol injection
Preparation having the same uses and effects of iron dextran but injected intra muscularly.
f injection de sorbitol-dextran
d Dextransorbitolinjektion
i iniezione di sorbitolo-destrano
e inyección de sorbitolo-dextran

1731 isaconitine
Benzaconine, picraconitine. Alkaloid of aconite formed from aconitine by loss of an acetyl group. It is capable of stimulating sensory nerve endings and therefore included in several liniments. Isaconitine is less toxic than aconitine.
f isaconitin
d Isaconitin
i isaconitina
e isaconitina

1732 isatin
Heterocyclic diketone, an orange-red solid substance that is soluble in hot water, less so in cold water, and is obtained by oxidation of indigo and synthetically starting from the alpha-iso nitroacetanilide. In its lactamic form, isatin is used in analytical chemistry as a reagent.
f isatine
d Isatin
i isatina
e isatina

1733 isoagglutinin
Each of the antibodies causing isoagglu-

tination, i.e. the phenomenon of the agglutination of serum with the red corpuscles of the same series.
f isoagglutinine
d Isoagglutinin
i isoagglutinina
e isoaglutinina

1734 isoaminile
Cough suppressant agent mainly used in the treatment of dry obstinate and irritative coughs. It is either taken on its own or in the form of linctus. It has no sedative or analgesic effects, nor does it depress respiration. It is particularly indicated in chronic pharyngitis and acute and chronic tracheobronchitis.
f isoaminile
d Isoaminil
i isoaminile
e isoaminilo

1735 isoamylacetate
Pear oil. Ester of acetic acid. Colourless liquid characterized by a pear smell and taste. It is soluble in water and in organic solvents and finds employment in the extraction of penicillin.
f acétate d'isoamyle
d Isoamylacetat
i isoamile acetato
e acetato de isoamilo; éster isoamilacético

1736 isoamyl alcohol
The alcohol which represents the largest proportion of the fusel oil of sugar fermentation. Optically inactive.
f alcool isoamylique
d Isoamylalkohol
i alcool isoamilico
e alcohol isoamílico

1737 isoamyl nitrite
Ester of the nitrous acid obtained from the acid by esterification with isoamyl alcohol. A yellow, hyaline, fruit-smelling, pungent-tasting inflammable liquid, soluble in alcohol and ether. It is used in medicine as a vasodilator and antispasmodic.
f nitrite de isoamyle
d Isoamylnitrit
i isoamile nitrito
e nitrito de isoamilo

1738 isoamyl valerianate
Or isoamyl valerate. Ester of the valerianic acid obtained by direct esterifica tion of this acid with isoamyl alcohol. It is a clear, apple-smelling liquid which is used in medicine as a sedative.
f valérianate de isoamyle
d Isoamylvalerianat
i isoamile valerianato
e valerianato de isoamilo

1739 isoascorbic acid
An acid which displays about one 20th of the vitaminic activity of the ascorbic acid and is used, generally in the form

of a sodium salt, as an antioxidant in foods.
f acide isoascorbique
d Isoascorbinsäure
i acido isoascorbico
e ácido isoascórbico

1740 isobromindione
Uricosuric (uricosuria is the excretion of uric acid in the urine) agent derived from phenindione. It has no anticlotting action but is capable of inhibiting the reabsorption of uric acid. It is used for the treatment of gout and hyperuricemic conditions. Its action is prompt, potent and reversible and can be inhibited by salicylates. Diarrhoea and skin rashes may occur as side effects. The drug must be used with caution as there exists the danger of precipitating the formation of renal calculi.
f isobromindione
d Isobromindion
i isobromindione
e isobromindiona

1741 isocarboxazid
Antidepressant agent of the hydrazine group of the monoamine oxidase inhibitors. The drug is used for the treatment of patients suffering from states of depression. There is no clear evidence that the drug attains the effects of electroshock therapy in severe cases but it is undoubtedly useful and indicated whenever a physical pathological conditions enhances the risk of such therapy. The drug is equally indicated in the depressive phase of manic-depressive states, involutionary psychosis and chronic depression caused by schizophrenia or some chronic debilitating disorders. See also phenezine. Side effects include anaemia, hepatitis, peripheral oedemas and cutaneous affections.
f isocarboxazide
d Isocarboxazid
i isocarbossazide
e isocarboxazida

1742 isocitric acid
Citric acid, in its isomerization, is subject to dehydration and thus becomes cis-aconitic, takes up a molecule of water and forms the isocitric acid. This is found in the leaves of several plants and in blackberry juice. It occurs in the cytric acid cycle of the metabolism.
f acide isocitrique
d Isozitronensäure
i acido isocitrico
e ácido isocítrico

1743 isoconazole
Therapeutic agent indicated in the treatment of fungal and protozoal vaginal infections. Nausea and a metallic taste in the mouth may occur as side effects.
f isoconazole
d Isoconazol
i isoconazolo
e isoconazolo

1744 isoephedrine
The stereoisomer (displaying an isomerism in which atoms are linked in the same order but differ in their spatial arrangement) of ephedrine, an alkaloid capable of stimulating the myocardium.
f pseudoéphédrine
d Isoephedrin
i pseudoefedrina
e isoefedrina

1745 isoetharine
Bronchodilator indicated in the treatment of bronchospasm occurring in bronchial asthma, chronic bronchitis and emphysema. It is a sympathomimetic agent displaying beta-adrenergic activity. Tachycardia, precordialgia, dizziness and hypotension may be observed as adverse effects.
f isoétharine
d Isoetharin
i isoetarina
e isoetarina

1746 isoflurane
Potent anaesthetic used by inhalation in major surgery and characterized by analgesic properties. Though capable of strengthening the effects of muscle relaxants, it is less likely to sensitize the heart to catecholamines.
f isoflurane
d Isofluran
i isoflurano
e isoflurano

1747 isoleucine
Amino acid containing two asymmetrical carbon atoms and present in most common proteins. It is an essential amino acid in diet for the maintenance of health.
f isoleucine
d Isoleucin
i isoleucina
e isoleucina

1748 isolysergic acid
An acid which is similar to the alkaloid ergotinine (q.v.) of the ergot group. It occurs in colourless crystals.
f acide isolysergique
d Isolyserginsäure
i acido isolisergico
e ácido isolisérgico

1749 isometheptence
Sympathomimetic spasmolytic agent with adrenaline-like actions and side effects. It is mainly used in the symptomatic treatment of migraine as it appears to be capable of constricting dilated blood vessels which are bound to cause very strong headache. The drug also finds employment for the alleviation of intestinal spasms, biliary colic, dysmenor-

rhoea and peptic ulcer. Hypertension may set in as a side effect.

f isométheptène
d Isomethepten
i isometeptene
e isometepteno

1750 isoniazid
Isonicotinic acid hydrazide. A compound displaying a potent activity against the Mycobacterium tuberculosis. Its action is bacteriostatic and bactericide. It is used in the prophylaxis and therapy of tuberculosis, especially pulmonary tuberculosis. The drug has also proved effective against streptomycin-resistant micro organisms. Neurovegetative disorders, polyneuritis, insomnia and psychic excitation are among the most frequent adverse effects.

f isoniazide
d Isoniazid
i isoniazide
e isoniacida

1751 isonixine
Antiphlogistic and analgesic agent used in the treatment of rheumatoid arthritis, arthrosis, lumbago, neuritis, tendinitis etc. It also finds employment in surgery, urology, gynecology and dentistry.

f isonixine
d Isonixin
i isonissina
e isonixina

1752 isopentaquine
Antimalarial drug acting on gametocytes and, combined with quinine, on the exoerythrocytic stage of the malarial parasite.

f isopentaquine
d Isopentaquin
i isopentachina
e isopentaquina

1753 isophane insulin
Very slow-acting insulin used in the treatment of mild to moderate diabetes mellitus. Some allergic reactions may be occurring as side effect. It is a sterile suspension of insulin containing the isophanic equivalent of protamine zinc chloride.

f isophane insuline
d Isophan-Insulin
i insulina isofano
e insulina isofano

1754 isophthalic acid
Dicarboxylic aromatic acid, isomer of the phthalic and terephthalic acids. It derives from the benzene by substitution of two hydrogen atoms in meta position with as many carboxylic groups. They occur as white crystals, not easily soluble in water, which are prepared by the oxidation of the m-xilene.

f acide isophtalique
d Isophthalsäure

i acido isoftalico
e ácido isoftálico

1755 isoprenaline
A beta adrenoreceptor agonist (a synthetic sympathomimetic amine). It is used chiefly in the treatment of bronchial asthma. The drug does not affect cardiac activity and blood pressure. Some cardiac disorders, e.g. ventricular fibrillation and tachycardia, may occur as side effects.

f isoprénaline
d Isoprenalin
i isoprenalina
e isoprenalina

1756 isopromethazine
Therapeutic agent used in the treatment of various allergies.

f isoprométazine
d Isopromethazin
i isoprometazina
e isoprometazina

1757 isopropamide
A peripheral-mechanism parasympatholytic agent mainly used in the symptomatic treatment of the gastroduodenal ulcer, gastrointestinal colic, spastic gastritis, hyperchlorhydria and biliary colic. Its action is both rapid and long-lasting. Dryness in the mouth, mydriasis, constipation and some accommodation disorders may occur as side effects. Generally used in its iodide form.

f Isopropamide
d Isopropamid
i isopropamide
e isopropamida

1758 isosorbide dinitrate
Coronary vasodilator with a peripheral mechanism of action, used in insufficiencies of the coronary circulation marked by inevitable cardiac disorders. In angina pectoris, it acts like nitroglycerin. Its action is prompt and generally lasts up to four hours. Methenoglobinaemia, gastric disorders, nephrotoxicity if the drug is injected and frequently nausea may occur as adverse effects.

f dinitrate d'isosorbide
d Isosorbiddinitrat
i isosorbide dinitrato
e dinitrato de isosorbido

1759 isosorbide mononitrate
Coronary vasodilator used for the prophylaxis of angina pectoris. Being an active metabolite of isosorbide dinitrate it is not subject to further metabolism and its effect is thus more predictable.

f mononitrate d'isosorbide
d Isosorbidmononitrat
i isosorbide mononitrato
e mononitrato de isosorbido

1760 isotonic
In chemistry, relating to solutions mark-

ed by the same osmotic pressure. In medicine, isotopic serums are those aqueous solutions of salts and proteic or medicinal substances with an osmotic pressure and a hydrogenionic concentration that is analogous to that of the circulating blood plasma. Such serums are widely used for intravenous injections and for phleboclysis in the most diverse conditions, e.g. shock, acute anaemia, grave intoxications etc., for which they are prepared with adequate substances, e.g. salts, proteinic substances (amino acids), sugars etc.

f isotonique
d isotonisch
i isotonico
e isotónico

1761 isotretinoin

A vitamin A derivative used for the treatment of severe antibiotic-resistant acne. It probably acts directly on the sebaceous glands in the skin to bring about a reduction of sebum production. Pronounced dryness of the skin, membranes and conjunctivae and sometimes nausea, cephalalgia, pains in the joints, teratogenesis and biochemical evidence of liver damage may follow the treatment as adverse effects. The drug should not be prescribed in the course of pregnancy and in the presence of some liver or kidney disease.

f isotrétinoïne
d Isotretinoin
i isotretinoina
e isotretinoina

1762 isovaleric acid

Carboxylic fatty acid present in considerable quantities in the roots of valerian and angelica and also occurring in decomposed cheese and the sweat of the feet. In certain diseases, it is also found in urine. It is used medicinally as a sedative in hysteria and in neurotic states.

f acide isovalérianique
d Valeriansäure
i acido valerianico
e ácido valeriánico

1763 isoxsuprine

Vasodilator especially active at the cerebral level and generally used for the relief of vascular-insufficiency-related symptoms almost always associated with arteriosclerosis and cerebral ischaemia. The drug is also used for the relaxation of uterine muscles.

f isoxsuprine
d Isoxsuprin
i isossuprina
e isoxuprina

1764 ispaghula

The ripe seed of the Plantago ovata, a herb growing in India and in Iran. It contains mucilage and swells in water or in contact with water to form a gelatinous mass which is used in diarrhoea as a demulcent and, generally, in the treatment of intestinal atony. Also known as spongel.

f ispaghul
d indisches Flohkraut
i ispaghul; piantaggine d'India
e llantén de India

1765 itramin tosylate

2-nitrate-ethylamine toluene-p-sulphonate. Coronary vasodilator used for the treatment of coronary insufficiency and angina pectoris. Its action is rapid and long-lasting.

f tosylate d'itramine
d Itramintosylat
i itramina tosilato
e tosilato de itramina

J

1766 jaborandi
Name of several plants of the families
Rutaceae and Piperaceae, characterized
by diaphoretic and salivant properties.
Also called maranham.
f jaborandi
d Jaborandi
i jaborandi
e jaborandi; jaborandi de Marañon

1767 jaborandi leaves
The small leaves of the Brazilian shrub
Pilocarpus microphyllus, which contain
approximately 0.5% of an alkaloid called
pilocarpine (q.v.) on which the main
therapeutic action of the drug depends.
Being reputed to promote the growth of
hair, the extract, or tincture, obtained
is added to hair tonics.
f feuilles de jaborandi
d Jaborandiblätter
i foglie di jaborandi
e hojas de jaborandi

1768 jalap
The dried tubers of the Convolvulus
jalapa or Ipomoea purga which contains
up to 12% of a resin which resembles in
its action that obtained from ipomoea
(genus of tropical and subtropical climbing herbs and shrubs whose roots or
tubers contain a cathartic principle),
yet differs from it in its chemical constitution. The powdered drug and the tincture prepared from it are used as purgatives.
f jalap
d Jalape
i gialappa
e jalapa

1769 jalapin
The portion of the jalap resin which is
precipitated from an alcohol solution by
the addition of ether and generally produced in pills used as purgative.
f jalapine
d Jalapin
i gialappina
e jalapina

1770 jalap resin
The resin contained in the dried tubers
of jalap and used for its cathartic properties. It is obtained from jalap by
extraction with alcohol.
f résine de jalap

d Jalapenharz
i resina di gialappa
e resina de jalapa

1771 Jamaica quassia wood
The wood obtained from some plants of
tropical America, which, once reduced
into powder, is used for the preparation
of decoctions and tonics.
f bois de quassia; quassia de la Jamaïque
d Jamaica-Quassiaholz
i legno quassio
e leño de cuasia de Jamaica

1772 jambul
Or jambolan plum. The Syzygium jambu-
lanum, a tree growing in tropical Asia,
whose bark is used for medicinal preparations.
f jambul; jamboulier
d Jambulbaum
i jambul
e jambul

1773 jasmine
Any of several shrubs of temperate and
warm regions, whose flowers are used
for the extraction of an essence used in
pharmaceutical preparations. The dried
roots and rhizomes of the yellow jasmin
which grows in South America, constitute the gelsenium used in pharmacy.
f jasmin
d Jasmin
i gelsomino
e jazmín

1774 jasmone
Liquid ketone derived from cyclopentene.
Peppermint oil. Aromatic carminative,
mild antispetic agent mainly used in the
treatment of meteorism. Also used in association with laxatives to avoid abdominal pains.
f essence de menthe poivrée
d Pfefferminzöl
i menta essenza; olio essenziale di menta
e esencia de menta piperita

1775 Java tea
The dried leaves of an East Indian mint
which are used to produce a powerful
diuretic agent.
f thé de Java; barbiflore
d Javatee; indischer Nierentee
i the di Giava; baffo di gatto
e té de Java; ortosifón

1776 jecoris extractum cardiacum
Preparation obtained from the liver of healthy calves and used as a therapeutic drug to regulate the myocardial function by its action on the cardiac metabolism. It is a vasodilator, increases the contractility of the myocardium and improves the cardiac rhythm. It is used in the treatment of acute or chronic diseases of the coronary arteries, asthenia and insufficiency affecting the heart, myocardosis and conduction disorders.
f extrait cardioactive de foie
d Jecoris Extractum cardiacum; herzwirksamer Leberextract
i estratto cardioattivo di fegato
e extracto cardioactivo de hígado

1777 jecoris extractum (per os)
Preparation obtained from fresh or frozen liver of healthy mammals by aqueous extraction followed by deproteinization by thermal coagulation and successive filtration. The extract must contain the hepatic thermostabile extraction fraction and be soluble in water and partially soluble in alcohol to 70%. The preparation may be used in its liquid form or reduced to a dry state by means of particular processes. The jecoris extractum is an antianaemic agent acting in accordance with its content of vitamin B_{12}, folic acid and other haemopoietic factors and used in the treatment of pernicious and macrocytic anaemia resistant to cyanocobalamin and hydroxycobalamin. In addition to a detoxifying action, the drug also protects the liver.
f jecoris extractum (per os)
d Jecoris extractum (per os)
i jecoris extractum (per os)
e jecoris extractum (per os)

1778 jecoris extractum purificatum (ad injectionem)
See definition of jecoris extractum (per os).
f jecoris extractum purificatum (ad injectionem)
d Jecoris extractum purificatum (ad injectionem)
i jecoris extractum purificatum (ad injectionem)
e jecoris extractum purificatum (ad injectionem)

1779 jecoris extractum totum (ad injectionem)
See definition of jecoris extractum (per os). The injection may be somewhat painful.
f jecoris extractum totum (ad injectionem)
d Jecoris extractum totum (ad injectionem)
i jecoris extractum totum (ad injectionem)
e jecoris extractum totum (ad injectionem)

1780 Jerusalem artichoke root
The root of a perennial plant growing in North America and all over Europe, from which inulin can be extracted.
f racine de topinambour
d Topinamburknolle; Erdbirnenknolle
i radice di topinambur
e raíz de aguaturma; cotufa

1781 Jerusalem coaslip
Better known as lungwort or pulmonary. A lichen, widely distributed both in Europe and America, formerly thought to be useful in the treatment of bronchitis.
f herbe aux poumons; herbe-coeur; sauge de Jérusalem
d Lungenkraut; blaue Schlüsselblume; Frauenmilchkraut
i polmonaria; erba dei polmoni; salvia di Gerusalemme
e pulmonaria manchade

* Jesuit bark   s. cinchona bark

1782 josamycin
Antibiotic obtained from cultures of Streptomyces narbonensis (josamyceticus), effective against sensitive strains of staphylococcus, streptococcus haemolyticus, diplococcus pneumoniae and several types of influenza.
f josamycine
d Josamycin
i josamicina
e josamicina

1783 juice
The fluid contents of animal cells and flesh, which can be extracted.
f jus; suc
d Saft; Pressaft
i succo
e jugo

1784 jujube
Mucilaginous and sweet berries from which a paste was prepared in the past for the manufacture of lozenges.
f jujube
d rote Brustbeere; Judendornbeere
i giuggiola
e azufaifa

1785 juniper
Evergreen shrub whose fruits are widely used in veterinary practice. The juniper oil which is extracted from them (a distillate) is chiefly used as a diuretic and a urinary antiseptic. It also finds employment as a carminative in meteorism, in colic and in lumbago.
f genévrier; genièvre
d Wacholder; Machandel
i ginepro
e enebro

1786 juniper oil
Juniperi aetheroleum. Volatile oil obtained by distillation of the ripe berries of juniper (Juniperus communis). It is a colourless or yellowish liquid

marked by a characteristic odour (a
penetrating turpentine-like smell) and
a pungent taste. Its principal constituents are pimene, camphene, cadinene
terpineol and juniper camphor. It is
used in medicine as a diuretic and
urinary antiseptic agent, as well as a
carminative in meteorism and in colic.

f essence de genièvre
d Wacholderöl
i essenza di ginepro
e esencia de enebro

K

1787 kainic acid
Anthelmintic agent used in the treatment of infections caused by the common roundworm to obtain a better effect out of santonin (a bicyclic sesquiterpenoid ketonic lactone found in santonica).
f acide kaïnique; acide chénique
d Kaininsäure
i acido cainico
e ácido caínico

1788 kaladana
The seeds of the Ipomoea hederacea, a climbing plant growing in India. For its activity and uses, see jalap.
f kaladane
d Kaladan
i caladana
e caladana

1789 kalii aminosalicylas
Potassium aminosalicylate. Antibacterial agent used in association with isoniazide or streptomycin in the treatment of pulmonary and other types of tuberculosis. Fever, diarrhoea, nausea accompanied by vomiting, albuminuria, blood dyscrasia, goiter and hypothyroidism may occur in case of prolonged administration.
f aminosalicylate de potassium
d Kaliumaminosalicylat
i potassio aminosalicilato
e aminosalicilato de potasio

1790 kalii bromidum
Potassium bromide. Sedative whose action is due to the bromine and which is indicated in the treatment of nervous hypersensitivity. Respiratory depression may be caused by the drug. Bromine acne may occur as a side effect.
f bromure de potassium
d Kaliumbromid; Bromkalium
i potassio bromuro
e bromuro potásico

1791 kalii canrenoas
Potassium canrenoate. Aldosterone-antagonist diuretic agent which inhibits the water and sodium reabsorption and the potassium depletion in the distal renal tubules. It is especially indicated in the treatment of oedemas of cardiac origin, liver cirrhosis and some nephrosis-related syndromes. The drug also finds employment in cardiovascular surgical operations, in intestinal paresis occurring after an operation, as an adjuvant in the treatment of arrhythmias, acute intoxications caused by tricyclic antihypertensives, hypokalaemia, respiratory insufficiency and cardiac decompensation. Nausea accompanied by vomiting, general exhaustion, a diminished sexual potency in men and hirsutism as well as cycle disorders in women may occur as side effects.
f canrénoate de potassium
d Kaliumkanrenoat
i potassio canrenoato
e canrenoato de potasio

1792 kalii chloras
Potassium chlorate. Oxidizing agent used as an antiseptic in skin and mucous membrane disorders. Particularly indicated in disorders of the oral mucosa. The drug displays a mild bactericide action. Its absorption may cause renal alterations and methaemoglobinaemia, i.e. the presence of methaemoglobin (haemoglobin) in the blood.
f chlorate de potassium
d Kaliumchlorat; chlorsaures Kalium
i potassio clorato
e clorato potásico

1793 kalii chloridum
Potassium chloride. Therapeutic compound capable of releasing potassium by dissociation. The potassium thus released acts on the cardiac, muscular and renal activity. It is indicated in the treatment and the prophylaxis of hypokalaemia, diabetic acidosis and infantile diarrhoea. It is also used as a diuretic and in the Menière's disease. Some cardiac disorders may occur as an adverse effect.
f chlorure de potassium
d Kaliumchlorid
i potassio cloruro
e cloruro potásico; cloruro de potasio

1794 kalii citras
Potassium citrate. Therapeutic agent capable of enhancing the potassium content of the body and indicated in potassium-deficiency syndromes. It has diaphoretic, diuretic and febrifugal properties and is excreted as the carbonate thus rendering the urine alkaline. It is mainly administered in cystitis, gout and enuresis caused by over-acid urine.

It is further used to prevent crystalluria in sulphonamide therapy.
f citrate de potassium
d Kaliumcitrat
i potassio citrato; citrato potassico
e citrato potásico

1795 kalii clorazepas
Potassium clorazepate. Benzodiazepine--related tranquillizer indicated in all anxiety states. The drug should be used with great caution in aged patients. Drowsiness, sometimes ataxia and hypotension may occur as adverse effects.
f clorazépate monopotassium
d Kaliumclorazepat
i potassio clorazepato
e clorazepato de potasio

1796 kalii ferrocyanidum
Potassium ferrocyanide. Soluble yellow compound used as a laboratory reagent. It has diaphoretic properties and was used in the past in night perspiration caused by phthisis.
f ferrocyanure de potassium
d Kaliumferrocyanid
i potassio ferrocianuro
e ferrocianuro de potasio

1797 kalii gluconas
Potassium gluconate. The potassium salt of gluconic acid which is rapidly absorb ed and well tolerated when administered orally to prevent and treat potassium deficiency in patients undergoing diuretic therapy, as well as a therapy based on digitalis compounds or corticosteroids.
f gluconate de potassium
d Kaliumgluconat
i potassio gluconato
e gluconato de potasio

1798 kalii glycerophosphas
Potassium glycerophosphate. Therapeutic agent used like other glycerophosphates in the treatment of nervous diseases and in conditions of general debility.
f glycérophosphate de potassium
d Kaliumglycerophosphat
i potassio glicerofosfato
e glicerofosfato de potasio

1799 kalii hydrogenotartras
Potassium bitartrate. Mild saline purgative less active than sodium and magnesium sulphate. The drug is particularly indicated in case it is advisable to soften the faeces and thus facilitate their passage.
f bitartrate de potassium; tartrate acide de potassium
d Kaliumbitartrat; weinsaures Kalium
i bitartrato potassico; potassio tartrato acido
e bitartrato potásico

1800 kalii jodidum
Potassium iodide. Soluble crystalline compound characterized by diuretic and expectorant properties. The drug is partially excreted by the bronchial glands and it thus renders the bronchial mucus less viscid. It is indicated in the treatment of chronic catarrhs and bronchitis. It is also capable of absorbing recently formed fibrous tissue and is therefore used in the later stages of syphilis to enhance the effectiveness of antisyphilitic drugs. It is also used in the treatment of actinomycosis, sporotrichosis and blastomycosis, but large doses are required for the purpose. In some circulatory diseases, it finds employment as an analgesic agent. Also used as a liniment for external use in case of enlarged glands.
f iodure de potassium
d Kaliumjodid
i potassio ioduro
e yoduro potásico

1801 kalii nitras
Potassium nitrate. Potassium salt used in the treatment of hypokalaemia occurring as an adverse effect of some diuretics or corticosteroids. In large doses, the drug may cause irritation of the kidney and of the gastrointestinal tract.
f nitrate de potassium; salpètre; nitre
d Kaliumnitrat; Kalisalpeter; Salpeter
i nitrato di potassio; salnitro; nitro
e nitrato potásico; nitro; salitre

1802 kalii perchloras
Potassium perchlorate. A compound capable of interfering with the iodine-binding mechanism of the thyroid gland. Owing to this property, it has been used in the past in the treatment of thyrotoxicosis.
f perchlorate de potassium
d Kaliumperchlorat
i potassio perclorato
e perclorato potásico

1803 kalii permanganas
Potassium permanganate. Disinfectant and oxidizing agent particularly used in the treatment of infections of the genito-urinary apparatus, as a deodorant and for vaginal irrigation. Owing to its oxidizing action, potassium permanganate finds employment as an antidote for several alkaloids.
f permanganate de potassium
d Kaliumpermanganat
i potassio permanganato; camaleonte minerale
e permanganato potásico

1804 kalii sorbas
Potassium sorbate. The potassium salt of 2,4-hexadienoic acid characterized by antibacterial and antifungal properties and used as a preservative agent.
f sorbate de potassium
d Kaliumsorbat
i potassio sorbato
e sorbato potásico

1805 kallidinogenase
Enzyme secreted by the salivary glands. A physiological vasodilator which, at the level of the ischaemic tissue, releases kallidin which is a hypotensive alphaglobulin. Owing to its action as a hypocholesteraemic agent and blood fluidifier, the drug is used in the treatment of peripheral vascular circular disorders, e.g. endo-angiitis and thromboangiitis obliterans, of cardiovascular circulatory complaints and, more generally, hypotension, severe wounds and ulcers. Precordial pains and cardiopalmus may occur as a side effect.
f callidinogénase
d Kallidinogenase
i callidinogenasi
e calidinogenasa

1806 kamala
The hairs and glands of the fruits of a tree (Mallotus philippinensis) growing in India, Australia and Malaysia, which are used as a cathartic and taenicide agent.
f glandes de kamala
d Kamaladrüsen; Mallotusdrüsen
i ghiandole di camala
e glándulas de rotlera; camala

1807 kamala tree
The tree (see kamala) from which capsules are gathered, which yield a reddish cathartic powder containing rottlerin and used as a vermifuge, particularly in veterinary practice.
f kamala
d Kamalabaum
i camala
e camala

1808 kanamycin
Antibiotic produced by Streptomyces kanamyceticus. Its sulphate has proved effective, as a polybasic antibiotic, against a large number of aerobic Gram-positive bacteria, e.g. strains of Staphylococcus and Mycobacterium tuberculosis as well as against Gram-negative bacteria, e.g. Bacterium coli, Aerobacter, Salmonella, Shigella, Klebsiella pneumoniae etc. As a therapeutic drug, it is used in the treatment of infections of the respiratory tract and the urogenital and digestive apparatus. It also finds employment in surgery. Allergic reactions and respiratory disorders may occur as side effects.
f kanamycine
d Kanamycin
i kanamicina
e kanamicina

1809 kaolin
China clay, argilla. Natural hydrated aluminium silicate which is found as a fine deposit and, once purified, is used in medicine as an adsorbent to be applied externally as a dusting powder

and orally in the treatment of diarrhoea owing to its capacity of increasing the faecal bulk and slowing down peristalsis.
f kaolin; bol blanc
d Kaolin; weisser Ton
i caolino; argilla lavata
e caolin; arcilla blanca

1810 kavain
Or kawain. Delta-lactone of the 5-hydroxy-3-metoxy-7-phenyl-2.6-heptadioenoic acid. Therapeutic agent which may be used in neurology. It is present in a beverage drunk on festive occasions in Polynesia to heigthen excitement. Addiction may cause extreme debility and the loss of use of the legs.
f kawaïne
d Kawain
i kawaina
e kawaina

1811 kava-kava
The rhizome of the Piper methysticum, a tropical plant whose action is similar to that of pepper. It is used to be administered in the form of a liquid extract as a diuretic and an antiseptic agent.
f kava-kava
d Kawa
i kava-kava
e kava-kava

1812 kavatin
Or methysticin. The liquid extract obtained from the rhizome of the kava-kava and used in Polynesia to heighten sensations and increase excitement. Its prolonged use may lead to addiction and cause nervous disorders.
f méthysticine
d Methysticin
i metistina
e methistina

1813 kebuzone
Active phenylbutazone metabolite having the same therapeutic effects of the latter but displaying much less toxicity. The drug is used as an analgesic, antipyretic and antiphlogistic agent in the treatment of rheumatism, both articular and extraarticular, phlebitis, gout and degenerative arthropathies. Water and salt retention may occur as side effects.
f kébuzone
d Kebuzon
i chebuzone
e cebuzona

1814 kellah
Or picktooth. Annual herb growing in Mediterranean countries. Its fruits are used therapeutically as tonic and digestive agents.
f visnage; herbe aux cure-dents
d Khellakraut; Zahnstöcherkraut
i bisnaga; pastricciano; visnada
e biznaga; viznaga

1815 kelp
Mass of growth of seaweeds whose ashes, the red wracks, are used for the extraction of their iodine content. Such process was very common in Spain, Scotland and Normandy.
f varech
d Meergras; Seetang
i fuco; varech
e fuco; hierba de mar

1816 keracyanin
Cyaninoside chloride. Vitamin factor P marked by an accelerating effect on the regeneration of rhodopsin. It is mainly used to improve accommodation to darkness particularly with a view to relieve the stress of night driving.
f kéracyanine
d Keracyanin
i keracianina
e keracianina

1817 keratin
Any of several sulphur-containing fibrous proteins which derive from the surface ectoderm (of which they form the chemical basis) constituting the horny layer of the skin. Keratins are therefore the principal components of hair, nails, feathers and scales. They are insoluble in most solvents and, thus differing from collagen and other proteins, are not digested by enzymes of the gastrointestinal tract. The elastic properties of the fibres are due to them. Keratins are rich in amino acid and are frenquently used for the preparation of pills, capsules etc. so as to allow the latter to pass through the stomach unaltered.
f kératine
d Keratin; Hornstoff
i cheratina
e queratina

1818 kermesic acid
Acid which is found in the red dye that is extracted from the dried bodies of various scales of the kermes oak. A dyestuff, possibly the oldest known to man, producing a red colour.
f acide kermésique
d Kermessäure
i acido chermessico
e ácido quermesínico

1819 kerosene
Or kerosine. Inflammable hydrocarbon not as volatile as petrol, generally obtained by the distillation of petroleum and used, among other things, as a diluent for insecticides and emulsions.
f kérosène; lampant
d Kerosin; Leichtpetroleum
i cherosene
e keroseno; queroseno

1820 ketamine

Non-barbituric parenteral anaesthetic characterized by analgesic properties when administered in subanaesthetic doses. Its action is rapid. Great caution is required in its use as it may cause some psychotic effects, e.g. hallucinations, which may however occur less frequently if a tranquillizer is associated. Some cardiac disorders may be observed as side effects.
f kétamine
d Ketamin
i chetamina
e cetamina

1821 keto acid
Or ketonic acid. Any compound which is both a ketone and an acid, i.e. an organic acid which displays in its structure both the carboxyl group –COOH and the divalent ketone group, =CO.
f acide cétonique
d Ketonsäure
i acido chetonico
e ácido quetónico

1822 ketobemidone
Analgesic agent acting through a central mechanism and used in the treatment of hepatic and renal colics, neuritis, postoperatives pains, as well as acute pains caused by angina pectoris, childbirth and fractures.
f cétobémidone
d Ketobemidon
i chetobemidone
e cetobemidona

1823 ketoglutaric acid
Either of the two crystalline keto derivatives of glutaric acid. It is an intermediate compound formed in the citric acid cycle, i.e. the chain of metabolic reactions in which the citric acid is the intermediary. Also known as oxoglutaric acid.
f acide cétoglutarique
d Ketoglutarsäure
i acido cetoglutarico
e ácido cetoglutárico

1824 ketone
Any of a class of organic compounds (acetone) which are characterized by a carbonyl group attached to two carbon atoms that are usually present in hydrocarbon radicals or in single bivalent radicals. They are similar to aldehydes but are considerably less reactive.
f cétone
d Keton
i chetone
e quetona

1825 ketone body
Either acetone or aceto-acetic acid or ß-hydroxybutyric acid occurring in the blood and urine of diabetics owing to the incomplete breakdown of fatty acids, a condition known as ketosis and some-

times developing following the administration of ether. Also known as acetone
body.
f corps cétonique
d Ketonkörper
i corpo chetonico
e cuerpo cetónico

1826 ketoprofen
Non-steroid antiinflammatory, analgesis
and antipyretic agent characterized by
a rapid absorption and mainly used
in rheumatoid arthritis, coxarthrosis,
spondylarthrosis and bursitis. The drug
is also indicated in the treatment of
sprains, contusions and luxations.
Nausea and sometimes anorexia may occur as side effects if the drug is administered orally.
f kétoprofène
d Ketoprofen
i chetoprofene
e cetoprofeno

1827 ketotifen
Antiallergic agent. An antiasthmatic
with antianaphylactic action capable
of blocking allergic mechanisms. The
drug is mainly used in the prophylaxis
of bronchial asthma, asthmatic conditions caused by hay-fever, allergic bron
chitis, rhinitis and some allergic dermatoses. Dryness of the mouth and vertigo
may occur as side effects.
f kétotifène
d Ketotifen
i chetotifene
e cetotifeno

1828 khellin
A smooth-muscle relaxant and coronary
dilator used in the treatment of angina
pectoris and bronchial asthma. It is
the active principle of Ammi visnaga
and generally administered in conjunction with papaverine.
f khelline
d Khellin
i chellina
e quelina

1829 khus-khus
Or vetiver. Aromatic grass whose roots
yield a useful essential oil of a yellow-
brown colour which contains sesquiterpenes.
f kus-kus; chiendent des Indes
d Vetivergras; indische Narde
i vetiver
e vetiver; ivarancusa

1830 kidneywort
One of the plants or herbs, e.g. navel-
wort, which were, and in some measure
still are, used for the treatment of
kidney diseases.
f gobelet; cotylet ombiliqué
d Nabelkraut
i erba bellica; orecchio d'abate
e conchelo; ombril de Venus

1831 kiku oil
Essential oil extracted from the Chrysanthemum segctum (kiku in Japanese) constituted by camphene, coumarin, esters
of angelic acid etc. It is either colourless or pale green and is used as a
therapeutic substance exactly as chamomile.
f essence de kikou
d Kikuöl
i essenza di kiku; olio di kiku
e aceite de kiku

1832 kino
Or kino gum. A tannin-containing dried
juice or extract obtained from several
tropical trees, in particular the trunk
of an East Indian tree, the Pterocarpus
marsupium, which is used as an astringent in diarrhoea.
f gomme kino
d Kino; Malabarkino
i gomma kino
e goma kino

1833 kitasamycin
An antibiotic isolated from cultures of
Streptomyces kitasatoensis. The drug is
effective against Gram-positive and
Gram-negative microorganisms, in particular when responsible for infections
affecting the respiratory tract and the
gastrointestinal tract. It is also indicated in the treatment of biliary, menin
geal and cutaneous infections.
f kitasamycine
d Kitasamycin
i chitasamicina
e leucomicina

1834 knight's spur
Or larkspur. Cultivated annual plant
(delphinium) yielding an acetic tincture
which may be used against infestation
by ectoparasites, in particular lice.
f dauphinelle
d Ackerrittersporn
i erba del cardinale; erba cornetta; cap-
puccio
e calcitrapa

1835 knotty rhatany
Or rhatany. The dried root of shrubs
growing in Central and South American
and used locally as an astringent.
f rhatanhia du Pérou
d Peru-Ratanhia; Rhatanhia
i retania del Peru
e ratania

1836 kojic acid
Crystalline phenolic-type water-soluble
toxic antibiotic derived from gamma-
-pyrone by fermentation, e.g. of glucose
with moulds. Chemically, a hydroxy
(hydroxymethyl)-pyrone produced by
Aspergilli (a genus of fungi of the
Ascomycetes family), which is mildly
active against several bacteria.
f acide kojique

d Kojisäure
i acido kojico
e ácido kójico

1837 kosam
Small evergreen shrub widely growing in South East Asia and Northern Australia. Its intensely bitter fruit (its seed) is still used in local medicine, particularly in the treatment of various types of diarrhoea and dysentery.
f kosam
d Brucea; Makassar
i brucea
e brucea

1838 kosso
Or brayera. The dried pistillate flowers of a tree growing in Abyssinia, which are used as an anthelmintic.
f cousso; fleurs de cousso
d Kosoblumen; Kussoblumen
i cusso; fiori di cusso
e flores de couso

1839 kynurenic acid
Crystalline decomposition product of tryptophane occurring in the urine of dogs and some other animals.
f acide cynurénique
d Kynurensäure
i acido cinurenico
e ácido quinurénico

1840 kynurenine
Amino acid occurring in the urine of dogs and several other animals as one of the normal products of tryptophan metabolism. It is capable of forming kynurenic acid and other products.
f cynurénine
d Kynurenin
i cinurenina
e cinurenina

L

1841 labdanum
Oleoresin exudating from the leaves of various species of Cistus in the form of red-brown masses characterized by a pleasant odour. It is a fixative agent which is used in several chemical compounds.
f ladanum
d Ladanharz; kandisches Ladanum
i ladano
e ládano

1842 labetalol
Alpha and beta adrenergic blocking agent. Antihypertensive agent because of the beta-blocking action on the peripheral arteries. A reflex increase of cardiac rhythm is thus prevented. The drug should not be used in the presence of atrioventricular block and caution is required in the course of general anaesthesia. Cephalalgia, cramps, nasal congestion and epigastric pains may occur as side effects.
f labétalol
d Labetalol
i labetalolo
e labetalolo

1843 lachesine
A chemical compound; colourless, odourless, bitter and water-soluble crystals having a midriatic action that is analogous, yet less intense, than atropine.
f lachésine
d Lachesin
i lachesina
e laquesina

1844 lacmoid
Litmus-resembling violet-blue dye that is obtained by the action of nitrites on resorcinol. The compound is used as an indicator in titration.
f lacmoïde; bleu de résorcinol
d Lacmoid; Resorcinblau
i blu di resorcinolo
e lacmoide; azul de resorcinol

1845 lactalbumin
The soluble heat-coagulable protein of the albumin class present in milk. In accordance with the order of sedimentation of the whey treated with an ultracentrifuge, two albumins can be identified, alpha and beta. The alpha-lactalbumin is the most abundant protein in the serum of human milk, while the beta one is more abundant in the serum of cow's milk Lactalbumin is used to increase the proteic value of food.
f lactalbumine
d Laktalbumin
i lattoalbumina
e lactalbúmina

1846 lactase
The digestive ferment which is contained in the gastric juices and is characterized by its capacity of splitting a lactose molecule into one of glucose and one of galactose.
f lactase
d Laktase
i lattasi
e lactasa

1847 lactate
Salt or ester of the lactic acid. The salts are generally water-soluble and, less easily, alcohol-soluble. The most important are the lactates of calcium and iron which are used in medicine.
f lactate
d Laktat; milchsaures Salz
i lattato
e lactato

1848 lactic acid
Hydrocarboxylic acid containing an asymmetrical carbon atom and thus existing in two optically active forms and in one racemic form. It occurs in sour milk, in the blood and tissues of animals, frequently in cheese and in the fermentation products of carbo-hydrates. It is formed in the muscular tissue following strong contractions when there is a condition marked by an inadequate supply of oxygen.
f acide lactique
d Milchsäure
i acido lattico
e ácido láctico

1849 lactic ferment
Ferment produced by milk-infecting bacteria, which converts lactose into lactic acid. Lactic ferments act as intestinal bacteriotherapeutic agents by acidifying the intestinal system and reestablishing the balance between alkaline putrescent flora and acid-fermentative and vitamin-synthesizing flora. Lactic ferments are

used in the prophylaxis and therapy of various disorders of the gastrointestinal apparatus and ensuing cutaneous manifestations, of pathological conditions affecting the intestinal flora and of dysvitaminosis caused by antibiotics and chemotherapeutic agents.

f ferment lactique; lactobacillus culture
d Milchferment
i fermento lattico
e fermento láctico

1850 lactobionic acid
A syrupy acid which is obtained by the oxidation of lactose, q.v.

f acide lactobionique
d Laktobionsäure
i acido lattobionico
e ácido lactobiónico

1851 lactoflavin
Riboflavin, vitamin B_2. A flavin, i.e. water-soluble yellow pigment of the lyochrome class originally found in milk and subsequently in various other foods and animal tissues. It possesses a vitamin B_2 activity. As it can be obtained from other substances than milk, the term is no longer used, and "riboflavine" is preferred.

f lactoflavine; riboflavine
d Laktoflavin; Riboflavin
i lattoflavina; riboflavina
e lactoflavina; riboflavina

*** lactogenic hormone   s. prolactine**

1852 lactoglobulin
Protein, more specifically called ß-1, occurring in the whey of milk, in particular cow's milk. Albeit in lower concentration, it also occurs in the milk of other mammals. There are two different types of lactoglobulin, A and B. The presence of the one rather than the other type, or both together, is determin ed genetically.

f lactoglobuline
d Laktoglobulin
i lattoglobulina
e lactoglobulin

1853 lactol
Either a hydroxyderivative of a lactone, or the ester of the lactic acid with naphthol, used in medicine as an antiseptic.

f lactole
d Laktol
i lattolo
e lactolo

*** lactomycin   s. demeclocycline**

1854 lactone
A generic term denoting the internal esters of the oxycarboxylic acids, i.e. formed by the esterification of two groups, an alcoholic one and a carboxylic one, belonging to the same molecule. Some

lactones occur in nature and are part of physiologically active compounds or antibiotics.

f lactone
d Lakton
i lattone
e lactona

*** lactone of the glucuronic acid   s. glucurolactone**

1855 lactophosphate
Any salt of lactic and phosphoric acids combined with the same base, e.g. a mixture of a lactate and a phosphate.

f lactophosphate
d Laktophosphat; Phospholaktat
i lattofosfato
e lactofosfato

1856 lactose
Milk sugar. Disaccharide constituted by the union of a molecule of glucose with one of galactose and existing in two isomeric forms, alpha and beta. It occurs in the milk of mammals and thus also called milk sugar. Lactose is hydrolyzed to galactose and glucose by lactase (q.v.) and is used in medicine as a diuretic and a purgative as well as a powder base.

f lactose; sucre de lait
d Laktose; Laktobiose; Milchzucker
i lattosio; zucchero di latte
e lactosa; lactobiosa; azúcar de leche

1857 lactulose
Therapeutic agent capable of reducing ammoniaemia and of acting as a mild laxative. It splits when it reaches the colon forming fatty acids having a low molecular weight capable of reducing the faecal pH. The reduction of the ammoniaemia occurs because of the diminish ed production and absorption of NH and the increase of its elimination. The cathartic effects occurs because of the osmotic effect and the stimulation of the peristalsis. The drug is also used in the treatment of portal-systemic encephalopathies, hyperammoniaemia, hepatic coma and cirrhosis of the liver, as well as chronic constipation. In paediatrics, it finds employment as a corrective of artificial feeding and in enteric disorders caused by Escherichia coli.

f lactulose
d Laktulose (Kohlenhydrat)
i lattulosio
e lactulosa

*** ladanum   s. labdanum**

1858 lady's hair
Or Venus hairfern. A fern of the Polipodiaceae family growing in all warm countries. Its leaves are used to treat catarrh of the respiratory tract and as a diaphoretic agent.

f adiante; cheveux de Vénus

d Frauenfarn; Venushaar
i capelvenere; adianto nero; erba fonta-
nina
e capilaria; adianto negro

1859 lady's mantle
Or lion's foot. Plant of the genus
Rosaceae growing on high ground and
whose leaves are used in medicine as an
astringent agent.
f alchemille; pied de lion
d Alchimistenkraut; Frauenmantel; Löwen-
klau
i alchemilla; erba stella; piede di leone
e alquimila; pie de león

1860 lady's thistle seed
Or St. Mary's seed, blessed thistle seed.
The seed of an annual pubescent herb,
which is used in medicine for its tonic,
stomachic and febrifuge properties that
are due, at least partly, to a bitter
glycosidic principle known as cnicin.
f semence de chardon Marie
d Marienkorn; Stechkorn
i seme di cardo santo; seme di cardo ma-
riano
e fruto de cardo de Maria; fruto de cardo
lechar

1861 lady's tresses
Or ladies' tresses. Orchid of the genus
Spiranthes grown in England (Spiranthes
autumnalis) which is reputed to have
aphrodisiac properties.
f spiranthe automnale
d Drehling; Wendelähre
i spiranto
e espiranto

1862 laevocystine
Sulphuretted amino acid marked by a
disintoxicating action owing to the presence of sulphydryl groups utilized by
the liver in the processes of transsulphu
ration. The cutaneous trophic action
which is also capable of stimulating the
cartilaginous metabolic processes is due
to the sulphur contained in an easy-to-
-exploit form. The drug is used in hypo
proteinaemia, disorders of the liver,
poisoning by heavy metals, arthrosis,
some forms of dermatosis and polyarthritis.
f lévocystine
d Lävocystin
i levocistina
e levocistina

1863 laevoglutamide
The mono-amide of the glutamic acid (a
common constituent of protein and an
important metabolic intermediate), i.e.
a biological active form of the latter.
It acts physiologically on the metabolism
of the nerve cells and is used as a
psychotherapeutic drug in the treatment
of mental strain, loss of memory, intellective retardation in children and
senile depression.

f lévoglutamide
d Lävoglutamid
i levoglutamide
e levoglutamida

1864 laevopropicillin
Laevorotatory isomer of semisynthetic
propicillin, acting as antibiotic. It is
quite effective in the treatment of infections caused by penicillin-sensitive
micro-organisms. It is resistant to inactivation caused by gastric juices.
f lévopropicilline
d Lävopropicillin
i levopropicillina
e levopropicilina

1865 laevopropoxyphene
Antitussive agent marked by a central
action and used in the symptomatic treat
ment of cough associated with acute and
chronic respiratory disorders. Nausea
and vertigo are often occurring as
adverse effects.
f lévopropoxyphène
d Lävopropoxyphen
i levopropossifene
e levopropoxifeno

1866 laevothyroxine sodium
Synthetic laevorotatory thyroxine. It is
indicated for the treatment of various
forms of hypothyroidism accompanied by
growth problem and metabolism disorders.
Cumulative effects may be expected in
case of prolonged treatment. Also hypertension may occur as a side effect.
f lévothyroxine sodique
d Natriumlävothyroxin
i levotirossina sodica
e levotiroxina sódica

1867 laevulinic acid
Gamma-ketoacid which is obtained by
heating a hexose (galactose, fructose
etc.) with concentrated hydrochloric acid.
A white solid crystalline soluble in
water and in alcohol. The acid has
also been found in the products of the
breakdown of the nucleic acid of the
thymus.
f acide lévulinique
d Lävulinsäure
i acido levulinico
e ácido levulínico

1868 laevulose
Laevorotatory D-fructose generally obtain
ed by hydrolysis either of inulin from
dahlia tubers, or from sucrose or the
Jerusalem artichoke. It is a monosaccharide metabolized and converted into
hepatic glycogen more rapidly than glucose even when there is some liver
damage or in the absence of insulin. In
fact, laevulose displays a tonic, detoxifying protective action for the liver as
it is an immediate donor of energy for
the functional activity of the hepatic
cell and improves hepatocellular respira-

tion and basal metabolism. It is used
parenterally in various pathological conditions borne of carbohydrate insufficien
cy also in diabetic patients. It is also
used in combination with amino acids
owing to its positive protein balance.
The drug is mainly indicated in the
treatment of acute and chronic liver
diseases, toxicosis, diabetes, asthenia
etc., as well as a sweetener for food
and pharmaceutical preparations. Overdose may cause abdominal discomfort and
diarrhoea.
f lévulose
d Lävulose; Schleimzucker
i levulosio
e levulosa

1869 laminaria
Genus of seaweeds belonging to the
Brown Algae. Various species, in particular the Laminaria cloustonii or hyper-
borea, yield alginic acid and alginates.
Their stalks are kept in alcohol; once
in contact with an aqueous liquid, they
swell owing to the absorption of water
by the mucilage contained in the muci-
parous canals and are used in surgery
to dilate cavities and particularly in
obstetrics to dilate the cervix uteri
(the neck of the womb).
f laminaire; baudrier
d Laminarie; Riementang
i laminaria
e laminaria

1870 laminaria stalks
Ten cm long stalks of laminaria, of a
cylindrical shape, which are conserved
in alcohol before use (see the definition
of laminaria).
f tiges de laminaire
d Laminariestifte
i stipiti di laminaria; candelette di laminaria
e estipites de laminaria

1871 laminated tablet
Or layer tablet, sandwich tablet. Medicinal tablet consisting of various layers
of different therapeutic substances pressed together.
f comprimé multicouche; comprimé stratifié
d Mehrschichttablette
i compressa stratificata
e comprimido estratificado

1872 lanatoside
Term denoting the glycosides contained
in the extract of the Digitalis lanata.
It is a cardioactive glycoside displaying
a positively inotropic and a negatively
chronotropic action. It is often preferred
to other digitalis drugs as it is less
cumulative. It is indicated in the therapy of congestive cardiac insufficiency,
also when this is accompanied by conduc
tion disorders and shows some prefibril-
lation conditions, and more generally of
heart insufficiency both in children and

in elderly patients. Overdosage may
cause gastrointestinal disorders as well
as rhythm and conduction disorders.
f lanatoside
d Lanatosid
i lanatoside
e lanatósido

* **lanitop   s. methyldigoxin**

1873 lanoline
Wool fat. A purified, fat-like substance
obtained from the wool of the sheep and
widely used in creams for topical use.
It is mainly used, apart from cosmetics,
to promote the absorption of drugs contained in a cream.
f lanoléine; lanoline; graisse de laine
d Lanolin; Wollfett
i lanolina; grasso di lana
e lanolina; manteca de lana

1874 lantana
Genus of tropical, often half-climbing
shrubs of the Verbenaceae family, from
which a lotion is derived that is used
in South Africa to treat wounds and
sores. Also known as red sage.
f lantane
d Camarakraut; Wandelröschen
i lantana
e lantana; caraquito colorado

1875 larch agaric
Or purging agaric. A mushroom growing
on larch trees and containing agaricin,
a poisonous principle used in popular
medicine for the treatment of night
perspiration caused by tuberculosis.
f agaric blanc; bolet du mélèze
d Lärchenschwamm; Purgierschwamm
i agarico bianco; agarico del larice
e agárico blanco; hongo de alerce

1876 larch turpentine
Better known as Venice turpentine. The
turpentine which is obtained from the
bark of the larch and is used in medicine as an ingredient of liniments.
f térébenthine de Venise; térébenthine du
mélèze
d venetianisches Terpentin; Lärchenter-
pentin
i trementina di Venezia; trementina di
larice
e trementino de Venecio

* **largomycin   s. metacycline**

1877 laricic acid
Principle extracted from the bark of the
larch and is used in medicine as an
expectorant.
f acide laricinique; acide larixinique
d Laricinsäure
i acido laricinico
e ácido laricínico

1878 L-asparginase
A cytotoxic enzyme obtained from bac-

terial cultures and used in the treatment of neoplastic diseases. Nausea and vomiting, neurotoxicity and bone marrow depression may occur as side effects.

f L-asparginase
d L-Asparginase
i L-asparginasi
e L-asparginasa

1879 laudanidine
An alkaloid of opium ($C_{20}H_{15}NO_4$). White crystals. The laevorotatory form of laudanine.

f laudanidine
d Laudanidin
i laudanidina
e laudanidina

1880 laudanin
Alkaloid of opium in which it is present by 0.005%. It is extracted from the alkaline mother liquor of the morphine. Prismatic crystals soluble in benzene, less so in alcohol, optically inactive and poisonous. Used in medicine for its convulsive action.

f laudanine
d Laudanin
i laudanina
e laudanina

1881 laudanosine
Opium alkaloid. White needle-shaped crystals (methyltetrahydropapaverine) characterized by a bitter taste. It is soluble in alcohol and ether, insoluble in water. Very poisonous. It displays a physiological action which is slightly narcotic.

f laudanosine
d Laudanosin
i laudanosina
e laudanosina

1882 laudanum
A tincture prepared from raw opium with water, wine or alcohol and various other adjuvants by maceration. It contains one percent of anhydrous morphine and displays the antispastic and analgesic properties characteristic of opium.

f laudanum
d Laudanon
i laudano
e láudano

1883 laurate
Ester and salt of the lauric acid, q.v.

f laurate
d Laurat
i laurato
e laurato

1884 laurel oil
Essential oil obtained by distillation in a stream of steam of the leaves of Laurus nobilis. It is a clear yellow liquid, smelling like laurel and turpentine and with a sweetish taste. It is soluble in alcohol and ether and mainly

consists of pinene, eucalyptol, small quantities of eugenol, linalool, geraniol, partly free and partly esterified by fatty acids. It is used in medicine as a stimulant and antiseptic agent.

f huile de laurier
d Lorbeeröl
i olio essenziale di lauro
e aceite de laurel

1885 lauric acid
Crystalline fatty acid occurring in the form of its glycerol esters in coconut oil, palm-kernel oil and in the berries of the European laurel. It is a constituent of saturated fatty acid. It is used for the preparation of the laurates, of the lauroyl chloride, of the laurylic acid, of the n-dodecylamine etc. Also called laurostearic acid.

f acide laurique
d Laurinsäure; Laurostearinsäure
i acido laurico
e ácido láurico

1886 laurocerasus
The leaves of the cherry laurel which can yield hydrocyanic acid on hydrolysis when they are fresh. The solution obtained from them, i.e. the cherry--laurel water, is used as a mild sedative and carminative.

f eau distillée de laurier cerise
d Kirschlorbeerwasser
i acqua distillata di lauroceraso
e agua de laurel cerezo

1887 laurodoxine
Lauroyldiacetylpyridoxine. Dermatological therapeutic agent used in the treatment of seborrhoea of the scalp with or without alopecia), of the face and trunk, seborrhoic acne, pityriasis sicca and steatoides of the scalp.

f laurodoxine
d Laurodoxin
i laurodossina
e laurodoxina

1888 laurotetanine
An alkaloid occurring in the bark of the Litsea latifolia; a crystalline, colourless powder, soluble in alcohol, with a physiologic action similar, though milder, to that of the strychnine.

f laurotétanine
d Laurotetanin
i laurotetanina
e laurotetanina

1889 lauryl alcohol
Crystalline compound made by reduction of ethyl laurate or by the hydrolysis and subsequent hydrogenation of coconut oil. Used for its detergent and disinfectant properties.

f alcool laurique
d Laurylalkohol
i alcool laurico
e alcohol láurico

1890 lauryl mercaptan
A mixture of various isomers which appears as a colourless or yellowish liquid marked by a characteristic odour. It is water-insoluble but soluble in organic solvents. It is frequently used in the synthesis of pharmaceutical products.
f laurylmercaptan
d Laurylmercaptan
i laurilmercaptano
e laurilmercaptano

1891 lauryl pyridinum chloride
Compound of quaternary ammonium, soluble in water and marked by germicide properties.
f chlorure de lauryle et pyridium
d Laurylpyridiumchlorid
i cloruro di laurile e piridinio
e cloruro de laurilo y piridinio

1892 Lauth's violet
Thionine. Purplish stain frequently used in microscopy owing to its affinity for acidic constituents of cells and intercellular matrices, e.g. mucopolysaccharides. Also used to stain the Nissl substance of nerve cells. It is a dark--green powder soluble in water.
f violet de Lauth; thionine
d Lauth's Violett; Thionin
i violetto di Lauth; tionina
e violeta de Lauth; tionina

1893 lavender
Widely cultivated Mediterranean plant from whose white or pale-violet flowers an oil is extracted by distillation in a stream of steam, which is used for medicinal purposes.
f lavande; nard d'Italie
d Lavendel
i lavanda; spigo nardo
e lavándula; espliego

1894 lavender oil
Essential oil obtained by distillation in a stream of steam of the lavender flowers. It is a colourless to yellowish liquid of an aromatic slightly bitter taste and a pleasant odour. It is used in medicine as a stimulant.
f essence de lavande
d Lavendelöl
i essenza di lavanda
e esencia de alhucema; esencia de lavándula

1895 laxative
A mild cathartic agent characterized by a light and mild action, i.e. increasing intestinal peristalsis. Laxatives may be of a vegetable nature, or synthetic.
f laxatif
d Laxativum; Abführmittel
i lassativo
e laxante

1896 lead acetate
Or sugar of lead. White crystalline soluble substance characterized by astringent properties.
f acétate de plomb; sucre de plomb; sel de Saturne
d Bleiacetat; Bleizucker
i acetato di piombo; zucchero di piombo
e acetato de plomo; azúcar de Saturno

1897 lead borate
Salt of boric acid. A white powder, insoluble in water and soluble in diluted nitric acid, prepared by treating boric acid and lead hydrate and used in some pharmaceutical products.
f borate de plomb
d Bleiborat; borsaures Blei
i borato di piombo
e borato de plomo

1898 lead carbonate ointment
Ointment consisting of a white amorphous compound and used as an astringent.
f onguent de sous-acétate de plomb
d Bleisalbe
i pomata saturnina
e pomada de subacetato de plomo

1899 lead tetraethyl
Oily, colourless, pleasantly-smelling, strongly toxic liquid, easily absorbed by the organism through inhalation of its vapour and also through direct contact with the skin. The ensuing intoxication (lead poisoning) affects mainly the nervous system. Lead tetraethyl is sometimes used in organic syntheses.
f plomb-tétraéthyle
d Tetraethylblei
i tetraetil-piombo
e tetraetil-plomo

1900 leafcup
A tall weedy herb characterized by a strong scent and wide thin leaves and panicled corymbs of yellowish or white flower heads. The herb has febrifuge properties.
f polymnie
d amerikanische Bärenklau
i polimnia
e polimnia

1901 leafy spurge
One of several plants of the family Euforbaceae, some of which are slightly poisonous. The leafy spurge (euphoria esula) displays febrifuge properties.
f euphorbe ésula; embranchée
d Eselswolfsmilch
i euforbia esula
e euforbia ésula

1902 lecithin
Group of phosphatides constituted by glycerol, phosphoric acid, choline and two residues of fatty acid. Lecithins are present in all animal and plant cells as they are essential components, together with some other phosphatides, of the cellular membrane. They represent a

source of choline for the nerve cell, participate in the blood coagulation and are important for the transport of the sodium and potassium ions and of the amino acids. Owing to their properties as a bioplastic agent, lecithins are used in medicine as a general restorative, particularly in the course of convalescence, and in nervous and physical exhaustion.

f lécithine
d Lezithin
i lecitina
e lecitina

1903 lectin
Substance occurring in the seeds of some plants and characterized by the property of specifically reacting to certain erythrocytic agglutinogens.

f lectine
d Lektin
i lectina
e lectina

1904 ledol
Ledum camphor. Crystalline sesquiterpenoid alcohol found in the oil which is obtained from the leaves and flowering tops of the marsh tea and used for the destruction of vermin and parasites.

f camphre de lédon
d Ledöl
i canfora dell'inbrentina
e alcanfor de ledo

1905 ledum
Marsh tea or wild rosemary. A big shrub, growing in Europe and in Asia characterized by narrow leaves yielding an infusion which is used for killing vermin and parasites.

f lédon des marais; romarin sauvage
d Sumpfporst; wilder Rosmarin
i rosmarino silvestre
e ledo palustre

1906 leech
Any member of the class Hirudinea, some species of which, in particular the Hirudo medicinalis, were applied locally in the past in order to remove small quantities of blood.

f sangsue
d Blutegel
i sanguisuga
e sanguijuela

1907 lemon balm oil
Melissa oil. Yellowish or brownish liquid characterized by a strong smell of verbena and containing citral (an aldehyde occurring in lemon oil) as well as traces of geraniol, citronella, limonene and dipentene. It was used in the past as a carminative agent.

f essence de mélisse
d Melissenöl
i essenza di melissa; essenza di citronella
e esencia de melisa

1908 lemon oil
The essential oil obtained from the lemon peel. A pale-yellow or greenish liquid having a characteristic odour and a strongly aromatic, slightly bitter taste. Citral and D-limonene are its principal constituents. It is widely used for its flavouring properties and has also found employment as a carminative agent.

f essence de citron
d Zitronenöl; Citronenöl
i essenza di limone
e esencia de limón

1909 leptandra root
Veronicastrum. The root of Veronica virginica, a North American plant, which is used as a cathartic agent.

f rhizome de véronique de Virginie
d Leptandrawurzelstück
i radice di veronica di Virginia
e raíz de verónica de Virginia

1910 leptazol
Pentamethylenetetrazole. Drug acting as respiratory and medullary stimulant which was used in the past to treat narcotic poisoning, in particular by barbiturates, as well as in cases of collapse during anaesthesia. In association with ephedrine, the drug is also used in the treatment of bronchial asthma. Also known as pentetrazole or pentamethazol.

f pentétrazol
d Pentetrazol
i leptazolo; pentetrazolo
e pentetrazol

1911 lespedin
A disintoxicating agent used in the treatment of renal and extrarenal hyperazotaemia.

f lespédin
d Lespedin
i lespedin
e lespedin

1912 leucine
An amino acid which is essential in human diet and is found in most protein hydrolisates. It may also be found together with thyrosine in urinary deposits, but is usually broken down into aceto-acetic acid in the metabolism.

f leucine
d Leucin
i leucina
e leucina

1913 leucinocaine
Local anaesthetic whose effects are more potent and lasting than those of procaine. It has anticholinergic and antispasmolytic properties and is used to attain anaesthetic block and spinal anaesthesia. It is frequently associated with adrenaline.

f leucinocaïne
d Leucinocain

  ı leucinocaina
  e leucinocaina

1914 leucocyanidol
Vitamin factor C capable of promoting the action of ascorbic acid endothelially with a capillary-protection action. It is used in the treatment of haemorrhages caused by capillary fragility, idiopathic purpura and hypertension-related purpura, diabetes, hepatopathies, disorders due to excessively prolonged treatment with corticosteroids or antivitamin K, haemorrhoids, varices, oedemas etc.
  f leucocyanidol
  d Leucocyanidol
  ı leucocianidolo
  e leucocianidolo

* **leucodinine-B** s. mequinol

1915 levallorphan
Narcotic antagonist capable of reducing or eliminating the typical effects of morphine as well as of drugs displaying a similar pharmaceutical action. Unless associated with some other suitable drug, levallorphan may cause respiratory depression.
  f lévallorphane
  d Levallorphan
  ı levallorfano
  e levalorfano

1916 levamisole
Anthelmintic agent mostly used in veterinary practice but also in man in cases of ascaridiasis and ankylostomiasis. Persistent nausea and a general feeling of malaise may occur as side effects.
  f lévamisole
  d Levamisol
  ı levamisolo
  e levamisolo

1917 levant wormseed
Santonica. Medicinal drug obtained from the buds of the European wormwood and used as an anthelmintic agent.
  f armoise de Barbarie
  d Wurmkraut; Wurmbeifuss
  ı santonico
  e santónico; cina

1918 levant wormseed oil
Yellow essential oil obtained from the flowers of Levant wormseed, used as an anthelmintic agent.
  f essence d'armoise maritime
  d Meerstrandbeifussöl
  ı olio d'assenzio marittimo
  e aceite de ajenjo marino

1919 levarterenol
Sympathomimetic agent, a precursor of adrenaline (noradrenaline). A vasoconstrictor characterized by a physiological mechanism. The drug is also a hypertensor and a cardiac stimulant.
  f lévartérénol

  d Levarterenol
  ı levarterenolo
  e levarterenolo

1920 levodopa
Or laevodopa. An amino acid which is converted in the body into dopamin, a neurotransmitter substance which is deficient in the Parkinson's disease. It potentiates the action of the sympaticomimetics but its effect on tremor is not as strong as that of anticholinergic drugs. Hepatic and gastro-intestinal as well as psychiatric disorders may occur as adverse effects which can however be reduced by combination with peripheral inhibitors of dopamine synthesis.
  f lévodopa
  d Lävodopa
  ı levodopa
  e levodopa

1921 levomepromazine
Tranquillizer of the phenothiazine group acting at subcortical level and used in the treatment of neuroses, psychoses, anxiety, mental agitation, insomnia etc. Also used to soothe pain. Orthostatic hypotension may occur as side effect.
  f lévomépromazine
  d Lävomepromazin
  ı levomepromazina
  e levomepromazina

1922 levomethiomeprazine
Chemotherapeutic compound, a depressant of the central nervous system, used as a sedative and antiemetic agent.
  f lévométhioméprazine
  d Lävomethiomeprazin
  ı levometiomeprazina
  e levometiomeprazina

1923 levomethorphan
A very potent synthetic analgesic based on the morphine skeleton. L-3-methoxy--N-methylmorphinan.
  f lévométhorphane
  d Lävomethorphan
  ı levometorfano
  e levometorfano

1924 levophacetoperane
Psychostimulant used in depressive states, oligophrenia, asthenia etc. It also finds employment in cases of children showing serious difficulties of adaptation and in obesity.
  f lévophacétopérane
  d Lävophacetoperan
  ı levofacetoperano
  e levofacetoperano

1925 levorphanol
Narcotic analgesic whose action is similar to that of morphine but more reliable and prolonged. It is mostly used to soothe the severe pains caused by biliary and renal colics, myocardial infarction, traumas and terminal dis-

eases. It is also used in surgery as a preoperative technique and to combat postoperative pains. It can be associated with barbiturates. General malaise, dizziness, respiratory depression and hypotension may occur as side effects.

f lévorphanol
d Levorphanol
i levorfano; levorfanolo
e levorfanolo

1926 licebane
Or staphisagria, louseseed. The ripe seed of the stavesacre, which contain delphinine, used to destroy head lice.

f herbe aux poux; staphisaigre
d Läusekraut; Stephanskraut
i erba dei pidocchi; stafisagria; tusano
e hierba pipjera; estafisagria

1927 lichen
Complex thallophytic plant constituted by the association of an alga and a fungus and growing on solid substances, e.g. rocks or the bark of a tree. It consists of a branching thallus which is not differentiated into stem and leaves but appears crustose and fruticous or foliaceous. It contains algal gonidia in a meshwork of fungal hyphae and may be considered a source of food and a dye for some pharmaceutical preparations.

f lichen
d Flechte
i lichene
e liquen

1928 lichenin
Lichen starch, moss starch. A polysaccharide contained in various types of lichens in which it occurs as a reserve substance or as a constituent of the cell walls. It is very similar to the cellulose, in particular the hydrocellulose and, like the latter, it can be transformed into glucose by total hydrolysis and into collobiose by partial hydrolysis. Thus differing from the cellulose, it can easily be saccharified by some enzymes.

f lichénine
d Flechtenstärke; Lichenin
i lichenina; amido dei licheni
e liquenina, fécula de liquen

1929 licorice
The Glycyrrhiza glabra and the juice which is extracted from its roots. The plant is a perennial herb growing in Mediterranean regions. Its dried root, having a gummy texture and a sweetish rather astringent flavour is used for the preparation of pharmaceutical drugs. The extract and the syrup of licorice is also used for its emollient, bechic and sweetening properties. Also known as liquorice.

f réglisse
d Süssholzstrauch; Lakritze
i liquirizia

e regaliz

1930 licorice extract (spissum)
Extract of licorice (q.v.) used in the treatment of peptic ulcers.

f extrait de réglisse
d Süssholzextrakt
i estratto di liquirizia
e extracto de regaliz

1931 licorice root
The dried roots and stolons of the Glycirrhiza glabra. Spasmolytic and antacid agent used as an expectorant and a demulcent. It is mildly purgative and diuretic. Also indicated as a coadjuvant in cortisone therapy. The aglycone (the compound remaining when a glycoside is hydrolysed and the sugar is extracted) of the glycyrrhizine (glycyrrhizic acid, q.v.) which occurs in non-hydrolysed licorice roots displays an antiphlogistic action and is indicated in the treatment of gastrointestinal ulcers.

f racine de réglisse
d Süssholzwurzel
i radice di liquirizia
e raíz de regaliz

1932 lidocaine
Amide-type local anaesthetic, probably the principal one of series. The drug stabilizes the neural membrane and checks the transmission of nerve impulses close to the site of its application. It is metabolized in the liver and excreted through the kidneys. It is mainly used to produce local or regional anaesthesia by infiltration techniques and intravenous regional anaesthesia by peripheral nerve block techniques, as well as by central neural techniques including epidural and caudal block. It is used to control cardiac arrhythmias and in open-heart surgery. It is also used in cases of premature ventricular contractions in patients who have suffered a myocardial infarction.

f lidocaïne
d Lidocain
i lidocaina
e lidocaina

1933 lidoflazine
Preventive anti-anginal agent capable of potentiating the action of adenosine and adenosine triphosphate by acting on the smaller coronary arteries and thus increasing the flow of the blood. The drug is indicated in the treatment of angina pectoris, coronary insufficiency and post-infarction conditions. Headache and tinnitus may occur as side effects.

f lidoflazine
d Lidoflazin
i lidoflazina
e lidoflazina

1934 lignaloe oil
Or linaloe oil. An essential oil used as

a source of linalool, q.v. and used to flavour some medicinal preparations. It is an almost colourless, rather pleasantly smelling liquid soluble in alcohol.
f essence de linaloès
d Lokariöl; Linaloeöl
i olio di linaloe
e esencia de lináloe

1935 lime
The fruit of the Citrus acida used as an antiscorbutic agent owing to its considerable vitamin C content.
f tilleul
d Linde
i tiglio
e tilo

1936 lime flower water
Or linden flower water. The liquid which is obtained from the lime flowers and is used as a tonic in the treatment of vitamin C deficiency.
f eau de tilleul
d Lindenblütenwasser
i acqua di tiglio
e agua de tilo

1937 lime milk
Emulsion-like solution of lime used in the treatment of intestinal disorders, particularly in very young children.
f lait de chaux
d Kalkmilch
i latte di calce
e lechada de cal

1938 lime syrup
Syrup consisting of calcium hydroxide and used in the treatment of diarrhoea in infants.
f sirop de chaux
d Kalksirup
i sciroppo di calce
e jarabe de cal

1939 limette oil
Lime oil, limetta oil. Essential oil obtained from limes and used especially as a flavouring agent in the preparation of medicinal drugs.
f essence de limette; huile de limette
d Limettenöl; Cedratöl
i essenza di cedro
e esencia de cidra

1940 lime water
Calcium hydroxide solution of slaked lime (the powder that is obtained by the action of water on quicklime) in water used as an antacid in the treatment of diarrhoea, particularly in infants.
f eau de chaux
d Kalkwasser
i acqua di calce
e lechada de cal

1941 limonene
Widely distributed naturally occurring monocyclic terpene, whose dextrorotatory

form, the carvene, is a colourless, lemon-smelling and principal constituent of the essential oils of lavender, lemon rind, orange, bergamot and caraway. The laevorotatory form of limonene is found with L-pinene in pine needle and fir cones and in the essential oils of peppermint and spearmint. A racemic form, the dipentene, occurs naturally in turpentine, citronella oil and cubed oil. It may be obtained synthetically from isoprene.
f limonène
d Limonen
i limonene
e limoneno

1942 linalol
Tertiary alcohol occurring in two optically active modifications of which the dextrorotatory one is also known as coriandrol. It is a colourless liquid, soluble in alcohol and ether, contained in the essential oils of linaloe, coriandrol, nux vomica, ylang-ylang, lavender etc. It is easily isomerized by the action of the acids and is used for the preparations of various medicinal drugs. Also called linalool.
f linalol
d Linalool
i linalolo
e linalool

1943 lincomycin
Antibiotic active against penicillin-resistant Gram-positive bacteria, in particular the Diplococcus pneumoniae, the Streptococcus pyogenes (group A), the Streptococcus viridans and the Bacillus anthracis. The drug is used in the treatment of various infections, e.g. sinusitis, otitis, mastoiditis, osteomyelitis, furunculosis and other cutaneous infections. Rashes, diarrhoea and pseudomembranous colitis may occur as adverse effects.
f lincomycine
d Lincomycin
i lincomicina
e lincomicina

1944 linoestrenol
A potent sex hormone (progestogen) with an action similar to that of progesterone. It has a contraceptive effect without symptoms of virilization. Also used as an adjuvant agent in the course of anticoagulant therapy. It is sometimes associated with mestranol in the treatment of uterine bleeding caused by physiological dysfunctions, of endometriosis and cancer of the breast.
f linestrol
d Lynöstrol
i linestrolo
e linestrolo

1945 linoleate
Ester and salt of the linoleic acid con-

taining the monovalent radical
$C_{17}H_{31}COO-$.

f linoléate
d Linoleat
i linoleato
e linoleato

1946 linoleic acid

Or linolic acid. Fatty acid which is an
essential component of the vitamin F and
occurs as a glyceride in many natural
fatty oil, e.g. linseed oil, soja oil, sun
flower oil etc. It is a yellow oily liquid
insoluble in water, used in the preparation of emulsions. Its function in the
organism is probably the promotion of
the utilization of lipids. It is presumed
that in the organisms linoleic acid is
transformed into arachidonic acid, an
unsaturated fatty acid present as a
glycerol in lecithin and cephalin, which
is necessary for normal growth and very
important in hepatic fat metabolism.

f acide linoléique
d Linolsäure; Linolensäure
i acido linoleico
e ácido linoleico; ácido linólico

1947 linolenic acid

Unsaturated fatty acid occurring as a
glyceride in linseed oil, soja oil etc.
It is an oily, colourless, water–insoluble liquid, soluble in alcohol and
ether and displaying siccative properties.
It is one of the essential fatty acids for
the human organism.

f acide linolénique
d Linolensäure
i acido linolenico
e ácido linolénico

1948 linseed

The seeds of the flax containing between
30 and 40% of fixed oil together with
mucilage and proteins. Its infusion is
used therapeutically as a demulcent
(relating to a substance capable of protecting the mucous membranes and soothing irritation).

f graine de lin
d Leinsamen
i seme di lino ; linosa
e linaza

1949 linseed cake

The residue that is left when oil is
extracted or expressed from linseed,
principally used as a feed for cattle but
also for therapeutic purposes.

f pain de lin
d Leinkuchen; Flachskuchen
i panello di lino
e torta di lino

1950 linseed meal

Infusion prepared with the seeds of the
flax (Linum usitatissimum) which is widely cultivated in all continents. It contains 30 to 40% of fixed oil together
with mucilage and protein, and is main-

ly used as a demulcent drink in the
treatment of inflammatory diseases of the
gastrointestinal tract. It is also used
against burns and some dermatoses. The
linseed oil and other fats of the same
nature are, in fact, vitamin F, whose
deficiency in the otherwise normal diet
may be at the origin of psoriasis, eczema etc. Also called crushed linseed.

f farine de lin
d Flachsmehl
i farina di lino; lino seme
e harina de linaza

1951 linseed poultice

A poultice consisting of crushed linseed
and used to apply warmth in bronchitis
and other disorders of the respiratory
tract.

f cataplasme de farine de lin
d Leinmehlumschlag
i cataplasma di farina di lino
e cataplasma de linaza

1952 liothyronine

Synthetic thyroid hormone, a laevorotatory physiologically active derivative of
the triiodothyronine. It is used in the
treatment of hypothyroidism associated
with hypometabolism and of male and
female sterility of a hypothyroid origin.
Its absorption is rapid and its elimination occurs in the urine within 8 to 12
hours.

f liothyronine
d Liothyronin
i liotironina
e liotironina

1953 lipocaic

Lipotropic preparation from the pancreas
used for the absorption, mobilization,
utilization and excretion of lipids and
lipoids. The drug is recommended in the
treatment of steatosis, steatorrhoea,
initial atheromatosis, hypercholesterolaemia, disorders in the lipid metabolism.
It is also sometimes used as a coadjuvant in the therapy of diabetes mellitus.

f lipocaic
d Lipocaic
i lipocaic
e lipocaic

1954 lipopolysaccharide

Liposaccharide, component of O antigen
of Gram-negative bacteria, in particular
endobacteria, e.g. Escherichia coli and
Salmonella. Lipopolysaccharides can be
injected intravenously, but extreme
caution is required as they produce
shock leucopenia followed by leucocytosis and pyrexia. They also act as antigen adjuvants.

f lipopolysaccharide
d Lipopolysaccharid
i lipopolisaccaride
e lipopolisacárido

1955 lippia

Large genus of tropical American herbs, shrubs and small trees of the Verbenaceae family, whose crushed flowers are used for some therapeutical preparations.
l lippia; aloyse citronnée
d Lippienkraut
i cedrina; limonaria; erba Luisa
e yerba dulce; regaliz de Cuba

1956 liquefied phenol
Solution consisting of 80% phenol and water, used in lotions, mouthwashes and gargles for the disinfection of the oral cavity.
f phénol aqueux; phénol liquéfié
d flüssige Karbolsäure; verflüssigtes Phenol
i fenolo liquido
e fenol líquido

1957 liquid amber
Or liquidambar. Genus of trees of the Hamamelidaceae family, some of which, in particular the Liquidambar orientalis is a source of storax, a grey brown fragrant liquid which contains resin, styrene and cinnamic acid and is used therapeutically as an expectorant.
f liquidambar; copalme; styrax liquide; thymiane
d Styrax-Balsam; amerikanischer Storax; flüssiger Storax
i storace liquido
e ámbar líquido; bálsamo estoraque; goma dulce

1958 liquid liver extract
Aqueous extract of liver containing the antipernicious-anaemia principle (see jecoris extractum purificatum ad injectionem) and is so prepared in relation to its strength that 56 ml is equivalent to 0.4535 kg of fresh mammalian liver. The drug is administered orally but is at present rarely prescribed as a purer preparation suitable for injections is now available.
f extrait de foie buvable
d flüssiger Leberextrakt
i estratto di fegato liquido
e extracto de hígado bebible

1959 liquor ammonii caustici
Concentrated aqueous solution of ammonium hydroxide with a marked caustic action when not diluted. If diluted, generally to a ratio of 10, it may be used as a respiratory stimulant as its action is very similar to that of ammonium carbonate and ammonium anisate liquor.
f eau ammoniacale
d wässerige Ammoniaklösung
i ammoniaca
e agua amoniacal

1960 liquorice juice
The juice that is obtained from Glycyrrhiza glabra and other species of the same plant and is characterized by a sweet active principle, glycyrrhizin. It is used in powder form or as a liquid as a mild expectorant.
f suc de réglisse
d Lakritzensaft; Süssholzsaft
i succo di liquirizia
e jugo de regaliz; zumo de regaliz

1961 lithium carbonate
Neuroleptic agent used in the prophylaxis and treatment of manic-depressive disorders. Grave toxic disorders may occur in overdosage. It is advisable to discontinue the administration of the drug in case of anorexia, nausea, diarrhoea and tremors.
f carbonate de lithium
d Lithiumcarbonat
i litio carbonato
e carbonato de litio

1962 lithium citrate
Antidepressant agent used in the treatment and prophylaxis of acute manic states and of unipolar and bipolar manic-depressive psychoses. Intense drowsiness, polyuria, difficulty of concentration and neuromuscular disorders may occur as side effects.
f citrate de lithium
d Lithiumcitrat
i litio citrato; citrato di litio
e citrato de litio

1963 lithium gluconate
An important tranquillizer used in the treatment of manic-depressive psychoses, recurrent depression crises, involutive-type depression and in the prophylaxis of cyclothymia (a mild manic-depressive insanity).
f gluconate de lithium
d Lithiumgluconat
i litio gluconato
e gluconato de litio

1964 lithium glutamate
Antidepressant drug used in the treatment of manic and hypomanic states and in the prophylaxis of occasional manic and depressive episodes of manic-depressive psychosis. The specific biochemical mechanism of the lithium ion in its action in manic crises is not yet absolutely clear.
f glutamate de lithium
d Lithiumglutamat
i litio glutamato
e glutamato de litio

1965 lithium salts
Generally administered as carbonate or citrate, it yields lithium ions which substitute for sodium in excitable tissues and reduce the level of brain catecholamines. It is used in the prophylaxis and treatment of mania and depression. The effects of the drug must be carefully monitored in the presence of cardiac or renal diseases. Vomiting, diarrhoea,

polyuria causing confusion and fits may occur as adverse effects.
f sels de lithium
d Lithiumsalze
i sali di litio
e sales de litio

1966 lithium sulphate
Long-acting neuroleptic agent used in the treatment of manic and hypomanic attacks, and in the prophylaxis of manic-depressive psychoses as well as periodic psychoses. It is also indicated in the treatment of some premenstrual syndromes.
f sulfate de lithium
d Lithiumsulfat
i litio solfato
e sulfato de litio

1967 lithocolic acid
Biliary acid which is present, together with cholic, deoxycolic and chenodeoxycolic acids, in the bile of man and ox.
f acide lithocholique
d Lithocholsäure
i acido litocolico
e ácido litocólico

1968 litmus
Blue colouring matter which is prepared from lichens by a suitable treatment with ammonia and contains azolitmin, erythrolitmin and erythrolein. It is used as an indicator, red in acids and blue in alkalis.
f tournesol
d Lakmus
i tornasole
e tornasol

1969 livelong
Orpine. Glabrous Eurasian sedum bearing terminal cymes of reddish purple flowers, used in the past in popular medicine.
f grand orpin; grassette; orpin reprise
d grosse Fetthenne; knolliges Steinkraut
i erba da calli; fava grassa
e hierba callera; telefio

1970 liver of sulphur
Sulphurated potash. Mixture mainly consisting of potassium polysulphides and potassium thiosulphate, obtained by heating sublimed sulphur and potassium carbonate as liver-brown lumps becoming yellowish and decomposing in air. The mixture thus obtained is used in medicine principally in the treatment of skin diseases.
f foie de soufre
d Schwefelleber
i fegato di zolfo
e hígado de azufre

1971 liverwort
The lichen Peltigera canina which displays demulcent properties and is used in popular medicine to treat liver complaints. Hepatica.

f hépatique des fontaines; lichen étoilé
d Brunnenleberkraut; Steinleberkraut
i fegatella
e fegatella

1972 lobelanidine
Minor alkaloid constituent of lobelia, the dried herb of which contains a mixture of alkaloids of which the most important is lobeline, q.v. It has an expectorant action and is also used in the treatment of spasmodic asthma and bronchitis. Administered in large doses, it acts as an emetic, but the treatment must be very carefully monitored as it may cause paralysis of the medulla.
f lobélanidine
d Lobelanidin
i lobelanidina
e lobelanidina

1973 lobelanine
Chemical compound ($C_{22}H_{25}NO_2$) obtained from lobeline by oxidation of the secondary alcoholic group. It is next in importance to lobeline.
f lobélanine
d Lobelanin
i lobelanina
e lobelanina

1974 lobelia
The leaves and flowers of the Lobelia inflata which are gathered towards the end of blossoming and dried. It is a therapeutic agent capable of stimulating the respiratory centre and mainly used in asphyxia of the newborn and asphyxia caused by poisoning. Its antiasthmatic action is probably due to its content of isolobinine and its derivatives. Gastric inflammation may follow excessive doses.
f lobélie
d Lobelie
i lobelia
e lobelia

1975 lobeline
Or inflatine. Alkaloid obtained from the Lobelia inflata. Needle-like colourless crystals, soluble in organic solvents, less soluble in water. It is a derivative of piperidine which, owing to its pharmacological property of electively exciting the respiratory centre at the dosis of 3 to 10 mg, is used parenterally against asphyxia of the newborn, in case of apnoea or dyspnoea caused by nerve lesions and of poisoning by drugs exerting a depressive action on the bulbar centres, e.g. morphine.
f lobéline
d Lobelin; Inflatin
i lobelina
e lobelina; inflatina

1976 lofepramine
Tricyclic antidepressant of the imipramine series, which is metabolized to

desipramine after absorption. It is mostly used for the treatment of endogenous
mild depression accompanied by neurovegetative disorders. Dizziness and blurred vision, copious perspiration and
urine retention may occur as side effects.
f lofépramine
d Lofepramin
i lofepramina
e lofepramina

1977 lomustine
Cytotoxic, antimycotic, alkylating agent
acting by causing the enzymatic block
of the DNA synthesis. It is used in the
treatment of cancer and particularly of
primary and secondary brain tumours,
as well as bronchopulmonary, genital
and bone tumours. The drug is also indicated in the therapy of melanomas and
haematosarcomas. Nausea, vomiting and
ulcers of the digestive tract may occur
as side effects. There is also the danger
of haematopoietic depression.
f lomustine
d Lomustin
i lomustina
e lomustina

1978 long pepper
Hot pepper plant displaying exceedingly
pungent elongated red or yellowish
fruits which are the main source of
cayenne pepper. It is applied externally
in the form of a plaster or medicated
lint for counter-irritation.
f poivre long
d langer Pfeffer; brasilianischer Pfeffer;
Stangenpfeffer
i pepe lungo
e pimentero largo

1979 loperamide
Antidiarrhoeal agent characterized by
a marked antiperistaltic action and used
in the symptomatic treatment of acute
diarrhoea caused by food poisoning,
viral infection and sometimes antibiotics.
It is also indicated in the therapy of
chronic diarrhoea due to organic lesions,
intestinal motility disorders and fermentation disorders. Some abdominal pains
may occur as a side effect.
f lopéramide
d Loperamid
i loperamide
e loperamide

1980 lorajmine
Antiarrhythmic agent characterized by a
latent-phase action and representing the
inactive form of ajmaline (amber crystalline alkaloid obtained from trees or
shrubs of the genus Rauwolfia) which is
regenerated at the myocardial receptors
and has a stronger and more lasting
effect. The drug is capable of increasing the coronary flow and is used in
sinusal tachycardia, in extrasystoles,
in paroxysmal atrial or ventricular

tachycardia and in atrial fibrillation.
f lorjamine
d Lorjamin
i loraimina
e loraimina

1981 lorazepam
Benzodiazepin tranquillizer indicated for
the treatment of all anxiety states and
nervous tension, insomnia, psychoneuroses, acute feeling of distress and
schizophrenia. It is also used as a
sedative prior to surgical operations and
as a coadjuvant with other psychotropic
drugs. Drowsiness, fatigue, blurred vision, intestinal disorders and sometimes
ataxia may occur as side effects.
f lorazépam
d Lorazepam
i lorazepam
e lorazepam

1982 lotus
Name of the genus Zizyphus which may
be divided into three main categories,
i.e. the Zizyphus lotus, the Zizyphus
spina-Christi and the Celtis australis.
Various parts of these and their fruit
are used in popular medicine as astringent and to treat dysentery and fevers.
f lotier
d Lotos
i loto
e loto

1983 louseseed
Or stavesacre seed. The seed of the
stavesacre, a Eurasian larkspur, which
contain delphinine, are potently emetic
and cathartic and are used in Eurasia
as a fish poison. The term is also used
for staphisagria, whose seeds are very
toxic and used to kill lice in the hair.
f graine de staphisaigre; graine de capu-
cine
d Läusesamen; Stephanskraut
i stafisagria; grano di stafisagria
e semilla de estafisagria; semilla de
albarraz

1984 lousicide
Any substance or preparation capable of
destroying lice.
f pédiculicide
d Läusemittel; Entläusungsmittel
i pediculocida
e pediculocido

1985 lovage herb
Stout glabrous herb growing in Southern
Europe and characterized by a fairly
strong aromatic odour. Its roots and
leaves have been used in popular medicine as a diuretic, carminative and
febrifuge agent.
f feuilles de livèche
d Liebstöckelkraut
i erba di levistico
e hierba de ligústico

1986 lovage root oil
Essential oil obtained by treating the roots of the lovage herb and used as a carminative and a stimulant.
f essence de racine de livèche
d Liebstöckelwurzelöl
i essenza di levistico
e esencia de ligústico

* **love apple   s. tomato**

1987 loxapine
Tranquillizer mainly used in the treatment of schizophrenia. Extrapyramidal disorders, tachycardia, frequent blood--pression alterations, nausea, dyspnoea etc. may occur as side effects.
f loxapine
d Loxapin
i lossapina
e loxapina

1988 luffa
Genus of tropical climbing plant, two species of which are used in medicine, as they contain, particularly in their ripe fruits, a bitter juice acting as a cathartic agent.
f luffe amère
d bittere Luffa
i luffa amara
e lufa

1989 lungwort lichen
European herb with small blue flowers used in popular medicine for the treatment of cough and bronchitis. Also known as oaklungs.
f crapaudine; lichen pulmonaire
d Lungenflechte; Grubenflechte
i lichene polmonario
e liquen pulmonario; sticta

1990 lupine
Genus of herbs of the Leguminosae family. The Lupinus albus has been cultivated since times immemorial and used in popular medicine for its alleged property of acting as an anthelminthic and diuretic agent.
f lupin; lupin blanc
d Lupine; Hasenkleebohne
i lupino; lupino bianco
e altramuz

1991 lupulin
A powder of a yellow-brownish colour consisting of the glandular hair detached from the fruits of the hop and used as a mild hypnotic and a tonic agent.
f lupuline
d Lupulin
i lupulina
e lupulina

1992 lupulus
The fruit of the common hop, which contains as essential oil having sedative properties. An infusion prepared on the basis of its bitter principle is used as a tonic agent.
f houblon
d Hopfen
i luppolo
e lúpulo

1993 lutein
A mixture of xanthophyll (70%) and zeaxanthine (30%) widely diffuse in the yolk of the egg, in green and yellow leaves, in many yellow flowers and in the grain of mais. It also occurs in the follicle in the course of luteinization. The term is also loosely used to denote the luteinizing hormone.
f lutéine
d Lutein
i luteina
e luteina

1994 luteinizing hormone
The interstitial cell stimulating hormone, i.e. a hormone of protein-carbohydrate composition obtained from the anterior lobe of the pituitary gland. It stimulates the development of the corpora lutea in the female and the secretion of progesterone together with the follicle-stimulating hormone. In the male, it stimulates the development of interstitial tissue in the testis and the secretion of testosterone.
f hormone lutéinisante; gonadotrophine B
d luteinisierendes Hormon; Prolan B
i ormone luteinizzante; prolan B
e hormona luteinizante; gonadotrofina B

* **luteinizing   hormone-releasing   hormone
s. gonadorelin**

1995 lutenurin
Antibacterial, antimycotic and antitrichomonal agent mainly used in the treatment of vaginitis and cervicitis.
f luténurine
d Lutenurin
i lutenurina
e lutenurina

1996 lututrin
Protein-similar substance extracted from the corpora lutea, the bodies found in the ovary after, rupture of the graafian follicles, of the sow. Used in the treatment of gynecological disorders.
f lututrine
d Lututrin
i lututrina
e lututrina

1997 lymecycline
Tetracycline-L-methylenelysine. Antibiotic of the tetracycline group. It is characterized by a broad-spectrum action and indicated in the treatment of infections caused by Gram-positive and Gram-negative microorganisms. The association of the drug with lysine promotes its absorption and diminishes its toxicity.
f lymécycline

    d Lymecyclin
    i limeciclina
    e limeciclina

1998 lypressin
Opotherapeutic hormone. Synthetic post-hypophysis hormone having an antidiuretic and antihaemorrhagic action. It is indicated in the treatment of diabetes insipidus and of haemorrhages, particularly in the oesophagus and the stomach.
    f lypressine
    d Lypressin
    i lipressina
    e lipresina

1999 lysergic acid
Monocarboxylic acid containing the heterocyclic nucleus of the indole, obtained, as its amide, from ergotic alkaloids by hydrolysis. Some synthetic derivatives of the lysergic acid, namely the diethylamide and the monoethylamide, have revealed considerable psychomimetic properties because their administration causes the development of the conditions which are observed in psychoses, e.g. schizophrenia. Owing to these properties, lysergic acid is now sometimes used in psychotherapy. Unfortunately, owing to its very characteristics, the drug is widely abused as a hallucinogen.
    f acide lysergique
    d Lysergsäure
    i acido lisergico
    e ácido lisérgico

2000 lysergide
Hallucinogen which is not used therapeutically but mainly for its psychedelic effects, above all for the very considerable alterations of visual perception. The consciousness of the individual, as well as his awareness, are not subject to great alterations, but serious mental confusion and disorders, changes in the personality and conditions which are sometimes diagnosed as psychotic diseases may be diagnosed.
    f lysergide
    d Lysergid
    i lisergide
    e lisergido

2001 lysidine
Condensation product of ethylene and ethylidene diamine apparently displaying a solvent action on uric acid and therefore used in the treatment of gout. Methylglyoxalidine.
    f lysidine
    d Lysidin
    i lisidina
    e lisidina

2002 lysine
Dyamino-hexanoic acid, i.e. an amino acid which is essential for animal nutrition. It is found in casein, gelatin and in some proteins.
    f lysine
    d Lysin
    i lisina
    e lisina

2003 lysine acetylsalicylate
Soluble salt of acetylsalicylic acid used intravenously, intramuscularly and orally as a potent, long-lasting analgesic and antiphlogistic agent. The drug is indicated in the treatment of acute articular rheumatism, degenerative arthropaties and muscular pains.
    f acétyle salicylate de lysine
    d Lysin-Acetylsalicylat
    i lisina acetilsalicilato
    e lisina acetilsalicilato

2004 lysine ibuprofen
Lysine alpha-p-isobutylphenylpropionate. A non-steroid analgesic, antiphlogistic and antipyretic agent rapidly absorbed from the gastrointestinal tract owing to its high degree of solubility in water. It is used in the treatment of phlogistic and painful processes of the osteoarticular system, e.g. lumbago, sciatica and rheumatoid arthritis. Intestinal haemorrhages may occur as an adverse effect.
    f ibuprofen lysine
    d Lysin-Ibuprofen
    i ibuprofen lisina
    e ibuprofen lisina

2005 lysine-vasopressin
Or syntopressin. A synthetic compound and the form in which the posterior pituitary hormone is administered as a nasal spray in the treatment of diabetes insipidus. The compound stimulates pituitary ACTH output in the presence of hypothalamic damage which does not affect pituitary glands. It also stimulates the output of the growth hormone and is used to differentiate hypothalamic from pituitary function.
    f lysine-vasopressine
    d Lysin-Vasopressin
    i lisina vasopressina
    e lisina vasopresina

2006 lysozyme
A basic protein occurring in egg white and in biological secretions, e.g. tears, saliva, as well as in the latex of some plants. It acts as a mucolytic enzyme and against the capsules of various bacteria. It is therefore indicated in the prophylaxis and therapy of bacterial and viral infections, blood-clotting disorders and (as an analgesic) pains caused by cancer.
    f lysozime
    d Lysozym
    i lisozima
    e lisozima

2007 lysuride

An antiserotonic agent used in the treatment of severe migraine, particularly in
patients suffering from hypertension.
f lysuride
d Lysurid
i lisuride
e lisurido

M

2008 mace
The dried arillode of the seed of the nutmeg shaped like a cup in its basal part and then divided into several more or less anastomosed lobes. When fresh its colour is a vivid green. It is used like nutmeg as an aromatic spice and a flavouring agent. An essential oil is obtained from mace, a mobile, either colourless or pale yellow liquid marked by a burning taste and soluble in alcohol, ether and chloroform and containing pinene, dipentene and myristine.
f macis; arille de la noix muscade
d Macis; Muskatenblüte
i macis; arillo di noce moscata
e macis; arilo de nuez moscada

2009 Macedonian parsley
Annual or biannual herb of the family Umbelliferae, widely cultivated, particularly in southern Europe for its finely dissected leaves which are frequently used for culinary and medicinal purposes.
f ache de rochers; persil de Macédoine
d macedonische Petersilie
i petrosello macedonico
e perejil de Macedonia

2010 Madagascar periwinkle
Cultivated woody herb with opposite entire leaves and large flowers used to treat various diseases in some tropical countries. There is no scientific basis explaining its effectiveness.
f pervenche de Madagascar
d Soldatenblume
i vinca di Madagascar
e flor del príncipe; domínica

2011 madder
The root of the madder plant used in the past because of its content of alizarin in the form of the glycoside ruberythric acid. It was considered useful as a tonic and astringent.
f garance
d Krapp
i robbia
e granza

2012 madia
Or melosa, tarweed. Genus of stick herbs of the family Compositae growing in North America and in Chile. Its heads, marked by deeply grooved bracts and by an unpleasant odour, are used for the production of an essential oil which is somewhat similar to walnut oil and can be used in food.
f madia
d Madi; Melosa
i madia; melosa
e madia; melosa

2013 madwort
Or squincy wort, quincy wort. A cress of the genus Lobularia yielding a roundish fruit. It was used in medicine as a diuretic.
f aspérule à l'esquinance; petite garance; rubiole
d Ackermeier; kleiner Waldmeister
i asperula dei campi; palloncino
e hierba de la esquinancia; bregandia

2014 mafenide
A sulphonamide used in the treatment of burns and infected wounds, generally applied as a powder a solution or an ointment as it is too easily soluble and too rapidly excreted for oral use. Thus differing from the usual sulpha drugs, mafenide is not inhibited by p-amino benzoic acid. A mild histolytic action may occur as side effect.
f mafénide
d Mafenid
i mafenide
e mafenido

2015 magenta
Or magenta red. A red dyestuff consisting of a mixture of pararosaniline, rosaniline and other amine dyes, which is used as a stain in histology and bacteriology. Better known as fuchsine or basic fuchsine.
f fuchsine; rouge d'aniline
d Fuchsin; Anilinrot
i fucsina; magenta; rosanilina
e fucsina; rojo de anilina

2016 magnesia milk
Or milk of magnesia, magnesia cream. Magnesium hydroxide mixture. Suspension in water of hydrated magnesium oxide. It contains just over 8% of $Mg(OH)_2$, and is used as an antacid and mild cathartic agent.
f lait de magnésie
d Magnesia-Milch
i latte di magnesia

e leche de magnesia

2017 magnesium
Silver-white bivalent metallic element, highly malleable and ductile, widely occurring in nature always in combination in minerals, e.g. magnesite, dolomite, olivine, olivite, spinel etc., in sea and mineral waters and both in animals and plants. Many of its derivatives, e.g. acetate, bromide, citrate, phosphates etc. are used in medicine.
f magnésium
d Magnesium
i magnesio
e magnesio

2018 magnesium alginate
The magnesium salt of alginic acid (a polysaccharide obtained from seaweeds and kelps) used as an emulsifying and thickening agent in antacid preparations aiming at combating acid reflux.
f alginate de magnésium
d Magnesiumalginat
i magnesio alginato
e alginato de magnesio

2019 magnesium aluminium hydroxide
An antacid agent marked by a rapid, though transient, buffer effect on gastric hyperacidity. The drug finds very useful employment in several acidity--caused stomach disorders and is indicated in the symptomatic treatment of peptic ulcer, oesophagitis and chronic or acute gastric or duodenal diseases due to inflammation or irritation of the mucous membrane.
f hydroxyde de magnésium et aluminium
d Magnesium- aluminiumhydroxyd
i magnesio alluminio idrossido; magaldrato
e hidróxido alumínico-magnésico

2020 magnesium antacids
Those magnesium salts which are used either by themselves or with other compounds to neutralize gastric acids in the treatment of peptic ulcers. They are sometimes combined with aluminium antacids when large doses of the drug are bound to act as laxatives. They are not easily absorbed but there is the danger of toxic magnesium blood levels in renal disorders.
f antacides de magnésium
d Magnesium - antazide; Magnesium - säureneutralisierendes Mittel
i antacidi di magnesio
e antiácidos de magnesio

2021 magnesium ascorbate
Salt of magnesium which is well tolerated in the parenteral magnesium therapy. Its action, which resembles that of curare, is muscle-relaxant and sedative thus finding employment in the treatment of spasms, circulatory disorders, the pathological conditions which follow cranial traumas, epilepsy, tetanus,

eclampsia, more generally neurovegetative disorders. Overdosage may cause respiratory depression.
f ascorbate de magnésium
d Magnesiumascorbat
i magnesio ascorbato
e ascorbato de magnesio

2022 magnesium bicarbonate solution
A clear colourless liquid (fluid magnesia) effervescing upon warming and used as an antacid and a mild laxative.
f solution de bicarbonate de magnésie
d Magnesiumbicarbonatlösung
i soluzione di bicarbonato di magnesio
e solución de bicarbonato de magnesio

2023 magnesium carbonate
Either a white, odourless and near-tasteless white granular powder (heavy magnesium carbonate) mainly used as an antacid agent, or a light, odourless and near-tasteless white powder (light magnesium carbonate) characterized by antacid and mildly laxative properties. Both forms of magnesium carbonate are used in the treatment of gastrosuccorrhoea accompanied by repeated belching and gastric pains.
f carbonate de magnésium; carbonate de magnésie
d Magnesiumcarbonat; kohlensaures Magnesium
i magnesio carbonato; magnesia alba
e carbonato magnésico

2024 magnesium chloride
A mild saline purgative capable of reducing water absorption from the intestine and of stimulating the secretion of enteric juice thus increasing the intestinal bulk and promoting peristalsis. Also used as a choleretic. It is also used in solution for peritoneal dialysis and haemodialysis.
f chlorure de magnésium
d Magnesiumchlorid; Chlormagnesium
i magnesio cloruro
e cloruro de magnesio

2025 magnesium clofibrate
Ethyl-methylpropionate used to reduce serum cholesterol and triglycerides in the treatment of hypercholesterolaemia. Among its adverse effects, the most frequent is nausea, but vomiting, diarrhoea, meteorism and hepatomegalia may also occur.
f clofibrate de magnésium
d Magnesiumclofibrat
i magnesio clofibrato
e clofibrato de magnesio

2026 magnesium glutamate hydrobromide
Anti-dystonic, sedative and hypnotic agent mainly used in the treatment of psychosomatic illnesses, neuroses, hypnagogic and hypnopompic disorders and, more generally, of states of nervous tensions. The drug is also indicated for

the treatment of behavioural disorders, sleep-walking, difficulty in learning and erethistic conditions in children.
f hydrobromure de glutamate de magnésium
d Magnesiumglutamathydrobromid
i magnesio glutamato bromidrato
e glutamato bromhidrato de magnesio

2027 magnesium glycerophosphate
White amorphous powder characterized by a bitter taste and used in the past as a component of preparations that were termed nerve tonics but later abandoned for its total lack of a useful action.
f glycérophosphate de magnésium
d Magnesiumglycerophosphat
i magnesio glicerofosfato
e glicerofosfato magnésico

2028 magnesium hydroxide
Odourless, amorphous, tasteless powder, hardly soluble in water and used as an antacid in hyperchlorhydria and in the treatment of peptic ulcer. The drug is also used as a mild laxative for children as it does not form carbon dioxide but produces magnesium chloride. It is an antidote for mineral acids and arsenic.
f hydroxyde de magnésium; hydrate de magnésie
d Magnesiumhydroxyd
i idrossido di magnesio
e hidróxico magnésico

2029 magnesium lactate
Therapeutic drug used in the treatment of hypomagnesaemia (too little magnesium in the blood plasma) due to stress or inadequate absorption, with or without neuromuscular hyperexcitability.
f lactate de magnésium
d Magnesiumlaktat
i magnesio lattato
e lactato de magnesio

2030 magnesium oxide
Antacid agent prepared in the form of heavy magnesium oxide, used also as a mild laxative, and light magnesium oxide. Both find employment in the therapy of conditions that are characteristic of gastrosuccorrhoea (the persistent secretion of gastric juice in abnormally large quantities).
f oxyde de magnésium; magnésie
d Magnesiumoxyd
i ossido di magnesio; magnesia
e oxido magnésico; magnesia

2031 magnesium peroxide
White powder having mild antiseptic properties mainly used in mouth-washes and tooth-pastes. The antiseptic properties are due to the release of nascent oxygen, i.e. oxygen in active form when it is free from a compound.
f peroxyde de magnésium; superoxyde de magnésium
d Magnesium peroxyd; Magnesiumsuperoxyd
i perossido di magnesio; magnesio diossido
e peróxido magnésico

2032 magnesium pyroglutamate
See definition of magnesium lactate.
f pyroglutamate de magnésium
d Magnesiumpyroglutamat
i piroglutammato di magnesio
e piroglutamato magnésico

2033 magnesium salicylate
Antiphlogistic and analgesic agent used in the treatment of mild osteoarthritis and rheumatoid arthritis. Nausea and epigastric disorders may occur as adverse effects.
f salicylate de magnésium
d Magnesiumsalicylat
i magnesio salicilato
e salicilato magnésico

2034 magnesium stearate
Mixture, in varying proportions, of magnesium stearate, i.e. the magnesium salt of stearic acid, whose main constituents are stearic and palmitic acids, and magnesium palmitate, containing not less than 3.8% and not more than 5% of the desiccated substance. A powder displaying an excellent adhesiveness to skin surfaces and thus used in various protective creams and ointments applied against inflammation-causing substances. Magnesium stearate is also used as a lubricant in the preparation of tablets.
f stéarate de magnésium
d Magnesiumstearat
i magnesio stearato
e stearato magnésico

2035 magnesium sulphate
Commonly known as Epsom salts. A saline cathartic agent. A colourless crystalline compound marked by a bitter taste and easily soluble in warm water, less so in cold water. It is not easily absorbed by the digestive tract and thus acts as a potent and rapid purgative. Hypotension may sometimes occur as a side effect.
f sulfate de magnésium; sel anglais
d Magnesiumsulfat; Bittersalz
i solfato di magnesio; sale inglese
e sulfato magnésico

2036 magnesium trisilicate
Tasteless, odourless white powder displays a good adsorbent and antacid property. Fairly large amounts of the drug may be administered as it does not cause alkalosis.
f trisilicate de magnésium
d Magnesiumtrisilikat
i magnesio trisilicato
e trisilicato magnésico

2037 mahaleb cherry tree
A small and slender European cherry having all-white and pleasantly smelling flowers in racemes and small inferior

fruit from which a medicinal cordial can
be obtained.
f cerisier mahaleb; bois joli
d Steinweichsel; Felsenkirsche
i ciliegio canino
e cerezo de Santa Lucia

2038 maize oil
The oil that is obtained (by expression)
from the fruits of the maize plants. It
consists of glycerides, principally of
oleic and linoleic acids and is used
therapeutically as a dietetic supplement
for patients suffering from the consequences of malnutrition.
f huile de maïs; huile de germes de maïs
d Maisöl
i olio di mais; olio di germi di grano
turco
e aceite de maíz

2039 malachite green
Triphenylmethane basic dye prepared
from benzaldehyde and dimethylaniline
used for the production of organic pigments, biological stains and, in medicine, as an antiseptic agent.
f vert malachite
d Malachitgrün
i verde malachite
e verde malaquita

2040 malate
Salt or ester of the malic acid, obtained
by substitution of the hydrogen atom
with one or both its carboxylic groups,
respectively with atoms of a metal or
hydrocarbon radicals. The salts of the
malic acid are transformed into salts of
the fumaric acid at a temperature of
approximately 200°C.
f malate
d Malat
i malato
e malato

2041 male fern oil
Oleoresin obtained from a European and
North American fern and used in medicine to expel tapeworms.
f essence de fougère mâle
d Farnkrautwurzelöl; ätherisches Farnöl
i essenza di felce maschio
e esencia de helecho macho

2042 maleic acid
Ethylenedicarboxylic acid, an oleofine
acid formed by the dehydration of malic
acid or by catalytic oxidation of benzene or naphthalene. It is a stereo-
-isomer of fumaric acid which is found
in combination of cetraric acid. It is
used, among other things, in the preparation of modified essential oils.
f acide maléique
d Maleinsäure
i acido maleico
e ácido maleico

2043 male orchis

Salep plant. Genus of orchidaceous
plants some of which are used for medicinal purposes. The male orchis yields
the salep (dried tubers) which is used
medicinally as a demulcent agent.
f orchis mâle; salep; scrotum de chien
d Saleporchis; Knabenkraut
i orchidea; satirione; testicolo di cane
e satirión

2044 maleylsulphadiazole
A sulphonamide of the thiazole series
chiefly acting on the intestinal flora
owing to its very limited absorbability.
It is thus used for the treatment of
specific intestinal disorders, in particular infections affecting the colon.
f maléylsulfatiazole
d Maleylsulfatiazol
i maleilsulfatiazolo
e maleilsulfatiazolo

2045 malic acid
Hydroxy dicarboxylic acid occurring in
three optically isomeric forms; the
L-form is found in various plant juices,
in particular apples, grapes and
rhubarb, and is formed as an intermediate in the Krebs cycle.
f acide malique
d Apfelsäure
i acido malico
e ácido málico

2046 mallow leaves
The leaves of a plant of the family
Malvaceae, which, together with the
flowers of the same plant, are used in
popular medicine for the preparation of
infusions having an expectorant action
and also for enemas.
f feuilles de mauve
d Malvenblätter
i foglie di malva
e hojas de malva

2047 malonate
Salt or ester of malonic acid obtained
by substitution of the hydrogen atom of
one or both carboxylic groups of the
acid itself with atoms of a metal or,
respectively, hydrocarbon radicals.
Malonate compounds are used in medicine
as demineralizers and/or intermediate
products in the synthesis of some pharmaceutical preparations, in particular
barbiturates.
f malonate
d Malonat
i malonato
e malonato

2048 malonic acid
Bicarboxylic acid found in several
plants and in particular in the juice of
beets. It is prepared by synthesis starting from a monochloroacetic acid which
is transformed into a cyanoacetic acid
and the latter into malonic acid. The
two hydrogen atoms of the $-CH_2-$ group

can be substituted with the same number
of sodium or potassium atoms and the
products derived therefrom have a consi
derable importance in many organic
syntheses for the preparation of barbituric and hypnotic agents and for vitamins. It inhibits the transformation of
succinic acid into fumaric acid, a rever
sible reaction which is extremely important for the intermediate metabolism.

f acide malonique
d Malonsäure
i acido malonico
e ácido malónico

2049 malt
A material consisting of grain, in particular barley, which has been softened
by steeping it in water and allowed to
germinate so as develop the enzyme dias
tase. It is owing to this enzyme and to
the carbohydrates it contains that it
develops digestive and nutritive properties. It has also been used in the treatment of wasting disease, e.g. cholera
of infants, and tuberculosis.

f malt
d Malz; Gerstenmalz
i malto
e malta

2050 malt extract
The preparation which is obtained from
the grains of barley caused to germinate
artificially. It is especially used for its
nutritive properties.

f extrait de malt
d Malzextrakt
i estratto di malto
e extracto de malta

2051 maltose
Malt sugar. Crystalline dextrorotatory
fermentable reducing disaccharide sugar
which can be extracted from malt or
germinated barley following the action
of the diastase. An intermediate product
in metabolism, in brewing and distilling, it is mainly used in food as a
sweetening agent and in biological culture media.

f maltose; sucre de malt
d Maltose; Malzzucker
i maltosio
e maltosa; azúcar de malta

2052 malt vinegar
Product that is obtained by fermentation
without distillation of an infusion of
barley malt or of cereals whose starch
has been converted by malt.

f vinaigre de malt
d Malzessig
i aceto di malto
e vinagre de malta

2053 mammae extractum
Preparation (total extract, lipoid extract
etc.) obtained from the mammary gland
of healthy animals. It is a mammary

opotherapeutic agent used to relieve
uterine inflammation and congestion. Its
hormonostimulating action is mainly
directed to the mammary cell. It also
finds employment in the treatment of
hyperpolymenorrhoea, dismenorrhoea,
menometrorrhagia, metritis, uterine fibroma and inadequate milk secretion.

f extrait mammaire
d Mammaextrakt
i estratto mammario
e extracto mamario

2054 manaca
The dried root of a shrub growing in
Brazil and the West Indies, which has
been used in popular medicine for the
treatment of rheumatism and syphilis.
It is also known as vegetable mercury.

f manaca; mercure végétal
d Manaka
i manaca
e manaca

2055 manchineel
Or mancinel. Poisonous tropical American
tree, the Hippomane mancinella, which
contains in all its parts a blistering
and poisonous milky juice.

f manchinelle
d Manschinellenbaum; Manzanillabaum
i manzanillo; manzaniglio
e manzanillo

2056 mandarin oil
Or tangerine oil. Fragrant yellow essential oil obtained from the peel of mandarin oranges. It is rich in vitamins
and mainly used for flavouring.

f essence de mandarine
d Mandarinenöl
i essenza di mandarino
e esencia de mandarina

2057 mandelic acid
Acid occurring in apples and pears,
once used in tooth-cleaning tablets. It
now finds employment as an astringent
agent in skin therapy.

f acide mandélique
d Mandelsäure
i acido mandelico
e ácido mandélico

2058 mandragorine
Alkaloid found in the root of the mandra
gora and having a mydriatic and antispasmodic action that is analogous to
that of atropine. It is hardly ever used
for therapeutic purposes.

f mandragorine
d Mandragorin
i mandragorina
e mandragorina

2059 mandrake
Herb growing in southern Europe and
North Africa, having ovate leaves and
whitish or purple flowers followed by
fruits which were believed to have aphro

disiac powers. Its root was used in the past to promote conception and also as a cathartic or a narcotic. Also called mayapple.

f podophylle
d Podophyll; Maiapfel
i podofillo
e manzana de mayo; podófilo

2060 manganate
Salt which contains the bivalent group $MnO_4=$; it derives from the manganic anhydride which is not known in its free state. In a neutral or acid solution, manganates liberate manganic acid which is converted into manganese dioxide and permanganate. Manganates are oxidizing substances, not frequently used as such, but fairly important as intermediates in the preparation of permanganates.

f manganate
d Manganat
i manganato
e manganato

2061 manganese
Grey-white polyvalent metallic element occurring in nature and present in many foods and in the tissues of plants and animals. Traces of manganese are essential for the normal development of the body and for the right functioning of the body's enzyme system. It is equally essential for the formation of haemoglobin. It is employed for therapeutic purposes in the form of permanganates which have fairly potent germicidal properties.

f manganèse
d Mangan
i manganese
e manganeso

2062 manganese butyrate
Therapeutic agent used in the treatment of boils, carbuncles and, more generally, for staphylococcal cutaneous infections.

f butyrate de manganèse
d Manganbutyrat
i manganese butirrato
e butirato de manganeso

2063 manganese chloride
Therapeutic agent used in the treatment of staphylococcal infections, mainly by hypodermic injection. Minute quantities of the drug are sometimes included in iron preparations as a trace element in the therapy of microcytic anaemia.

f chlorure de manganèse
d Manganchlorid
i manganese cloruro
e cloruro de manganeso

2064 manganese dioxide
Insoluble compound used in its precipitate form together with iron salts in the treatment of anaemia.

f bioxyde de manganèse
d Mangandioxyd; Mangansuperoxyd
i biossido di manganese
e dióxido de manganeso

2065 manganese glycerophosphate
Compound occasionally used as a tonic.

f glycérophosphate de manganèse
d Manganglycerophosphat
i manganese glicerofosfato
e glicerofosfato de manganeso

2066 manganese hypophosphite
Constituent of compound syrup of hypophosphites used as a tonic in the course of convalescence and, more generally, in debilitated conditions. It is sometimes employed as a trace element with iron in the treatment of microcytic anaemia.

f hypophosphite de manganèse
d Manganhypophosphit
i ipofosfito di manganese; manganese ipofosfito
e hipofosfito de manganeso

2067 manganese peptonate
Constituent of solution of iron peptonate (a combination of pepsin with a metallic salt) with manganese, consisting of iron peptonate with manganese chloride, q.v. It is used orally in the treatment of microcytic anaemia.

f peptonate de manganèse
d Manganpeptonat
i manganese peptonato
e peptonato de manganeso

2068 manganese saccharate
Yet another manganese compound used in the treatment of microcytic anaemia.

f saccharate de manganèse
d Mangansaccharat
i manganese saccarato
e sacarato de manganeso

2069 manganese sulphate
Haematinic agent supposed to increase the effect of iron sulphate in the treatment of iron-deficiency anaemia.

f sulfate de manganèse
d Mangansulfat
i manganese solfato
e sulfato de manganeso

2070 manganic acid
Acid which does not exist in the free state but only as salt (see manganate).

f acide manganique
d Mangansäure
i acido manganico
e ácido mangánico

2071 mangosteen
Or mangostane. The fruit of an East India tree whose pericarp is used as an astringent.

f mangoustan
d indische Mangostane
i garcinia purpurea
e garcinia purpúrea

2072 mango tree
The Mangifera indica of the Anacardiaceae cultivated in tropical countries, particularly in India and Malaya, for its fibrous-fleshy fruit (mango) that has a diuretic and laxative action. Also used as an astringent.
f manguier
d indischer Mangobaum
i mango
e mango; mancho

2073 manna
The juice or exudate, rather sweetish in taste, which is obtained by means of an incision in the bark of the manna ash tree. It contains up to 50% of mannitol (hexahydric alcohol very effective as a potent diuretic) and is characterized by a mild laxative action.
f manne; manne en larmes
d Manna
i manna
e mana

2074 mannitol
Osmotic diuretic agent capable of inhibiting the reabsorption of water normally accompanying sodium reabsorption from kidney tubule. It is mainly used in case of a danger of renal failure and in the fluid overload which does not respond to other diuretics. The drug increases the circulating blood volume. It is also used in the treatment of oedema and endocranial and intraocular hypertension.
f mannitol; mannite
d Mannitol
i mannitolo; zucchero di manna; mannite
e manitol; azucar de mana; manita

2075 mannitol hexanitrate
Coronary vasodilator, mainly at the level of the vasal smooth muscles. It is frequently used in the treatment of angina pectoris.
f hexanitrate de mannitol
d Mannitolhexanitrat
i esanitrato di mannitolo
e hexanitrato de manitol

* **mannitol nitrogen mustard   s. mannomustine**

2076 mannomustine
A mannitol complex of mustine, q.v.; a cytotoxic agent particularly used in the treatment of lymphatic leukaemia and of reticulosarcomata. It also finds employment in the therapy of solid neoplasias.
f mannomustine
d Mannomustin
i mannomustina
e manomustina

2077 mannose
Aldohexose occurring as a glycoside in various plants, as a unit in compound saccharides and in mucoproteins. It is a stereoisomer of glucose and has very little importance as a nutrient agent.
f mannose
d Mannose
i mannosio
e manosa

2078 mannosulphan
Tetramesylmannitol. Alkylating and cytostatic agent used for the palliative treatment of neoplasias of the haemopoietic and reticuloendothelial systems. Also used in the treatment of ovarian, gastric, rectal and urogenital tumours independently of surgical and radiotherapeutic procedures.
f mannosulfan
d Mannosulfan
i mannosulfan
e manosulfan

2079 maprotiline
Tetracyclic antidepressant whose pharmacological properties are similar to those of the tricyclic antidepressant drugs and particularly to imipramine. It has no anticholinergic effects and is used in the treatment of depression caused by psychic or somatic disorders. Blurred vision, precipitation of glaucoma, retention of urine, constipation, cardiac arrhythmias and sometimes convulsions may occur as side effects.
f maprotiline
d Maprotilin
i maprotilina
e maprotilina

2080 marigold flowers
The flowers of a group of plants having composite flower formations. One of them, the Calendula officinalis is characterized by resolvent and stimulant properties. Its dried florets are also used as a mild aromatic and diaphoretic.
f fleurs de souci
d Goldblumen; Ringelblumenblüten
i fiori di calendola
e flores de caléndula

* **marihuana   s. hemp**

2081 marking nut
The nut of a tree growing in tropical America, as big as a chestnut, whose shell is rich in an oil that is used locally in medicine. Also known as varnish tree.
f anacardier oriental
d ostindischer Elephantenläusebaum
i anacardio orientale
e anacardo oriental

2082 marshmallow
European perennial herb also growing in the eastern US, with a dense velvety pubescence, near-oval leaves and pink racemose flowers. Its mucilaginous root has been used in the past in the treatment of cough and inflammation of the oral membrane and of the throat.

f guimauve; mauve blanche
d Eibisch
i altea; bismalva
e altea; malvavisco

2083 marshmallow root
The root of the marshmallow herb from which a therapeutic substance is extracted, used in the past to treat cough and inflammation of the mouth and of the throat.
f racine de guimauve
d Eibischwurzel; Altheewurzel
i radice di altea
e raíz de altea

2084 marsh marigold
European and North American swamp herb characterized by bright yellow flowers and gathered in early spring for its nutrient value.
f souci d'eau; pacoteur
d Sumpfranunkel; Butterblume
i farferugine; calta
e calta; hierba centella

2085 marsh willow weed
Or water willow. A type of willow which thrives in swamps and very wet areas and is used for its mucilagenous and astringent properties.
f épilobe des marais
d Sumpfweidenröschen
i epilobio palustre
e epilobio palustre

2086 master root oil
Perennial herb growing in various European countries and cultivated in the past in the belief that it possesses considerable healing powers for several illnesses.
f essence d'impératoire
d Meisterwurzelöl
i essenza d'imperatoria
e aceite de imperatoria

2087 mastic
Aromatic resinous exudation from the tree Pistacia lentiscus, containing resenes, resin acid and an essential oil. It occurs in lustrous and transparent tears from incisions in the tree.
f mastic
d Mastix
i mastice
e almáciga

2088 mate bush
A shrub growing wildly in Brazil and cultivated in Argentina. It contains tannin, a very small percentage of caffeine and aneurine. It is used as an infusion in South America for its medicinal properties, particularly as a stimulant.
f maté
d Matepflanze
i mate; acquifoglia mate
e arbusto mate

2089 matico
Term used to denote both the brush Piper angustifolium and its leaves. It grows in South America and its leaves are oblong, pointed and leathery, characterized by an agreable odour and an aromatic taste. They are used in medicine as a haemostatic agent and in the treatment of urethral infections and internal haemorrhages.
f matico
d Matiko; Matikobaum
i matico
e matico; yerba del soldado

2090 matico oil
Essential oil extracted from the matico leaves; a yellow brownish liquid containing asarin and methyleugenol and soluble in alcohol and ether. It is used in medicine as an astringent and as a stimulant of the genito-urinary tract.
f essence de matico
d Matikoöl
i essenza di matico
e esencia de matico

2091 maxillae extractum
Maxillary extract. Preparation obtained from the maxillary tissue of embryos or very young animals. A specific opotherapeutic agent used in dentistry. It has both a stimulant and plastic activity in functional deficiencies of the masticatory apparatus and is therefore indicated as an adjuvant agent in the treatment of inflammations of the periodontal membrane, gingivitis, dentine hypersensitivity, caries and jaw fractures.
f extrait maxillaire
d Maxillarextrakt
i estratto mascellare
e extracto maxilar

2092 maypop
Climbing perennial passion flower (also known as wild passion flower) having large flowers followed by yellow edible, yet somewhat tasteless berries which may be used in medicine as a mild sedative.
f grenadille; passiflore
d Passionsblume
i passiflora selvatica
e granadilla; pasionaria

2093 mazaticol
Anticholinergic agent used in the treatment of the Parkinsonian syndrome caused by neuroleptics or by arteriosclerosis.
f mazaticol
d Mazaticol
i mazaticolo
e mazaticolo

2094 mazindol
Anorectic indole derivative having central stimulant properties. It is used in the treatment of obesity. Tachycardia accompanied by a rise in blood pressure, restlessness, insomnia, fever and tremors

may occur as side effects.
f mazindole
d Mazindol
i mazindolo
e mazindolo

2095 mazipredone
Rapid-acting glucocorticoid used in the treatment of shock caused by traumas, severe scalds, surgical operations, intoxications, anaphylaxis, transfusion, infarction, drug-allergy reactions, toxaemia, acute adrenal insufficiency etc. Dyspepsia, ulcers, haemorrhages, hyperglycaemia, osteoporosis, greater proneness to infections, oedema, hypertension and a negative nitrogen balance have been observed as adverse effects.
f maziprédone
d Mazipredon
i mazipredone
e mazipredona

2096 mead
Alcoholic drink, also known as honey wine, obtained by the fermentation of an aqueous solution of honey with malt and yeast. Sometimes used medicinally as a mild tonic and stimulant. Metheglin.
f hydromel vineux
d Honigwein; Met
i idromiele; vino di miele
e vino de miel

2097 meadow buttercup
Or tall buttercup. Perennial widely diffused buttercup with a short and thick rootstock and petioled rosette leaves from which an infusion is obtained that can be used medicinally.
f bouton d'or acre; jauneau; renoncule des prés
d scharfer Hahnenfuss; brennender Hahnenfuss
i ranuncolo di palude
e botón de oro; ranúncolo

2098 mebendazole
Broad-spectrum anthelmintic particularly indicated in the treatment of roundworm, more generally against nematodes and cestodes. It must be used very carefully in children under two.
f mébendazole
d Mebendazol
i mebendazolo
e mebendazolo

2099 mebeverine
Antispasmodic agent mainly acting on the muscle. The drug is also used as an analgesic. Spastic colitis, irritable colon, gastritis, spastic constipation and gastric ulcers can also be treated with the drug.
f mébévérine
d Mebeverin
i mebeverina
e mebeverina

2100 mebhydrolin
Antihistaminic agent used in the treatment of several allergic conditions, e.g. urticaria, itch, eczematous dermatitis, hay fever, rhinitis. It also finds employment in the treatment of allergic drug reactions. Nausea and occasionally dizziness may occur as side effects.
f mebhydroline
d Mebhydrolin
i mebidrolina
e mebidrolina

2101 mebolazine
Andostrane-derived anabolic steroid without androgenic action. It promotes proteic synthesis and reduces the consumption of nitrogen. It also increases body weight. Some hirsutism has been observed in women as a side effect.
f mébolazine
d Mebolazin
i mebolazina
e mebolazina

* **mebubarbital**   s. pentobarbital

* **mebumal**   s. pentobarbital

2102 mebutamate
Antihypertensive agent acting centrally, i.e. at the level of the vasomotor control centres in the brain stem and vasomotor tracts in the spinal cord. It is mainly indicated in the treatment of hypertensive conditions that are principally due to some nervous disorder.
f mébutamate
d Mebutamat
i mebutamato
e mebutamato

2103 mebutizide
Sulphonamide-group, benzothiadiazine derivative. A diuretic agent capable of increasing the urinary excretion of water and salt without causing hypokalaemia. The effect sets in within 12 hours and generally lasts for 48 hours. It is particularly indicated in the treatment of oedema and water retention of any kind. Cardiac disorders due to loss of potassium may occur as adverse effects.
f mébutizide
d Mebutizid
i mebutizide
e mebutizida

2104 mecamylamine
Antihypertensive; a ganglion-blocking drug mainly indicated in the treatment of hypertension of neurogenous origin or caused by endocrine-gland disorders. It is generally used in association with reserpine.
f mécamylamine
d Mecamylamine
i mecamilamina
e mecamilamina

2105 **meclocycline**
Broad-spectrum tetracycline to be used topically for the prophylaxis and treatment of superficial cutaneous infections caused by Gram-positive and Gram-negative bacteria.
f méclocycline
d Meclocyclin
i meclociclina
e meclociclina

2106 **meclofenamic acid**
An antiinflammatory agent.
f acide méclofénamique
d Meclofenamsäure
i acido meclofenamico
e ácido meclofenámico

2107 **meclofenoxate**
Cerebral stimulant. A psychotherapeutic agent acting both as sympathomimetic and as a sedative. Owing to its soothing effect, it is used to relieve states of tension, anxiety and depression. It is also used to combat intoxications caused by neurotropic drugs, as well as hypophyseal dyskinesia and some neuralgias.
f méclofénoxate
d Meclofenoxat
i meclofenossato
e meclofenoxato

2108 **mecloqualone**
A non-barbituric sedative and hypnotic agent capable of rapidly promoting sleep without affecting general conditions on awakening. Intestinal disorders may occur as a side effect.
f mécloqualone
d Mecloqualon
i mecloqualone
e mecloqualona

2109 **mecloralurea**
Tranquillizer used to combat states of anxiety, neurovegetative dystonia and depression. Nausea and lipothymia may occur as side effects.
f mécloralurée
d Mecloralurea
i mecloralurea
e mecloralurea

2110 **meclozine**
Antihistamine and antiemetic agent used for the prevention and treatment of seasickness, the Menière's syndrome, nausea and vomiting. A mild degree of dizziness may occur as a side effect.
f méclozine
d Meclozin
i meclozina
e meclozina

2111 **mecobalamin**
An oral haematopoietic agent, synthetic cyanocobalamin derivative used in the treatment of vitamin B_{12} deficiency, e. g. anaemias, megaloblastic blood dyscrasia, neuropathies etc.
f mécobalamine
d Mecobalamin
i mecobalamina
e mecobalamina

2112 **meconate**
Salt or ester of meconic acid. The term also denotes some addition products of the meconic acid with basic organic compounds, e.g. the meconate of morphine and narcotine which displays a narcotic and sedative action.
f méconate
d Mekonat; mekonsaures Salz
i meconato
e meconato

2113 **meconic acid**
Dibasic oxyacid which can be extracted from opium (it contains 4 to 6% of it). It crystallizes with either one or three molecules of water. Its solution appear blood-red when combined with ferric chloride.
f acide méconique
d Mekonsäure
i acido meconico
e ácido mecónico

2114 **meconium**
Opium. A dense, viscous, brown-greenish substance marked by a slightly acid reaction. It is present in the intestine of the human fetus as from the fourth or fifth month of uterine life and is subsequently expelled in varying quantities of 80 to 100 g in the first two or three days of life. It consists of biliary secretion as well as secretion from the stomach, guts and pancreas, intestinal epithelium and amniotic liquid ingested by the fetus.
f méconium
d Mekonium
i meconio
e meconio

* **mecysteine**   s. **methylcysteine**

2115 **medazepam**
Benzodiazepin-related tranquillizer used to soothe anxiety and in the treatment of emotional disorders associated with neurovegetative affections. Among its various side effects (s. diazepam) drowsiness occurs with considerable frequency.
f médazépam
d Medazepam
i medazepam
e medazepam

2116 **medibazine**
Therapeutic agent used as a sedative for the heart. It is a coronary vasodilator and thus used against angina pectoris. The increase of the coronary blood flow is not accompanied by any alteration of the normal sinus rhythm. The drug is particularly indicated in post-infarction

therapy.
f médibazine
d Medibazin
i medibazina
e medibazina

2117 medifoxamine
Anxiolitic and antidepressant drug with analgesic action. It is used in the treat ment of states of anxiety and depression, particularly when occurring in paediatrics and gyniatrics. Also indicated in neurovegetative disorders.
f médifoxamine
d Medifoxamin
i medifossamina
e medifoxamina

2118 medrogestone
Synthetic progestogen (any substance which is identical to that of progesterone) derived from hydroprogesterone. The drug is used in habitual abortion, in amenorrhoea, oligomenorrhoea as well as functional sterility. The drug has also been used as an antineoplastic agent, in particular in the treatment of breast, uterus and prostata tumours. It seems to act as a masculinization agent for the foetus.
f médrogestone
d Medrogeston
i medrogestone
e medrogestona

2119 medroxyprogesterone
Synthetic derivative of progesterone, a progestetional steroid used as a contraceptive by three monthly injections accompanied by the daily oral administration of stilboestrol for a week each month. The drug is also used in the treatment of habitual or threatened abortion and infertility, associated with an oestrogen, e.g. ethinyl estraoediol, and in the treatment of hormone-imbalance-caused tumours. In some instances it is indicated against neoplasms of the breast, testis, uterus and endometrium. Thrombo-embolic disorders have been observed as side effects.
f médroxyprogestérone
d Medroxyprogesteron
i medrossiprogesterone
e medroxiprogesterona

2120 medrylamine
Antihistaminic agent used for the prophylaxis and the treatment of several allergic disorders, e.g. rhinitis, drug allergies and contact dermatosis. Some respiratory depression has been observed as a side effect.
f médrylamine
d Medrylamin
i medrilamina
e medrilamina

2121 medrysone
Locally administered antiphlogistic agent;

a corticosteroid derived from hydroxy-progesterone and used in the treatment of allergic conjunctivitis, scleritis and other affections of the eye.
f médrysone
d Medryson
i medrisone
e medrisona

2122 medullae ossium rubrae extractum
Extract of the red medulla of the bones. A highly purified hydrosoluble extract obtained from deproteinized bone marrow with the addition of malic acid and dissolved in a sterile physiological solution with 2% of benzylic acid added. It is an opotherapeutic substance displaying a balancing action on pituitary-adrenal axis and a protective action on the func tional activity of the collagen tissue. It is particularly indicated for the treat ment of arthritis, gout, sciatica, osteo-arthritis etc.
f extrait de moelle osseuse rouge
d roter Knochenmarkextrakt
i estratto di midollo osseo rosso
e extracto de medula osea roja

2123 medullae renalis extractum
Extract of renal medulla (the deep part of the kidney, which does not contain glomeruli). Total hydroglyceric extract of adrenal medulla (the internal part of the suprarenal gland containing glandular chromaffin cells permeated by the venous sinusoid) obtained from healthy animals. It contains the specific hormone that is elaborated by the chromaffin cells, i.e. the epinephrine. It is an opotherapeutic agent having a physiologically stimulant action capable of substituting the activity of the suprarenal gland medulla. Its action is very similar to that of epinephrine, but is more gradual and stable.
f extrait de moelle rénale
d Nierenmarkextrakt
i estratto di midollo renale
e extracto de medula renal

2124 mefenamic acid
Analgesic, antiphlogistic and antipyretic agent. It is not related to morphine and used to soothe pain which is not toc severe. Rashes, diarrhoea and blood dys crasia have been observed as side effects.
f acide méfénamique
d Mefenaminsäure
i acido mefenamico
e ácido mefenámico

*** mefenamide s. orphenadrine**

2125 mefenorex
A compound capable of diminishing the appetite for food and used in the treatment of simple obesity, as well as of obesity associated with diabetes, hypertension and arthrosis. It is also indicat

ed against post-partum obesity and meno
pausal obesity. One of its principal side
effects is insomnia.
f méfénorex
d Mefenorex
i mefenorex
e mefenorex

2126 mefruside
Sulfonamidic diuretic having an action
which lasts from 6 to 10 hours and used
in the treatment of cardiac, renal and
hepatic oedema as well as pregnancy-
-caused oedema. It acts as a hypotensive agent. Gastrointestinal disorders
may occur as side effect.
f méfruside
d Mefrusid
i mefruside
e mefrusido

2127 mefuralazine
Dihydroxymethylfuratrizine.   Antiseptic
agent active against Gram-positive and
Gram-negative microorganisms and chiefly used in the treatment of infections of
the urinary tract. Anorexia, nausea and
diarrhoea can be expected as side effects.
f méfuralazine
d Mefuralazin
i mefuralazina
e mefuralazina

2128 megestrol
Progestetional   steroid considerably more
potent   than   progesterone.   It   blocks
ovulation and is used in the treatment
of   amenorrhoea,   oligomenorrhoea   and
uterine haemorrhagies of a functional
origin. It does not affect the water-salt
metabolism. It is also indicated as palliative or co-therapeutic agent in the
treatment of endometrial carcinoma, as
an oral contraceptive (with ethinyloestra
diol), and also for the treatment of
hirsutism, female sexual infantilism and
of malignant neoplasms of the breast
and endometrium. The side effects are
similar to those of progesterone.
f mégestrol
d Megestrol
i megestrolo
e megestrolo

2129 meglumine
Organic base used for the preparation
of salts of iodinated organic acids employed as a contrast medium in radiological tests. Its action is of short dura
tion. Particularly used for arteriography,
pyelography, cholangiography and for
preoperative diagnosis.
f méglumine
d Meglumin
i meglumina
e meglumina

2130 meglumine amidotrizoate
Iodinated contrast medium particularly

used for urography and angiography and
for   the   radiographic   examination   of
splenoportography (the demonstration of
splenic   and   portal   veins   by   injecting
the radioopaque medium into the spleen),
sinus and fistulous tracts, arthrography,
hysterosalpingography, lymphography etc.
Allergic   reactions   may   be   observed   as
a side effect.
f amidotrizoate de méglumine
d Megluminamidotrizoat
i meglumina amidotrizoato
e meglumina amidotrizoato

2131 meglumine antimoniate
Antiprotozoal antimonial agent characterized by a lower degree of toxicity than
other antimonial compounds. It is rapidly eliminated through the kidneys and
especially used in the treatment of visceral leishmaniasis (kala-azar) and post-
-kala-azar dermal leishmaniasis. A febrile reaction may occur at the beginning of the therapy.
f antimoniate de méglumine
d Megluminantimoniat
i meglumina antimoniato; meglumina stibiato
e meglumina antimonato; meglumina esti-
  biato

2132 meladrazine
Antispasmodic agent acting on the synap
tic junction between two intermediate
neurons (a polysynaptic inhibitor) and
used in the treatment of nervous disorders characterized by spastic syndrome.
The drug is also used in the treatment
of lumbago, torticollis and some neurogenous diseases.
f méladrazine
d Meladrazin
i meladrazina
e meladrazina

2133 melarsonyl potassium
Trypanosomicide used in the treatment
of trypanosomiasis and filariasis. Peripheral neuritides may occur as a side
effect. The drug should not be administered in cases of epidemic influenza.
f mélarsonyl potassique
d Melarsonylkalium
i melarsonil potassico
e melarsonil potásico

2134 melarsoprol
A trivalent organic arsenical combined
with dimercaprol (BAL) on the premise
that this latter compound diminishes the
toxicity of arsenic. The drug is used in
the treatment of human trypanosomiasis
which has reached an advanced state.
f mélarsoprol
d Melarsoprol
i melarsoprolo
e melarsoprolo

2135 melilotic acid
An acid which is naturally occurring

in yellow clover. It is an aromatic oxyacid found in colourless crystals, which can be transformed into the corresponding lactone by heating.
f acide mélilotique
d Melilotsäure
i acido melilotico
e ácido melilótico

* **melin s. rutin**

2136 melissa
Melissa officinalis, a perennial herb, approximately 1 m. tall, having axillary clusters of small flowers with bilabiate calyx, exserted corolla and divergent anther lobes. Its leaves are used for the preparation of pharmaceutical products, for the distilled water and the spirit of melissa, as well as other compounds, all characterized by nerve-stimulating and antispastic properties.
f mélisse
d Melisse
i melissa
e melisa

2137 melissic acid
Monobasic fatty acid found in some varieties of wax, in particular, combined with palmitic and stearic acids, in bees-wax. It also occurs in some vegetables, e.g. sugar cane, cotton etc.
f acide mélissique
d Melissinsäure
i acido melissico
e ácido melísico

2138 melitose
Disaccharide occurring in the Australian manna. The term is frequently used to denote raffinose.
f mélitose
d Melitose
i melitosio
e melitosa

2139 melitracen
Tricyclic antidepressant with strong anticholinergic properties and a mild sedative action. The drug is a good stimulant agent of the psychomotor activity when administered in small doses and a sedative in cases of anxiety, emotional and neuropsychic tension and agitation when administered in larger doses. The drug is frequently employed in checking the risk of suicide. Its action is rapid especially if compared with various other thymoleptic preparations.
f mélitracène
d Melitracen
i melitracene
e melitraceno

2140 mellitic acid
Benzene-hexacarboxylic acid. A crystalline acid occurring in the form of its aluminium salt. Used in preparations for the treatment of juvenile diabetes.

f acide mellique; acide mellitique
d Mellitsäure; Honigsteinsäure
i acido mellitico
e ácido melítico

* **melperone s. methylperone**

2141 melphalan
Cytotoxic drug capable of inhibiting the nucleic acid biosynthesis and used in the treatment (almost always palliative) of multiple myeloma, lymphosarcoma and various solid tumours, e.g. melanoma.
f melphalan
d Melphalan
i melfalan
e melfalan

2142 menadiol
Synthetic vitamin K, as effective as the natural vitamins K1 and K2. It is mainly used in the treatment of haemorrhagic episodes, particularly when occurring in the newborn. It is also employed for essential clotting factors. Haemolytic anaemia and icterus may occur as side effects.
f ménadiol
d Menadiol
i menadiolo
e menadiolo

2143 menadiol sodium sulfate
The soluble salt of the synthetic vitamin K, an antihaemorrhagic agent used for the treatment of haemorrhages caused by obstruent (obstruction-causing) icterus and by insufficiency of the liver. The drug is also indicated in pre- and post-operative treatment of patients suffering from some hepatopathy, haemoptysis (the expectoration of bright-red blood from the lungs or bronchi and trachea) and sprue (a tropical disease). Gastric disorders are frequently occurring as side effect.
f sulfate sodique de ménadiol
d Menadiolnatriumsulfat
i menadiolo sodio solfato
e sulfato sódico de menadiolo

2144 menadione
Synthetic vitamin K3, as active as the natural one and mainly used in the treatment of conditions borne of hypoproteinanaemia as it reduces the coagulation time of the blood. More generally, it is indicated in all haemorrhagic conditions. Haemolytic anaemia and icterus may occur as side effects.
f ménadione
d Menadion
i menadione
e menadiona; menaftona

2145 menadione sodium bisulphite
Menadione derivative, soluble in water, mainly used to promote the production of essential clotting factors, i.e. the concentration in the plasma of the vari-

ous factors of blood coagulation. Essentially, vitamin K3. It is chiefly used in the treatment of haemophilia and other haemorrhagic conditions.

f   bisulfite sodique de ménadione
d   Menadionnatriumsulfit
i   menadione sodio bisolfito
e   bisolfito sódico de menadiona

2146 menadoxime

A vitamin K analogue mainly used in haemorrhagic conditions in which a low level of prothrombin has been diagnosed.

f   ménadoxime
d   Menadoxim
i   menadossima
e   menadoxima

2147 menatetrenone

Vitamin K2 indicated in the treatment of haemorrhages caused by a low prothrombin concentration and of haemorrhages affecting the newborn. Bronchospasms, cyanosis, sometimes a peripheral circulatory collapse may occur as side effects.

f   ménatétrénone
d   Menatetrenon
i   menatetrenone
e   menatetrenona

2148 menotropin

Preparation obtained from urine in the menopausal and post-menopausal stage. Its active principle is the follicle-stimulating hormone which is capable of stimulating the growth and the maturation of the ovarian germinal follicles. It also contains the luteinizing hormone which is capable of stimulating the interstitial cells. As a therapeutic agent, it is indicated in the treatment of primary and secondary amenorrhoea, anovular (relating to uterine haemorrhage independently of ovulation) cycles and sterility, both in males and females, due to dysendocriniasis. An abnormal stimulation of the ovaries accompanied by pains, ascites, oliguria, hypotension etc. may occur as side effects.

f   ménotropine
d   Menotropin; humanes follikelstimulierendes Hormon
i   menotropina
e   menotropina

*   menthae piperitae aetheroleum s. jasmone

2149 menthol

Colourless crystalline compounds characterized by a strong peppermint-like odour and an aromatic taste whose immediate warm sensation is followed by a cold one. It is obtained from various mint oils or prepared synthetically from thymol. It is soluble in alcohol, olive oil and volatile oils. Menthol is used therapeutically as a very strong antiseptic in the treatment of coryza, rhinitis and tracheitis, and as mild local

anaesthetic in dentistry. Also used in the form of ointment or liniment in muscular rheumatism.

f   menthol; alcool mentholique
d   Menthol; Pfefferminzkampfer
i   mentolo; canfora di menta
e   alcanfor de menta; alcohol mentólico

2150 mepacrine

Odourless, crystalline, bitter-tasting powder of a vividly yellow colour used in the past for the treatment of malaria. Its hydrochloride displays a potent action on the parasites of the asexual cycle of the erythrocytic phases in all three species of Plasmodium. It is a schizonticide and an anthelmintic and is also used for the treatment of intestinal amoebiasis. The drug was widely used for the prophylaxis and suppression of malaria during the last world war. It has also been employed in the treatment of certain heart diseases and of rheumatism. Headache and gastrointestinal disorders may occur as side effects. Also known as quinacrine.

f   quinacrine
d   Mepacrin; Chinacrin
i   chinacrina; mepacrina
e   mepacrina; quinacrina

*   mepazine s. pecazine

2151 mepenzolate bromide

Parasympatholytic agent. The drug has good antispasmodic properties and is mainly used in the treatment of ulcerative colitis, diarrhoea, infectious enterocolitis and other intestinal disorders. It acts like atropine and has no direct spasmolytic action on smooth muscles.

f   bromure de mépenzolate
d   Mepenzolatbromid
i   mepenzolato bromuro
e   bromuro de mepenzolato

2152 mephenesin

A compound tending to cause a temporary paralysis of striped muscle superficially not dissimilar to that provoked by curare but developing in an entirely different way. Its site is found in the basal ganglia, in the brain stem and the thalamus. The drug inhibits some types of muscular tremor –and it is therefore indicated in parkinsonism and acute alcoholism– but its action has only a short duration. Though less efficient than curare, it has been used as a muscular relaxant in surgery. It also is administered orally for the treatment of spastic, hypertonic and hyperkinetic conditions. Hypotension and sometimes collapse may occur as adverse effects.

f   méphénesine
d   Mephenesin
i   mefenesina
e   mefenesina

2153 mephenoxalone

Tranquillizer with central mechanism of action particularly used as an adjuvant agent in the treatment of some forms of muscle spasms occurring in arthrosis, lumbago etc.

f méphénoxalone
d Mephenoxalon
i mefenossalone
e mefenoxalona

2154 mephentermine

Trimethylphenetylamine. A vasoconstrictor and sympathomimetic agent displaying alpha- and beta-adrenergic activity and thus used (by topical application) as a vasal decongestant in rhinitis and (by systematic application) as an antihypotensive and cardiokinetic agent. Insomnia and tension may occur as side effects.

f méphentermine
d Mephentermin
i mefentermina
e mefentermina

2155 mephenytoin

An anticonvulsant agent which is both structurally and pharmacologically related to hydantoin and is indicated in the treatment of epilepsy and also, though with milder effects, of petit mal. It has a milder hypnotic action than phenytoin but a greater potential toxicity.

f méphénytoïne
d Mephenytoin
i mefenitoina
e mefenitoina

2156 mepivacaine

Amide-related local anaesthetic mainly used for infiltration, nerve blocking and epidural and spinal anaesthesia. The drug is indicated when anaesthetics containing adrenaline are not advisable. Also used in gynecology. The drug may cause hypotension, depression of the central nervous system and convulsions.

f mépivacaïne
d Mepivacain
i mepivacaina
e mepivacaina

2157 meprobamate

Chemical compound. Colourless crystals of a bitter taste, not easily soluble in water and stable in weakly acid or alkaline solution. The drug is used as a tranquillizer, albeit a minor one, characterized by a selective action on the hypothalamus and the spinal cord. It is principally used in the treatment of neuroses, alcoholism and various functional disorders. Small doses are advised for the elderly and for debilitated patients. In any case a careful monitoring of the effects is recommended especially because of the danger that pre-existing symptoms, e.g. anxiety, insomnia, anorexia, withdrawal reactions etc. might reemerge. The drug displays considerable side effects, in particular, in addition to gastro-intestinal disorders, dizziness accompanied by hypotension and hepatic drug metabolism with the danger of drug interactions. Overdosage may induce coma and respiratory depression.

f méprobamate
d Meprobamat
i meprobamato
e meprobamato

2158 meprylcaine

Topical anaesthetic agent mainly used in surgery and dentistry. Not to be used for patients suffering from myasthenia gravis and low plasma-cholinesterase concentrations. Allergic reactions, vertigo, nausea and vomiting may occur as side effects.

f méprylcaïne
d Meprylcain
i meprilcaina
e meprilcaina

2159 mepyramine maleate

Antihistaminic agent used in anaphylactic and allergic conditions, in particular hay fever and urticaria. It has a comparatively short duration and very rarely causes sedation.

f maléate de mépyramine
d Mepyraminmaleat
i mepiramina maleato
e maleato de mepiramina

2160 mequinol

Melanogenesis inhibitor indicated for the topical treatment of skin areas suffering from hyperpigmentation caused by drugs, cosmetics, vitiligo and, at times, post-traumatic and post-operative consequences. Also used in the treatment of Addison's melanoderma, chloasma, ephelides and Riehl's melanosis.

f méquinol
d Mequinol
i mechinolo
e mequinolo

2161 mequitazine

Antihistamine indicated as an antiemetic and, more generally, in the treatment of allergic reactions. No antipsychotic effects have been observed, but its sedative action renders it useful as a hypnotic in children and in pre-operative medication. It must be used with great caution in liver diseases and epilepsy. Postural hypotension, blurred vision, cholestatic jaundice and deposits in the lens and cornea are possible side effects. Overdosage may cause coma.

f méquitazine
d Mequitazin
i mequitazina
e mequitacina

2162 merbaphen

Mercuric derivative of the diethylmalonyl
urea; a crystalline, white, odourless
powder soluble in water and insoluble
in alcohol and ether. It has a diuretic
action and was used in the past for the
treatment of syphilis either by itself or
in association with arsenobenzene.
f merbaphène
d Merbaphen
i merbafene
e merbafeno

2163 merbromine
A topical antibacterial and disinfectant
agent generally used for the treatment
of comparatively small cutaneous lesions
and/or infections. The drug may display
a mild irritant action.
f merbromine
d Merbromin
i merbromina
e merbromina

2164 mercaptamine
Drug sometimes used for the treatment
of radiation sickness and of diseases
which may occur following treatment with
isotopes.
f mercaptamine
d Mercaptamin
i mercaptamina
e mercaptamina

2165 mercaptan
Family of organic compounds which are
sulphur analogues to the alcohols and
phenols but contain sulphur instead of
oxygen. Mercaptans are weak acids
which form salts known as mercaptides
(the metallic derivatives).
f mercaptan
d Mercaptan
i mercaptano
e mercaptano

* mercaptoalanine   s. cysteine

2166 mercaptomerin
A mercurial diuretic agent mainly used
for the treatment of peripheral and pulmonary oedemas caused by congestive
cardiac insufficiency, nephroses, glome-
rulonephritis, cirrhosis of the liver as
well as obstruction of the porta hepatis.
f mercaptomérine
d Mercaptomerin
i mercaptomerina
e mercaptomerina

2167 mercaptopurine
A cytotoxic drug particularly used in
the treatment of leukaemia. More generally, as it inhibits nucleoprotein synthesis,
it finds employment in many neoplastic
diseases. Also very useful in the therapy of chronic myelocytic leukaemia.
f mercaptopurine
d Mercaptopurin
i mercaptopurina
e mercaptopurina

* 4-mercaptopyrazole 3.4-D1 pyrimidine   s.
tysopurine

2168 mercurial
Relating to a pharmaceutical preparation
which contains mercury, e.g. mercurial
ointment (30% mercury in a fatty excipient).
f mercuriel
d Quecksilber-
i mercuriale
e mercurial

2169 mercuric benzoate
White crystalline compound, hardly soluble in water but easily soluble in a
solution of ammonium benzoate, used in
the past, either orally, intramuscularly
or by urethral injection for the treatment of syphilis and gonorrhoea.
f benzoate mercurique
d Quecksilberbenzoat
i benzoato di mercurio
e benzoato mercúrico

2170 mercuric chloride
Topical antiseptic and disinfectant mainly used for the disinfection of non-dete-
riorating objects and, albeit very rarely
and with caution, hands and skin completely free of any lesion. A weak solution is sometimes used as astringent.
Also known as corrosive sublimate.
f chlorure mercurique
d Quecksilberchlorid
i cloruro mercurico
e cloruro mercúrico

2171 mercuric cyanide
Colourless, transparent, poisonous crystals which darken when exposed to light
and are soluble in water and alcohol.
It is prepared by causing mercury oxide
to react with hydrocyanic acid at a
raised temperature and under pressure.
It is used in medicine as a rather potent antiseptic and antiluetic agent.
f cyanure mercurique
d Quecksilbercyanid
i cianuro mercurico
e cianuro mercúrico

2172 mercuric iodide
Used in the past as an antiluetic agent,
it may find some employment as an antiseptic for surgical instruments and other
objects.
f iodure mercurique
d Jodquecksilber; Quecksilberjodid
i ioduro mercurico
e yoduro mercúrico

2173 mercuric nitrate
Compound, $Hg(NO_3)_2 \cdot H_2O$, occurring in
colourless crystals or in the form of
white deliquescent powder. It is decomposed by heating and used in medicine
for its antiseptic properties.
f nitrate mercurique
d Quecksilbernitrat

i nitrato mercurico
e nitrato mercúrico

2174 mercuric oxide (red)
Or red precipitate. Once regarded as an
antiluetic, it is now sometimes used as
an antiseptic ointment. It should not
be used in the presence of nephropathies
or in conjunction with chloride, bromide
and iodide.
f oxyde mercurique rouge; précipité rouge
d Hydrargyrum oxydatum rubrum
i ossido mercurico rosso; precipitato rosso
e óxido mercúrico rojo; precipitado rojo

2175 mercuric oxycyanide
A compound that is obtained by the
interaction of mercuric oxide and an
excess of mercuric cyanide. It is a
strong antiseptic agent but less irritant
than mercuric chloride and therefore
used for the irrigation of the conjunctiva in cases of ophthalmia affecting the
newborn. It is also used for the treatment of conjunctivitis, for irrigation of
the urinary bladder and for the sterilization of surgical instruments.
f oxycyanure mercurique
d Quecksilberoxycyanid
i ossicianuro mercurico
e oxicianuro mercúrico

2176 mercuric salicylate
A compound that is obtained by boiling
a solution of salicylic acid with yellow
mercuric oxide. It is used in medicine,
either in the form of powder or as an
ointment, for its antiseptic and antisyphilitic properties.
f salicylate mercurique
d Quecksilbersalicylat
i salicilato mercurico
e salicilato mercúrico

2177 mercuric sulphate (yellow)
A compound that is obtained when water
is mixed with mercuric sulphate. It is
used in medicine as a quick-acting emetic and, in the form of ointment, for the
treatment of seborrhoea capitis and ring-
worm.
f sulfate mercurique
d Merkurisulfat; Quecksilbersulfat
i solfato mercurico
e sulfato mercúrico

2178 mercuric sulphide
A compound which occurs as a brilliant
red powder and was used in the past,
in the form of an ointment, for the treatment of chronic skin diseases. Also called Chinese red.
f sulfure mercurique
d Quecksilbersulfid
i solfuro mercurico
e sulfuro mercúrico

2179 mercurobutol
Topical disinfectant agent indicated
against bacteria and protozoa. It is

generally used in gynecology, surgery,
dermatology and otorhinolaryngology.
Some mild irritation has been observed
as side effect.
f mercurobutole
d Merkurobutol
i mercurobutolo
e mercurobutolo

2180 mercurochrome
Water-soluble phthalein dye which was
once used as a mild antiseptic.
f mercurochrome
d Merkurochrom
i mercurocromo
e mercurocromo

2181 mercurophylline
Combination of the organic mercurial
ß-methoxy-gamma-hydroxy mercuropropyla-
mide of trimethylcyclopentane dicarboxylic
acid with theophylline (mercurophylline
sodium). It is a quick-acting mercurial
diuretic agent. Also used in the treatment of peptic ulcer.
f mercurophylline
d Merkurophyllin
i mercurofillina
e mercurofilina

2182 mercurous chloride (mild)
Drug used orally as cathartic agent,
particularly in veterinary medicine. It
is also used parenterally as an antiluetic. Renal lesions have been observed
following prolonged use.
f chlorure mercureux
d Merkurochlorid
i cloruro mercuroso
e cloruro mercurioso

2183 mercurous iodide (yellow)
Compound resulting from the interaction
of mercurous nitrate and potassium
iodide. It is principally used in the
form of ointment for the treatment of
chronic cutaneous diseases and enlarged
glands. It has also found employment in
the treatment of syphilis in its later
stages.
f iodure mercureux
d Merkurojodid
i ioduro mercuroso
e yoduro mercurioso

2184 mercurous oxide
Black powder (Hg_2O) soluble in acids
and insoluble in water, which can be
prepared by treating the mercurous nitrate with an alkaline hydrate. Also
known as black mercury oxide.
f oxyde mercureux
d Merkurooxyd; Quecksilberoxyd
i ossido mercuroso
e óxido mercurioso

2185 mercury
An element occurring naturally as the
sulphide in the ore cinnabar. Many
pharmaceutical preparations for thera-

peutic use contain small doses of strongly diluted mercury as this is a general
poison of the protoplasm. It is owing to
this property that so many inorganic
compounds are employed for external use
as disinfectant agents.
f mercure
d Quecksilber
i mercurio
e mercurio

2186 mercury and potassium iodide
Complex salt formed by the combination
of mercuric iodide and potassium iodide.
Being a potent germicide, it is used
for the treatment of cutaneous infections
and for the disinfection of surgical
instruments.
f iodure de mercure and potassium
d Quecksilber- und Kaliumiodid
i ioduro di mercurio e potassio
e yoduro de mercurio y potasio

2187 mercury bichloride
Mercuric chloride, corrosive sublimate.
Water-soluble white crystalline powder.
Strong solutions of the compound, though
caustic to tissues, are used as disinfectants. Weak solutions may be used as
antiseptic and astringent lotions. Caution is required in its use owing to the
danger of mercury poisoning.
f chlorure mercurique
d Merkurichlorid; Quecksilberchlorid
i cloruro mercurico
e cloruro mercúrico

2188 mersalyl
Organic mercurial diuretic agent capable
of depressing the active reabsorption of
sodium and chloride by the kidney
tubules. It is particularly indicated in
the treatment of fluid retention. Caution
is indicated owing to the danger of an
excessive loss of sodium and chloride.
The drug must be injected intramuscular
ly. Gastro-intestinal disorders and skin
rashes may occur as side effects.
f mersalyl
d Mersalyl
i mersalil
e mersalil

2189 mescaline
The alkaloid contained in the peyote
(Echinocactus lewinii) and occurring in
colourless crystals fairly soluble in
water. It has a toxic action on the central nervous system and strong hallucinogen properties. Its oral ingestion or
parenteral administration induces grave
neurovegetative disorders, e.g. nausea,
vomiting, mydriasis, and a schizophrenic
-like symptomatology characterized by
vivid colourful hallucinations.
f mescaline
d Mescalin
i mescalina
e mescalina

* **mescopolamine methyl sulphate   s. scopolamine methyl sulphate**

2190 mesna
Mucolytic and secretory agent principally used to protect the mucosa of the
bladder from the irritant effect of cytotoxic drugs, e.g. cyclophosphamide,
ifosfamide, which are activated by metabolism in the liver and mainly excreted
in the urine. The drug is also used as
a nasal spray in the treatment of nasal
obstruction caused by excessive secretions and for nebulization for the treatment of asthma, mucoviscidosis, bronchial emphysema etc. The nasal spray
may cause some inflammation of the
mucous membrane and nebulization cough
and spasms in asthmatic patients.
f mesna
d Mesna
i mesna
e mesna

* **mesocaine   s. trimecaine**

* **mesorgydin   s. lysuride**

2191 mesoridazine
Phenothiazine neuroleptic indicated in
the treatment of schizophrenia, psycho-
neurosis and cerebral diseases for its
strong tranquillizing properties. Hypotension, extrapyramidal disorders, nausea and vomiting, drowsiness may occur
as side effects. When administered by
parenteral injection, the patient must be
kept lying down for some time to avoid
the danger of a sudden drop in blood
pressure.
f mésoridazine
d Mesoridazin
i mesoridazina
e mesoridacina

2192 mesotartaric acid
Optically inactive form of the tartaric
acid. Such inactivity derives from an
internal symmetry of the molecule which
leads to intramolecular balance and thus
to the annulment of the rotatory action.
Together with the racemic tartaric acid
it is formed by heating with water or
mineral acids of the d-tartaric acid, or
by oxidation of various compounds (e.g.
maleic acid, inactive erythrol etc.). It
occurs in colourless, water-soluble crystals.
f acide mésotartrique; acide antitartrique
d Mesoweinsäure
i acido mesotartarico
e ácido mesotartárico

2193 mesterolone
Male sex hormone marked by actions,
applications and effects, both positive
and adverse, similar to those of testosterone, q.v.
f mestérolone
d Mesterolon

ı mesterolone
e mesterolona

2194 mestranol
Sex hormone. An oestrogenic agent generally used in conjunction with various
progestogens as oral contraceptive for
its capacity to inhibit ovulation. Also
used for haemorrhagic metropathies and
dysmenorrhoea (the pain in the back
and lower abdomen at the time of the
menses). Gastro-intestinal disorders may
occur as side effects.
f mestranol
d Mestranol
ı mestranolo
e mestranolo

2195 mesulphen
Dimethyldiphenylene disulphide. Yellow
viscous oil which contains approximately
25% of sulphur in organic combinations.
It is a topical scabicide and antipruritic agent and thus used in the treatment
of itching dermatoses and eczemas as
well as acne. It is particularly effective
when applied to hairy regions of the
body.
f mésulfène
d Mesulfen
ı mesulfene; mesulfen
e mesulfen

2196 mesuximide
Succimide-related anticonvulsant agent
less toxic than other antiepileptics deriv
ed, both structurally and pharmacologically, from barbiturates and/or hydantoin. The drug is used in the treatment
of petit mal.
f mésuximide
d Mesuximid
ı mesuccimide
e mesuximida

2197 metabisulfite
Or pyrosulfite. Any of the pyrosulfites
of various metals and particularly the
potassium metabisulfite. Colourless crystals which are obtained by saturating
with sulphurous trioxide a solution of
potassium bisulfite or heating the latter
to about 100°C. It is slowly soluble in
water and its solution acts like the bisulfite and thus capable of displaying
a reducing and antiseptic action.
f pyrosulfite
d Metabisulfit; Pyrosulfit
ı metabisolfito; pirosolfito
e metabisulfito; pirosulfito

2198 metaborate
Salt of the metaboric acid with monovalent Me metal. Metaborates are unstable in solution and by the action of
carbonic acid decompose into the corresponding tetraborates.
f métaborate
d Metaborat
ı metaborato

e metaborato

2199 metacresol
Isomeric cresol present in the cresol
that is obtained from coal tar. Its use
as a disinfectant has proved more effective than ortho- or paracresol and is
generally prepared either as a soapy so
lution or as an emulsion as its solubility in water is negligeable. Parachloro-
metacresol (the halogen derivative) displays an approximately ten times greater
activity and a comparatively lesser degree of toxicity. It is present in many
disinfectant solutions.
f métacrésol
d Metacresol
ı metacresolo
e metacresolo

* **metacresolsulphonic acid-formaldeyde** s.
polycresolsulphonate

* **metadiazine** s. **pyrimidine**

* **metamphetamine** s. **desoxyephedrine**

2200 metanilic acid
Crystalline sulfonic acid which is isomeric with sulfanilic acid and formed by
sulfonating nitrobenzene and subsequent
reducing. It is used as an intermediate
for azo-dyes.
f acide métanilique
d Metanilsäure
ı acido metanilico
e ácido metanílico

2201 metantimoniate
The salt of antimonic acid in which Me
is a monovalent metal. Alkaline metanti-
moniates are not easily soluble in water;
the sodium one is a white powder used
for the constitution of pharmaceutical
preparations. The potassium metantimoniate, in a mixture with the free acid,
is used in medicine as an expectorant
(diaphoretic antimony).
f métantimoniate
d Metantimoniat
ı metantimoniato
e metantimoniato

2202 metaphosphoric acid
Acid which may regarded as a derivative of phosphoric acid by the elimination of a water molecule. Its salts are
important in medicine (sodium, calcium
etc.).
f acide métaphosphorique
d Metaphosphorsäure
ı acido metafosforico
e ácido metafosfórico

2203 metaraminol
Sympathicomimetic agent whose alpha and
beta effects are similar to those of
adrenaline. Its hydrogen tartrate is a
vasopressor marked by a prolonged action. It is also a hypertensive and car-

diokinetic agent and as such is indicated to combat hypotension in emergencies.
The drug considerably improves the circulation of the blood in the regions of
the heart, cerebrum and kidneys. Headache, nausea, vomiting and dizziness
may occur as side effects.
f métaraminol
d Metaraminol
i metaraminolo
e metaraminolo

2204 metasaccharic acid
Dibasic acid isomeric with mannosaccharic acid (the acid which results from
further oxidation of mannose) and obtain
ed by the oxidation of mannitol (a hexa
hydric acid occurring in several plants,
e.g. sugar cane, celery etc.).
f acide métasaccharique
d Metazuckersäure
i acido metasaccarico
e ácido metasacárico

2205 metazocine
Therapeutic agent used as a narcotic
analgesic.
f métazocine
d Methazocin
i metazocina
e metazocina

2206 metescufylline
Capillary-protecting agent used in the
treatment of pathological conditions and
consequences of venous stasis (the slowing down or outright stoppage of the
flow of blood in a vein), varices, phlebitis, varicose ulcers, haemorrhoids,
gangrene, erythrocyanosis etc.
f métescufylline
d Metescufyllin
i metescufillina
e metescufilina

2207 metformin
Synthetic oral antidiabetic capable of
increasing the use of glucose by peripheral tissues. It is mainly used in the
milder forms of diabetes and in senile
diabetes. Some degree of anorexia has
been observed as a side effect.
f metformine
d Metformin
i metformina
e metformina

2208 methabarbital
Barbituric agent characterized by a long
action and anticonvulsant properties,
which is especially indicated in the
treatment of epilepsy (both grand mal
and petit mal and myoclonic seizures).
The drug is particularly useful for patients who do not positively react to
other anticonvulsants. The drug has
proved very effective when administered
to very young children especially when
the condition is the consequence of some
brain damage. Cardiovascular disorders,

the neutralization of the contraceptive
activity of oral oestro-progestational
agents etc. may occur as side effects.
f methabarbital
d Methabarbital
i metabarbitale
e metabarbital

2209 methacholine chloride
Therapeutic compound which can be
administered by mouth or injection and
has the muscarinic effects of acetylcholine and a greater stability. It is
capable of stimulating the parasympathetic nerve receptors reducing the heart
rate and blood pressure and dilating
the peripheral blood vessels. It is there
fore used in the treatment of paroxysmal
tachycardia, the Raynaud's syndrome,
scleroderma (diffuse sclerosis of the
skin), vasospasms in the limbs and
glaucoma. Nausea, vomiting, ptyalism,
dyspnoea etc. may occur as side effects.
f chlorure de méthacholine
d Methacholinchlorid
i metacolina cloruro
e cloruro de metacolina

2210 methacycline
Methacycline hydrochloride. Tetracycline-
-related broad-spectrum antibiotic acting
against Gram-positive and Gram-negative
bacteria, protozoa, spirochetes, as well
as some viruses. The drug is indicated
for the treatment of infections of the
respiratory, urinary, gastrointestinal
tracts, bones and joints as well as soft
tissues. Peak plasma levels are attained
within 4 hours and the effect generally
persists for 12 hours.
f méthacycline
d Methacyclin
i metaciclina
e metaciclina

2211 methadone
Synthetic narcotic analgesic agent used
to relieve pain. Its action is similar to
that of morphine but it causes less seda
tion and respiratory depression. It is
also used in the control of withdrawal
symptoms for narcotic addiction. It also
finds employment for the relief of chronic pains in terminal diseases. Drowsiness may occur as side effect.
f méthadone
d Methadon
i metadone
e metadona

2212 methadyl acetate
Analgesic agent synthesized from methadone hydrochloride. It displays a delayed morphine-like action. Its use is only
rarely recommended.
f acétate de méthadyle
d Methadylacetat
i metadil acetato
e acetato de metadilo

2213 **methahexamide**
Hypoglycaemic agent of the sulphanylurea series used in the treatment of
diabetes occurring in adults independent
ly of the presence of obesity. Hypoglycaemia and allergic muco-cutaneous reac
tions may occur as side effects.
f méthahexamide
d Methahexamid
i metaesamide
e metaesamida

2214 **methallenestril**
Non-steroid synthetic sex hormone with
an action on the uterine membrane that
is similar to that of oestrone but much
weaker (approximately 10 times). It is
indicated for the treatment of menopause
and ovulation disorders, of cancer of
the prostata and of osteoporosis. Withdrawal bleeding, breast swelling in the
male, water retention, the stimulation
of tumours, nausea, as well as arterial
and venous thrombosis may occur as side
effects.
f méthallènestrile
d Methallenestril
i metallenestrile
e metalenestrilo

2215 **methamivan**
Diethylamide of the 3-ethoxy-4-hydroxy-
benzoic acid. Cardiovascular and respira
tory analeptic agent capable of increasing the activity of the heart by stimulat
ing the respiratory and vasoconstrictor
centres in the medulla oblongata and by
a cardiac positive inotropic action and
a cardiac negative chronotropic action.
The drug is used in the treatment of
both chronic and acute cardiac and
respiratory insufficiency, dyspnoea,
hypotension and circulatory collapse. It
is also indicated in the treatment of
intoxications and as a postoperative
analeptic agent.
f méthamivan
d Methamivan
i metamivan
e metamivan

2216 **methamphazone**
Analgesic and antirheumatic agent (4-
-amino-6-methyl-2-phenyl-3-pyridazone),
without any noteworthy adverse effects.
f méthamphazone
d Methamphazon
i metamfazone
e metamfasona

2217 **methamphepramone**
Metabolic agent acting on the hypophysis
and the diencephalon with anorexic effects. It is indicated in the control and
treatment of obesity due both to lack of
physical exercise and immoderate diet,
as well as to psychosomatic disorders.
f métamphépramone
d Methamphepramon
i metamfepramone

e metamfepramona

2218 **methamphetamine**
Methylamphetamine derivative acting as
a sympathicomimetic amine. A cardio-,
angio- and psychostimulant agent charac
terized by peripheral actions, i.e. the
rising of systolic and diastolic blood
pressure, dilation of the bronchi and
the stimulation of respiration. The drug
displays all the clinical properties of
amphetamine. Side effects include restleness, insomnia, dizziness, tremor and
headache. Gastro-intestinal, allergic and
endocrine disorders have also been observed.
f méthylamphétamine
d Methylamphetamin
i metamfetamina
e metamfetamina

2219 **methampicillin**
Semi-synthetic antibiotic agent derived
from penicillin and indicated in the
treatment of infections due to penicillin-
resistant microorganisms. The high-
-plasma levels of the drug last for
about 12 hours. Some allergic reactions
have been observed as side effects.
f méthampicilline
d Methampicillin
i metampicillina
e metampicilina

2220 **methandienone**
Synthetic anabolic agent capable of stim
ulating protein synthesis, increasing
body weight, improving appetite and,
more generally, supporting the process
of convalescence. It is particularly indi
cated in the treatment of osteoporosis,
various types of haemorrhages and of
albuminuria.
f méthandiénone
d Methandienon
i metandienone
e metandienona

2221 **methandriol**
Synthetic anabolic steroid capable of
promoting protein biosynthesis and used
to support the process of growth and
increase body weight. It is also indicated as a palliative in the treatment of
carcinoma of the breast.
f méthandriol
d Methandriol
i metandriolo
e metandriolo

* methandrostenolone   s. methandienone

2222 **methane disulphonic acid**
Methionic acid. An acid which proves
important in organic synthesis.
f acide méthionique
d Methionsäure; Methandisulfonsäure
i acido metionico
e ácido metiónico

2223 methane sulfonate
Or methionate. Salt of the methansulfonic
acid, i.e. CH_3SO_3 Me, where Me is a
monovalent metal. The methanesulfonates
are generally crystalline and are used
as intermediates in organic syntheses
and as catalyzers.
f méthanesulfonate; méthionate
d Methionat; methansulfonsäures Salz
i metansolfonato; mesilato
e metansulfonato; mesilato

2224 methaniazide
Chemotherapeutic agent derived from isoniazid and chiefly used in the prophylaxis and treatment of acute pulmonary
tuberculosis. It is less toxic than isoniazid. Gastro-intestinal disorders and
excitability may be observed as side effects.
f méthaniazide
d Methaniazid
i metaniazide
e metaniazida

2225 methanol
Methyl alcohol. Colourless, neutral liquid characterized by a penetrating
odour, soluble in water, ether, ethyl
alcohol etc. It is used as raw material
for the preparation of formaldeyde. Very
toxic. Also known as wood alcohol.
f méthanol; alcool méthylique
d Methanol; Holzgeist; Methylalkohol
i metanolo; alcool metilico
e metanol; alcohol metílico; alcohol de
madera

2226 methansulfonic acid
Sulfonic acid derived from methane,
CH_3SO_3H (methylsulfonic acid). Its salts
are more important than the free acid
and some derivatives, e.g. chloride,
amides etc., are used as catalyzers in
alkylation reactions.
f acide méthansulfonique
d Methansulfonsäure
i acido metansolfonico
e ácido metansulfónico

2227 methanteline
Parasympathetic therapeutic agent capable of acting on the parasympathetic
nerve receptors like atropin but combining a ganglionic blocking action like
tetraethylammonium chloride. It is mainly used for the treatment of peptic ulcer,
pylorospasm, spastic colitis and disorder
of the biliary tract, as well as hyperemesis gravidarum, pacreatitis and uri-
nary-bladder spasms. Mydriasis, dysuria
and restlessness are the most frequent
adverse effects of the drug which can
be administered orally or parenterally
by intramuscular or intravenous injection.
f méthanteline
d Methantelin
i metantelina
e metantelina

2228 methaphenilene
Antihistamine agent which is very effective and has a low incidence of adverse
effects. It is generally used in the treatment of allergic states, e.g. hay fever,
chronic vasomotor rhinitis, angioneurotic
oedema and any reaction due to antibiotics.
f méthaphénylène
d Methaphenilen
i metafenilene
e metafenileno

2229 methapyrilene
Chemical compound. The hydrochloride
of the base is an antihistamine agent
of limited capacity, i.e. capable of reducing the sensitivity of the tissues to
histamine. The drug is indicated in the
treatment of allergic symptoms associated
with hay-fever, serum sickness, rhinitis,
chronic urticaria and, more generally,
any allergy to drugs. Some nervous
depression and occasionally anorexia
may occur as adverse effects.
f méthapyrilène
d Methapyrilen
i metapirilene
e metapirileno

2230 methaqualone
Quinazoline-related hypnotic and sedative drug whose effects are similar to
those of barbiturates although it appears to act on a different central nervous system site from the site of barbiturates and glutethimidum. Its action,
if compared to that of barbiturates, is
potent and prolonged. It is generally
used to combat insomnia due to physical
pain or emotions. It is also a spasmolytic.
f méthaqualone
d Methaqualon
i metaqualone
e metaqualona

2231 methaxalone
Centrally-acting muscle relaxant agent
mostly indicated for the relief of pain
accompanied by muscular spasm, e.g.
sprains, fractures and traumatic lesions
of the tendons and the ligaments. Drowsiness, nausea, vomiting and in some
cases blood dyscrasia have been observed as side effects.
f méthaxalone
d Methaxalon
i metassalone
e metaxalona

2232 methazolamide
Diuretic agent with immediate action or
the distal tubules by inhibiting carbonic
anhydrase and the reabsorption of electrolytes. It is particularly used in the
treatment of glaucoma. Drowsiness may
occur as a side effect.
f méthazolamide
d Methazolamid

i metazolamide
e metazolamida

2233 methdilazine
Antihistaminic agent capable of reducing
the sensitivity of the tissues to histamine and used in the treatment of allergic symptoms which are associated
with dermatoses and oedemas of an allergic nature. Drowsiness may occur as
a side effect.
f methdilazine
d Methdilazin
i metdilazina
e metdilacina

2234 methenamine
US synonym for hexamine, an antiseptic
agent used topically and for urinary infections. The drug acts by releasing
formaldehyde in an acid medium either
in the stomach or in the urine and is
particularly effective against gram-negative bacteria. Some degree of albuminuria may occur as side effect.
f méthénamine
d Methenamin
i metenamina
e metenamina

2235 methenolone
Or metenolone. Synthetic anabolic steroid
capable of stimulating protein biosynthesis and of retaining nitrogen. It
improves the general physiological condition, accelerates convalescence, ba-
lances loss of weight and promotes the
development of osseous tissue.
f méthénolone
d Methenolon
i metenolone
e metenolona

2236 methicillin
Semi-synthetic penicillin characterized
by its resistance to inactivation by
penicillase. It is specifically used for
the treatment of penicillase-producing
Staphylococcus aureus infections. Intramuscular injection makes it possible to
reach peak blood concentrations in 30
to 60 minutes. Topical histolytic action,
anaemia, hyperazotaemia, haematuria
etc. may occur as side effects.
f méthicilline
d Methicillin
i meticillina
e meticilina

2237 methicotinium iodide
Methylic ester of nicotinic acid methyl
iodide. A iodonicotinic organic compound
marked by a stimulant action on metabolism and the blood-clarifying action
of iodine plus, in addition, the peripheral and cerebral vasodilating action
of nicotinic acid but with a more gradual and longer effect. It is mainly
indicated in the treatment of arteriosclerosis, hypertension, angiospasms and,
more generally, pathological conditions
which are caused by a defective circulation.
f méthyl iodométhylnicotinate
d Methicotiniumjodid
i meticotinio ioduro
e yoduro de meticotinio

2238 methimazole
Or thiamazole. Powerful antithyroid
agent capable of inhibiting the conversion of iodide into iodine, iodination of
thyrosine and the formation of diiodo-
tyrosine. The drug is recommended for
the treatment of hyperthyroidism, juvenile thyrotoxicosis and menopausal disor
ders. It may be used by itself or in
association with radioactive iodine. It
reduces the basal metabolism. Allergic
cutaneous reactions and aplastic anaemia
may occur as side effects.
f méthimazol
d Methimazol; Thiamazol
i metimazolo; tiamazolo
e metimazol

2239 methionine
An amino acid that is an essential constituent of the diet. It occurs in protein
hydrolisates and is especially important
in nutrition as it is one of the principal sources of methyl groups necessary
for the biological methylation processes
of the body. Methionine may be used as
an antidote in severe cases of poisoning
caused by paracetamol as it reduces the
concentration of a toxic metabolite. The
adenosyl methionine acts as an important
methyl donor, e.g. in the synthesis of
choline phospholipids and in the synthesis of catecholamine, noradrenaline.
f méthionine
d Methionin
i metionina
e metionina

2240 methionine S^{35}
Radioactive sulphur-containing agent,
whereby the presence and concentration
of methionine can be identified and as-
sessed in tissues.
f méthionine S^{35}
d S^{35}-Methionin
i metionina S^{35}
e metionina S^{35}

2241 methiosulphon iodide
Methylmethioninsulphon iodide. Iodo-sul-
phorated organic compound capable of
combining the metabolism-promoting action of iodine with the catalyzing, desensitizing and resolving action of sulphur. The drug is particularly indicated
in the treatment of chronic and/or acute
arthropathies. Patients susceptible to
iodine may display some idiosyncratic
disorders as a side effect.
f iodure de méthiosulfon
d Methiosulfonjodid
i metiosolfonio ioduro

e yoduro de metiosolfonio

2242 methiosulfonium chloride
Vitamin U. Liver-protecting and anti-ul cer agent. Its transmethylating and transulfurant activity produces an antinecrotic and lipotropic effect for the hepatic cell and a regenerating effect on the mucous membrane of the gastro-intestinal tract. The drug is therefore indicated in the treatment of diseases of the liver, of cholecystopathies and peptic ulcers. Some nausea and a foul-smelling breath may occur as side effect.
f chlorure de méthiosulfonium
d Methiosulfonchlorid
i metiosolfonio cloruro
e cloruro de metiosulfonio

2243 methisazone
Thiosemicarbazone-related chemotherapeutic drug, which has proved effective in the prophylaxis and the treatment of the vaccine virus. Some teratogenic action has been observed as a side effect.
f méthisazone
d Methisazon
i metisazone
e metisazona

2244 methixene
Parasympatolytic agent indicated in the treatment of Parkinson's disease (paralysis agitans) owing to its capacity to reduce tremor. It is also used as a spasmolytic agent in the treatment of spastic gastro-intestinal disorders. The drug is also employed as an antihistaminic.
f méthixène
d Methixen
i metisene
e metiseno

2245 methocarbamol
Muscle relaxant mainly used for the relief of muscle spasm and pain. Its action is similar to that of mephesin, q.v. It is indicated in the treatment of acute skeletal muscle hyperactivity caused by neuro-musculo-skeletal disorders, e.g. fibromyositis. It has also been used, as a coadjuvant, in the treatment of tetanus.
f méthocarbamol
d Methocarbamol
i metocarbamolo
e metocarbamolo

2246 methocidin
Broad-spectrum antibiotic which is obtained from the splitting-off tyrothricin. It acts as a bactericide and is used as disinfectant for various pathological cutaneous disorders. Also used in otorhinology. Some histolytic action has been observed as a side effect.
f méthocidine
d Methocidin

i metocidina
e metocidina

2247 methohexital
Or methohexitone. A very quick-acting barbiturate hypnotic chiefly used for the induction of anaesthesia. Respiratory disorders, e.g. some depressions, have been observed as side effects.
f méthohexitone; méthohexital
d Methohexital
i metoesital
e metohesitalo

* **methohexitone s. methohexital**

2248 methonium
Term denoting organic bases which are members of a polymethylene bis series. Pentamethonium and decamethonium are homologues and the salts of their bases are highly active in blocking autonomic ganglia. The salts of the decamethonium base are used as muscle relaxants owing to their property of producing neuromuscular block.
f méthonium
d Methonium
i metonio
e metonio

2249 methophenazine
Phenothiazine neuroleptic drug indicated in the treatment of psychic and motor disturbances, above all acute anxiety and irritability, paranoic crises, paranoid schizophrenia, catatonic condition etc. Also employed in paediatrics, gynecology and oncology. Extrapyramidal syn dromes have been observed as side effects.
f méthophénazine
d Methophenazin
i metofenazina
e metofenazina

2250 methoserpidine
Reserpine-derived drug used as a hypertensive. No associated neuropsychic depression has been noted and the required effects, although slow at the beginning of the treatment, last for a good many days. Essential hypertension, as well as hypertension accompanied by heart or kidney diseases may be treated by this drug either by itself or in combination with diuretics. Side effects are negligeable.
f méthoserpidine
d Methoserpidin
i metoserpidina
e metoserpidina

2251 methotrexate
Cytotoxic drug capable of antagonizing folic acid and used in the treatment of grave cases of leukaemia, particularly in children. It is also used in chorionic carcinoma and in lymphosarcoma. Alopecia, stomatitis, oligospermia, liver

toxicity, folate-deficient anaemia, bone marrow depression and skin reactions are possible adverse effects. The drug should be avoided in pregnancy.

f méthotrexate
d Methotrexat
i metotressato
e metotrexato

* **methotrimeprazine s. laevomepromazine**

2252 methoxamine
A sympathomimetic agent capable of stimulating alpha-adrenoreceptors so as to cause constriction of blood vessels and an increase in blood pressure. It is mainly used to balance hypotension in the course of anaesthesia and in the treatment of acute peripheral circulatory collapse. Arrhythmia marked by a considerable slowing of heart rate and excessive blood-pressure rise may easily occur as side effects.

f méthoxamine
d Methoxamin
i metossamina
e metoxamina

2253 methoxsalen
Xanthotoxin. Therapeutic agent which increases the skin pigmentation that is caused by exposure to ultraviolet rays increasing the production of melanin. The drug is used in the treatment of vitiligo and for sun-tanning. Gastric disorders, e.g. diarrhoea and nausea, insomnia, depression and photosensitization may occur as side effects.

f méthoxsalen
d Methoxsalen
i metossalene
e metoxsaleno

2254 methoxyflurane
A potent, non-inflammable and non-explosive volatile anaesthetic marked by a high boiling point. It is used for the induction and maintenance of anaesthesia, as well as for the production of analgesia. If used in the course of labour and delivery, the drug may be rendered more potent by barbiturates and curare derivatives. Administered in large doses, the drug may cause some renal damage; in fact, it should not be used in case of renal dysfunction. Respiratory depression and hypotension may occur as side effects.

f méthoxyflurane
d Methoxyfluran
i metossiflurano
e metoxiflurano

2255 methoxyphenamine
Sympathomimetic agent derived from phenylethylamine and used, for its bronchodilating properties, in the treatment of asthma, allergic rhinitis and allergic skin disorders. Some palpitation and insomnia may occur as side effects. The drug should not be used for patients suffering from hypertension.

f méthoxyphénamine
d Methoxyphenamin
i metossifenamina
e metoxifenamina

2256 methoxypsoralen
5-methoxypsoralen. Drug that is used topically to exclude sun rays which may prove harmful. Photosensitivity may occur as side effect. It is thought that the drug may increase the incidence of skin cancer.

f méthoxypsoralen
d Methoxypsoralen
i metossipsoralene
e metoxipsoraleno

2257 methyclothiazide
A thiazide diuretic (s. bendroflumethiazide) characterized by direct action on the proximal renal tubules by a mechanism very similar to that of chlorine- and fluorine-compounds, though more potent in its effects. It is indicated in the treatment of oedema due to some heart diseases, hepatopathies, nephrosis, hypertension, obesity etc. A reduction of blood potassium and myasthenia may occur as adverse effects.

f méthyclothiazide
d Methyclothiazid
i meticlotiazide
e meticlotiazida

2258 methyl acetanilide
A chemical compound which finds use in medicine as an antipyretic and analgesic agent.

f méthylacétanilide
d Methylacetanilid
i metilacetanilide
e metilacetanilida

2259 methylacetic acid
Propanoic or propionic acid. Saturated fatty acid which is obtained from the reduction of lactic acid. It also occurs in fermentation and in the conversion of glycogen into lactic acid in muscle. It is used as a fungicide in the form of ointment or dusting powder.

f acide méthylacétique; acide propionique
d Methylessigsäure; Propionsäure
i acido metilacetico; acido propionico
e ácido metilacético; ácido propiónico

2260 methyl acetoacetate
Colourless liquid soluble in alcohol used as an ingredient of solvent mixtures and in organic syntheses.

f acétate de méthyle
d Methylacetat; essigsaures Methyl
i metilacetato
e metilacetato; acetato de metilo

2261 methylacetylsalicylate
Colourless, odourless crystals insoluble in water, which decompose with alkalis.

It is used in the treatment of rheumatism, as a febrifuge and analgesic.
f méthylacétylsalicylate
d Methylacetylsalicylat
i metilacetilsalicilato
e acetilsalicilato de metilo

2262 **methylaesculetin**
Methylated derivative of the aesculetin (crystalline lactone obtained by hydrolysis of aesculin, the glycoside that is obtained from the inner bark of the horse chestnut and the roots of the yellow jessamine) but more potent than the latter in strengthening the resistance of the capillaries. It is widely used in medicine in the form of sodium salt of the disulphuric ester.
f méthylesculétine
d Methyläskuletin
i metilesculetina
e metilesculetina

2263 **methyl alcohol**
Methanol. Colourless liquid marked by a characteristic odour and a burning taste. It is obtained by the destructive distillation of wood or synthesized from water gas. Methyl alcohol depresses the central nervous system and the products of its toxic oxidation cause optic atrophy and acidosis. It cannot be administered internally and is only used as a solvent.
f alcool méthylique; méthanol
d Methylalkohol; Methanol
i alcool metilico; metanolo
e alcohol de metilo; metanol

* **methyl aldehyde   s. formaldehyde**

2264 **methylaminonaphthol**
Vitamin K5 displaying antihaemorrhagic properties and used in the treatment of hypothrombinaemia-related haemorrhagies. Perspiration, repeated flushing, tachycardia and sometimes cyanosis may occur as side effects.
f méthylaminonaphtol
d Methylaminonaphthol
i metilaminonaftolo
e metilaminonaftolo

* **methylamphetamine   s. methamphetamine**

2265 **methylanthraquinone**
A compound occurring in tiny white needles, soluble in ether and benzene and not easily in water. It is prepared by methylation of the anthraquinone with methyl alcohol in the presence of sulphuric acid and is widely used in organic syntheses.
f méthylanthraquinone
d Methylanthrachinon
i metilantrachinone
e metilantraquinona

2266 **methylbenzethonium chloride**
Principally an antiinfective agent. A

quaternary ammonium effective against gram-positive and gram-negative microorganisms. It is widely used to rinse bedding and underwear of incontinent patients with a view to avoiding ammoniacal dermatitis. Also used to check dermatitides, miliaria rubra and intertrigo on the perianal area. Hypokalaemia and myasthenia may occur as side effects.
f chlorure de méthylbenzéthonium
d Methylbenzethoniumchlorid
i metilbenzetonio cloruro
e cloruro de metilbencetonio

2267 **methyl bromide**
Colourless, volatile liquid of a burning taste and a chloroform-like odour. Inflammable. It forms a crystalline hydrate with cold water, and can be mixed with ordinary organic solvents. It is an intoxicating agent electively acting on the central nervous system and causing cerebellar and labyrinthic disorders which may prove permanent. The treatment against the intoxication is symptomatic.
f bromure de méthyle
d Methylbromid; Brommethyl
i metilbromuro
e bromuro de metilo

2268 **methyl butesalicylate**
Methyl-salicylate derivative, odourless, capable of being easily absorbed by the tissues and quickly turning into methyl salicylate and diethylacetic radicals, i. e. two sedative compounds. The drug is indicated for the relief of rheumatic and arthritic pains, as well as pains caused by frostbite or by excessive physical stress or traumas. Some revulsive reaction may be observed as side effect.
f butésalicylate de méthyle
d Methylbutesalicylat
i butesalicilato di metile
e butesalicilato de metilo

2269 **methyl carbonate**
Colourless, pleasantly smelling liquid which can be mixed with alkalis and acids. It is insoluble in water but soluble in the ordinary organic solvents. It finds employment in several organic syntheses.
f carbonate de méthyle
d Methylcarbonat
i metilcarbonato
e metilcarbonato; carbonato de metilo

2270 **methylcellulose**
Methyl ether of the cellulose containing approximately 30% of methoxyls. A laxative agent also used to treat functional disorders of the intestine.
f méthylcellulose
d Methylcellulose
i metilcellulosa
e metil celulosa

2271 methyl chloride
Colourless non-corrosive gas prepared by distilling methyl alcohol with hydrochloric acid either in a gaseous phase at high temperature (300 to 450°C) or in an aqueous medium at low temperature (100 to 150°C) in the presence of a metal chloride acting as catalyzer. It has considerable narcotic properties but its use is restricted owing to the risk of hepatic damage.
f chlorure de méthyle
d Methylchlorid
i metilcloruro
e cloruro de metilo

2272 methylchromone
3-methylchromone. Analgesic and antispasmodic agent mainly used in the treatment of angina pectoris and, more generally, spastic conditions. Some nausea accompanied by vomiting may occur as side effect.
f méthylchromone
d Methylchromon
i metilcromone
e metilcromona

2273 methylclothiazide
Diuretic agent chiefly used in the treatment of fluid overload and for the control of high blood pressure. The drug acts by diminishing sodium reabsorption in the kidneys.
f méthylclothiazide
d Methylclothiazid
i metilclotiazide
e metilclotiazida

2274 methylcysteine
Cysteine derivative used as a decongestant and mucolytic agent marked by a physiological mechanism of action. It promotes processes of repair and restoration and is used especially in the treatment of acute and chronic respiratory diseases, mucopurulent rhinitis, bronchitis and laryngotracheitis. The drug is also used for prophylactic purposes.
f mécystéine; méthylcystéine
d Methylcystein
i metilcisteina
e metilcisteina

2275 methyldigoxin
Cardioactive glycoside characterized by a positively inotropic action and by a bradycardiac effect. A quick and complete absorption is effected by the intestine. The drug is mainly used in the treatment of acute and chronic cardiac insufficiency.
f méthyldigoxine
d Methyldigoxin
i metildigossina
e medigoxina; metildigoxina

2276 methyldopa
Antihypertensive agent capable of inhibiting the biosynthesis of norepinephrine

and thus useful in the treatment of hypertension. Sedation, depression, fluid retention, impotence and haemolytic anaemia may occur as side effects.
f méthyldopa
d Methyldopa
i metildopa
e metildopa

2277 methylephedrine
Sympathomimetic agent. Cardiovascular and respiratory arousal-causing drug used in the treatment of hypotension and respiratory incompetence.
f méthyléphédrine
d Methylephedrin
i metilefedrina
e metilefedrina

* **methylergol carbamide   s. lysuride**

2278 methylergometrine
The name of D-1-hydroxy-2-butylamide of D-lysergic acid, an ergot-related alkaloid obtained from the dried sclerotium of the Claviceps purpurea. It produces rhythmical contractions of the uterus, in particular of the puerperal uterus and finds a very useful employment in case of post-partum haemorrhage and menorrhagia. Its effect is extremely rapid and lasts for 4 to 8 hours.
f méthylergométrine
d Methylergometrin
i metilergometrina; metilergonovina
e metilergometrina; metilergonovina

2279 methyl ether
Colourless gas sometimes used in medicine for its anaesthetic properties.
f méthyle d'éther
d Äthermethyl
i metiletere
e éter de metilo

2280 methyl eugenol
Compound occurring in oleum myrciae. It is insoluble in water and soluble in alcohol and ether. Used medicinally for its aromatic properties.
f eugénol de méthyle
d Methyleugenol
i metileugenol
e eugenol de metilo

2281 methyl furoate
Colourless liquid, tending to yellowish if exposed to light, almost insoluble in water but soluble in alcohol and ether. It is used as a solvent and in organic syntheses.
f furoate de méthyle
d Methylfuroat
i metilfuroato
e metilfuroato; furoato de metilo

* **methyl-gag   s. mitoguazone**

2282 methylheptaminol
Sympathomimetic and cardiotonic agent

capable of promoting contractility of the heart and of improving coronary circulation. It is therefore used in the treatment of heart disorders, in particular insufficiency, spasms and inability of the heart to fulfil its task.

f méthylheptaminole
d Methylheptaminol
i metileptaminolo
e metilheptaminolo

2283 methylhexaneamine
Sympathomimetic and vasoconstricting agent used in the treatment of nasal congestion. Headache and, albeit very rarely, tremors may occur as side effects.

f méthexamine
d Methylhexanamin
i metessamina
e metilhexanoamina

2284 methyl hydroxybenzoate
White crystalline substance easily soluble in water and used to hinder bacterial growth in solutions and/or emulsions of pharmaceutical preparations. Also used with the propyl analogue in solution for eye drops.

f hydroxybenzoate de méthyle
d Methylhydroxybenzoat
i metilidrossibenzoato
e hidroxibenzoato de metilo

2285 methyl iodide
Colourless liquid becoming brownish when exposed to light, soluble in alcohol and ether but insoluble in water. It can be prepared by distillation of the product that is obtained through a reaction between methyl alcohol, sodium iodide and sulphuric acid. It is used in medicine as topical anaesthetic, in organic syntheses, in the microscopic technique and as a reagent. It has also found employment as a vesicant.

f iodure de méthyle
d Methyljodid
i ioduro di metile; metilioduro
e yoduro de metilo

* **methylisooctenylamine   s. isometheptene**

2286 methyl nicotinate
A vasodilator and rubefacient agent used in ointments and creams to be applied topically for the relief of rheumatic and arthritic pains.

f nicotinate de méthyle
d Methylnicotinat; Methylnikotinat
i metile nicotinato
e nicotinato de metilo

2287 methyloxybenzoate
Compound of whose possible isomers the most notable is the one with the substituting groups in positions 1 and 4 (para); a crystalline, white, odourless powder having a slightly bitter taste, soluble in alcohol and ether. It has an antiseptic and antifermentative action and is therefore used for the conservation of pharmaceutical and alimentary products.

f méthyloxybenzoate
d Methyloxybenzoat
i metilossibenzoato
e metiloxibenzoato

2288 methyl paraben
Methylparahydroxybenzoate, methylhydroxybenzoate. A white crystalline compound, easily soluble in water, mainly used to inhibit bacterial growth in solutions or emulsions of drugs. The substance is also used for eye drops with the propyl analogue in solution.

f hydroxybenzoate de méthyle
d Methylhydroxybenzoat
i metilidrossibenzoato
e benzoato de metilhidroxi

2289 methylpentynol
Tertiary alcohol acting as hypnotic and sedative and characterized by a rapid but short-lasting general depression of the central nervous system. As it does not cause apathy or drowsiness it is used to soothe anxiety and tension, also in children and aged patients. It is also indicated for the treatment of insomnia. Normal doses may cause some belching; large doses, or overdosage, are bound to bring about symptoms which resemble inebriation with alcohol.

f méthylpentynol
d Methylpentynol
i metilpentinolo
e metilpentinolo

2290 methylperone
Butyrophenone-related neuroleptic agent, a sedative and myorelaxant used in the treatment of disorders characterized by psychomotor agitation, confusional states also in senility and epilepsy. Some alterations in blood pressure and cardiac frequency may occur as side effects.

f méthylpérone
d Methylperon
i metilperone; melperone
e metilperona; melperona

2291 methylphenidate
Methylic acid of the alpha-phenyl-2-piperidinacetic acid. Analeptic agent capable of stimulating the central nervous system and (not specific) the respiratory system. More generally, an antidepressant drug used in the treatment of stress, mental fatigue and psychic neuroses. The agent also finds employment in the treatment of disorders caused by the menopause and by senescence. Insomnia may easily occur as side effect, in which case the dose must be reduced. Some allergic conditions have also been observed.

f méthylphénidate
d Methylphenidat
i metilfenidato

e metilfenidato

2292 methylphenobarbital
Anticonvulsant and sedative agent especially indicated in the treatment of epilepsy and, in small doses, of petit mal. It also has an antispasmodic action and is frequently used instead of phenobarbital, although the transition from one drug to the other must be carried out gradually, usually in the course of ten days.
f méthylphénobarbital
d Methylphenobarbital
i metilfenobarbitale
e metilfenobarbitalo

2293 methylphenylquinolinecarboxylic acid
Methyl derivative of cinchophen used as an analgesic antipyretic and in the treatment of sciatica, lumbago and gout.
f acide méthylphénylquinolinecarboxylique
d Methylphenylquinolincarbonsäure
i acido metilfenilquinolincarbossilico
e ácido metilfenilquinolincarboxílico

2294 methylprednisolone
Corticosteroid (see also prednisolone) displaying a mainly glucocorticoid action and used in the treatment of rheumatism, arthritis, collagenosis, phlogistic and allergic dermatopathies, traumatic ophthalmopathies, anaemia and acute or chronic leukaemia. The drug is also employed in the treatment of multiple sclerosis, sarcoidosis, pulmonary fibrosis, adrenocortical insufficiency, tuberculosis and autoimmune haemolytic anaemia. Various side effects, above all of a gastrointestinal nature, have been observed. A withdrawal syndrome may be observed in cases of sudden interruption of the therapy.
f méthylprednisolone
d Methylprednisolon
i metilprednisolone
e metilprednisolona

2295 methylrosanilium chloride
Methyl rosaniline chloride. Anthelmintic agent mainly used in the treatment of oxyuria (the infestation with a nematode worm). It probably acts as a myoneural blocking for oxyuris (a threadworm) without being an oxyurifuge. Gastro-intestinal disorders may occur as a side effect.
f chlorure de méthylrosaniline
d Methylrosanilinchlorid
i metilrosanilina cloruro
e cloruro de metilrosanilina

2296 methyl salicylate
Oily, colourless or yellowish or reddish liquid having a strong aromatic odour, fairly soluble in alcohol, soluble in ether and acetic acid and much less so in water. It can be prepared by heating salicylic acid and methyl alcohol in the presence of a dehydrating agent, e.g. sulphuric acid, by distillation from the leaves of Gaultheria procumbens or the bark of Betula lenta. The compound has a local irritant and rubefacient action, and is used as a topical application to sooth inflammation of the joints. It is also used as a liniment in non-articular rheumatism. A mild sense of revulsion may be felt as side effect.
f salicylate de méthyle; acide gaulthérique
d Methylsalicylat; Salicylsäuremethylester
i metilsalicilato; salicilato di metile
e salicilato de metilo; metilsalicilato

2297 methylsulphuric acid
Acid sulphate of methyl; an oily liquid, soluble in anhydrous ether, mildly soluble in alcohol and water. It is mainly used in organic syntheses and as a chemical aggressive agent.
f acide méthylsulfurique
d Methylschwefelsäure
i acido metilsolforico
e ácido metilsulfúrico

2298 methyltestosterone
Sex hormone. A synthetic male hormone acting physiologically and used in the treatment of hypogonadism, sterility, frigidity and oligospermia, as well as in the treatment of prostatic hypertrophia and cirrhosis. Excess fluid and water retention, the growth of prostate tumours and virilization in women may occur as side effects.
f méthyltestostérone
d Methyltestosteron
i metiltestosterone
e metiltestosterona

2299 methylthionine chloride
Methylene blue. A mild antiseptic agent used in the treatment of infections of the urinary tract. It may also be used intravenously to combat poisoning caused by metahaemoglobinaemic drugs, and orally to combat metahaemoglobinaemia. Inflammation of the urinary tract and gastro-intestinal disorders have been observed as side effects as a consequence of high doses of the drug.
f chlorure de méthylthionine
d Methylthioninchlorid
i metiltioninio cloruro
e cloruro de metiltionina

2300 methylthiouracil
Synthetic antithyroid agent capable of inhibiting the biosynthesis of the thyroid hormone by hindering the iodination of tyrosine. The drug is used for the treatment of hyperthyroidism, in particular Basedow's disease, thyrotoxicosis associated with a pathological condition of the cardiovascular system and of the central nervous system. A possible greater production of thyrotropic hormone by the drug may lead to hypertrophy of the thyroid.

f méthylthiouracile
d Methylthiouracil
i metiltiouracile
e metiltiouracilo

* **methyluracil   s. thymine**

2301 methypranol
Beta-adrenergic blocking agent mainly used in the treatment of coronary incompetence and sclerosis, stenocardia, neuro vegetative diseases of the heart and blood hypotension. The drug must be administered with great caution to patients with recent myocardial infarction. Nausea with vomiting, diarrhoea, paraesthesia of the limbs and vertigo may occur as side effects.
f méthypranol
d Methypranol
i metipranolo
e metipranolo

2302 methyprylone
Glutethimide-related hypnotic sedative, mainly used as a tranquillizer. The drug is also used to intensify the action of other tranquillizers. Vertigo accompanied by nausea may occur as side effect.
f méthyprylone
d Methyprylon
i metiprilone
e metiprilona

2303 methysergide
A serotonin antagonist used in the preventive treatment of severe headache caused by vasomotor disorders. The drug should not be administered in pregnancy or to patients suffering from angina pectoris, thrombosis and, more generally, vascular lesions. Nausea, dizziness, abdominal cramp and psychiatric disorders may occur as side effects. The prolonged use of the drug may cause retroperitoneal fibrosis and ensuing impairment of renal function.
f méthysergide
d Methysergid
i metisergide
e metisergido

2304 metiazinic acid
A non-steroid synthetic antiphlogistic agent characterized by analgesic, antipyretic and anti-bradykinin properties. It is used in the treatment of rheumatism, arthritis and arthrosis, gout, neurologic and muscular pains, post--traumatic pains and phlebitis. A burning sensation while urinating as well as gastrointestinal disorders may occur as side effects.
f acide métiazinique
d Metiazinsäure
i acido metiazinico
e ácido metiazínico

2305 meticrane
Thiazide diuretic agent acting selective-

ly on the proximal renal tubules. It is principally indicated in the treatment of oedemas and essential hypertension. A reduction of erythropoiesis and hyperaldosteronaemia may occur as side effects.
f méticrane
d Meticran
i meticrano
e meticrano

* **metiperidol   s. moperone**

2306 metirosine
An antihypertensive agent. Specifically, an enzyme inhibitor capable of blocking the synthesis of catecholamines by the adrenal gland. The drug is used in the treatment of hypertension due to the adrenal tumour phaeochromocytoma (chro maffinoma mainly occurring in the adrenal medulla) where the increased blood pressure is caused by excessive production of catecholamines. Diarrhoea, sedation and a degree of hypersensitivity may occur as side effects.
f métirosine
d Metirosin
i metirosina
e metirosina

2307 metmorfin
Oral antidiabetic agent capable of augmenting the use of glucose by the peripheral tissues. It suppresses appetite and is thus indicated in overweight patients.
f metmorfin
d Metmorfin
i metmorfin
e metmorfin

2308 metochalcone
Choleretic and cholagogue agent used in the treatment of functional liver incompetence, biliary dysfunction and biliary stasis. The drug also promotes the digestion of fats and is sometimes used in the therapy of nervous disorders due to functional hepatopathies.
f méthochalcone
d Metochalcon
i metocalcone
e metocalcona

2309 metoclopramide
Antiemetic agent with dopamine antagonist actions in the brain and peripheral effects on the gastro-intestinal tract where it acts by promoting motility and thus improving intestinal transit. It is principally used in the treatment of nausea and vomiting and as an adjunct to the X-ray exalination of the intestine. Drowsiness and some involuntary movements may occur as side effects.
f métoclopramide
d Metoclopramid
i metoclopramide
e metoclopramida

2310 metolazone
Diuretic agent used in the treatment of
moderate hypertension, oedemas and pre-
-eclampsia (a condition occurring in
pregnancy and considered a precursor
of eclampsia, i.e. an attack of convulsions). Nausea accompanied by vomiting,
diarrhoea or constipation, dizziness and
hypotension may occur as side effects.
f métolazone
d Metolazon
i metolazone
e metolazona

* **metophenazate s. methophenazine**

2311 metopimazine
Phenothiazine derivative acting electively on the bulbar centre of vomiting. The
drug has a very considerable anti-apo-
morphine activity, greater than that of
chlorpromazine and prochlorpromazine.
Normal doses do not cause sedative or
neuroleptic action. The drug is mainly
used in the symptomatic and preventive
treatment of nausea and vomiting (e.g.
seasickness). Drowsiness may occur as
side effect.
f métopimazine
d Metopimazin
i metopimazina
e metopimacina

2312 metopon
Or methopon. Dihydromethylmorphinone,
a modified morphine compound. Its hydro
cloride occurs in the form of colourless
crystals and is used as an analgesic
agent. Its action is followed by collateral effects which are milder than those
produced by morphine.
f métopon
d Metopon
i metopone
e metopona

2313 metoprolol
Beta-adrenoceptor blocking drug used in
the treatment of hypertension and angina
pectoris.
f métoprolol
d Metoprolol
i metoprololo
e metoprololo

2314 metrisilol
Sodium monomethylsilanol orthohydroxy-
benzoate. Therapeutic agent capable of
regenerating connective tissues and used
in rheumatology for the treatment of
osteoporosis (especially of the aged),
postmenopausal osteoporosis and osteoporosis due to adverse effects of cortisone derivatives. The drug is also used
for the treatment of arteritis and athero
sclerosis, cerebral ischaemia, Raynaud's
disease (cyanosis of the limbs caused by
a degree of temperature which would not
affect a healthy individual), ischaemia-
-related disorders of the coronary arte-

ries and microcystic and macrocystic
mastoses. The drug is generally adminis
tered intravenously.
f métrisilol
d Metrisilol
i metrisilolo
e metrisilolo

2315 metrizamide
Contrast medium which is used in lumbar
myelography. Nausea, vomiting and head
ache may occur as side effects.
f métrizamide
d Metrizamid
i metrizamide
e metrizamida

2316 metronidazole
Antimicrobial agent effective against
trichomonas, anaerobic bacteria, Vincent's organisms (Borrelis vincenti and
Fusiformis fusiformis) gardiasis and
amoebiasis. It is a synthetic drug of
the nitroimidazole series and is widely
used in the treatment of diseases caused
by the above-mentioned microorganisms,
as well in the treatment of septicaemia,
cerebral abscesses, pelvic cellulitis and
infections following surgical lesions in
gynecology. Nausea, vomiting, hypersensitivity reactions and a metallic taste
in the mouth are the most frequent side
effects.
f métronidazole
d Metronidazol
i metronidazolo
e metronidazolo

2317 metyrapone
Powerful aldosterone inhibitor mainly
used as a diuretic agent in the treatment of oedemas caused by hyperaldosteronism, as well as antidote and diagnostic test. Gastro-intestinal disorders
may occur as side effects.
f métyrapone
d Metyrapon
i metirapone
e metirapona

2318 mexenone
Sunburn-protecting agent capable of
absorbing ultraviolet light and very use
ful in the treatment of sun-induced erythema and photodermatosis.
f méxénone
d Mexenon
i mesenone
e mexenona

2319 mexiletine
Cardiac antiarrhythmic agent indicated
in the treatment of ventricular arrhythmias. In the presence of grave disorders
of atrioventricular conduction, caution
and accurate monitoring are required.
Nausea, vomiting, drowsiness, tremors
and hypotension are possible side effects
of the drug.
f mexilétine

d Mexiletin
i mexiletina
e mexiletina

2320 mezereum bark
The bark of the branches of the Daphne
mezereum growing in several Eurasian
countries. It is used as a rubefacient
and vesicant agent. It was also used
internally, but its administration is no
longer admitted owing to its tendency
to cause nephritis, gastroenteritis etc.
The powder is a strong sneezing agent.
f écorce de mézéréon
d Kellerhalsrinde; Seidelbastrinde
i corteccia di mezereo; pepe montano
e corteza de mezereón

2321 mezlocillin
Penicillin antibiotic marked by a broad
spectrum of activity against gram-negative microorganisms, in particular
Pseudomonas and Proteus. It is injected.
The drug has proved useless against
penicillase-producing bacteria. Hypersensitivity reactions have been observed as
side effcts. Also encephalopathy accompanied by convulsions if the drug is
administered intrathecally or in excessive doses.
f mezlocilline
d Mezlocillin
i mezlocillina
e mezlocilina

2322 mianserin
Tetracyclic antidepressant agent with
actions and uses similar to those of imipramine (q.v.) but without its peripheral autonomic side effects. It is indicated in the treatment of depression symptoms and particularly endogenous depression. The drug is not advisable, or
requires great caution, in patients suffering from schizophrenia.
f miansérine
d Mianserin
i mianserina
e mianserina

2323 miconazole
Antifungal and antibacterial agent mainly used topically for skin infections.
The drug has a broad spectrum (it includes dermatophytes, yeasts and various mycetes) and is capable of a bactericidal activity against bacilli and gram
-positive cocci. It is specifically used
for the treatment of primary and secondary vulvovaginal infections (in the
form of cream).
f miconazole
d Miconazol
i miconazolo
e miconazolo

2324 midecamycin
Antibiotic active against Gram-positive
and Gram-negative bacteria and mainly
used in the treatment of infections due

to susceptible bacteria of the respiratory
and urinary tracts. It also finds employ
ment in the treatment of ear, nose and
throat infections, of osteomyelitis and
periostitis. Gastric disorders may occur
as side effect.
f midécamycine
d Midecamycin
i midecamicina
e midecamicina

2325 midodrine
Sympathomimetic indicated in the treatment of arterial hypotension and orthostatic disorders. Bradycardia and disorders of the urinary function in the
presence of hypertrophy of the prostate
may occur as adverse effects. The drug
should be used with caution in patients
suffering from hypertension.
f midodrine
d Midodrin
i midodrina
e midodrina

2326 milfoil flowers
Or yarrow flowers. The Achillea mille-
folium, a herb growing on grazing and
uncultivated ground and used as a bitter and astringent agent.
f fleurs de millefeuille
d Tausendblattblumen;   Scharfgarbenblüten
i capolini di millefoglie
e sumidades de milenrama

2327 milfoil oil
Or yarrow oil. Essential oil constituted
by terpene monocyclic compounds, e.g.
limonene, bicyclic compounds, e.g.
pinene, and sesquiterpene bicyclic compounds and acids. It has tonic and
stomachic properties.
f essence de millefeuilles
d Feldgarbenöl; Scharfgarbenöl
i olio essenziale di millefoglie
e esencia de milenrama

2328 milk
The fluid that is secreted by mammals
to feed their young. It is a near-perfect
food as it contains fat, protein and
carbohydrate as well as the essential
inorganic salts and vitamins. The term
is also used to describe any fluid, in
particular vegetable fluid, which resembles milk.
f lait
d Milch
i latte
e leche

* **millefolii herba   s. achillea**

2329 mimosine
Chemical compound constituting the optic
ally active form of the eugenol, the
phenolic compound occurring in clove
oil and oil of cinnamon. It is present
in several vegetables.
f mimosine

d Mimosin
i mimosina
e mimosina

2330 mineralcorticoids
Steroid hormones derived from cholesterol capable of increasing the reabsorption of Na^+, Cl^-, and HCO_3^- by the kidney which leads to an increase in blood volume and blood pressure.
f minéralo-corticoïdes
d Mineralkortikoide; Mineralokortikoide
i mineralosterocorticoidi
e mineralocorticoides

2331 mineral fat
Any substance which is similar to fat but cannot be saponified or grow rancid. Mineral fats consist of solid or semisolid mixtures of hydrocarbons, e.g. petrolatum, paraffin, ceresin.
f graisse minérale
d Mineralfett
i grasso minerale
e grasa mineral

2332 mineral oil
Liquid product of mineral origin within the viscosity limits set our for oils. In medicine, it is generally regarded as a synonym for liquid paraffine, i.e. a mild laxative. Owing to its capacity to promote the granulation and regeneration of the epithelial and connective tissues, a particular form of mineral oil is used as a topical cicatrizant agent. When administered orally for its cathartic action, anal inflammation may occur.
f huile minérale
d Paraffinöl
i olio minerale; olio di paraffina
e aceite mineral

2333 mineral water
The water that is obtained from a mineral spring, i.e. a source of water characterized by a high mineral content. Some mineral water qualities are thought to be of some therapeutic value.
f eau minérale
d Mineralwasser; Heilwasser
i acqua minerale
e agua mineral

2334 minium
A scarlet powder which is obtained by roasting lead carbonate in air. It is used as a pigment and as a drier. Also known as Paris oxide, red lead, lead tetroxide.
f minium; oxyde rouge de plomb
d Bleirot; Minium; Mennige; rotes Bleioxyd
i minio; ossido di Parigi; ossido rosso di piombo
e minio; rojo de plomo; óxido rojo de plomo

* **mirbane oil   s. nitrobenzol**

2335 mistletoe

Parasitic plant which grows on tree, particularly apple, plum, oak and poplar. The leaves and to a lesser extent the branches contain a glutinous substance, the viscin, gum and tannin. Both the European and the American species display pharmacological properties presumed to be antispasmodic, vasodilator and oxytocic (childbirth-promoting).
f gillon; gui de chêne
d Mistel
i vischio
e visco; muérdago

2336 mithramycin
Cytotoxic antibiotic once used in the treatment of cancer, in particular testicular tumours, and at present to reduce hypercalcaemia caused by a malignant disease. The drug may be used for only a few days and cause thrombocytopenia, haemorrhages, liver necrosis, etc.
f mithramycine
d Mithramycin
i mitramicina
e mitramicina

2337 mitobromitol
Alkylating agent characterized by a neoplastic action and chiefly used in the therapy of chronic myeloid leukaemia and polycythaemia. Marked anaemia, anorexia, thrombocytopenia and renal lithiasis may occur as side effects.
f mitobromitol
d Mitobromitol
i mitobromitolo
e mitobromitolo

2338 mitoguazone
Cytotoxic agent used in the treatment of acute lymphoblastic and granulocytic leukaemia and chronic myeloid leukaemia undergoing a pathological process. The drug may be administered with great caution. Throat disorders, up to obstruction of the respiratory tract, hypoglycaemia and intestinal ulcers may occur as adverse effects.
f mitoguazone
d Mitoguazon
i mitoguazone
e mitoguazona

2339 mitomycin
Cytotoxic antibiotic agent mainly used in the treatment of gastro-intestinal and breast cancer. It is possible that the drug cause an arrest of the mitotic division at the metaphase stage and act by inhibition of protein and DNA synthesis. It causes delayed effects on the bone marrow, lung fibrosis and kidney damage. It may also cause mucous membrane toxicity, diarrhoea, vomiting, vision disorders, fatigue, oedema and thrombophlebitis.
f mitomycine
d Mitomycin

ı mitomicina
e mitomicina

2340 **mitopodozide**
Antineoplastic agent marked by an antimitotic action and mainly used in the
treatment of cancers, often in association with surgery, radiotherapy, alkyat-
ing agents and various other antimitotic
drugs. Nausea and diarrhoea, restleness,
fever, anorexia, alopecia, vertigo and
hypotension may occur as side effects.
The drug should be used with caution
in patients suffering from hepatic or
renal disorders or from some heart disease.
f mitopodozide
d Mitopodozid
ı mitopodozide
e mitopodozida

2341 **mitotane**
Cytotoxic drug marked by a specific action on the adrenal cortex and used in
the treatment of inoperable functional
and non-functionable adrenal cortical
cancer. Gastro-intestinal disorders and
depression of the central nervous system
accompanied by vertigo and drowsiness
are the principal side effects of the
drug.
f mitotane
d Mitotan
ı mitotano
e mitotano

2342 **mitozandrone**
Antineoplastic agent capable of acting
against resting-phase tumours and a-
gainst those undergoing DNA synthesis.
The drug has thus proved useful in both
slow and rapid growing tumours. It is
used in the treatment of breast cancer
at an advanced stage. Bone-marrow suppression, alopecia, gastro-intestinal and
hepatic disorders, as well as heart failure may occur as adverse effects.
f mitozandrone
d Mitozandron
ı mitozandrone
e mitozandrona

2343 **mocaya oil**
Or mocaya butter. The non-volatile oil
which is obtained from fruits of some
species of Acrocomia. It is identical
with coconut oil and is used as a basis
for pharmaceutical suspension of drugs.
f huile d'acrocoma
d Mocaya-Öl; Macauba-Öl
ı olio di acrocoma
e aceite de acrocoma

2344 **mofebutazone**
Butazone derivative used as an antirheumatic and anti-inflammatory agent owing
to its antipyretic and analgesic action.
Some gastro-intestinal disorders may occur as side effects.
f mofébutazone

d Mofebutazon
ı mofebutazone
e mofebutazona

* **mogadon s. nitrazepam**

2345 **molasses**
Or treacle. The dense, light to dark-
-brown viscid syrup that is separated
from raw sugar in the various processes
of sugar manufacture. It contains 40 to
50% saccharose and used, among other
things, for the production of ethyl alcohol.
f mélasse
d Melasse
ı melassa
e melote

2346 **molindone**
Neuroleptic agent marked by an antipsychotic action, used in the treatment of
schizophrenia, particularly in the case
of patients who do not respond to phenothiazines, butyrophenones and thioxanthene. Dizziness, drowsiness, sometimes
euphoria, nausea, tachycardia and ortho
static hypotension may occur as side
effects. The drug should not be used
when the patient is in a comatose state.
f molindone
d Molindon
ı molindone
e molindone

2347 **molsidomine**
Hypotensive agent particularly active on
the systolic and pulse pressure thus
reducing the heart's effort through an
action which resembles that of nitrites.
The drug is used in the prophylaxis and
therapy of angina pectoris, cardiopathies due to hypotension, infarction
and the strong pains that are caused by
incompetence of the coronary arteries.
As it occurs with many vasodilators,
headache may easily be caused shortly
after ingestion of the drug. Also known
as morsydomine.
f molsidomine
d Molsidomin
ı molsidomina
e molsidomina

2348 **molybdate**
A salt of molybdic acid, in which Me is
a monovalent metal. The solution of the
polymolybdate is used as a reagent of
the phosphoric ion.
f molybdate
d Molybdat
ı molibdato
e molibdato

2349 **molybdic acid**
Crystals which are white when anhydrous and yellow in the hydrate form.
Molybdic acid forms both normal salts
(the molybdates) and complex, mainly
acid, salts. It is also capable of form-

ing various compounds with phenols, sugars, tannins and organic acids. The molybdic trioxide (or anhydride) is used in the preparation of various compounds, as a reagent in analytical chemistry and in the production of catalyzers.
f acide molybdique
d Molybdensäure
i acido molibdico
e ácido molíbdico

2350 monesia bark
The bark of a tree growing in Brazil (Chrysophyllum glycyphloeum), rich in tannin and used medicinally as an expectorant and an astringent.
f écorce du Brésil; guaranhem
d Monesiarinde
i corteccia di monesia
e corteza de monesia

2351 monobenzone
Therapeutic agent capable of inhibiting the production of melanine and used in the treatment of hyperpigmentation of the skin in the course of pregnancy. It is also employed in the treatment of Addison's disease and in generalized melanodermatides. Depigmentation, as well as inflammation and some sensitivity reaction may occur as side effects. Prolonged exposure to sunlight and contact with eyes should be avoided.
f monobenzone
d Monobenzon
i monobenzone
e monobenzona

2352 monoethanolamine phosphate
Phosphorylcholamine. Phosphoretted (of a compound formed with phosphorus) tonic particularly used in cerebral and hepatic states of debility. The drug may be regarded as a precursor in the synthesis of acetylcholine and as such capable of acting as a stimulant. Gastro-intestinal disorders may occur as side effects.
f phosphate d'éthanolamine
d Monoethanolaminphosphat
i etanolammina fosfato; fosforilcolammina
e fosfato de etanolamina

2353 monophosphothiamine
Neurotrophic(nutrient of the nervous system) and antineuritic agent capable of improving cellular metabolism, in particular in the cerebral region. The drug should be administered with great caution in the presence of heart disease.
f monophosphothiamine
d Monophosphothiamin
i tiamina monofosfato
e monofosfato de tiamina

* monoxychlorosene   s. oxychlorosene

2354 moperone
Tranquillizing and sedative agent of the butyrophenone series. It is used in the treatment of manic psychosis and hallucinatory states, schizophrenia and oligophrenia. Extrapyramidal dysfunctions may occur as adverse effects.
f mopérone
d Moperon
i moperone
e moperona

2355 moquizone
True choleretic agent producing a considerable and prolonged increase of biliary flow and an equally considerable excretion of solid bile constituents. Being a potent hypocholesterolaemic agent without cholecystokinetic effect, it is used in the treatment of both organic and functional hepatobiliary disorders, as well as bad digestion caused by incompetence of the liver, hypercholesterolaemia and atherogenic hyperlipidaemia. Meteorism, diarrhoea and flush may occur as side effects.
f moquizone
d Moquizon
i mochizone
e moquizona

2356 morazone
Antiphlogistic, antirheumatic agent marked by antipyretic and analgesic action. Gastro-intestinal disorders may occur as side effects.
f morazone
d Morazon
i morazone
e morazona

2357 morclofone
Non-narcotic antitussive agent principally used in the treatment of cough caused by laryngotracheitis, bronchitis, pneumonia and bronchopneumonia, whooping cough etc.
f morclofone
d Morclofon
i morclofone
e morclofona

2358 morinamide
Antituberculous agent indicated for the treatment of pulmonary and extrapulmonary tuberculosis which has proved resistant to more potent drugs. It is generally administered in association with other antituberculous drugs as it does not produce chemoresistance. Some allergic dermatosis has been observed as side effect.
f morinamide
d Morphazinamid
i morinamide; morfazinamide
e morinamida

2359 moroxydine
Therapeutic agent capable of checking the growth of viruses and used in the treatment of many viral infections, e.g. herpes zoster, varicella etc. Gastro-intestinal disorders may occur as side effects.

f moroxydine
d Moroxydin
i moroxidina
e moroxidina

2360 morpheridine
Morpholinoethylnorpethidine. A chemical compound used in medicine as a narcotic analgesic.
f morphéridine
d Morpheridin
i morferidina
e morferidina

* **morphetylbutine   s. promolate**

2361 morphine
The principal alkaloid of opium (up to 10-12%) and of other plants of the Papaveraceae. It may be regarded as a phenanthrene derivative with two hydroxylic groups and the tertiary nitrogen atom bound to methyl group. It occurs in colourless prisms of a bitter taste, is hardly soluble in water, ether, benzene and chloroform, easily soluble in absolute alcohol and in alkalis and acids. It is obtained from the opium by a complex process of extraction and purification. In medicine, rather than the free base, one uses the salts which morphine easily forms with acids. Its principal action is on the central nervous system, causing a mixture of depression and stimulation. Being a strong analgesic, the drug is mainly used for the soothing of pain. The respiratory centre is easily depressed by morphine, even if administered in small doses. Large doses are required to depress the motor cortex. Other important side effects must be taken into account when using the drug. A euphoric state with consequent probable addiction may accompany the physiological effects.
f morphine
d Morphin
i morfina
e morfina

2362 morphine hydrochloride
Alkaloid present in opium (10 to 12%). Extremely active and used as a potent analgesic agent. It generally does not affect the respiratory and cardiovascular systems and has a rather mild effect on the smooth-muscle organs. It displays a narcotic and stupor-causing action. The drug presents the danger of dependence (s. morphine).
f chlorhydrate de morphine
d Morphinhydrochlorid
i morfina cloridrato
e hidrocloruro de morfina

2363 morpholine
Tetrahydrooxazine. Colourless liquid with an alkaline reaction, soluble in water and used as a solvant and in several chemical syntheses.
f morpholine; tétrahydrooxazine
d Morpholin
i morfolina; tetraidro-ossazina
e morfolina; tetrahidrooxazina

2364 morpholine salicylate
Retard-type analgesic, antirheumatic and antipyretic agent less toxic than sodium salicylate and easily excreted in the urine. It is generally used to relieve headache, fever and the physical discomfort which accompany colds and influenza, as well as neuralgias and neuritis--related pains.
f salicylate de morpholine
d Morpholinsalicylat
i morfolina salicilato
e salicilato de morfolina

2365 morpholinethylrutoside
Water-soluble derivative of rutin having the same action and effects of Vitamin P. It checks capillary fragility thus displaying a cardiotonic action. The drug has also found employment as a diuretic, antiphlogistic and hypotensive agent, and is principally used in the treatment of cerebral, ocular and coronary disorders.
f morpholinéthylrutine
d Morpholinethylrutin
i morfolinetilrutoside; etossazorutoside
e morfolinetilrutosida

2366 morpholinylethylmorphine
Morphine derivative resembling codeine in its action and principally used as a sedative in the treatment of cough caused by various disorders of the upper respiratory tract.
f morpholinyléthylmorphine
d Morpholinylethylmorphin
i morfoliniletilmorfina
e morfoliniletilmorfina

2367 morrhuic acid
Non-saturated acid extracted in small quantities from cod-liver oil and used in the treatment of varicose veins. The term also denotes a mixture of fatty acids obtained from the same source.
f acide morrhuique
d Morrhuinsäure
i acido morruico
e ácido morruico

2368 morsuximide
Anticonvulsant agent used in the treatment of epilepsy (both petit and grand mal) and, more generally, of psychomotor disorders. The drug is almost always administered together with some other antiepileptic agent. Gastric disorders, nausea, vertigo and some allergic reactions may occur as side effects.
f morsuximide
d Morsuximid
i morsuccimide
e morsuximida

2369 motherwort
A herb growing on rocky ground and displaying axillary whorls of small purple flowers which are used in popular medicine for the treatment of circulatory disorders.
f agripaume; cardiaire; queue de lion; cardiaque
d Herzheil; Löwenschwanz
i cardiaca; coda di leone
e agripalma; mano de Santa Maria

2370 mountain flax
Or senega, senega snakeroot, rattle senega root. Herb of the polygalaceae family including some 475 species. The most important of these species is the Virginian senega growing in North America, which reaches a height of up to 30 cm. and has a short ramified root and small greenish flowers. The roots are used for the extraction of a drug which finds employment in medicine for its expectorant properties.
f polygala de Virginie
d Klappenschlangenwurzel; Senega
i poligala virginiana
e polígala de Virginia

2371 mountain sage
Or wood sage, garlic sage. The dried herb Teucrium scordonia used in medicine as an antidiaphoretic agent.
f sauge de montagne
d Bergsalbei
i salvia alpina; assenzio alpino
e salvia de montaña

2372 mouthwash
The aqueous solution of a medicament for cleaning the mouth or for gargling. Collutorium.
f bain de bouche
d Mundwasser
i collutorio
e colutorio bucal

2373 moxastin
Antihistamine used in the treatment of several allergic reactions, e.g. bronchial asthma, Ménière's disease, some dermatites etc. The drug is also used in preanaesthesia. Moxastin is also used as an antiemetic in the treatment of nausea and vomit caused by motion, e. g. seasickness, in the course of pregnancy and of Ménière's disease and, in larger doses, to relieve anxiety states and treat certain psychoses.
f moxastine
d Moxastin
i moxastina; mefenidramina
e moxastina; mefenidramina

2374 moxaverine
Meteverine. A papaverine-like antispasmodic agent less toxic and more effective than papaverine. It is generally indicated in the treatment of all spastic states of the smooth muscles of hollow organs,

of angina pectoris, cardiac asthma, migraine etc. The drug, administered by endovenous injection, is also used in the treatment of arterial embolism (clotting). The possibility of respiratory block must be taken into consideration.
f métévérine
d Moxaverin
i mossoverina
e moxoverina

2375 moxestrol
Synthetic oestrogenic hormone used in the treatment of menstrual disorders due to follicular insufficiency, functional metrorrhagia, menopause, vaginal atresia related to menopause etc. The drug has also been used in case of threatened abortion, inhibition of lactation and post-partum haemorrhages. Nausea and vomiting, as well as water retention may occur as side effects. Patients having a personal or family history of mammary or genital-tract neoplastic diseases should not be treated with this drug.
f moxestrol
d Moxestrol
i moxestrolo
e moxestrolo

2376 mucic acid
Bycarboxylic acid derived from the galactose by oxidation of both the aldeydic and the primary alcoholic groups. It is obtained by oxidation with nitric acid of the lactose or of the pentosanes, or by oxidation of the galactose obtained by hydrolysis of the sawdust of wood. It is an epimer of the saccharic and mannosaccharic acids; a white crystalline powder, hardly soluble in water and insoluble in alcohol. It is used in organic synthesis and as a substitute of the tartaric acid. Also known as saccharolactic acid.
f acide mucique; acide saccharolactique
d Muzinsäure; Schleimsäure
i acido mucico; acido saccarolattico
e ácido múcico; ácido sacaroláctico

2377 mucilage
Rubber-like substance occurring in many plants and having the function of absorbing and holding water by swelling. It contains proteins and polysaccharides as well as uronides.
f mucilage
d Pflanzenschleim
i mucillagine
e mucílago

2378 mucin
Glucoproteides whose prosthetic group is constituted by the mucoitinsulphuric acid. Its main functions are the yielding of highly viscous solutions, the capacity not to undergo coagulation when heated and to be precipitated by acetic acid. Mucins lubricate and protect the tissues against chemical and bac-

terial agents. They are secretion products of particular glandular epithelia
and represent the principal component
of the synovial liquid, of the mucus
and catarrh.
f mucine
d Muzin
i mucina
e mucina

2379 mucoitinsulphuric acid
Hypothetical mucopolysaccharide occurring in mucus. Actually, it has been
observed that the mucous secretions contain a mixture of mucopolysaccharides
and glycoproteins in addition to a variable quantity of neutral polysaccharides.
f acide mucoïtinsulfurique
d Mucoitinschwefelsäure
i acido mucoitinsolforico
e ácido mucoitinsulfúrico

2380 muconic acid
A bicarboxylic acid. Colourless water-
-soluble crystals, the product of the
oxidation of benzene occurring in the
urine following the administration of
benzene or benzenic derivatives.
f acide muconique
d Mukonsäure
i acido muconico
e ácido mucónico

2381 mucopolysaccharides
Hexosamine-containing polysaccharides.
They occur either alone or together with
proteins, as mucins in mucoid secretions.
f mucopolysaccharides
d Mucopolysaccharide
i mucopolisaccaridi
e mucopolisacaridas

2382 mucopolysaccharid polysulphate
Heparin-like anticoagulant and antiinflammatory agent capable of inhibiting
thrombin and thromboplastin and promoting fibrinolysis. The drug also acts as
an antiexudative agent, stimulates blood
and lymph circulation locally, as well
as the metabolism of the connective
tissue. It is generally indicated in
case of venous insufficiency and its consequences, e.g. phlebitis, thrombophlebitis, ulcus cruris, varices etc.
f polysulfate de mucopolysaccharide
d Mucopolysaccharidpolysulfat
i mucopolisaccaride polisolfato
e polisulfato de mucopolisacarida

2383 multialkylpeptide
Polypeptidic agent capable of slowing
down the bacterial growth by an alkylating and antimetabolic action. The drug
is widely used in the treatment of the
haemolymphoblastoses, e.g. reticulosarcoma, lymphosarcoma, multiple myeloma,
both acute and chronic leukaemia and
polycythaemia. It is equally indicated
in the therapy of solid tumours and to
inhibit recidivity. Reversible alopecia,

leukopenia and thrombocytopenia may
occur as side effects.
f multialkylpeptide
d Multialkylpeptid
i multialchilpeptide
e multialcoilpeptida

2384 mupirocin
Previously called pseudomonic acid. Antibiotic used in the form of ointment in
the treatment of bacterial skin infections.
f mupirocine
d Mupirocin
i mupirocina
e mupirocina

2385 muramic acid
Amino sugar formed by the union of the
glycosamine with the D-lactic acid
through an ethereal bond. It occurs
copiously in the wall of the bacterial
cells as component of a polymer in
which molecules of muramic acid alter-
nate with molecules of N-acetylglyco-
samine.
f acide muramique
d Muraminsäure
i acido muramico
e ácido murámico

* muramidase   s. lysozime

2386 murexine
Choline-related chemical compound occurring in various species of mollusca. It
is easily hydrolysable and its salts
(chloride, picrate etc.) are stable. It
has an action similar to that of curare
and therefore employed in therapeutic
treatments as a myorelaxant agent.
f murexine
d Murexin
i muressina
e murexina

2387 muscarine
Alkaloid occurring in Amanita muscaria,
a poisonous mushroom. Its action is
similar to that of postganglionic stimulation of parasympathetic nerves and
thus to that of acetylcholine. Muscarine,
thus displaying a good miotic action,
increases the tone and intestinal peristalsis, stimulates respiration when administered in low doses and inhibits it
at strong doses, causes the slowing
down of the heart rate up to the arrest
in diastole and a notable peripheral
vasodilation with consequent fall of the
arterial pressure. All these various actions are inhibited by the atropine.
f muscarine
d Muscarin
i muscarina
e muscarina

2388 muscle relaxant
Or myorelaxant. Any pharmaceutical
drug which induces the diminution of the
tone and eventually the loss of reflexes

and paralysis of the skeletal muscles without affecting the superior nervous activity. This latter property distinguishes myorelaxants from general anaesthetics.

f myorelâchant; relâchant musculaire
d Muskelrelaxans; Muskelentspannungsmittel
i miorilassante
e miorelajante; relajante muscular

2389 musculi extractum

Muscle extract. Opotherapeutic agent marked by its capacity of converting energy derived from food into tissue and that of regulating metabolism. It is generally used in denutrition and convalescence states, as well as in organic debility, difficult pregnancies and lactation.

f extrait de muscle
d Muskelextrakt
i muscolo estratto
e extracto de músculo

2390 mussena bark

The bark of the Akbizzia anthelmintica, a tree growing in eastern Africa, which displays an anthelmintic action.

f écorce de moucéna; écorce de mousséna
d Massenarinde; Basenarinde
i corteccia di mussena
e corteza de musena

2391 mustard

Plant of the genus Brassica whose seeds are widely used for therapeutic purposes.

f moutarde
d Senf
i senape
e mostaza

2392 mustard bath

A bath to which a mixture of black and white mustard is added to obtain a counter-irritant so as to increase the flow of blood through the skin and consequently of sweating.

f bain de moutarde
d Senfbad
i bagno di senape; bagno senapizzato
e baño de mostaza

2393 mustard flour

The ground seeds of black and white mustard without their seed-coats. Its ingestion increases the flow of saliva and gastric juices. In case of poisoning, it is also used as an emetic. It also finds employment in the form of poultice to sooth irritation.

f farine de moutarde
d Senfmehl
i farina di senape
e harina de mostaza

2394 mustine hydrochloride

Cytotoxic drug used in the treatment of neoplastic diseases. It has a similarity with irradiation but it acts more rapid-

ly. Administered in small doses, its action is almost exclusively applied to blood-forming organs and, by the inhibition of mitotic division, it succeeds in destroying the precursors of the red and white blood cells in the bone marrow and causing necrosis of the germ cells in the limph nodes. Agranulocytosis, anaemia and thrombocytopenia purpura may occur with prolonged administration. Nausea, vomiting, thrombosis, anorexia, diarrhoea etc. are possible adverse effects.

f hydrochlorure de mustine
d Mustinhydrochlorid
i mustina idrocloruro
e hidrocloruro de mustina

2395 mycophenolic acid

Acid isolated in the crystalline state from Penicillium glaucum and capable of inhibiting the development of the Bacillus anthracis (the anthrax-causing bacillus). It is the first example of an antibiotic prepared with fungi.

f acide mycophénolique
d Mychophenolsäure
i acido micofenolico
e ácido micofenólico

2396 mycose

Or trehalose. Carbohydrate ($C_{12}H_{22}O_{11}$) crystallizing with two molecules of water and occurring in manna, in several varieties of fungi and in the oil of ergot.

f mycose; tréhalose
d Mycose; Mutterkornzucker
i micosio; trealosio
e micosa; trehalosa

*** myprozin s. natamycin**

2397 myralact

N-(hydroxyethyl) tetradecylammonium lactate. Antiseptic agent marked by an antiprotozoal action and mainly used in the treatment of vaginal infections caused by Trichomonas or Mycetes, e.g. vaginitis, leukorrhoea-associated vulvitis etc.

f myralact
d Myralact
i miralact
e miralact

2398 myristic acid

Saturated aliphatic acid occurring as an ester in many animal and vegetable fats in human milk and in butter. It is also found in the subcutaneous layer of the human body up to a content of 1%. It is used in the preparation of some pharmaceutical drugs.

f acide myristique
d Myristinsäure
i acido miristico
e ácido mirístico

2399 myristica oil

Nutmeg oil. The oil which is obtained

by distillation from nutmeg. Colourless or pale yellow, nutmeg-tasting liquid chiefly consisting of D-camphene and acting as a carminative agent. When the oil is obtained by hot expression, it is employed in plasters (expressed nutmeg oil).

f essence de myristique
d Myristikaöl
i olio di miristica
e esencia de mirística

2400 myristicin

Aromatic ether characterized by a penetrating odour occurring in the nutmeg oil. It is a yellowish liquid which can also be obtained synthetically.

f myristicine
d Myristicin
i miristicina
e miristicina

2401 myristin

Ester of the glycerol with three molecules of myristic acid, occurring in nutmeg and coconut butter, in spermaceti and in other fats.

f myristine
d Myristin
i miristina
e miristina

2402 myronate

The salt of the myronic acid, a substance of acid nature whose potassium salt constitutes the sinigrin, a glucoside occurring in black mustard seeds and capable of yielding allylisothiocyanate

and glucose on hydrolysis.

f myronate
d Myronat
i mironato
e mironato

2403 myrrh

Or gum myrrh. Oleo gum resin displaying antiseptic action. It is mainly used in mouth washes and with purgatives as a carminative agent.

f myrrhe
d Myrrhe
i mirra
e mirra

2404 myrtecaine

Analgesic and antispastic agent used in the treatment of gastritis, gastroduodenal ulcers, oesophagitis and any intestinal disorders caused by drugs.

f myrtecaïne
d Myrtecain
i mirtecaina
e mirtecaina

2405 myrtol

Fraction obtained from the volatile oil of Dutch myrtle and indicated as an antiseptic and a stimulant to the pulmonary and urogenital mucous membranes. It is used in the treatment of bronchitis, cystitis, pyelitis, silicosis, rhinitis and pneumomycosis.

f myrtol
d Myrtol; Myrtenölkampfer
i mirtolo; cantoro di mirto
e mirtol; alcanfor de mirto

N

2406 nabilone
Cannabis derivative displaying antiemetic and anxiolytic properties. It acts on the opiate receptors in the brain so as to inhibit the transmission of vomiting impulses. It is also employed against nausea and vomiting caused by cytotoxic agents. Drowsiness accompanied by headaches, postural hypotension and some abdominal pains are the most frequent side effects.
f nabilone
d Nabilone
i nabilone
e nabilona

2407 nabumetone
Non-steroid, antiphlogistic and analgesic agent mainly indicated in the treatment of osteoarthritis and rheumatoid arthritis. Although the drug is in itself a rather mild inhibitor of prostaglandin synthesis, it is capable of metabolizing in the liver to form a more active compound. Gastro-intestinal disorders are thus less probable if compared to other agents of the same class, but it should be avoided or carefully monitored if the presence of peptic ulcers appears in the anamnesis. Diarrhoea, headache, dizziness and sedation are possible adverse effects.
f nabumétone
d Nabumeton
i nabumetone
e nabumetona

* **nadine** s. nicotinamide adenine dinucleotide

2408 nadolol
1-(tert-butylaminol)-3-(5,6,7,8-tetrahydro -cis-6,7-dihydroxy-1-naphthyl)oxyl-2-propanol. A beta-adrenoceptor blocking drug used in the treatment of angina pectoris, arrhythmias, hyperthyroidism, headaches and states of anxiety. It is also indicated to prevent recurrence of myocardial infarction. Bronchoconstriction, sleep disorders, cold hands and feet and cardiac failure may occur as side effects.
f nadolol
d Nadolol
i nadololo
e nadololo

2409 nadoxolol
Antiarrhythmic agent capable of reducing cardiac activity by a probable depressive effect on the myocardial cell. The drug is principally used in the treatment of cardiac arrhythmias, fibrillation and flutter, frequent extrasystoles, paroxystic ventricular tachycardia and those arrhythmias which follow myocardial infarction. The drug should not be used in cases of renal incompetence. Nausea and vomiting may occur as side effects.
f nadoxolol
d Nadoxolol
i nadossololo
e nadoxololo

2410 nafcillin
A penicillanic acid. Semisynthetic penicillin used in the treatment of respiratory infections as well as infections affecting skin, soft tissues and the urinary tract. The drug is also used in the treatment of suppurative osteomyelitis due to penicillin-resistant staphylococci and to staphylococci, pneumococci (bacterial genus in the family Lactobacillaceae) and streptococci which are sensitive to penicillin G. Itching, sometimes rashes, nausea, vomiting and diarrhoea may occur as side effects. The drug must be carefully monitored when administered in pregnancy.
f nafcilline
d Nafcillin
i nafcillina
e nafcilina

2411 nafiverine
Parasympatholytic and anticholinergic agent with action similar to that of papaverine and atropine. It is principally used in the treatment of gastro-intestinal, biliary and urogenital spastic conditions. Side effects include constipation and insomnia.
f nafivérine
d Nafiverin
i nafiverina
e nafiverina

2412 naftazone
Non-coagulant haemostatic agent marked by a good action on vasal walls. The drug is mainly used for the prophylaxis and actual treatment of haemorrhages and capillary fragility. Chemically,

1,2-naphthaquinone 2-semicarbazone.
f naftazone
d Naftazon
i naftazone
e naftazona

2413 naftidrofuryl
Cerebral and peripheral vasodilator which is capable of causing vasodilation by improving cellular metabolism and determining a relaxation of the muscle cells in the blood-vessel walls. Its action is carried out without affecting the systemic arterial pressure. The drug is used to increase the blood flow in the brain by a specific action at the level of the Krebs cycle and is thus indicated in the treatment of vascular encephalopathies, cerebral functional disorders and involution processes, and peripheral vascular diseases. Headaches and hypotension may occur as side effects.
f naftidrofuryl
d Naftidrofuryl
i naftidrofurile
e naftidrofurilo

2414 naftionine
4-amino-1-sodium naphthalensulphonate. Anticoagulant which can be administered orally, parenterally or rectally. The drug promotes the rate of fibrinogen and the number of blood platelets in circulation by a prompt and lasting action. It is therefore indicated in the treatment of haemorrhages in particular post--partum haemorrhages, metrorrhagia, haemoptysis, haematemesis (vomiting blood), retinal haemorrhages etc.
f naftionine
d Naftionin
i naftionina
e naftionina

2415 naftoclizine
An antitussive agent by the action of sodium dibunate and antihistaminic by the action of chlorcyclizine. It is indi-cated in the symptomatic treatment of coughs, especially when due to infections of the upper respiratory tract, e. g. laryngitis, acute and chronic bron-chitis, tracheitis, bronchiectasis (dilation of the bronchi) and emphysema. It is also sometimes used for the relief of allergic conditions. Some drowsiness and nervous depression may occur as side effects.
f naftoclizine; dibunate de chlorcyclizine
d Chlorcyclizindibunat; naftoclizin
i naftoclizina; clorciclizina dibunato
e naftoclizina; dibunato de clorciclizina

2416 nalidixic acid
1-ethyl-7-methyl-4-oxo-1,8-naphthyridine--3-carboxylic acid. Therapeutic agent used in the treatment of bacterial infections.
f acide nalidixique
d Nalidixinsäure

i acido nalidissico
e ácido nalidíxico

2417 nalorphine
N-allylnormorphine. Morphine derivative in which the N-methyl group has been replaced by an N-allys group. The drug is a narcotic antagonist, i.e. it antagonizes most of the actions of morphine thus proving an extremely valuable and effective antidote in the treatment of overdosage with morphine, pethidine, methadone and other related compounds. The drug may cause hallucinations and other mental disturbances. When adminis_tered to addicts, it causes severe withdrawal symptoms. It may safely be used in the newborn.
f nalorphine
d Nalorphin
i nalorfina
e nalorfina

2418 naloxone
Analgesic and narcotic antagonist acting as specific antidote to morphine. It is indicated for the treatment of respiratory depression caused by morphine. and of narcotic dependence. Must be used with caution when administered to pregnant women, children and babies.
f naloxone
d Naloxon
i nalossone
e naloxona

2419 nandrolone
19-norandrostenolone. Steroid hormone with tissue-building function, derived from androstane. Its effects are similar to those of testosterone, but its anabolic effects are greater than its androgenic effects. The drug is principally used in the treatment of conditions in which a potent tissue-building or protein-saving action is required, e.g. debilitating illnesses and carcinoma of the breast. The drug is not active by mouth. Some hirsutism in women has been observed as a side effect.
f nandrolone
d Nandrolon
i nandrolone; norandrostenolone
e norandrostenolona

2420 naphazoline
Imidazolin-series vasoconstrictor. A sympathomimetic agent with a marked alpha-adrenergic activity, particularly indicated for topical use on the nasal mucosa to attain decongestion, i.e. a reduced secretion and decrease of mucosal swelling. The drug is widely employ_ed in the treatment of cold, rhinitis, in particular allergic rhinitis, sinusitis etc. Episodes of tachycardia, hyperthyroidism and chronic rhinitis have been observed as side effects.
f naphazoline
d Naphazolin

l nafazoline
e nafazolina

2421 **naphthalene**
Crystalline aromatic hydrocarbon marked
by a characteristic odour and obtained
as a by-product in the manufacture of
coal gas. It is sometimes used in the
form of ointment in the treatment of
scabies and pediculosis and as a vermifuge.
f naphtalène; naphtaline
d Naphthalin
l naftalina
e naftalina; naftaleno

2422 **naphthalic acid**
Organic acid derived from naphthalene
by introduction of two carboxylic groups.
It is obtained by oxidation of the
acenaphthene.
f acide naphtalique
d Naphthalinsäure
l acido naftalico
e ácido naftálico

2423 **naphthalinic acid**
2-oxy-alpha-naphthoquinone. Colouring
substance that is isomeric of juglone;
also called lawsone as it is found in
the leaves of Egyptian henna.
f acide naphtalinique
d Naphthalinsäure
l acido naftalinico
e ácido naftalínico

* **naphthalinsulphonic acid** s. **naphtholsulphonic acid**

2424 **naphthenic acid**
Crystalline acid formed by oxidation of
acenaphthene. Naphthalic acids are
extracted in the form of an oily colourless liquid marked by a penetrating
odour and are generally used in industry as tensioactive agents.
f acide naphténique
d Naphthensäure
l acido naftenico
e ácido nafténico

2425 **naphthionic acid**
Naphthylamine-sulphonic acid, Tobias
acid. Acid obtained by the sulphonation
of alpha-naphthylamine and used in
medicine for the treatment of urinary
infections.
f acide naphtylaminosulfonique, acide de
Tobias
d Naphthylaminosulfonsäure
l acido naftilamminosolfonico
e ácido naftilaminosulfónico

2426 **naphthocaine**
Derivative of naphthoic acid whose hydro
chloride is marked by topical anaesthetic properties.
f naphtocaïne
d Naphthocain
l naftocaina

e naftocaina

2427 **naphthol**
Crystalline compound present in coal tar
in two isomeric forms: alpha- and beta-
-naphthol. The first is a substance having phenolic character and some antiseptic properties supposed to be more
potent than those of beta-naphthol, yet
are more irritant when administered internally. The substance is used for the
furfural test for sugars, for protein-con
taining guanidine groupings and to test
urine for indoxyl. Beta naphthol is a
more powerful antiseptic than phenol and
also less toxic. It is present in various
ointments and lotions indicated for the
treatment of scabies, psoriasis and
eczema. The beta-naphthol is also called
isonaphthol.
f naphtol
d Naphthol
l naftolo
e naftol

2428 **naphtholate**
Chemical compound derived from naphthol
by the substitution with a metal of the
hydrogen of the hydroxylic group. Sodium naphtholate has antiseptic properties; bismuth naphtholate is an intestinal antiseptic agent.
f naphtolate
d Naphtholat
l naftolato
e naftolato

2429 **naphthol green**
Ferrous derivative of the sodium salt of
the 1-nitrous-2 oxynaphthalic-6-sulphonic
acid; a dark-green powder, insoluble in
water, used in microscopy.
f vert naphtol
d Naphtholgrün
l verde naftolo
e verde naftol

2430 **naphtholphthalein**
Naphthol derivative of the phthalein;
crystals not easily soluble in water
and used as a pH indicator.
f naphtolphtaléine
d Naphtholphthalein
l naftolftaleina
e naftolftaleina

2431 **naphtholsulphonic acid**
Isomeric-named monosulphonic acid mainly used as a dyestuff intermediate but
also in the manufacture of synthetic
drugs.
f acide naphtolsulfonique
d Naphtholsulfonsäure
l acido naftolsolfonico
e ácido naftolsulfónico

2432 **naphthopyrine**
Loose compound of beta-naphthol and
phenazone. It is used in medicine as an
antiseptic.

f naphtopyrine
d Naphthopyrin
i naftopirina
e naftopirina

2433 **naphthoquinone**
The alpha- or 1,4-isomer is the parent
of the synthetic vitamin-K analogues.
Both vitamin K and vitamin K_2 are derivatives of 2-methyl-1,4-naphthaquinone
which is, in fact, a potent K vitamin.
f naphtoquinone
d Naphthoquinon
i naftochinone
e naftoquinona

2434 **naphthoresorcin**
Naphthalene derivative containing two
hydroxylic groups in position 1 and 3.
Water-, alcohol-. ether-soluble crystals
which are obtained from the naphthalin-
disulphonic acid. It is used as an antiseptic.
f naphtorésorcine
d Naphthoresorcin
i naftoresorcina
e naftoresorcina

* **naphthylaminemonosulphonic acid s.
naphthionic acid**

* **naphthylamine-sulphonic acid s. naphthionic acid**

2435 **naphthylpararosaniline**
Dye substance used in the past for the
treatment of malignant growths in animals. The drug has also been used in
an attempt to cure malignant epithelial
growths in humans, yet without positive
results.
f naphtylpararosaniline
d Naphthylpararosanilin
i naftilpararosanilina
e naftilpararosanilina

2436 **naphthylsalicylate**
Chemical compound having intestinal
antiseptic and antirheumatic properties.
f acide naphtylsalicylate
d Naphthylsalizilat
i naftilsalicilato
e naftilsalicilato

2437 **naproxen**
Non-steroid anti-phlogistic and analgesic
agent marked by a prompt and long-lasting action. It is particularly indicated
in the treatment of chronic polyarthritis,
especially when evolutive, as it has a
corticosteroid-saving effect. Also used
in the treatment of gout, non-articular
rheumatism, neuralgias and, more generally, in those painful conditions which
follow a trauma or surgery. Overdosage
symptoms may be observed in patients
treated with anticoagulants. Insomnia
and sometimes thrombocytopenia may occur as side effects.
f naproxène

d Naproxen
i naprossene
e naproxeno

2438 **narceine**
Narcotine-related alkaloid of the isoquinoline group. It is found in opium
and displays properties that are similar
to those of papaverine. It is rarely
used for therapeutic purposes owing to
its side effects.
f narcéine
d Narcein
i narceina
e narceina

2439 **narcobarbital**
5-(2-bromoallyl)-5-isopropyl-1-methylbarbi-
turic acid. Hypnotic and sedative barbiturate, very short-acting. It is administered intravenously for surgical operations and in order to induce general
anaesthesia. Great caution is required
in the case of patients suffering from
bronchial asthma and circulatory disorders. Some allergic reactions have been
observed as side effects.
f narcobarbital
d Narcobarbital
i narcobarbitale
e narcobarbital

2440 **narcotic**
Relating to any drug or agent which
induces stupor or torpor. Also relating
to any substance which produces narcosis or general anaesthesia.
f stupéfiant; narcotique
d Betäubungs-; narcotisch
i narcotico
e narcótico

2441 **natamycin**
Antifungal and antiprotozoal wide-spec-
trum antibiotic agent mainly used in the
treatment of fungal infections of the
skin and of the vagina. It may be used
in association with other antibiotics or
corticosteroids. Owing to its poor absorbability, it is not used orally. Gastric irritation may, in fact, be caused.
Skin rashes may occur as side effect.
f natamycine
d Natamycin
i natamicina
e natamicina

2442 **natrii acetrizoas**
Sodium acetrizoate. Contrast medium
chiefly used in urography, retrograde
pyelography, nephrography, angiocardio-
graphy etc. Some topical reactions, headaches and hypotension may occur as side
effects.
f acétrizoate de sodium
d Natriumacetrizoat
i sodio acetrizoato
e acetrizoato sódico

2443 **neat's-foot oil**

Pale yellow fatty oil which is obtained
by boiling the feet and shinbones of
cattle and is used as a very fine lubricant.
f huile de pied de boeuf
d Klauenfett; Rinderklauenfett
i grasso di piede di bue
e grasa de pie de buey

2444 nebularine
Purine riboside. Chemical compound
$(C_{10}H_{12}N_4O_4)$ present in some varieties
of mushrooms. Colourless crystals, fairly
soluble in water and characterized by
tuberculostatic (arresting the growth of
the tubercle bacillus) and antimitotic
properties.
f nébularine
d Nebularin; Purin-Ribosid
i nebularina
e nebularina

2445 nedocromil
Therapeutic agent used in the preventive
treatment of asthma and bronchitis with
reversible obstruction of the airways.
The drug is capable of blocking allergic
mechanisms. Headache and nausea may
occur as side effects.
f nédocromil
d Nedocromil
i nedocromil
e nedocromil

2446 nefopam
Centrally acting analgesic agent marked
by a mode of action only vaguely known
but not related to narcotics. It is used
to alleviate acute and chronic pains, in
particular pains caused by orthopaedic
conditions, dental operations, wounds,
neoplasiae and, more generally, surgery.
Drowsiness, headaches, malaise, lack of
appetite, a dry mouth, copious perspiration and depression may occur as side
effects.
f néfopam
d Nefopam
i nefopam
e nefopam

*** nembutal   s. pentobarbital**

2447 neoarsphenamine
Therapeutic agent of the arsenobenzene
series, used in the treatment of primary
and secondary syphilis. The drug inhibits the sulphydryl group and cellular
oxidation.
f néoarsphénamine
d Neoarsphenamine
i neoarsfenamina
e neoarsfenamina

2448 neocarzinostatin
Antineoplastic agent used in the treatment of acute leukaemia and of carcinoma of the pancreas and stomach.
Leukopenia, anorexia accompanied by
nausea and diarrhoea are possible side

effects.
f néocarcinostatine; néocarzinostatine
d Neocarzinostatin
i neocarcinostatina
e neocarcinostatina

2449 neomycin
Broad-spectrum antibiotic produced by
Streptomyces fradiae. Neomycin is the
result of several components of a basic
nature, soluble in water and active
against gram-positive bacteria but not
against mycetes and viruses. Also effective against gram-negative bacilli, e.g.
E.coli. Nausea accompanied by vomiting,
diarrhoea, nephrotoxicity accompanied
by oliguria, albuminuria and haematuria
are among the side effects of the drug.
f néomycine
d Neomycin
i neomicina
e neomicina

*** neomycin B   s. framycetin**

2450 neopyrithiamine
Pyrithiamine. Chemical compound corresponding to thiamine (or vitamin B_1) in
which the thiazolic nucleus has been
substituted by a pyridinic nucleus. It
is characterized by properties which are
antagonistic to those of thiamine. In
fact, symptoms of thiamine deficiency in
experimental animals are cured by
thiamine.
f pyrithiamine
d Neopyrithiamin; Pyrithiamin
i piritiammina
e piritiamina

2451 neostigmine
The dimethylcarbamic ester of 3-hydroxy-
phenyltrimethylammonium methyl sulphate.
Neostigmine is a synthetic alkaloid mark
ed by effects that are similar to those
of physostigmine (q.v.) in so far as it
inhibits cholinesterase (enzyme present
in several animal tissues and especially
abundant in the nervous tissue). It is
thus capable of potentiating the action
of acetylcholine at the motor end-plate
and is therefore used in anaesthesia in
order to counteract the effects of curare-
-like compounds.
f néostigmine
d Neostigmin
i neostigmina
e neostigmina

2452 neostigmine bromide
See neostigmine and neostigmine methylsulphate. Parasympathomimetic agent
capable of inhibiting or blocking acetylcholine esterases. The drug stimulates
intestinal motility and is therefore indicated in the treatment of atonic constipa
tion. It is characterized by an immediate effect on the smooth muscles with
cholinergic innervation. It is also used
in myasthenia gravis (muscle dysfunction

and abnormal fatigability) when its action may be intensified by ephedrine hydrochloride. Gastro-intestinal disorders accompanied by pains, abnormal salivation, visual disturbances, myosis etc. may occur as side effects.

f bromure de néostigmine
d Neostigminbromid
i neostigmina bromuro
e bromuro de neostigmina

2453 neostigmine methylsulphate

Compound similar to neostigmine bromide and used when administration by injection is desirable or required. Its reversible anticholesterasic activity (on the skeletal muscles and, albeit more rarely, on the smooth muscles) explains its use in the treatment of myasthenia gravis, paralytic ileus (the intestinal obstruction caused by paralysis of muscles, also known as adynamic ileus), acute congestive glaucoma, retention of urine following a surgical operation and for the elimination of calculi in the urinary tract etc. Gastro-intestinal disorders, often accompanied by pains, myosis, bradycardia and abnormal bronchial secretion may occur as side effects.

f méthylsulfate de néostigmine
d Neostigminmethylsulfat
i neostigmina metilsolfato
e metilsulfato de neostigmina

* **neptazane**   s. methazolamide

2454 neroli flowers

Orange flowers from which a saturated solution is obtained (neroli oil), which is used as a pharmaceutical preparation and in toilet water.

f fleurs d'oranger
d Neroliblüten; Orangenblüten
i fiori d'arancio
e flores de naranja; flores de azahar

2455 netilmicin

Aminoglycoside antibiotic whose spectrum resembles that of neomycin. It is particularly active against Pseudomonas aeruginosa (bacterial genus known as "blue pus"). Ototoxicity and nephrotoxicity may occur as side effects.

f nétilmicin
d Netilmicin
i netilmicin
e netilmicin

2456 nettle-leaved bellflower

Throatwort; a herb growing in Europe and used in the past to treat sore throat.

f gantelée; gant de Notre-Dame; ortie bleue
d Halskraut; Nesselglockenblume
i ortica turchina; campanula guantana
e dedalera

2457 neuramide

Sterile colloidal solution of proteolytic enzyme obtained from the glandular layer of the fresh stomach of a pig. It is an analgesic, neurotrophic and anti-inflammatory agent which has been found to be capable of an antiviral action in the treatment of herpes zoster, herpes genitalis, radiculitis, neuraxitis and, more generally, neuralgic and neuritic episodes. Also known as protamide.

f neuramide
d Neuramid
i neuramide
e neuramida

2458 neuraminic acid

Amino acid abundantly found in nature in the form of N-acetyl derivative. It is an important constituent of mucopolysaccharides, gangliosides and blood group substances.

f acide neuraminique
d Neuraminsäure
i acido neuroaminico
e ácido neuroamínico

2459 neurine

Amino alcohol, a compound corresponding to choline without a molecule of water. It is a syrupy substances, strongly poisonous, that is found among the ptomaines resulting from putrefaction of albumins. It is also found, though in minimal quantities, in the adrenal glands, urin and, in slightly greater quantities, in the brain where it is not preformed but derives from choline by the dehydrating action of some bacteria.

f neurine
d Neurin
i neurina
e neurina

2460 neutral copper acetate

Crystallized verdigris, crystals of Venus. Copper oxyacetate in needles of a somewhat indefinite composition. Its medicinal properties are similar to those of copper sulphate.

f acétate neutre de cuivre; cristaux de Venus; verdet cristallisé
d essigsaures Kupferoxyd
i acetato neutro di rame; verde eterno
e acetato de cobre

2461 neutral red

Toluylene red. Basic phenazine dye principally used as a biological stain and acid-base indicator.

f rouge neutre; rouge de toluylène
d Neutralrot; Toluylenrot
i rosso neutro
e rojo de toluileno

2462 ngai camphor

Camphor found in various essential oils, chemically a laevorotatory borneol, i.e. a substance resembling camphor in its therapeutic properties though its esters may display properties which are asso-

ciated with the particular acid radical,
e.g. the salicylate used in the treatment
of rheumatism.

f camphre de sombong; camphre ngai
d Ngaikampfer
i canfora di blumea
e alcanfor de blumea; alcanfor de ngai

2463 nialamide

Antidepressant agent of the hydrazine
group of the monoamino-oxidase inhibitors. Its activity and indications are
similar to those of phenelzine, q.v.
Widely used in the treatment of depressive and paranoid psychoses, cyclothym-
ic conditions and some psychosomatic
disorders. The drug is also used in
cases of mental retardation in children.

f nialamide
d Nialamid
i nialamide
e nialamida

2464 niaouli oil

Oil extracted from the leaves of a small
evergreen tree growing in the south-
western Pacific islands and containing
cineole. It is mainly used in the treatment of rheumatism, intestinal disorders
and toothache.

f huile de niaouli
d Niaulöl
i olio essenziale di niaouli
e aceite de niaouli

2465 niaprazine

Non-barbituric hypnotic and sedative
agent mainly indicated in the treatment
of sleep disorders and, in children, of
behaviour anomalies. Some drowsiness
may occur as side effect.

f niaprazine
d Niaprazin
i niaprazina
e niaprazina

2466 nicametate

A vasodilator acting through its intra-
organic scission into nicotinic acid and
diathylaminoethanol. The drug displays
a selective action on the blood vessels
and thus improves the circulation within
tissues. It is used in the treatment of
cerebral and cutaneous circulatory insufficiency.

f nicamétate
d Nikametat
i nicametato
e nicametato

2467 nicergoline

Vasifactive agent acting by a vasal
alpha-adrenolytic and myolitic mechanism
and by a stimulation of the metabolism
of tissues. It increases the peripheral
and cerebral blood flow and diminishes
vascular resistance without significantly
altering the systemic arterial pressure.
The drug is capable of enhancing oxygen and glucose absorption, thus stimu-

lating the cerebral energetic metabolism.
It is mainly indicated in the treatment
of acute and chronic cerebral vasculopathies, e.g. arteriosclerosis, thrombosis
etc. and of peripheral vascular diseases.
Also useful for the prevention of vasomotor headaches.

f nicergoline
d Nicergolin
i nicergolina
e nicergolina

2468 niceritrol

Vasodilator used in the treatment of
hyperlipaemia and arteriosclerosis, as
well as other peripheral and central
vascular disorders, e.g. intermittent
claudication, Raynaud's disease etc.
Gastro-intestinal disorders, cutaneous
hyperaemia, a mild diabetes and urticaria may occur as side effects. The
drug should not be administered in the
course of pregnancy.

f nicéritrol
d Niceritrol
i niceritrolo
e niceritrolo

*** nicetamide   s. niketamide**

2469 niclosamide

Anthelmintic agent particularly effective
in the treatment of tapeworm, characterized by a paralyzing and taenicide mechanism of action. The drug is not absorbed from the intestinal tract and thus
does not act on parasites that are present in muscles and/or in the lungs.
Side effects are practically nil because
of the lack of absorption.

f niclosamide
d Niklosamid
i niclosamide
e niclosamida

2470 nicoclonate

Hypolipaemic and hypocholesterolaemic
agent which acts by maintaining a constant blood level of nicotinic acid. It
is indicated in the treatment of atheromatous (relating to the process affecting
blood vessels with the formation of sub-
-intimal plaques) diseases which may or
may not be associated with alterations
of the lipidic and glucidic metabolism,
e.g. coronary insufficiency, cardiovascular disorders, hypertension and, more
generally, peripheral arteriopathies.

f nicoclonate
d Nikoclonat
i nicoclonato
e nicoclonato

2471 nicocodine

Analgesic, antitussive and narcotic
agent used in the symptomatic treatment
of cough. A codeine nicotinate.

f nicocodine
d Nikocodin
i nicocodina

e nicocodina

2472 nicofibrate
Vasodilator acting on the central and peripheral blood vessels. The drug also reduces the level of cholesterol and acts as an antihyperlipidaemic. It is used in the treatment of vascular disorders caused by arteriosclerosis and diabetes and whenever a reduction of serum cholesterol and total lipid levels is required.
f nicofibrate
d Nikofibrat
i nicofibrato
e nicofibrato

2473 nicofuranose
Peripheral vasodilator and hypotensive agent derived from nicotinic acid and especially used in the treatment of vascular spasms, thrombophlebitis and as a coronary vasodilator. Gastro-intestinal disorders may occur as side effect. The drug should not be administered in cases of severe cardiopathies.
f nicofuranose
d Nikofuranose
i nicofuranosio
e nicofuranosio

2474 nicofurate
Peripheral vasodilator and hypolipidaemic, hypocholesterolaemic agent generally used to improve blood circulation.
f nicofurate
d Nikofurat
i nicofurato
e nicofurato

2475 nicometamide
Hydroxymethylnicotinamide. Antiseptic agent of the biliary tract and a cholagogue. Its hepatotrophic action is due to the nicotinamidic radical. It is indicated in the treatment of hepatobiliary and enteric inflammations, cholecystopathies, peptic ulcer, acute enteritis, colitis and, more generally, gastro-intestinal disorders. Nausea accompanied by vomiting in patients suffering from hypochlorhydria, anorexia, headaches etc. may occur as side effects. Also known as nicotinylmethylamidum.
f nicométamide
d Nikometamid
i nicometammide
e nicometamida

2476 nicomethanol
Vasodilator and hypotensive agent used in the treatment of essential (functional) hypotension. Its action is mainly of a mechanical nature. Flush and the danger of circulatory collapse are possible side effects.
f nicométhanol
d Nikomethanol
i nicometanolo
e nicometanolo

2477 nicomol
Hypocholesterolamic agent indicated in the treatment of hypercholesterolaemia. It is a derivative of nicotinic acid. Flush, urticaria, diarrhoea, liver disorders and mild diabetes may occur as side effects.
f nicomol
d Nikomol
i nicomolo
e nicomolo

2478 nicomorphine
A narcotic analgesic marked by a prompt and lasting action and excreted in the urine. The drug is chiefly used to allay pains of various origin, e.g. post-operatory, neoplastic diseases-related, biliary spasms pains. The drug should not be administered in the presence of a severe cardiovascular incompetence.
f nicomorphine
d Nikomorphin
i nicomorfina
e nicomorfina

2479 nicotafuryl
Nicotinic-acid derived vasodilator chiefly used as a rubefacient (causing redness of the skin) and revulsive for the treatment of myalgias and arthrosis. It is also useful in the treatment of ischaemic conditions of the skin.
f nicotafuryle
d Nikotafuryl
i nicotafurile
e nicotafurilo

2480 nicotinamide
Amide of the nicotinic acid, with which it is widely present in animal and vegetable tissues. It has a vitaminic, antipellagra action and is rightly thought as the true vitamin PP (pellagra preventing) since the biological action of the vitaminic acid is probably conditioned to its transformation into nicotinamide. It is a constituent of two coenzymes performing important biological activities in so far as they act as acceptors and donors of hydrogen in the oxyreduction process. Both enzymes actively intervene by conditioning various biochemical and nutritional events.
f nicotinamide; niacinamide; vitamine anti pellagreuse
d Nikotinamid; Nikotinsäureamid; Antipellagravitamin
i nicotinammide; nicotammide; niacinammide
e nicotinamida; niacinamida; vitamina PP

2481 nicotinamide adenine dinucleotide
Antagonistic to alcohol and narcotic analgesics. It is a biologically active form of vitamin PP and one of the principal coenzymes of cell respiration. It is used in the treatment of several disorders of the tissular respiration occurring in cardiopathies, intestinal dysfunction and all vitamin PP indications.

f nicotine amide dinucléotide
d Nikotinamiddinucleotid
i nicotinamide adenina dinucleotide
e nicotinamida adenina dinucleotido

2482 nicotinate
Salt or ester of nicotinic acid which is
present in milk, meat, wheat, liver,
kidney, yeast and bread. A lack of
nicotinic acid in the diet leads to pellagra, a food-deficiency disease.
i nicotinate
d Nikotinat
i nicotinato
e nicotinato

2483 nicotine
Highly poisonous basic liquid alkaloid
constituting the active principle of tobacco. Nicotine darkens on exposure and
causes an acrid burning sensation in
the mouth. It has very little use in medicine and is chiefly used as an insecticide.
f nicotine
d Nikotin
i nicotina
e nicotina

2484 nicotinic acid
Carboxylic derivative of pyridine, which
is obtained by the oxidation of nicotine
or pyridine derivatives substituted, in
position 3, with quinoline. In small
quantities, it is present in all cells of
animal and vegetable organisms; in
larger quantities, in the liver, spleen,
yeast, milk and cereals. In all probabil_
ity, it is only found in the form of
amide which in its turn is bound to
other compounds, e.g. adenine. Together
with nicotinamide, nicotinic acid displays a vitaminic antipellagra property.
In large doses, it is sometimes used as
a vasodilator.
f acide nicotinique
d Nikotinsäure
i acido nicotinico
e ácido nicotínico

* **nicotinic acid amide   s. nicotinamide**

2485 nicotinyl tartrate
Peripheral vasodilator acting directly on
muscle in blood vessels and indicated
in the treatment of defective peripheral
blood circulation, e.g. chilblains and
Raynaud's syndrome. Flushing, tachycardia, gastro-intestinal disorders and
hypotension may occur as side effects.
f tartrate de nicotinyle
d Nikotinyltartrat
i nicotinile tartrato
e tartrato de nicotinilo

2486 nicoumalone
Anticoagulant which inhibits the synthesis of clotting factors by the liver. See
warfarin.
f nicoumalone

d Nicoumalon
i nicumalone
e nicumalona

2487 nifedipine
Vasodilator used in the treatment of
angina pectoris and hypertension. It
acts by blocking the influx of calcium
ions into vascular smooth muscles thus
reducing peripheral vascular resistance.
Flush, dizziness and asthenia may occur
as side effects. The drug should not be
administered in the course of pregnancy.
f nifédipine
d Nifedipin
i nifedipina
e nifedipina

2488 nifenalol
Sedative agent chiefly used in the treatment of angina pectoris and arrhythmias.
Its action does not depress the rate and
general activity (contraction) of the
heart. Insomnia and diarrhoea may occur as side effects.
f nifénalol
d Nifenalol
i nifenalolo
e nifenalolo

2489 nifenazone
Pyrazolone-related antiphlogistic, decongestant and anaesthetic agent mainly
used in the treatment of rheumatism, in
particular muscular pains, sciatica and
persistent headaches. Agranulocytosis
and gastric hypersecretion may occur as
side effects.
f nifénazone
d Nifenazon
i nifenazone
e nifenazona

2490 niflumic acid
Synthetic non-steroid analgesic and antiphlogistic agent used in the treatment
of rheumatism, both inflammatory and
degenerative, sciatic and scapulohumoral
neuralgias, gout, urinary disorders,
haemorrhoids, rhagades and, in association with antibiotics, of pneumonia and
bronchitis. The drug also finds employment in gynecology and dentistry.
f acide niflumique
d Niflumsäure
i acido niflumico
e ácido niflúmico

2491 nifuratel
Disinfectant agent used mainly in the
treatment of trichomoniasis and, more
generally, other forms of vulvovaginitis
due to lycetes, gram-positive and gram-
-negative micro-organisms.
f nifuratel
d Nifuratel
i nifuratel
e nifuratel

2492 nifurfuline

Nitrofurantoin-derived antibacterial agent particularly active against the pathogens of the urinary tract, Escherichia coli, enterobacter, proteus, staphylococci and enterococci. It is indicated in the treatment of cystitis, pyelonephritis, prostatitis, chronic or acute infections by colibacilli, enterococci and staphylococci. Also used as a prophylactic agent in the course of examination of the urinary tract.

f nifurfuline
d Nifurfulin
i nifurfulina
e nifurfulina

2493 nifuroxazide
5-nitrofurane-derived antiseptic agent particularly active against Escherichia coli, Shigella dysenteriae, Eberthelia typhosa and Streptococcus pyogenes. Also active against Streptococcus faecalis and Pseudomonas aeruginosa. The drug is used in the treatment of intestinal infec tions, infectious and acute diarrhoeas, food poisoning, salmonellosis, acute or chronic infectious colitis and bacillary dysentery.

f nifuroxazide
d Nifuroxazid
i nifurossazide
e nifuroxazida

2494 nifuroxime
Chemotherapeutic agent particularly active against candidiasis and trichomonia sis. It is mainly used in the treatment of both adult and very young women. The antibacterial action is rather mild. Nausea and diarrhoea may occur as side effects.

f nifuroxime
d Nifuroxim
i nifurossima
e nifuroxima

2495 nifurpipone
Nitrofuranic agent whose broad spectrum includes most gram-positive and gram- -negative bacteria causing infections of the urinary system. The drug is rapidly absorbed when taken by mouth and rapidly excreted in the urine. The drug is indicated for the prophylaxis and treatment of infections of the urinary tract and for those infections and febrile reactions which may be caused by catheterization or endoscopic examina- tions.

f nifurpipone
d Nifurpipon
i nifurpipone
e nifurpipona

2496 nifurprazine
Pyridazine-related topical antibacterial agent mainly used for the prophylaxis and treatment of pyodermatoses, ulcers, mastitis, bedsores, impetigo etc.

f nifurprazine

d Nifurprazin
i nifurprazina
e nifurprazina

2497 nifurtimox
Antiprotozoal agent used in the treatment of acute and chronic trypanosomiasis, in particular the American type, i.e. an acute febrile disease marked in its early stages by local swelling and anaemia and in its later stages by cardiac, nervous and mixoedematous symptoms. Insomnia, restlessness, nausea with vomiting, muscle tremors and neuritis may occur as side effects.

f nifurtimox
d Nifurtimox
i nifurtimox
e nifurtimox

2498 nifurtoinol
Nitrofuranic therapeutic agent whose broad spectrum includes gram-positive and gram-negative bacteria that are responsible for infections of the urinary tract. In its active form, it shows greater tolerance and a better excretion into the urine than nitrofurantoin, q.v. It is indicated in the treatment of bacterial acute, subacute and chronic infections of the urinary tract, as well as preoperative and postoperative urological prophylaxis. Also used for endoscopies which require the use of surgical instru ments. Sensitization reactions and haemo litic anaemia may be observed as side effects. Chemically, hydroxymethylnitrofurantoin.

f nifurtoïnol
d Nifurtoinol
i nifurtoinolo
e nifurtoinolo

2499 nigella oil
Fatty essential oil contained in high percentage in the seeds of Nigella sativa and marked by a particular, very penetrating odour. It is used in some countries as a therapeutic agent. It is soluble in alcohol.

f huile de nigelle; huile de cumin noir
d Schwarzkümmelöl
i olio di nigella
e esencia de comino negro; esencia de agenuz

2500 night-blooming cereus
Or sweet-scented cactus. Sprawling or climbing cactus with ribbed stems and yellow spines. Its flowers, which are large, white and strongly scented, are followed by yellow red-streaked fruit that is used for medicinal purposes.

f cactier; cactus à grandes fleurs
d Königin der Nacht
i regina della notte
e reina de la noche

2501 nigrosin
Generic name of basic dyes which are

obtained by melting together aniline, aniline hydrochloride, nitrobenzene and iron filings. The compound is used in neurological histology.

f nigrosine
d Nigrosin
i nigrosina
e nigrosina

2502 **niketamide**
Respiratory stimulant acting directly on respiratory centres in the brain. It is rarely used except to allay respiratory depression caused by severe chronic bronchitis. It should not be used in the treatment of respiratory depression caused by drug overdosage. Abnormal perspiration, convulsions, excitation, nausea and vomiting may occur as side effects.

f nicétamide
d Niketamid
i nicetammide
e nicetamida

2503 **nimorazole**
Antiprotozoal chemotherapeutic agent used in the treatment of gastro-intestinal and vaginal infections, as well as infections caused by Trichomonas vaginalis. Nausea may occur if alcohol is taken in the course of the treatment.

f nimorazole
d Nimorazol
i nimorazolo
e nimorazolo

2504 **ninhydrin**
Substance used as a very sensitive reagent to alpha-amino acids with which, by heating in aqueous solution, it yields a characteristic violet dye. Chemically, ninhydrin is either 1,2,3-indan-trione hydrate or triketo-hydrindene hydrate.

f ninhydrine
d Ninhydrin
i ninidrina
e ninhidrina

2505 **ninhydrin reaction**
Reaction of ninhydrin with amino acids or related amino compounds used for the colorimetric determination of amino acids, peptides or proteins by measuring the intensity of the blue to violet to red colour which results. It is also used for the quantitative determination of alpha-amino acids by measuring the amount of carbon dioxide that is produced.

f réaction de la ninhydrine
d Ninhydrinreaktion
i reazione della ninidrina
e reacción de la ninhidrina

2506 **nipplewort**
Or succory dock-cress. Slender branching annual herb characterized by small head of yellow flowers and so called for its therapeutic use for nipples.

f grageline; herbe aux mamelles
d Milchen; Ackerkohl

i lampsana; lassana; grespignolo
e lámpsana

2507 **niprofazone**
Pyrazolone derivative capable of strong analgesic action. Owing to its antipyretic and antiphlogistic activity, the drug is indicated for the relief of pains of neuralgic origin, e.g. toothache, sciatica etc. Also used for rheumatic pains as well as pains caused by traumas.

f niprofazone
d Niprofazon
i niprofazone
e niprofazona

2508 **niridazole**
Chemotherapeutic agent indicated in the treatment of infections caused by Dracunculus medinensis (guinea worm). Drowsiness, gastro-intestinal symptoms and headaches may occur as side effects.

f niridazole
d Niridazol
i niridazolo
e niridazolo

2509 **nitralinic acid**
Benzoquinone containing two hydroxylic and two nitric groups. Yellow, water-soluble crystals obtained by nitration of esters of the hydroquinone. They are used in analytical chemistry for the determination of alkaloids.

f acide nitralinique
d Nitralinsäure
i acido nitralinico
e ácido nitralínico

2510 **nitrate**
Salt or ester of nitric acid. Alkaline nitrates are easily soluble in water, but those of heavier metals are soluble only with difficulty.

f nitrate
d Nitrat
i nitrato
e nitrato

2511 **nitration**
The addition of a nitro group to a compound. The treatment of an organic compound with nitric or fuming nitric acid.

f nitration
d Nitrierung
i nitrazione
e nitración

2512 **nitrazepam**
Tranquillizer and hypnotic agent of the diazepinic series, indicated to combat insomnia in the elderly and in subjects affected by hyperemotivity. It acts as a pre-hypnotic and allows the resumption of interrupted sleep. It is also used as a pre-narcotic and was used in the past to allay myoclonic attacks. The drug should be avoided by patients suffering from respiratory insufficiency. Mental confusion and depression may be

caused as side effects in the elderly.
f nitrazépam
d Nitrazepam
i nitrazepam
e nitrazepam

2513 nitric acid
Aqua fortis. The most important of the
acids derived from nitrogen, present in
nature in the form of salts (nitrates).
Once diluted, nitric acid is an astringent, tonic and bile stimulating agent.
The fumic nitric acid consists of the
concentrated acid with oxides of nitrogen
in solution and acts as a strong caustic
agent.
f acide nitrique
d Salpetersäure; salpetrige Säure
i acido nitrico
e ácido nítrico

2514 nitric ester
Ester produced by the action of concentrated nitric acid on an alcohol.
f ester nitrique
d Ester der Salpetersäure
i estere nitrico
e éster nítrico

2515 nitride
Compound of one or more atoms of nitrogen with another element, e.g. lithium,
aluminium etc. It is capable of evolving
ammonia upon treatment with water.
f nitrure
d Nitrid
i nitruro
e nitruro

2516 nitrification
The chemical process in which ammonia
is oxidized (by bacteria) into nitrous
acid and subsequently into nitric acid.
More generally, the introduction of a
nitro group into a molecule.
f nitrification
d Nitrierung
i nitrificazione
e nitrificación

2517 nitrify, to
To treat with nitric acid. Specifically,
the process whereby certain bacteria capable of converting ammonia into nitrous
and nitric acids are qualified.
f nitrifier
d nitrieren
i nitrificare
e nitrificar

2518 nitrile
Any of the compounds marked by the
presence of the equivalent nitrogen
radical N= and which can be derived
from oxygen containing acids by total
removal of the elements of water from
their ammonium salts. Nitrile also denotes a compound which is characterized
by the presence of the cyanogen group,
can be derived from a carboxylic acid

or its amide and yields the acid on
complete hydrolysis. It is, in fact, an
organic cyanide.
f nitrile
d Nitril
i nitrile
e nitrilo

2519 nitrite
Salt of nitrous acid. Some nitrites, in
particular the organic alkyl nitrites,
have a vasodilating action associated
with a depressor effect on the motor
centres of the cord. They are therefore
used in the treatment of angina pectoris,
of intestinal disorders and of asthma
owing to their relaxing action on the
smooth muscles.
f nitrite
d Nitrit
i nitrito
e nitrito

2520 nitrobenzaldehyde
Chemical compound obtained by nitration
of benzaldehyde. It exists in the three
isomeric forms o-, m- and p- and is
used as an intermediate for drugs and
in organic syntheses.
f nitrobenzaldéhyde
d Nitrobenzaldehyd
i nitrobenzaldeide
e nitrobenzaldehido

2521 nitrobenzene
Aromatic hydrocarbon directly derived
from benzene by nitration. A yellow oily
liquid whose partial-reduction products
and total-reduction product are particularly important. The latter product is
aniline which is obtained by performing
the reduction in acid solution. Also
known as mirbane oil.
f nitrobenzène; nitrobenzol; essence de
mirbane
d Nitrobenzol; Mirbanöl
i nitrobenzolo
e nitrobenzol; nitrobenceno

2522 nitrobenzoic acid
Chemical compound used as an intermediate in the production of pharmaceuticals.
f acide nitrobenzoïque
d Nitrobenzoesäure
i acido nitrobenzoico
e ácido nitrobenzoico

2523 nitrobenzoyl chloride
Chemical compound having the same action of nitrobenzoic acid.
f chlorure de nitrobenzoyle
d Nitrobenzoylcyanid
i cianuro di nitrobenzoile
e cianuro de nitrobenzoílo

2524 nitrobenzoyl cyanide
Chemical compound having the same action as nitrobenzoic acid.
f cyanure de nitrobenzoyle

d Nitrobenzoylcyanid
i cianuro di nitrobenzoile
e cianuro de nitrobenzoilo

2525 nitrocellulose
Cellulose nitrate. Nitrate derivative of cellulose used to manufacture filters of a given porosity which find employment in molecular biology, in particular in nucleic acid hybridization experiments.
f nitrocellulose
d Nitrocellulose
i nitrocellulosa
e nitrocelulosa

2526 nitro-erythrol
Erythrityl tetranitrate. Chemical compound used in medicine in a solution with equal weight of lactose in the treat-ment of angina pectoris. It acts by reducing blood pressure.
f nitro-érythritol
d Nitroerythrit
i nitroeritrolo
e nitroeritritol

2527 nitroferricyanic acid
Acid used as the sodium salt (nitroprusside) to identify metaproteins as well as other compounds containing the thiol group -SH. The acid reacts with aceto-acetic acid and acetone in ketonuric urine.
f acide nitroferricyanique
d Nitroferricyansäure
i acido nitroferricianico
e ácido nitroferriciánico

2528 nitrofural
Topical antibacterial chemotherapeutic agent active against gram-positive and gram-negative microorganisms but inactive against mycotic infections and acne. The drug is used in the treatment and prophylaxis of ear, nose and throat disorders, as well as in dermatology, stomatology and ophthalmology.
f nitrofural
d Nitrofural
i nitrofurale
e nitrofurazona

2529 nitrofurantoin
Chemotherapeutic agent active against gram-positive and gram-negative microorganisms, in particular pathogens affecting the urinary tract (Pseudomonas aeruginosa has proved resistant to the drug). It is rapidly absorbed and excreted with urine in which it produces a yellow fluorescence. Bacteriologically, it is much more active in acid than in alkaline urine. It is indicated for the treatment of acute and chronic infections of the urinary system, e.g. urethritis, pyelonephritis, cystitis and in particular of those forms which prove resistant to antibiotics. Nausea, vomiting, headaches and some skin reactions may occur as side effects. Haemolytic anae-mia in G6PD-deficient subjects has been observed.
f nitrofurantoïne
d Nitrofurantoin
i nitrofurantoina
e nitrofurantoina

2530 nitroglycerin
Glyceryl trinitrate. Explosive ester obtained through the action of nitric and sulphuric acid on glycerol. It is used in medicine, either as alcoholic solution or in tablets as a heart stimulant. Its action is similar to that of ethyl nitrite which was administered in the past in cases of angina pectoris. Commonly known as trinitrine (trinitroglycerin). Headaches and dizziness may occur as side effects.
f nitroglycérine
d Nitroglycerin
i nitroglicerina
e nitroglicerina

2531 nitrohydrochloric acid
Nitromuriatic acid. Commonly known as aqua regia. In a diluted solution it is sometimes prescribed in the treatment of liver disorders.
f acide nitrochlorhydrique
d Nitrosalzsäure
i acido nitrocloridrico
e ácido nitroclorhídrico

2532 nitromersol
Antiseptic agent indicated for the disinfection of the skin before a surgical operation. Also used in the treatment of minor lesions of the skin, urethral irrigations and eye infections.
f nitromersol
d Nitromersol
i nitromersolo
e nitromersolo

* **nitropenthrite** s. pentaerithrityl tetranitrate

2533 nitrophenol
Compound formed by the progressive substitution of the hydrogen atoms of phenol with nitro- NO_2 groups. The term is also used to denote paranitrophenol, a crystalline compound which is obtained by the nitration of phenol and is used in the manufacture of several medicinal drugs. Owing to its antifungal activity, it is used topically to heal fungal infec-tions of the skin.
f nitrophénol
d Nitrophenol
i nitrofenolo
e nitrofenol

2534 nitrosorbide
Isosorbide dinitrate. Vasodilator mainly indicated for the coronary arteries and characterized by aperipheral mechanism of action. The drug is used for the prevention and treatment of angina pectoris

and other cardiac disorders (see nitroglycerin). Its effects are almost immediate and last up to six hours. Methaemoglobinaemia (the presence of methaemoglobin in the blood), gastro-intestinal disturbances and vomiting may occur as side effects.
f nitrosorbide; dinitrate d'isosorbide
d Nitrosorbitdinitrat
i isosorbide dinitrato
e nitrato de isosorbido

2535 nitrosulphathiazole
Very active sulphonamide having important therapeutic properties.
f nitrosulfathiazol
d Nitrosulfathiazol
i nitrosulfatiazolo
e nitrosulfatiazol

2536 nitrosylsulphuric acid
Acid nitrosylsulphate. Compound obtained by action of the nitrous anhydride on concentrated sulphuric acid. It is used in organic chemistry in the denitrification of weak basic amines.
f acide nitrosylsulfurique
d Nitrosylschwefelsäure
i acido nitrosilsolforico
e ácido nitrosilsulfúrico

2537 nitrous acid
Acid known only in aqueous solution, whereas its salts are stable (the nitrites). At temperatures above the ordinary the solutions of nitrous acid decompose yielding nitric acid and liberating nitrogen oxide which is used in anaesthesia with an effect that goes from analgesia to complete anaesthesia according to its concentration.
f acide nitreux
d salpetrige Säure
i acido nitroso
e ácido nitroso

2538 nitrous oxide
Inhalational anaesthetic rather weak in its action and a strong analgesic. Generally used in dental and obstetric practice for a light anaesthesia with at least 30% oxygen. Also known as laughing gas.
f oxyde nitreux; gaz hilarant
d GMI-Stoff; Stickoxydul; Lachgas
i ossido nitroso; gas esilarante
e óxido nitroso; gas hilarante

2539 nitroxoline
Broad-spectrum chemotherapeutic agent active against gram-positive and gram--negative microorganisms. It is rapidly absorbed and eliminated in the urine. It is especially used in the treatment of infections of the upper and lower urinary tract, e.g. cystitis and pyelonephritis, and of the prostate. It is advisable to administer the drug together with vitamin B6.
f nitroxoline

d Nitroxolin
i nitrossolina
e nitroxolina

2540 nizatidine
Gastric histamine receptor blocking agent used in the treatment of peptic ulcers and gastric hyperacidity, oesophageal reflux and prophylaxis of gastro-intestinal bleeding in patients who are seriously ill. Diarrhoea, dizziness and breast enlargements in males, as well some depression of the central nervous system may occur as side effects.
f nizatidine
d Nizatidin
i nizatidina
e nizatidina

2541 nomifensine
Antidepressant agent. Uses and effects are similar to those of imipramine, q.v. However, the drug has been withdrawn owing to its somewhat dangerous side effects, including haemolytic anaemia.
f nomifensine
d Nomifensin
i nomifensina
e nomifensina

2542 nonoxinol
Non ionic surfactant (capable of reducing surface tension) used as a dissolving, dispersive agent or as support for iodine. Each nonoxinol is followed by a figure indicating the approximate number of the oxyethylenic group present in it. Nonoxinol "9" is used for the preparation of vaginal ointments as a spermicide and for the treatment of vaginitis. A burning sensation of the vagina or the penis may occur as side effect.
f nonoxinol
d Nonoxinol
i nonossinolo
e nonoxinolo

2543 noracymethadol
Morphine-like narcotic analgesic indicated in the treatment of very severe pains, e.g. pains occurring in neoplastic diseases. The drug is also used to combat morphine addiction. Drowsiness and mental confusion may appear as side effects.
f noraciméthadol
d Noracymethadol
i noracimetadolo
e noracimetadolo

2544 noradrenaline
Norepinephrine. Cathecolamine hormone secreted with adrenalin in the suprarenal medulla and sympathetic nerve endings together with epinephrine in response to a low blood glucose level. Both norepinephrine and epinephrine stimulate the mobilization of glycogen and triacylglycerol by triggering the cAMP- (cyclic adenosin monophosphate)

-cascade. A further important function of noradrenaline and epinephrine is their capacity to inhibit the uptake of glucose by muscle.

f noradrénaline; norépinéphrine
d Noradrenalin
i noradrenalina; norepinefrina
e noradrenalina; norepinefrina

2545 noramidopyrine methanesulphonate sodium
Analgesic, antipyretic and antirheumatic agent more potent than aminopyrine, indicated for the relief of pains of various origin, in particular persistent headaches. Haemolytic anaemia and other blood dyscrasias may occur as adverse effects.

f méthanesulfonate sodique de noraminopyrine
d Noramidopyrine-Natriummethansulfonat
i noramidopirina metansulfonato sodico
e metansulfonato sódico de aminopirina

* **norandrosterone   s. nandrolone**

2546 norclostebol
Anabolic steroid capable of promoting the retention of nitrogen and of stimulating protein biosynthesis. It is thus indicated for the treatment of physiological weakness and loss of weight.

f norclostébol
d Norclostebol
i norclostebol
e norclostebol

2547 norcodeine
N-dimethylcodeine. A narcotic analgesic agent generally used to alleviate pain.

f norcodéine
d Norcodein
i norcodeina
e norcodeina

2548 nordazepam
Dimethyldiazepam. Benzodiazepine-related tranquillizer marked by a prolonged action and used in the treatment of states of anxiety, emotional disturbances caused by organic diseases, stress, psychotic states and insomnia.

f nordazépam
d Nordazepam
i nordazepam
e nordazepam

* **norephedrine   s. noradrenaline**

* **norepinephrine   s. noradrenaline**

2549 norethandrolone
Synthetic androgen marked by anabolizing action and a lesser virilization than testosterone. Owing to its capacity of causing biosynthesis and reducing the consumption of proteins, it may be used as a tonic. It is indicated in the treatment of pathological thinness and osteoporosis. The drug should not be administered to young children.

f noréthandrolone
d Norethandrolon
i noretandrolone
e noretandrolona

2550 norethisterone
Sex hormone used in gynecology to treat premenstrual symptoms, metrorrhages and endometriosis (endometrial tissue present in abnormal situations in the body). It is also used in case of threatened abortion. In association with etinyl oestradiol, it is an oral contraceptive. Nausea and headaches may occur as side effects.

f noréthistérone
d Norethindron
i noretiniltestosterone
e noretiniltestosterona

2551 norethynodrel
Sex hormone similar to progesterone. It acts by inhibiting the secretion of pituitary gonadotropin either directly or on the hypothalamus. It is indicated in the treatment of primary and secondary amenorrhoea and of endometriosis. It is also used as a contraceptive. Nausea and vomiting may occur as side effects.

f noréthynodrel
d Norethynodrel
i noretinodrel
e noretinodrel

2552 norfenetrine
Sympathomimetic, cardiovascular analeptic indicated in the treatment of hypotension, i.e. essential, orthostatic, postoperative hypotension and those pathological and physiological conditions which may follow a shock. Arrhythmia in some hypersensitive patients may occur as a side effect.

f norfénétrine
d Norfenetrin
i norfenetrina
e norfenetrina

2553 norgestrel
Sex hormone acting as oral contraceptive, generally in association with ethynylestradiol, by preventing ovulation. The drug inhibits the output of gonadotropins from the pituitary gland and is indicated in the treatment of dysmenorrhoea and endometriosis. Multiple birth may occur as a side effect.

f norgestrel
d Norgestrel
i norgestrel
e norgestrel

2554 norgestrienone
Progestetional agent, orally active, acting by suppressing gonadotrophin secretion. It has been used, together with testosterone, as a male contraceptive (to maintain libido and secondary sexual characters) and in the treatment of primary and secondary menstrual disorders.

Gastro-intestinal disorders may occur as
side effect.
f norgestriénone
d Norgestrienon
i norgestrienone
e norgestrienona

2555 norlevopharnol
Hydroxymorphinan, a narcotic analgesic
agent.
f norlévopharnol
d Norlevopharnol
i norlevofarnol
e norlevofarnol

2556 normal saline
0.9% sterile solution of sodium chloride
in water. It is isotonic (of equal tension) with blood and used for injection
purposes.
f soluté physiologique; sérum physiologique
d physiologische Kochsalzlösung
i soluzione isotonica di cloruro di sodio;
siero artificiale
e suero artificial; solución fisiológica

2557 normethadone
Antitussive agent marked by a central
mechanism of action. Also a sympathomimetic agent acting as spasmolytic and
bronchodilator and indicated in the treatment of spasmodic cough and bronchitis.
Bradypnoea and reduction of expectoration may occur as side effects.
f norméthadone
d Normethadon
i normetadone
e normetadona

2558 normethandrone
Sex hormone indicated in the treatment
of amenorrhoea, dysmenorrhoea, premenstrual and intermenstrual syndrome and functional menometrorrhagia. Oedema, gynecomastia gastro-intestinal disorders and
leucorrhoea may occur as side effects.
f norméthandrone
d Normethandron
i normetandrolone; normetandrone
e normetandrolona

2559 nornicotine
Alkaloid ($C_9H_{12}N_2$) obtainable by de-
methylation of nicotine and present in
several varieties of tobacco. It is a
viscous, hygroscopic liquid marked by
a slight odour of amine. It is used as
an insecticide and is generally preferred
to nicotine as it is less volatile and
stabler.
f nornicotine
d Nornikotin
i nornicotina
e nornicotina

2560 norpipanone
Analgesic and antipyretic agent chiefly
used to allay rheumatic pains.
f norpipanone

d Norpipanon
i norpipanone
e norpipanona

* **nortriptyline   s. desmethylamitriptyline**

2561 noscapine
Alkaloid found in opium (isoquinoline
series) and used as an antitussive agent
marked by a central mechanism of action.
At normal doses, no modification of respiration, cardiac activity or intestinal
motility is caused. It is generally used
in the treatment of infections or inflammation of the upper respiratory tract.
Allergic reactions and a mild psychic
excitement may occur as side effects.
f noscapine
d Noscapin
i noscapina
e noscapina

2562 novobiocin
Bacteriostatic and bactericide antibiotic
characterized by an action which is similar to that of penicillin and erythromycin and used for infections by penicillin-resistant staphylococci. It is also
useful in the treatment of pertussis.
Allergic reactions, agranulocytosis and
thrombocytopenia, as well as a yellowish
coloration of the skin are possible side
effects.
f novobiocine
d Novobiocin
i novobiocina
e novobiocina

* **novocaine   s. procaine**

2563 noxiptyline
Tricyclic antidepressant having stabilizing effects on depressed-mood tone. Relief is observed 5 to 14 days from the
start of the treatment. The drug is indicated in endogenous involutive or reactive depression and in psychosomatic
disorders. See imipramine for side effects.
f noxiptyline
d Noxiptylin
i nossiptilina
e noxiptilina

2564 noxythiolin
Antifungal agent mainly used in the
treatment of vesical and vaginal infections caused by sensitive microorganisms,
e.g. Pseudomonas. It is also indicated
to combat mycetes and protozoa.
f noxythioline
d Noxythiolin
i noxitiolina
e noxitiolina

2565 nuclease
Enzyme capable of hydrolyzing a molecule. Nucleases can be specific for DNA
or RNA and for both single-stranded and
double-stranded molecules. Nucleases are

present in all cells and those that are
found in the intestinal lumen are of
pancreatic origin.
f nucléase
d Nuklease
i nucleasi
e nucleasa

2566 nucleic acids
Complex organic acids that are obtained
by precipitation from the nucleoproteins.
They consist of smaller units, the nucleotides, formed by a nitrogenous base,
a pentose and a molecule of phosphoric
acid. Their importance is fundamental
for the life of the cell.
f acides nucléiques
d Nukleinsäure; Kernsäure
i acidi nucleici
e ácidos nucleínicos

2567 nuclein
Intermediate product in the hydrolysis
of nucleoprotein. Protein and nucleic
acid are obtained by further hydrolysis.
In biochemistry, any of the various substances which are obtained from cell
nuclei after gastric digestion.
f nucléine
d Nuklein
i nucleina
e nucleina

2568 nucleoside
Glucoside resulting from the elimination
of the phosphoric acid from a nucleotide
and therefore appearing as the result
of the union of a pentose with a purinic
or pyrimidinic base. Nucleosides occur
in the organism as intermediate products
of the metabolism of the nucleic acid.
Some species of moulds and fungi are
capable of synthesizing nucleosides with
antibiotic activity.
f nucléoside
d Nukleosid
i nucleoside
e nucleósido

2569 nutmeg
The seed of Myristica fragrans of the
Myristaceae family, a large evergreen
tree growing in the Molucca islands, in
the Philippines, in Brazil etc. It is
used, like its fleshy aril (seed-coat)
which envelops it, to produce a volatile
oil principally consisting of terpenes
and myristicin, a solid fat and amylo-
dextrin for its stimulant and carminative
properties. Nutmeg may prove highly
toxic causing convulsions due to the
myristicin.
f noix muscade
d Muskatnuss
i noce moscata
e moscada; nuez moscada

*** nutmeg oil   s. myristica oil**

2570 nux vomica
The ripe dried seeds of Strychnos nux-
-vomica, a tree widely distributed in
the tropical regions of Asia. Two very
poisonous alkaloids can be obtained from
them, strychnine and brucine, used in
medicine for the preparation of drugs
principally administered in nervous diseases.
f noix vomique
d Brechnuss
i noce vomica
e nuez vómica

2571 nystatin
Antibiotic used in the treatment of Candida infections affecting skin and mucous membranes, in particular the mouth,
the alimentary tract and the vagina.
Also used in the prophylaxis and treatment of mycotic infections caused by
antibacterial and antibiotic agents. The
drug is not easily absorbed when taken
orally. In the latter case, nausea and
vomiting may occur.
f nystatine
d Nystatin
i nistatina
e nistatina

O

2572 oak bark
The dried bark of the smaller branches and young stems of the English oak which are collected in the spring from the trees. Quercitannic acid (amorphous phlobatannin oxidized to the oak red) and gallic acid are found in it. Medicinally, it is used as an astringent, either as a wash or an injection, in the treatment of haemorrhoids and leucorrhoea.
f écorce de chêne blanc
d Eichenrinde; Eichenlohe
i corteccia di quercia bianca
e corteza de roble blanco; corteza de encina

2573 oak moss
Or oakmoss. A lichen growing on various trees and taking up the shape of a small bush of a length of 15 to 20 cm. It is generally used in perfumery and in the manufacture of some medicinal drugs.
f mousse de chêne; beurre de mousse
d Eichenmoss
i musco quercino
e musgo de encina

2574 oat starch
Starch obtained from the oat seeds. It has a macromolecular structure constituted by several molecules of D-alpha (+) glucose bound together as in maltose.
f amidon d'avoine; farine d'avoine
d Haferstärke
i amido di avena
e almidón de avena; fécula de avena

2575 obidoxime chloride
Specific antidote to combat poisoning caused by alkylated phosphates of the parathion and malathion type.
f chlorure d'obidoxime
d Obidoximchlorid
i obidossima cloruro
e cloruro de obidoxima

2576 obstupefacient
Narcotic, soporific. Relating to any agent characterized by a stupefying effect on human beings.
f stupéfiant; obstupéfiant
d Betäubungsmittel; Obnarkoticum
i stupefacente; narcotico
e estupefaciente

2577 octacaine
Topical anaesthetic agent used for infiltration and conduction anaesthesia. It has a prolonged effect and little toxicity. It is frequently administered in association with a vasoconstrictor, e.g. adrenaline.
f octacaïne
d Oktacain
i ottacaina
e octacaina

2578 octamylamine
Spasmolytic and analgesic agent indicated in the treatment of spastic conditions of the digestive urinary, respiratory and genital tracts.
f octamylamine
d Oktamylamin
i ottamilammina
e octamilamina

2579 octatropine methylbromide
Atropine-resembling parasympatholytic agent used as anticholinergic and antispasmodic for the gastro-intestinal and biliary tracts. It is also indicated in the treatment of cardiospasms, pylorospasms and peptic ulcer. The drug is frequently used as it does not affect the coronary circulation and blood pressure. Tachycardia has been observed as a side effect.
f méthylbromure d'octatropine
d Oktatropinmethylbromid
i octatropina metilbromuro
e bromuro de octatropina

2580 octaverin
Therapeutic agent marked by spasmolytic properties which are similar to those of the alkaloids of belladonna.
f octavérine
d Oktaverin
i octaverina
e octaverina

2581 octopamine
Sympathicomimetic agent resembling noradrenaline in its effects and used therapeutically owing to its capacity to increase blood pressure. Its action is not as potent as that of adrenaline but longer lasting. It is indicated whenever a reduction of arterial pressure is observed, i.e. hypotension and vascular lability marked by a tendency to the

latter. Also used to alleviate conditions which are typical of convalescence.

f octopamine
d Oktopamin
i ottopamina
e octopamina

2582 octostanol

3-N-octilenolic dihydrotestosterone ether. Oenolic testosterone derivative practically without androgenic action and indicated as antiblastic (capable of slowing down bacterial growth) agent in the treatment of tumours due to hormonal diseases. Also used in the palliative treatment of mammary cancer.

f octostanol
d Oktostanol
i ottostanolo
e octostanolo

2583 octotiamine

Thioctothiamine. Vitamin B1 used in the treatment of the specific vitaminic deficiency.

f octotiamine
d Oktothiamine
i ottotiamina
e octotiamina

2584 octoxynol

Nonionic surfactant used as a spermicidal cream, i.e. a preparation capable of destroying spermatozoa.

f octoxynol
d Octoxynol
i octoxinol
e octoxinol

2585 oestradiol

Natural oestrogenic hormone (a steroid hormone) present in the urine of pregnant women. In small percentages, it is prepared by reduction of the oestrone. It derives from oestrane and contains, in addition to the three double bonds, two oxydrylic groups. The isomer with the oxydryl in trans- position is capable of inhibiting the function of the pituitary gland and that with the oxydryl in cis- position is used therapeutically as oestrogen, i.e. one of the five major classes of steroid hormones required for the development of female secondary sex characters.

f oestradiol; estradiol
d Östradiol
i estradiolo
e estradiol

2586 oestriol

One of the natural oestrogenic hormones found in the placenta and urine of pregnant women. Its activity is ten times less than that of oestrone, but its action more lasting and therefore its use suitable in particular pathological conditions. It is principally used in cases of oestrogen deficiency.

f oestriol; estriol

d Östriol
i estriolo
e estriol

2587 oestriol succinate

Oestrogen marked by activity inhibiting fibrolysis and protecting the capillary wall. It is indicated in the prophylaxis and the treatment of haemorrhages, in haemorrhagic diathesis and in primary and secondary thrombocytopenia.

f succinate d'estriol; succinate d'oestriol
d Östriolsuccinat
i estriolo succinato
e succinato de estriol

2588 oestrione

Oestrogenic hormone marked by oestrogenous activity and indicated in the treatment of ovarian insufficiency, primary or secondary amenorrhoea, puberal and menopausal disorders, lactation inhibition, senile vaginitis and kraurosis vulvae (distrophy of the epidermis and mucosa of the external feminine genitalia). At higher doses, the drug is also used in the palliative treatment of carcinomata affecting prostata or breast.

f estrone
d Östron
i estrone
e estron

2589 oil cake

The solid residue which is left after the extraction of most of the oil from various seeds (cotton, flax, soya beans etc.) and is subsequently ground and used as an important alimentary substance owing to its richness in proteins.

f tourteau
d Ölküchen; Pressküchen
i panello
e torta de borujo

2590 oil palm

Plant of the genus Elaeis yielding the fractionated palm kernel oil, a fat obtained by expression and used mainly as an ointment base.

f palme à huile
d Ölpalme
i palma oleifera
e palma oleífera; palma de aceite

2591 oily grain

Sesame seed. The small inversely ovate flat seed of sesame yielding a pale yellow semidrying fat used as a vehicle for several pharmaceutical preparations, in particular ointments and liniments.

f grain de sésame
d Sesamsamen
i seme di sesamo
e semilla de sésamo

2592 ointment

Generally, salve or unguent for application to the skin. Medicinally, a semisolid preparation with a base of fatty

or greasy material, e.g. petrolatum.
f onguent; pommade
d Salbe
i unguento; pomata
e unguento; pomada

2593 ointment base
The base of the preparation used as an
ointment. It may be fatty or a water-in-
-oil emulsion where absorption is indicat
ed or just soft paraffin where simple
protection is wanted.
f excipient pour pommades
d Salbengrundlage
i eccipiente per pomate
e excipiente para pomadas

2594 oleander
Or rosebay tree. Small evergreen tree or
shrub of the Apocinaceae family. Its
leaves and to a lesser degree its bark
and fruit contain glycosides which display a mild cardiokinetic action. However, owing to their toxicity, use is
very limited.
f oléandre; rosage; laurelle
d Oleander; Rosenlorbeer
i oleandro; leandro
e baladre; adelfa

2595 oleandomycin
Antibiotic isolated by B.A. Sabin et al.
active against most gram-positive and
some gram-negative bacteria, as well as
against some Rickettsiae, a few large
viruses (psicattosis and ornithosis) and
a protozoon (Entamoeba hystolitica).
Cross immunization between oleandomycin
and most of the other more widely known
antibiotics has not been shown.
f oléandomycine
d Oleandomycin
i oleandomicina
e oleandomicina

2596 oleandrin
Glycoside present in the leaves, bark
and unripe fruit of oleander; a white
crystalline powder, very bitter and insoluble in water, displaying diuretic
and cardiokinetic properties similar to
those of digitalin. It yields gitoxigenin
(a physiologically-active cardiac glycoside) on hydrolysis.
f oléandrine
d Oleandrin
i oleandrina
e oleandrina

* oleanol   s. oleanolic acid

2597 oleanolic acid
Terpenic hydroxy acid derived from a
pentacyclic triterpene containing a
hydroxylic group and a carboxylic group
and a double bond (without or almost
without reactivity). It is found in the
leaves of the olive tree, in the skin of
grapes, as saponine in the beetroot etc.
It is used in some pharmaceutical pre-

parations.
f acide oléanolique
d Oleanolsäure
i acido oleanolico
e ácido oleanólico

2598 oleate
Salt or ester of the oleic acid. Also,
a compound obtained by interaction between the oxide (or salt) of a metal and
olive oil. Copper oleate is used for its
fungicide properties and zinc oleate is
used in ointments.
f oléate
d Oleat
i oleato
e oleato

2599 oleic acid
Or oleinic acid. The fatty acid most
widely occurring in nature as it is
found in the form of glycerides in all
animal fats and in all vegetable oils,
especially olive and almond oil, as
olein, i.e. the glycerol ester. It is
insoluble in water but easily soluble in
ether, alcohol, chloroform and essential
oils. It is used for the manufacture of
some pharmaceutical preparations.
f acide oléique
d Ölsäure
i acido oleico
e ácido oleínico

2600 olein
The glyceride of the oleic acid. An oily
colourless liquid insoluble in water. It
is the principal constituent of the olive
oil and of many other animal and vegetable fats. Olein is particularly important in biochemistry.
f oléine
d Olein
i oleina
e oleina

2601 oleum iodatum
Iodized oil. Radioopaque contrast medium
used in uterosalpingography, bronchography, sialography (visualization of
the salivary glands) as well as visualization of fistulous passages. Allergic
reactions may occur as side effect.
f huile iodée
d Jodöl
i olio iodato
e aceite yodado

* oleum ricini   s. castor oil

2602 oligocyanidine
Therapeutic agent capable of reducing
the permeability of the capillaries with
antioedematous and antihaemorrhagic effects. It is indicated in the treatment
of exudative retinitis and of traumas of
the fundus oculi. Also used in contusion
-caused detachment of the retina.
f oligocyanidine
d Oligocyanidin

i oligocianidina
e oligocianidina

2603 oligomycin
Antibiotic produced from cultures of actinomyces marked by a selective fungicide property. It inhibits oxidative phos phorylation by its action on the mitochondrial ATPase system. The drug is indicated in the treatment of dermatitis.
f oligomycine
d Oligomycin
i oligomicina
e oligomicina

2604 oligosaccharide
Any of the saccharides containing a small number of constituent monosaccharide units, in particular disaccharides, trisaccharides and tetrasaccharides.
f oligosaccharide; oligoside
d Oligosacharid; Oligosaccharide
i oligosaccaride
e oligosacarido; oligósido

2605 olive kernel oil
Fatty substance contained in the seeds of olive; a yellowish-greenish liquid of a sweetish taste constituted by the glycerides of the oleic, palmitic and stearic acid. Generally, as the olives are pressed without a previous separation of the kernels, the kernel oil is mixed with that obtained from the pulp of the fruit, with which it forms the olive oil that is so called in commerce.
f huile de noyaux d'olive
d Olivenkernöl
i olio di noccioli di olive
e aceite de huesos de oliva

2606 olive oil
The fixed oil which is obtained by expression from the ripe fruit of the olive (Olea europaea). A pale yellow liquid marked by a vaguely sweetish taste. It consists of glycerides, chiefly of oleic acids and small percentages of palmitic, linoleic, stearic and myristic acids. It has demulcent and mildly laxative but above all very nutritious properties and is used in medicine in the treatment of gastric and duodenal ulcers in order to inhibit secretion of gastric juice. Also used externally as an emollient to protect inflamed surfaces.
f huile d'olive
d Olivenöl
i olio di oliva
e aceite de oliva

2607 olivomycine
Antineoplastic antibiotic used in the treatment of metastasized testicular tumours, tonsillar tumours both in situ and metastasized uterine chorioepithelioma, melanomas etc.
f olivomycine
d Olivomycin

i olivomicina
e olivomicina

2608 opianic acid
Carboxylic acid of the aromatic series; crystalline, colourless, water- and alcohol-soluble solid which is obtained by the splitting of narcotine. Also obtained from hydrastine by oxidation.
f acide opianique
d Opiansäure
i acido oppianico
e ácido opiánico

2609 opipramol
Tricyclic antidepressant agent indicated in the treatment of psychosomatic diseases in general, particularly conditions of anxiety and asthenia as it acts on cardiovascular, respiratory and endocrine neuroses. The drug is also capable of stabilizing the rate of cerebral serotonin and histamine. Considerable dryness of the mouth, drowsiness and hypotension may occur as side effects.
f opipramol
d Opipramol
i opipramolo
e opipramolo

2610 opium
The condensed juice obtained from the capsules of Papaver somniferum, a herbaceous plant widely cultivated in China, Japan, India and many other countries of Asia, Europe and the Americas. Owing to its high content of alkaloids, opium, in therapeutic doses, has an antispastic, analgesic, hypnotic and sedative action, the latter by reaching the respiratory and cough centres. Although it is still used, either by itself or in association with other drugs, e.g. Sydenham laudanum, Dower powder etc., it has been mostly replaced by morphine. Excessive doses may cause disorders of a varied entity which may affect the cardio-circulatory, respiratory and nervous functions and lead to coma and sometimes death.
f opium
d Opium; Meconium
i oppio
e opio

2611 opium extract
The condensed juice which is expressed from the opium capsules and used for various pharmacological preparations.
f extrait d'opium; extrait thébaïque
d Opiumextrakt
i estratto d'oppio
e extracto de opio

2612 opium powder
Or powdered opium. Dried and powdered opium adjusted to contain 10% anhydrous morphine by using a suitable diluent, e.g. lactose.
f poudre d'opium

d Opiumpulver
i polvere di oppio
e polvo de opio

2613 opium tincture
Laudanum; alcoholic extract of opium containing 1% weight in volume of anhydrous morphine, a suitable preparation for the administration of opium alkaloids. The camphorated tincture of opium (paregoric) contains 5% of tincture of opium representing 0.05% morphine, benzoic acid, camphor and anise oil and is used for the preparation of cough mixtures.

f teinture d'opium
d Opiumtinktur
i tintura di oppio
e tintura de opio

2614 opium tincture with saffron
Sydenham laudanum. Either of two opium preparations: a tincture of opium prepared with saffron or a solution of opium in aromatized sherry or diluted alcohol having the action and strength of ordinary laudanum.

f teinture d'opium safranée; laudanum de Sydenham
d safranhaltige Opiumtinktur; Laudanum Sydenham
i tintura di oppio crocata; laudano del Sydenham
e tintura de opio azafranada; láudano de Sydenham

2615 opium vinegar
Extract of opium, sugar and nutmeg prepared by maceration and percolation with dilute acetic acid. It was used in the past for the administration of opium.

f vinaigre d'opium
d Opiumessig
i aceto di oppio
e vinagre de opio

2616 opobalsam
Tolu balsam. Yellowish substance obtained from the trunk of Myroxylon balsamum having antiseptic and expectorant properties.

f baume de Tolu
d Tolubalsam
i balsamo di Tolu; opobalsamo di Tolu
e bálsamo de Tolú

2617 opopanax gum
Or opopanax resin. Odorous gum resin used medicinally in the past.

f opopanax; panais sauvage
d Opopanaxgummi; Heilwurzsaft
i opoponaco
e opopónaco

2618 opopanax oil
The essential oil which is obtained by distillation from the opopanax, and is used mostly in perfumery and, albeit rarely, in pharmacology.

f essence d'opopanax

d Opopanaxöl
i essenza di opoponaco; olio di opoponaco
e esencia de opopónaco

2619 orange
The edible fruit of the orange tree. The fresh peel of the bitter orange is used in medicine as a flavouring agent and for its considerable carminative properties. It contains an essential oil, the bitter glycoside aurantiamarin, hesperidin and vitamins A, B and C, though in minute quantities.

f orange
d Orange; Pomeranze; Apfelsine
i arancia
e naranja

2620 orange peel oil
The essential oil which is obtained from the orange peel, i.e. the bitter glycoside aurantiamarin and the glycoside hesperidin.

f essence d'écorce d'oranges
d Apfelsinenschalenöl
i essenza di arancia
e esencia de corteza de naranja

2621 orazamide
Precursor of purine and pyrimidine bases having a balancing effect on the proteic metabolism in various diseases of the liver and an antisteatogenous (active against fatty degeneration) and regenerating action of the hepatic parenchyma. It is therefore indicated in the treatment of chronic and acute hepatitis, cirrhosis of the liver and icterus.

f orazamide
d Orazamid
i orazamide
e orazamida

2622 orciprenaline
Beta-adrenoreceptor agonist indicated in the treatment of bronchial asthma owing to its capacity of favouring cardiac and respiratory activity. The drug is also used in some cardiac diseases due to defective conduction. Tachycardia, arrhythmia and tremor may occur as side effects.

f orciprénaline
d Orciprenalin
i orciprenalina
e orciprenalina

2623 orgotein
Group of soluble metalloproteins isolated from liver, red blood cells and other mammalian tissues. Antiinflammatory agent mainly indicated for the treatment of phlogistic oedemata, inflammation of the intestinal and urinary tracts caused by radiation therapy.

f orgotéine
d Orgotein
i orgoteina
e orgoteina

2624 origanum oil
Oily liquid of a yellow colour and a strongly aromatic odour, which contains phenols, e.g. thymol, sesquiterpenes and ethers. It is used pharmaceutically in various liniments.
f essence d'origan
d Origanumöl; Dostenöl
i essenza di origano; essenza di rigamo
e esencia de orégano

2625 ornidazole
Antiprotozoal agent derived from the 5 nitro-imidazole and quite effective against Trichomonas vaginalis, Entamoeba histolytica and Giardia intestinalis (a very common parasite of the small intestine). It is therefore used in the treatment of trichomoniasis, amoebiasis and gardiasis. Gastro-intestinal disorders, headaches and sometimes dizziness may occur as side effects.
f ornidazole
d Ornidazol
i ornidazolo
e ornidazolo

2626 ornipressine
Vasoconstrictor characterized by a specific action on veins and capillaries and used in the treatment of haemorrhages in surgery, gynecology and urology. Also indicated in otorhinolaryngology and in collapse therapy. Nausea, sometimes accompanied by vomiting, and abdominal pains may occur as side effects.
f ornipressine
d Ornipressin
i ornipressina
e ornipresina

2627 ornithine
Amino acid present in the proteins combined with guanidine so as to form the arginine. It is released by the proteic molecule through processes of putrefaction which split the arginine in urea and ornithine.
f ornithine
d Ornithin
i ornitina
e ornitina

2628 ornithine acetoglutarate
Ammoniaemia-antagonist detoxicant agent indicated in the treatment of acute, subacute and chronic hyperazotaemia, cirrhosis accompanied by severe liver disorders, hepatic coma and, more generally, the prevention of diseases caused by liver-affecting drugs.
f acétoglutarate d'ornithine
d Ornithinazetoglutarat
i ornitina aceglutato; ornitina acetoglutarato
e acetoglutato de ornitina

2629 ornithine carbamoylaspartate
Therapeutic agent capable of reducing

azotaemia and of protecting the liver. Indicated in the treatment of hyperazotaemia caused by encephalopathy and cirrhosis.
f carbamylaspartate de L-ornithine
d Ornithin-Carbamoylaspartat
i ornitina carbamilaspartato
e carbamilaspartato de L-ornitina

2630 ornithine transcarbamoylase
Enzyme capable of catalyzing a reaction whereby a carbamoyl group is transferred to ornithine to form citrulline, a crystalline amino acid formed as an intermediate in the conversion of ornithine into arginine.
f ornithine transcarbamoylase
d Ornithintranscarbamoylase
i ornitina transcarbamoilasi
e ornitina transcarbamoilasa

2631 orotic acid
Crystalline acid first found in milk. It is a growth factor for several micro-organisms and a precursor of the pyrimidinic nucleotides. In medicine, orotic acid is a therapeutic agent used as a choleretic, a hypocholesterolaemic and antisteatogenous and lithaemia-reducing drug in the treatment of hyperuricaemia, hyperlipaemia, hepatitis, cirrhosis and steatosis.
f acide orotique
d Orotsäure
i acido orotico
e ácido orótico

2632 orphenadrine
Parasympatholytic-antihistamine produced in the form of citrate and hydrochloride. The dihydrogen citrate of orphenadrine is chiefly used to relieve rigidity in Parkinsonism and the depression which often accompanies the advanced stage of this disease. The hydrochloride of orphenadrine has more or less the same action as the citrate but is considerably more soluble in water. Side effects are the same as those of atropine. In addition, some cases of aplastic anaemia and of difficulty to expel urine have been observed.
f orphénadrine
d Orphenadrin
i orfenadrina
e orfenadrina

2633 orris oil
Semisolid aromatic substance obtained from the roots of the orris (iris) by steam distillation. It consists mainly of myristic acid and irone (an odorous substance) and is used as a base for scents.
f beurre de violette; camphre d'iris
d Irisbutter; Veilchenwurzelöl
i olio d'iride; olio di giglio
e esencia de lirio

2634 orsellic acid

Or orsellinic acid. Crystalline aromatic
oxyacid contained in the form of depside
(one of the gallotannins) in several lichens. It is a solid, crystalline, colourless substance obtained by hydrolysis
of the lecanoric acid (a lichen acid).
f acide orsellique
d Orcellasäure
i acido orsellico
e ácido orcélico

2635 orthanilic acid
Sulphonic acid of aniline; crystals obtained by reduction of the o-nitrobenzen-
sulphonic acid and used in the preparation of some azoic dyes.
f acide orthanilique
d Orthanilsäure
i acido ortanilico
e ácido ortanílico

2636 orthocaine
Colourless, tasteless white crystalline
powder soluble in alcohol and ether,
less soluble in water. It is used as a
topical analgesic, preferably in the form
of powder owing to its low solubility.
f orthocaïne
d Orthocain
i ortocaina
e ortocaina

2637 orthoformic acid
Acid, $CH(OH)_2$, which is not known in
the free state but only in the form of
ester, i.e. the orthoformate. Also called
trihydroxymethane.
f acide orthoformique
d Trihydroxymethan
i acido ortoformico; triidrossimetano
e ácido ortofórmico; triidroximetano

2638 oryzanol
The active principle extracted from
Oryza sativa. Metabolic agent supposed
to activate the function of the hypothalamus and used in the treatment of
neurovegetative and endocrine disorders.
f oryzanol
d Oryzanol
i orizanol
e orizanol

2639 osalmid
Salicylic derivative marked by a choleretic action aiming at stimulating the
secretory activity of the liver by mobilizing (biliary elimination) cholesterol and
preventing, albeit indirectly, arteriosclerosis. It is indicated in the treatment of mild hepatic insufficiency characterized by cutaneous reactions and hepatic congestion accompanied by gastro-
intestinal disorders.
f osalmide
d Osalmid
i osalmide
e osalmida

2640 osmic acid

Hypothetical dibasic acid forming the
salts, i.e. the osmates. The term is
generally used to denote osmium tetroxide which is used for the fixation and
staining of fats in histological esperi-
ments.
f acide osmique
d Osmiumsäure
i acido osmico
e ácido ósmico

2641 osmium tetroxide
The most important of the osmium oxides,
obtained by treating osmium in its metallic state or in any of its combinations. It is used in biology for the
fixation of tissues in microscopic preparations as it is capable of staining fats.
f tétroxyde d'osmium
d Osmiumtetroxid
i tetrossido di osmio
e tetróxido de osmio

2642 ossis extractum
Preparation formed by organic and inorganic components of young animals. Opotherapeutic osteotrophic agent capable
of promoting osteogenesis. It is indicated in calcium deficiency and as an adjuvant in osteopathies, osteoporosis and
fractures. It has also been employed as
a prophylaxis for dental caries.
f extrait d'os
d Knochenextrakt
i estratto di ossa
e extracto de huesos

2643 otilonium bromide
Spasmolytic agent marked by a direct
action on the tonus and contractility of
the smooth muscles, in particular the
gastro-intestinal muscles. It is chiefly
used in the treatment of spastic conditions and functional dyskinesia (the
impairment of voluntary motion) of the
gastro-intestinal apparatus, e.g. gastritis, colitis enteritis and gastroduodenitis.
f bromure d'otilonium
d Otiloniumbromid
i ottilonio bromuro
e bromuro de otilonio

2644 ouabain
Strophantinum G. Highly toxic crystalline
steroidal glycoside obtained from the
seeds of an African shrub or from the
wood of trees of the genus Acocanthera.
It yields rhamnose ouabagenin (a cardiac hexahydroxy steroidal lactone) on
hydrolysis and is used similarly to digitalis.
f ouabaïne
d Ouabain
i ouabaina; ubaina
e ouabaina

2645 ouabaio
The wood or root of the tree Acocanthera
schimperi from which a crystalline glycoside is obtained, which is identical with

strophantin, a drug marked by actions which are similar to those of digitalis.

f ouabaio
d Ouabaio
i ouabaio; uabaio
e ouabaio

2646 ovalbumin
Water-soluble protein present in the white of the egg. It contains a well balanced quantity of essential amino acids and has therefore a high nutrient capacity. It also contains phosphorus and a small quantity of carbohydrates.

f ovalbumine; ovoalbumine
d Ovalbumin; Eialbumin
i ovalbumina
e ovoalbúmina; albúmina de huevo

2647 ovarii extractum
Preparation obtained from the ovarian tissue of healthy animals, containing all the active principles which are e-laborated by the gland itself. It is an ovarian opotherapeutic agent capable of stimulating the ovarian function and re-gulating all phenomena related to the female genital system. The drug is thus indicated in the treatment of ovarian infantilism, puberty and menopausal disorders, amenorrhoea, oligomenorrhoea and dysmenorrhoea as well as various disorders due to ovarian insufficiency or to dysfunction of ovary-related glands.

f extrait d'ovaire
d Eierstockextrakt
i estratto d'ovaia; ovaia estratto
e extracto de ovaio

2648 oxabolone cypionate
Anabolic steroid capable of stimulating the protein biosynthesis and of reducing the nitrogen consumption. It is indicated in the treatment of growth disorders marked by loss of weight, convalescence and diseases which are due to hepato-toxic drugs.

f cypionate d'oxabolone
d Oxaboloncypionat
i ossabolone cipionato
e cipionato de oxabolona

2649 oxacillin
Semisynthetic penicillin capable of resist ance to acids and penicillase. It is principally active against penicillin G--resistant pneumococci, staphylococci and streptococci. The haematic concentration to a peak level is attained after half or one hour. Allergic reactions may oc-cur as side effect.

f oxacilline
d Oxacillin
i ossacillina
e oxacilina

2650 oxaflumazine
Phenothiazine-derived neuroleptic drug with a spasmolytic, analgesic and mildly antihistaminic action. It is indicated in the treatment of some acute psychoses, e.g. paranoid and manic-depressive psychosis. Also schizophrenia and chron-ic hallucinatory psychoses may be treat-ed with this drug. It should not be used in the presence of parkinsonism, multiple sclerosis and epilepsy. Symp-toms of extrapyramidal dysfunction may occur as side effects.

f oxaflumazine
d Oxaflumazin
i ossaflumazina
e oxaflumazina

2651 oxalacetic acid
Bicarboxylic acid marked by two pos-sible tautomeric forms, a ketonic one and an enolic one. The latter is pre-valent and appears in two isomeric forms, i.e. cis- and trans-. Oxalacetic acid is an intermediate compound formed by reversible oxidation of malic acid, as in the metabolism of fats and carbo-hydrates, and in reversible transamina-tion reactions, as from aspartic acid. It must be noted that for each molecule of oxalacetic acid utilized in the reac-tion another one is regenerated in the course of the Krebs cycle. The acid also functions as a hydrogen transporter in oxido-reduction processes and promotes the development (biotinic action) of cer-tain strains of bacteria.

f acide oxalacétique
d Oxalessigsäure; Oxalazetat
i acido ossalacetico
e ácido oxalacético

2652 oxalate
Salt or ester of the oxalic acid, q.v.

f oxalate
d Oxalat
i ossalato
e oxalato

2653 oxalic acid
Bicarboxylic aliphatic acid widely diffus ed in algae, fungi, ferns etc. in the form of salts of calcium, potassium and magnesium. In the human organism, the acid is found in small quantities as a normal constituent of the blood, in which it represents a product of the intermediate metabolism participating in the Krebs cycle, of the urine, bile and faeces. It is present in greater quanti-ties in some pathological conditions, e.g. oxalic diathesis, nephrolithiasis etc.

f acide oxalique
d Oxalsäure
i acido ossalico
e ácido oxálico

2654 oxalosuccinic acid
Organic acid intervening as an interme-diate stage in the metabolism of fats and carbohydrates (the Krebs cycle).

f acide oxalsuccinique
d Oxalbernsteinsäure

ı acido ossalsuccinico
e ácido oxalsuccínico

2655 oxalylurea
Parabanic acid. Derivative of urea, forming salts which are easily converted into those of oxaluric acid, an oxalic monoureide appearing in normal human urine.
f oxalylurée; acide parabanique
d Parabansäure
ı ossalilurea; acido parabanico
e oxalilurea; ácido parabánico

2656 oxamate
Term denoting the salts or esters of oxamic acid. The esters are prepared by reaction of equimolecular quantities of ammonia and an acid ester of the oxalic acid.
f oxamate
d Oxamat
ı ossamato
e oxamato

2657 oxametacin
Non-steroid, antıinflammatory, analgesic and antipyretic agent marked by a general and topical high degree of tolerance. It is chiefly indicated in the treatment of rheumatoid arthritis, osteoarthritis and osteoarthrosis when there is inflammation all over the area, periarthritis, tenosynovitis, tendinitis and rheumatic myositis. Gastro-intestinal disorders in a mild form may occur as side effect.
f oxamétacine
d Oxametacin
ı ossametacina
e oxametacina

2658 oxamic acid
Monoamide of the oxalic acid; a crystalline colourless solid, insoluble in water, not easily soluble in alcohol, used in organic syntheses.
f acide oxamique
d Oxaminsäure
ı acido ossamico
e ácido oxámico

2659 oxamide
Diamide of the oxalic acid; white, needle-shaped crystals insoluble in water and mildly soluble in alcohol, which are obtained by saponification of the cyanogen or by heating the ammonium oxalate. Together with the oxalic and oxamic acids it is a terminal product of the oxidative decomposition of proteins.
f oxamide
d Oxamid
ı ossamide
e oxamida

2660 oxamniquine
Anthelminthic, anti-infective agent used in the treatment of schistosomiasis caus-ed by S. mansoni. Drowsiness, headaches, vertigo and transient fever may occur as side effects.
f oxamniquine
d Oxamniquin
ı ossamnichina
e oxamniquina

* **oxamphetamine   s. hydroxyamphetamine**

2661 oxandrolone
Anabolic agent particularly useful in the therapeutic treatment of psycho-physical debility, convalescence and those conditions, especially a lack of proteins, which may follow a surgical operation. Equally useful to balance excessive loss of weight due to some disease and, in geriatrics, to combat tissue atrophy.
f oxandrolone
d Oxandrolon
ı ossandrolone
e oxandrolona

2662 oxapium iodide
Spasmolytic and anticholinergic agent mainly used in the treatment of hyperkinesia (abnormally increased and purposeless, uncontrollable muscular movement) and painful spasms in chronic gastritis, gastroduodenal ulcer, colecysto pathies, urolithiasis etc. Constipation, sometimes diarrhoea may occur as side effects.
f iodure d'oxapium
d Oxapiumjodid
ı ossapio ioduro
e yoduro de oxapio

2663 oxatomide
Antihistamine capable of blocking histamine$_1$ receptors and the release of allergic mediators from mast cells. It is used in the treatment of various allergies but not asthma. Drowsiness and above all the potentiation of alcohol may occur as side effects.
f oxatomide
d Oxatomid
ı ossatomide
e oxatomida

2664 oxazepam
Anxiolytic tranquillizer of the benzodiazepine group chiefly indicated in the treatment of insomnia, cardiovascular, respiratory and gastro-intestinal disorders. The drug acts directly on the limbic system, thalamus and hypothalamus. In some instances, it has been used in the treatment of emotional crises and sexual neuroses. Drowsiness, ataxia, nausea, blurred vision etc. (see diazepam) may occur as side effects.
f oxazépam
d Oxazepam
ı ossazepam
e oxazepam

2665 oxazolam

Benzodiazepinic tranquillizer indicated in the treatment of emotional disorders, insomnia, neurosis and anxiety states. Drowsiness and occasionally ataxia may occur as side effects.

f oxazolam
d Oxazolam
i oxazolam
e oxazolam

2666 oxazoline
Heterocyclic compound, a dehydroxy derivative of oxazol. Three different types of it are possible but only the derivatives of the second one are known. Some of them are considered local anaesthetics.

f oxazoline
d Oxazolin
i ossazolina
e oxazolina

2667 ox bile extract
Organic extract indicated in the treatment of biliary deficiency. Although its use is still a matter of argument, there is no doubt that it increases bowel activity and thus relieves constipation.

f extrait de bile de boeuf; extrait de fiel de boeuf
d Oxengallenextrakt
i estratto di bile bovina; estratto di fiele bovino
e extracto de bilis de buey; extracto de hiel de toro

2668 oxedrine
Sympathomimetic agent with chiefly alpha-adrenergic effects. It is indicated in the treatment of orthostatic hypotension, cardiac adynamia, convalescence-related disorders, excessive stress and in the course of spinal anaesthesia.

f oxédrine
d Oxedrin
i ossedrina
e oxedrina

2669 oxedrine tartrate
Sympathomimetic agent marked by peripheral cardiovascular analeptic action. Though similar in its effects, oxedrine tartrate is less potent but more prolonged than epinephrine. It displays a constrictive activity on the arterioles and capillaries, as well as a positively chronotropic one. It is mainly indicated in the treatment of incompetence of the systemic circulation, of infectious diseases, senile heart, pathological conditions in the course of pregnancy etc. Arrhythmia and tachycardia may occur as side effects.

f tartrate d'oxédrine
d Oxedrintartrat
i ossedrina tartrato
e tartrato de oxedrina

2670 oxeladin
Antitussive agent acting on the cough centre of the central nervous system.

It is particularly useful in the treatment of spasmodic and dry cough. Though a sedative, it has no hypnotic action.

f oxéladine
d Oxeladin
i osseladina
e oxeladina

2671 oxerutine
Vitamin P factor with endothelium-protecting and antiphlogistic action on the capillaries, whose permeability to red-blood cells is thus reduced. It is mainly indicated, either by systemic or by topical route, in the treatment of pre-varicose conditions, chronic venous incompetence, varicose ulcers, haemorrhoids, periphlebitis, varicose ulcers and dermatides etc.

f oxérutine
d Oxerutin
i osserutina
e oxerutina

2672 oxethazaine
Or oxethacaine. Surface-active topical anaesthetic agent particularly indicated for a surface anaesthesia in the gastrointestinal region, for painful syndromes caused by ulcers in the stomach and duodenum, gastritis and colitis.

f oxétazaïne; oxétacaïne
d Oxethazain; Oxetacain
i ossetacaina
e oxetacaina

2673 oxetorone
Sedative, analgesic, antihistaminic, anti anaphylactic, antiemetic and vasodilator agent chiefly used in the treatment of headaches caused by vasomotor disorders. Drowsiness may occur as a side effect.

f oxétorone
d Oxetoron
i ossetorone
e oxetorona

2674 oxidized cellulose
Cellulosum oxydatum. Product obtained by a double process of regeneration and oxidation capable of attaining a uniform chemical and physical structure. Oxidized cellulose is a topical haemostatic agent capable of forming very rapidly an artificial clot when it comes into contact with blood at the site of a lesion. Haemorrhages can thus be checked in one to two minutes. It is completely absorbed by the tissues in less than seven days.

f cellulose oxydée; gaze oxydée
d Oxydationszellulose
i cellulosa ossidata; garza assorbibile
e celulosa oxidada; venda absorbable

2675 oxime
Any of the chemical compounds containing one or more =NOH groups. Oximes

are obtained by condensation of hydroxylamine with compounds including
the carboxylic group, i.e. aldehydes or
ketones.

f oxime
d Oxim
i ossima
e oxima

2676 oxitefonium bromide
Spasmolytic anticholinergic agent indicat
ed for the treatment of spastic crises in
the gastro-intestinal, urinary and biliary tracts. Also used to soothe broncho
spasms. Dryness of the mouth and my-
driasia may occur as side effects.

f bromure d'oxitéfonium
d Oxitefoniumbromid
i ossitefonio bromuro
e bromuro de oxitefonio

2677 oxolamine
Sedative mainly used in the treatment
of cough caused by infection in the upper respiratory tract. Indicated in acute
and chronic bronchitis. Gastro-intestinal
disorders may occur as side effect.

f oxolamine
d Oxolamin
i ossolamina
e oxolamina

2678 oxolinic acid
Antiproteus agent, i.e. an acid capable
of acting on proteus, a genus of the
Eubacteriales having four species, the
typical one of which is the Proteus vulgaris, one of the common bacteria of
putrefaction. One particular variety of
Proteus (Proteus X) has the property,
widely exploited for diagnostic purposes,
of being agglutinated by the serum of
patients suffering from eruptive Mediterranean (or Marseilles) fever.

f acide oxolinique
d Oxolinsäure
i acido ossolinico
e ácido oxolínico

2679 oxomemazine
Phenothiazinic antihistaminic agent mark
ed by sedative, moderatly analgesic and
antitussive action. It is indicated in the
treatment of acute and chronic allergic
conditions, e.g. itching dermatoses.

f oxomémazine
d Oxomemazin
i ossomemazina
e oxomemazina

2680 oxomethionine
Hydroxy amino acid, active methianine
metabolite. Its therapeutic action on the
liver is quite similar to that of methionine but considerably more soluble,
more easily tolerated and without coun-
terindications in the presence of some
dysfunction in the liver. Also indicated
in endogenous and exogenous toxicoses,
hepatitis and hepatonephrosis.

f oxométhionine
d Oxomethionin
i ossometionina
e oxometionina

2681 oxophenarsine
Logical successor of the arsenobenzene-
-related drugs for the treatment of syph
ilis and relapsing fever and preferably
administered orally as it is effective in
small doses.

f oxophénarsine
d Oxophenarsin
i ossofenarsine
e oxofenarsina

*** oxpentifylline   s. pentoxifylline**

2682 oxprenolol
Beta-adrenoreceptor blocking drug marked by a cardioanaleptic and cardioregu-
lating activity, as well as a myotropic
one. It is indicated in cases of arrhythmia characterized by supraventricular
and ventricular extrasystoles. Also used
as a bradycardiac agent in some anginoid conditions.

f oxprénolol
d Oxprenolol
i osprenololo; ossiprenololo
e oxprenololo

2683 oxybuprocaine
Local anaesthetic for topical use, active
on the sensitive nerve endings. It is
used in the form of solution in ophthalmology and in otorhinolaryngology in
case of painful lesions. It has roughly
the same systemic toxicity of tetracaine
(q.v.) but is less irritant when applied
to the conjunctiva. It also displays
bacteriostatic activity in the form of solution and has no adverse effect on the
pupil.

f oxybuprocaïne
d Oxybuprocain
i ossibuprocaina
e oxibuprocaina

2684 oxybutynin
Spasmolytic anticholinergic agent chiefly
used in the treatment of spasms in the
urinary tract, particularly vesical and
intestinal spasms. Urinary retention,
dryness of the mouth, reduced perspiration, tachycardia, palpitations, mydriasis etc. may occur as side effects.

f oxybutynine
d Oxybutynin
i ossibutinina
e oxibutinina

2685 oxychlorosene
Monoxychlorosene. Cytotoxic agent chiefly
used for the elimination of vital cancerous cells, especially when found in the
course of surgical operation. Also, as
a sodium salt, used for topical disinfection in surgery, urology and gynecology.

f oxychlorosène

d Oxychlorosen
i ossiclorosene
e oxicloroseno

2686 oxycinchophen
Antirheumatic, analgesic agent used in the treatment of chronic rheumatism. It has no effect in acute cases. Nausea and diarrhoea may occur as side effects.
f oxycinchophène
d Oxycinchophen
i ossicincofene
e oxicincofeno

2687 oxycodone
Codein derivative marked by antitussive properties acting through a central mechanism. In high doses it may act as narcotic. Addiction may be caused by a prolonged use.
f oxycodone
d Oxycodon
i ossicodone
e oxicodona

2688 oxycyanide
Compound of oxygen and cyanogen with an element or radical.
f oxycyanure
d Oxycianid
i ossicianuro
e oxicianuro

2689 oxygen
Colourless, odourless and tasteless gas, slightly soluble in water, constituting about one fifth of the atmosphere, half of the heart's crust and two thirds of animal and vegetable tissue. It is essential for every form of life. In medicine, it is chiefly used in hypoxaemia conditions due to inadequate pulmonary ventilation, impairment of gas transfer leading to low degree of oxygen saturation in the blood and consequent shortness of breath, chronic bronchitis, pulmonary oedema, circulatory incompetence, neonatal asphyxia, poisoning by carbon monoxide, excessive doses of barbiturates etc. High concentrations of oxygen may cause respiratory depression in patients suffering from inadequate ventilation and some damage to the pulmonary epithelium.
f oxygène
d Sauerstoff
i ossigeno
e oxígeno

2690 oxymesterone
Anabolic agent with very little, or no androgenic power. It stimulates protein biosynthesis and retains nitrogen. It is mainly used in the treatment of pathological loss of weight and osteoporosis. It does not cause hydrosaline retention. A mild virilization has been observed as side effect. The drug should not administered to children below fifteen.
f oxymestérone

d Oxymesteron
i ossimesterone
e oximesterona

2691 oxymetazoline
Sympathomimetic characterized by marked alpha-adrenergic effects and used on the nasal mucosa for the treatment of nasal congestion, rhinitis, acute sinusitis, laryngitis and pharyngitis.
f oxymétazoline
d Oxymetazoline
i ossimetazolina
e oximetazolina

2692 oxymetholone
Synthetic androgen marked by a potent anabolic action. It promotes protein biosynthesis and metabolic processes and is chiefly used for the treatment of protein inbalance, debility in the course of convalescence and, more generally, chronic debility in aged patients. Also employed to treat osteoporosis. A mild virilization has been observed as side effect.
f oxymétholone
d Oxymetholon
i ossimetolone
e oximetolona

2693 oxymorphone
Analgesic and narcotic agent marked by morphine-like activity. It is used to relieve mild to severe pains, in the preparation of the patient for surgical operations, and to allay post-operative and post-partum pains. Also employed for the relief of anxiety whenever dyspnoea, caused by pulmonary oedema and left ventricular failure, sets in. Drowsiness, nausea, itching, general malaise, delirium, headaches and myositis may occur as side effects.
f oxymorphone
d Oxymorphon
i ossimorfone
e oximorfona

2694 oxypendyl
Antiemetic with spasmolytic and anaesthetic effects used in the treatment of motion sickness (car-, sea- and air-sickness), hyperemesis gravidarum (the uncontrollable vomiting in pregnancy), intense nausea and vomiting caused by heart traumas and neurological disorders.
f oxypendyl
d Oxypendyl
i ossipendile
e oxipendilo

2695 oxypentifylline
Vasodilator indicated in the treatment of peripheral vascular diseases, e.g. intermittent claudication. The drug generally reduces blood pressure, lower blood glucose or cause gastro-intestinal disorders. Nausea, dizziness and hypo-

tension may also occur as side effects.
f oxypentifylline
d Oxypentifylline
i ossipentifillina
e oxipentifilina

2696 oxypertine
Tranquillizer chiefly used in the treatment of schizophrenia and other psychoses, as well as anxiety neuroses. Drowsiness, dizziness or hyperactivity may be observed as side effects, as well as agranulocytosis and aplastic anaemia.
f oxypertine
d Oxypertin
i ossipertina
e oxipertina

2697 oxyphedrine
Cardiotonic agent capable of increasing coronary circulation and cardiac contractile strength by its action on the beta-adrenergic receptors. It also improves the metabolism of the myocardium by the optimal exploitation of the energy substrates. It is therefore indicated in the treatment of angina pectoris, myocardial infarction particularly when symptoms of cardiac fatigue or insufficiency are diagnosed. When administered orally, there may occur a temporary loss of taste sensations.
f oxyphédrine
d Oxyphedrin
i ossifedrina
e oxifedrina

2698 oxyphenamate
Tranquillizer used in all forms of anxiety. Drowsiness, headaches, vertigo and nausea may occur as side effects.
f oxyphénamate
d Oxyphenamat
i ossifenamato
e oxifenamato

2699 oxyphenbutazone
Pyrazolone derivative indicated in the symptomatic treatment of acute gout and in rheumatoid arthritis when they fail to respond to less toxic drugs. See phenylbutazone for other uses and side effects. Agranulocytosis and aplastic anaemia, as well as intestinal haemorrhages may occur in the course of a prolonged administration.
f oxyphenbutazone
d Oxyphenbutazon
i ossifenbutazone
e oxifenbutazona

2700 oxyphencyclimine
Anticholinergic and antispasmodic agent whose effects may last up to 12 hours if administered orally. It is especially indicated in the treatment of peptic ulcer, spastic constipation and hyperacidity. Dryness of the mouth and urinary retention are possible side effects.
f oxyphencyclimine

d Oxyphencyclimin
i ossifenciclimina
e oxifenciclimina

2701 oxyphenisatin
Synthetic cathartic agent marked by a strong action on the peristalsis. Its acetate, administered orally in the form of tablet, acts almost exclusively on the colon. Abdominal pains, sometimes acute, may occur as side effect.
f oxyphénisatine
d Oxyphenisatin
i ossifenisatina
e oxifenisatina

2702 oxyphenonium bromide
Parasympatholytic agent with anticholinergic action and antispasmodic. It is used for adjunctive therapy in spasms of the gastro-intestinal, biliary and genitourinary tracts and in peptic ulcer. Dryness of the mouth and accommodation disorders may occur as side effects.
f bromure d'oxyphénonium
d Oxyphenoniumbromid
i ossifenonio bromuro
e bromuro de oxifenonio

2703 oxypinocamphone
Cardio-respiratory analeptic agent used in the treatment of chronic respiratory insufficiency in the course of bronchitis, bronchial dilatation and emphysema. The drug is also employed in pneumoconiosis, primary or secondary pulmonary fibrosis and pulmonary heart.
f oxypinocamphone
d Oxypinocamphon
i ossipinocamfone
e oxipinocamfona

2704 oxypolygelatin
Oxidized product of polymerized gelatin which has been used as a transfusion substitute for blood plasma.
f oxypolygélatine
d Oxypolygelatin
i ossipoligelatina
e oxipoligelatina

2705 oxypyrronium bromide
Spasmolytic anticholinergic agent indicated for the treatment of gastro-intestinal dystonia, hyperchlorhydria, gastritis, colitis and gastroduodenal ulcer. Some mydriasis has been observed as side effect.
f bromure d'oxypyrronium
d Oxypyrroniumbromid
i ossipirronio bromuro
e bromuro de oxipirronio

2706 oxyquinoline
Antiseptic and antimycotic agent used for the disinfection of the mucosas in gynecology, dermatology and stomatology. Also used topically on skin or vagina for minor infections or acne. It has deodorant and keratolytic properties and

may be used as a contraceptive.
f oxyquinoline
d Oxychinolin
i ossichinolina
e oxiquinolina

2707 oxytetracycline
Crystalline antibiotic isolated from the culture medium of Streptomyces rimosus. It is a bactericide marked by a wide range of activity resembling that of aureomycine but has no value in tuberculosis, leprosis and those diseases which are caused by filamentous fungi. It carries out its function by inhibiting protein biosynthesis.
f oxytétracycline
d Oxytetracyclin
i ossitetraciclina
e oxitetraciclina

2708 oxythiamine
Chemical compound obtained by introduction of an oxydrylic group into the molecule of thiamine. The hydrochloride forms needle-shaped crystals used as thiamine antagonists.
f oxythiamine
d Oxythiamin
i ossitiammina; ossitiamina
e oxitiamina

2709 oxytocic
Capable of stimulating or strengthening the contraction of muscles, in particular those of the uterus (see oxytocin).
f oxytocique
d oxytozisch; wehenanregend
i ossitocico
e oxitócico

2710 oxytocin
The fraction of the extract of the posterior lobe of the hypophysis cerebri capable of promoting and strengthening the contraction of the uterine muscles. It has an ecbolic action, not an abortive one. Hence its therapeutic use to stimulate and/or regulate the contractions of the uterus and to stop haemorrhages. Oxytocin also acts on the smooth muscles of the intestine, gall bladder, ureter and urinary bladder.
f oxytocine
d Oxytocin
i ossitocina
e oxitocina

2711 oxytoxin
Any agent or substance which is produced by the process of oxidizing a toxin.
f oxytoxine
d Oxytoxin
i ossitossina
e oxitoxina

2712 oxytriptane
Natural amino acid, a precursor of serotonin. It is chiefly indicated for the treatment of depressive state of melancholia of a confusional type, neurologic syndromes caused by deficiency of serotonin and, as an adjuvant, schizophrenia.
f oxytryptane
d Oxytryptan
i ossitriptano
e oxitriptano

2713 ozone
Allotropic form of oxygen present in the air at the average quantities of 0,02 mg per cubic metre at sea level and ten times as much at a height of 25 km. The absorption of ultraviolet rays is made possible by this layer of maximum concentration, which has a thickness of about 20 km. Ozone, when pure, is a highly unstable and reactive gas and used for sterilization of drinking water, air purification in ventilation and as a potent oxidizing and bleaching agent.
f ozone
d Ozon
i ozono
e ozono

P

2714 pabacidum

Aminobenzoic acid. A B-complex vitamin which is an essential metabolite of several bacteria and is antagonized by sulphonamides. It is used in the treatment of achromotrichia (the absence of pigment in the hair) and skin dyschromia; also as eupnoic in geriatrics, in sulphonamide intoxications and especially against acute articular rheumatism. Locally, it is used as a protective agent against ultraviolet radiations.

f papavérine monophosfadénine
d Monophosphadenin-Papaverin
i padefosfo; papaverina monofosfadenina
e padefosfo; papaverina monofosfadenina

2715 pacemaker

Small mass of specialized nervous tissue at the junction of the superior vena cava and the right atrium in the heart. It initiates and controls the rate and rhythm of the atria and of the ventricles. Therapeutically, the term is used to denote the artificial apparatus, i.e. an electric or electronic device aiming at keeping a regular and constant cardiac beat and rhythm whenever a lesion (generally of an atherosclerotic nature) affects the conduction system of the stimuli.

f pacemaker; stimulateur cardiaque
d Schrittmacher; Herzschrittmacher; Pacemaker
i pacemaker
e pacemaker; marcapaso

2716 palmitate

Any salt or ester of palmitic acid, q.v.

f palmitate
d Palmitat
i palmitato
e palmitato

2717 palmitic acid

Monocarboxylic aliphatic organic acid, a crystalline solid insoluble in water and soluble in alcohol. It is present in nature in oils and animal and vegetable fats mostly in the form of glycerides, as well as in beeswax as methyl ester. It is used for the production of cetyl alcohol (cetol), an alcohol which occurs in spermaceti and is employed in ointments for the treatment of eczema and pruritus.

f acide palmitique
d Palmitinsäure
i acido palmitico
e ácido palmítico

2718 palmitin

The pith of the palm tree. The tripalmitic ester of glycerol which is found in oils and is, together with olein and stearin one of the three basic glycerides of animal fats. It is used for the manufacture of soaps and some pharmaceutical preparations. Also known as tripalmitin.

f palmitine; tripalmitine
d Palmitin; Tripalmitin
i palmitina; tripalmitina
e palmitina; tripalmitina

2719 palmitoleic acid

Crystalline unsaturated fatty acid present in the form of glycerides in whale, seal, cod and other fish oils and producing palmitic acid on hydrogenation.

f acide palmitoléique
d Palmitoleinsäure
i acido palmitoleico
e ácido palmitoleico

2720 palm kernel oil

Fractionated palm-kernel oil. Fat obtained from the kernels of the fruit of the palm tree. It is used pharmaceutically for the preparation of some ointments. Also known as palm nut oil.

f huile de noyaux de palme
d Palmenkernfett; Palmenkernöl
i olio di noce di palma
e aceite de pepitas de palmas

2721 palm oil

Solid or semisolid red or brown-yellow edible oil expressed from the flesh of the fruit, in particular of the African oil palm and used widely in industry. Also known as palm butter.

f huile de palme; beurre de palme
d Palmfett; Palmöl; Palmbutter
i olio di palma; burro di palma
e aceite de palma; manteca de palma

2722 palm seed oil

The fat which is obtained by expressing the seeds of the palm fruit.

f huile de pépins de palme
d Palmsamenöl
i olio di semi di palma
e aceite de semillas de palma

2723 Palthe senna
Or Alexandria senna. The senna which is obtained from a cassia. Its dried leaflets are used in medicine as a purgative agent.
f séné d'Alexandrie; séné de Palthe
d Palthesenna
i sena alessandrina; sena della Palthe; sena d'Egitto
e sen de Alejandria; sen de Palthe

2724 pamaquine
Aminic derivative of the quinoline; a yellow, water-insoluble powder, the first antimalarial agent produced synthetically. It is considerably toxic, yet it has the advantage, in comparison with other malaria-fighting drugs, to destroy in small doses the gametic form of the Plasmodium malariae.
f pamaquine; plasmoquine
d Plasmochin; Pamachin
i pamachina; plasmochina
e pamaquina; plasmoquina

2725 pambacidum
Antifibrinolytic agent indicated in the treatment of haemorrhages caused by general or localized fibrinolysis. The drug should not be administered in the presence of thrombosis and embolia.
f pambacide; acide aminométhylbenzoïque
d Pambsäure; Aminomethylbenzoesäure
i pamba; acido aminometilbenzoico
e pambácido; ácido aminometilbenzóico

2726 Panama bark
Soapbark. The bark of the soapbark tree which, owing to its saponin content, produces a soapy lather and is used in emulsifying oils.
f écorce de Panama; écorce de Quillaya
d Panamarinde; Seifenrinde; Waschrinde
i corteccia di Quillaia; legno di Panama
e corteza de Quillaia; leño de Panama

2727 pancreatic proenzyme
Pancreatis cryodesiccatum. Lyophilized product of hog's pancreas extracted immediately after the death of the animal so as to keep fully unaltered the biological properties of the gland. It contains the enzymatic principles in the form of inactive proenzymes which are transformed into proteolytic and lipolytic ferments that are only active in the guts. In practice, the drug reproduces the conditions that resemble most to the physiological pancreatic secretion and is indicated in the treatment of primary or secondary pancreatic functional incompetence, chronic pancreatitis, total or partial exeresis of the pancreas, gastrectomy, duodenal fistula and, generally, digestive disorders.
f proenzyme pancréatique
d Pankreasproenzym
i pancreas liofilizzato
e pancreas liofilizado

2728 pancreatin
Mixture of enzymes obtained from hog and ox liver and principally consisting of a proteolytic (trypsin), amylolytic (amylopsin) and lipolytic agent. It is used in medicine as a digestant and, in particular, in the treatment of dyspepsia caused by pancreatic insufficiency.
f pancréatine
d Pankreatin
i pancreatina
e pancreatina

2729 pancuronium bromide
Non-depolarizing muscle relaxant acting on the neuromuscular joint thus inhibiting the transmission of the nerve impulse. The action is quick-acting and its duration good, both resembling those of d-tubocurarine. The drug is recommended when it is advisable to produce skeletal-muscle relaxation. It is also used as an adjuvant agent in surgery. It should not be administered in the presence of myasthenia gravis and severe renal or hepatic diseases.
f bromure de pancuronium
d Pancuroniumbromid
i pancuronio bromuro
e bromuro de pancuronio

2730 pantadephosphate
Coenzyme A. Regulator of the metabolism of lipides, protides and glucides at enzymatic cellular level. It promotes and maintains the defensive capacities of the organism and its capacity of metabolic adaptation.
f coenzyme A
d A-Coenzym
i coenzima A
e coenzima A

2731 panthenol
Active form of panthotenic acid capable of interfering with the metabolic processes promoting the biosynthesis of coenzyme A. It is used by itself or together with other vitamins for the disintoxication and protection of the liver.
f panthénol
d Panthenol
i pantenolo
e pantenolo

2732 panthesine
Leucinocaine. Topical anaesthetic with an action that is similar to that of lignocaine, q.v. It is used as injection in association with protoveratrines where its hypotensive effects are presumably a support in the treatment of severe hypertension. Nausea, vomiting, dizziness and some respiratory depression may occur as side effects.
f panthésine
d Panthesin
i pantesina
e pantesina

2733 panthetine
Biologically active derivative of pantothenic acid, a constituent of A-coenzyme that is essential in the acetylation process. It has a thrombocytopoietic, anabolizing and hepatoprotecting action and is indicated in the treatment of retarded or arrested growth, haemorrhagic episodes marked by thrombocytopenia, hepatic incompetence, inflammation of the skin and mucous membranes and psychastenia.
f panthétine
d Panthetin
i pantetina
e pantetina

2734 pantoic acid
Unstable dihydroxy acid usually occurring in the form of salts. It is a constituent of pantothenic acid, q.v.
f acide pantoïque
d pantoische Säure
i acido pantoico
e ácido pantoico

2735 pantothenate
Salt or ester of the pantothenic acid. Calcium pantothenate has a considerable interest as it is formed as an intermediate product in the synthesis of pantothenic acid by condensing pantolactone with a salt of the ß-alanine. Some esters of the pantothenic acid can be used as vitamin substitutes in animal nutrition.
f pantothénate
d Pantothenat
i pantotenato
e pantotenato

2736 pantothenic acid
Viscous oily acid formed by a molecule of ß-alanine with a carboamidic bond and a derivative of butyric acid, deoxybutylmethylic acid or pantoic acid. It belongs to the vitamin group and is essential for the growth of several animal and micro-organisms.
f acide pantothénique
d Pantothensäure
i acido pantotenico
e ácido pantoténico

2737 papain
Enzyme, also known as vegetable pepsin, which is found copiously in the papaya, from which it is extracted. A grey water-soluble powder marked by a very considerable proteolytic action especially in the pH field between 4 and 6.5. In medicine, it is mainly used as a protein digestant in the treatment of dyspepsia associated with hypochlorhydria.
f papaïne
d Papain
i papaina
e papaína

2738 papaverine

Alkaloid present in opium and obtained as a secondary product of morphine extraction. It can be prepared synthetically. It is a weak base, optically inactive, almost insoluble in water; it dissolves in sulphuric acid without coloration when pure and with a green-to-violet coloration when impure. Owing to its antispastic properties, it is used in medicine together with its salts, e.g. hydrochloride and sulphate.
f papavérine
d Papaverin
i papaverina
e papaverina

2739 papaverine codecarboxylate
Papaverine pyridoxal-5-phosphate. Vasodilator marked by a spasmolytic activity on the smooth vascular muscles and indicated in the treatment of vasospastic disorders associated to a defective circulation, neurosensorial disorders due to vascular insufficiency and defective coronary circulation. The drug is also used to treat hypertension, chronic nephropathies accompanied by renal incompetence, peripheral arteriopathies, acute cerebral vascular diseases, angina pectoris and myocardial infarction. Some arrhythmia may appear as side effect if the drug is injected intravenously.
f codécarboxylate de papavérine
d Papaverincodecarboxylat
i papaverina codecarbossilato
e codecarboxilato de papaverina

2740 papaverine hydrochloride
The most commonly used water-soluble salt of papaverine administered orally, subcutaneously or as a spray. (See definition of papaverine)
f chlorhydrate de papavérine
d Papaverinhydrochlorid
i papaverina cloridrato; idrocloruro di papaverina
e hidrocloruro de papaverina

2741 papaveroline
Therapeutic agent capable of regulating the cerebral and peripheral microcirculation. It displays a vasodilating action owing to its selective adrenolytic and myolytic activity at the level of the smooth vasal muscles. Also capable of improving the tone of the vasal walls, the mobilization of the blood from various sites and of diminishing peripheral resistances. Arterial pressure and heart output are not affected. The drug is indicated in the treatment of acute and chronic coronary insufficiency, as well as cerebral and peripheral defective circulation.
f papavéroline
d Papaverolin
i papaverolina
e papaverolina

* papayotin s. papain

2742 para-aminobenzoic acid
Acid present in yeast and liver. A member of the bios (the factor necessary for the growth of certain strains of yeast), it is considered essential for the growth microorganisms. It may be considered a bacterial vitamin and thus inhibited by sulphonamides. The acid is at times used as a component of vitamin mixtures but it has not yet been ascertained whether there is a deficiency disease in man. No side effects have been observed; occasionally, in case of large doses, skin rashes and hypoglycaemia have occurred.
f acide para-aminobenzoïque
d para-Aminobenzoesäure
i acido para-aminobenzoico
e ácido para-aminobenzoico

2743 para-aminohippuric acid
Crystalline acid chiefly used in the form of sodium salt for testing kidney function.
f acide para-aminohippurique
d para-Aminohippursäure
i acido para-aminoippurico
e ácido para-aminohipúrico

2744 para-aminosalicylic acid
Synthetic antituberculous agent generally administered as the sodium salt. It is the para isomer of aminosalicylic acid. Nausea, vomiting, hypothyroidism and hepatitis may occur as side effects.
f acide para-aminosalicylique
d para-Aminosalicylsäure
i acido para-aminosalicilico
e ácido para-aminosalicílico

2745 paracetaldehyde
Paraldehyde. Polymer of the acetic aldehyde, a centrally acting hypnotic agent marked by rapid and fairly prolonged effects. It is eliminated with the respiration and is chiefly used in the treatment of some psychopathies and convulsive syndromes.
f paraldéhyde
d Paraldehyd
i paraldeide
e paraldehído

2746 paracetamol
The active metabolite of phenacetin widely used as an analgesic and antipyretic agent. It has proved less toxic than phenacetin. The drug should not be administered to anaemic children and in the presence of hepatopathies and nephropathies. A low rate of haemoglobins and a trace of agranulocytosis have been observed as side effects.
f paracétamol
d Paracetamol
i paracetamolo
e paracetamolo

2747 parachlorometaxylenol
Chloroxylenol. Potent bacteriostatic agent particularly used in preparations generally known as "pine" disinfectants.
f parachlorométaxylénol
d Parachlorometaxylenol
i paraclorometassilenolo
e paraclorometaxilenolo

2748 Para cress
Genus of the Compositae tubuliflorae with 30 species, almost all of them growing in America. Some species are used against scurvy and others against tooth ache.
f abécédaire; cresson du Brésil; cresson de Para
d Husarenknopf; Parakresse; indianische Harnkraut
i spilanto; crescione di Para
e espilanto; berro de Para

2749 paradiazine
Or pyrazine; compound containing a diazine ring with the two nitrogen atoms para to each other. It has some use in medicine.
f paradiazine; pyrazine
d Pyrazin
i paradiazina; pirazina
e paradiazina; pirazina

2750 paradichlorobenzene
Insecticide widely used for its potent action against moth and furniture beetles.
f paradichlorobenzène
d Paradichlorbenzol
i paradiclorobenzene
e paradiclorobenceno

2751 paraffin oil
Oily colourless liquid formed by a mixture of hydrocarbons and obtained as a liquid residue during the preparation of paraffin. It is used pharmaceutically for the preparation of ointments, vaseline etc.
f huile de paraffine
d Paraffinöl
i olio di paraffina
e aceite de parafina

2752 paraflutizide
Chemical compound (6-chloro-3.4 dihydro-3-(p-fluorobenzyl)-2h-1,2,4-benzothiadia zine-7-sulphonamide-1,1-dioxide. The compound is medicinally used as a diuretic agent.
f paraflutizide
d Paraflutizid
i paraflutizide
e paraflutizido

*** paraform s. paraformaldehyde**

2753 paraformaldehyde
Polyoxymethylene containing between 90 and 98% of formaldehyde. A colourless mass characterized by the irritating smell of formaldehyde. It is used as an antiseptic.

f paraformaldéhyde; paraforme
d Paraformaldehyd; Trioxymethylen; Paraform
i paraformaldeide; paraformio
e paraformaldehído; paraformo

2754 paraglobin
Product of the denaturation of globin,
i.e. of the simple protein (the histone)
which constitutes the prosthetic group
of the haemoglobin. It is obtained every
time the globin is detached without due
precautions from the prosthetic group.
Its rotatory capacity is higher than that
of globin.
f paraglobine
d Paraglobin
i paraglobina
e paraglobina

2755 paraglobulin
Globulins isolated from plasma of horses
treated with cartilage and parathyroids.
Fibrinoplastin, serum globulin. A globulin which is present in blood cells,
blood   serum   and   in   many   connective
tissues.
f paraglobuline
d Paraglobulin
i paraglobulina
e paraglobulina

2756 paramethadione
Anticonvulsant agent generally indicated
in the treatment of petit mal when other
drugs have failed. As the drug may act
as a marrow depressant, accurate monitoring should be kept in the course of
its administration.
f paraméthadione
d Paramethadion
i parametadione
e parametadiona

2757 paramethasone
Corticosteroid hormone marked by principally glucocorticoid action indicated in
the treatment of rheumatic conditions,
e.g. rheumatoid arthritis, collagenosis,
inflammatory and allergic skin diseases,
grave   allergy   states   and   reactions,
traumatic   ophthalmopathies,   autoimmune
haemolytic   anaemia,   acute   and   chronic
leukaemia, Hodgkin's disease, ulcerous
colitis, multiple sclerosis and, associated with a mineralcorticoid agent, adreno
cortical insufficiency. Dyspepsia, gastro-
-intestinal   ulcers   and   haemorrhagies,
osteoporosis,   hypertension,   oedemas,
myopathies etc. may occur as adverse
effects.
f paraméthasone
d Paramethason
i parametasone
e parametasona

2758 paramorphine
Thebaine.   Opium   alkaloid,   very   toxic
and with very little narcotic action.
f paramorphine; thébaïne

d Paramorphin; Thebain
i paramorfina; tebaina
e paramorfina; tebaina

2759 parapenzolate bromide
Spasmolytic   agent   marked   by   an   anticholinergic   action   and   indicated   in   the
treatment   of   gastro-intestinal   ulcer,
gastritis, colitis and timpanism. Dryness
of the mouth and sometimes accommodation disorders may occur as side effects.
f bromure de parapenzolate
d Parapenzolatbromid
i bromuro di parapenzolato
e bromuro de parapenzolato

2760 parapropamol
Analgesic   used   to   combat   headaches,
toothache,   rheumatic   pains,   neuralgias,
lumbago and dysmenorrhoea.
f parapropamol
d Parapropamol
i parapropamolo
e parapropamolo

2761 paratartaric acid
Racemic acid, the optically inactive form
of tartaric acid which can be separated
into the dextro and laevo forms of which
it is a mixture. The acid is present in
grape juice.
f acide tartarique racémique
d Paraweinsäure; razemische Weinsäure
i acido tartarico racemico
e ácido tartárico racémico

2762 parathiazine
Antiemetic   used   for   the   prevention   and
the treatment of vomiting independently
of its origin and cause, e.g. the post-
-operative   vomiting   and   that   occurring
following the administration of narcotics.
It is also useful to combat seasickness
and neurovegetative dystonias. The drug
should not be administered in the course
of pregnancy. Drowsiness and occasionally some gastro-intestinal disorders may
occur as side effects.
f parathiazine
d Parathiazin
i paratiazina
e paratiazina

2763 parathyroid hormone
Parathormone,   the   polypeptide   hormone
produced by the parathyroid gland. The
drug   increases   the   plasma   calcium   by
mobilizing calcium from bone, diminishing calcium secretion with urine and
promoting its absorption from the gastro
-intestinal tract. It is therefore indicated in the treatment of conditions arising
from low plasma calcium (tetany). Abnormally high plasma calcium accompanied by extreme weakness, lethargy
and coma may occur as adverse effects.
f hormone parathyroïdienne; parathormone
d Nebenschilddrüsenhormon;   Parathormon;
   PTH
i paratormone; ormone paratiroideo

e parathormona; hormona paratiroidea

2764 pareira
The dried root of Chondodendron tomentosum, containing berberine (a tonic and antihaemorrhagic agent) which by methylation yields an isomer of the d-tubocurarine (q.v.). It has therefore been frequently used in chemical research on tubocurarine.
f pareira
d Pareira; Griedwurzel
i pareira
e pareira; butua

2765 parenzyme
Trypsine. The enzyme of the pancreatic juice which initiates the digestion of proteins. It hydrolyzes the peptidic bonds affecting the carboxylic group of the arginine and lysine; it is excreted by the pancreas in an inactive form as proenzyme of the trypsinogen. The latter is converted into trypsine by another enzyme (enterokinase) which is found in the intestinal juice.
f trypsine
d Trypsin
i tripsina
e tripsina

2766 paretoxycaine
Local anaesthetic used in the treatment of throat inflammation, e.g. tonsillitis, pharyngitis etc. No side effects have been noticed.
f parétoxycaïne
d Paretoxycain
i paretossicaina
e paretoxicaina

2767 pargyline
Monoamine oxidase inhibitor chiefly used in the treatment of fatigue of both mental and physical origin. It is marked by a central mechanism of action. Also indicated in the treatment of mild essential hypertension. Asthenia, vertigo, palpitations and constipation may occur as side effects.
f pargyline
d Pargylin
i pargilina
e pargilina

2768 parietis arteriosi extract
Preparation obtained from the arterial walls of healthy animals. It is a vasal opotherapeutic agent capable of controlling by physiological mechanism any dysmetabolism and/or alteration of the vasal walls of which it restores and improves the properties of lipid regulation, blood clarification, the stimulation of respiratory processes, the protection and hypotensive action of the vessels. It is therefore indicated in the treatment of arteriosclerosis, myocardial infarction, diabetic vasculopathies and hypertensive states.
f extrait de la paroi artérielle
d Arterienwandextrakt
i estratto di parete arteriosa
e extracto de la pared arterial

2769 paritary
Stone fern. Also called wall wort and wall pellitory as all seven species of the plant (family Urticaceae) grows on rocks and walls. The infusion and decoction of paritaria, especially the paritaria officinalis, have diuretic and cicatrizing properties.
f épinard des murailles; pariétaire; perce-pierre
d Mauerkraut; Glaskraut
i parietaria; muraiola; spaccapietra; vetriola
e parietaria

2770 paromomycin
Antibiotic produced by Streptomyces rimosus. The sulphate of paromomycin is a broad-spectrum antibiotic promptly active against Shigella, Escherichia coli, Salmonella, Proteus, Pseudomonas and various other intestinal organisms. As the drug is administered orally and not entirely absorbed, it is indicated in the treatment of intestinal infections, both acute and subacute, accompanied by diarrhoea and fever. Also used as a prophylactic agent in intestinal surgery.
f paromomycine
d Paromomycin
i paromomicina
e paromomicina

2771 paroxypropione
Pituitary gonadotropic hormone inhibitor indicated in the treatment of morbid syndromes caused by excessive secretion of thyrotropic hormone (mainly occurring in metabolic disorders, Basedow's disease etc.), gonadotropic hormone (mainly occurring in hyperfolliculunism), as well as pancreatic, corticotropic and other hormones. Heart disorders and drowsiness, especially if the drug is taken together with barbiturics, may occur as side effects.
f paroxypropione
d Paroxypropion
i parossipropione
e paroxipropiona

2772 parsalamide
Non-steroid antiinflammatory agent with analgesic, muscle-relaxant and sedative properties. The drug is mainly indicated in the treatment of inflammatory or degenerative rheumatism (osteoarthritis variously localized, myalgia of traumatic, arthritic and vascular origin, periphlebitis, thrombophlebitis etc.). Dyspepsia, gastralgia and nausea may occur as side effects.
f parsalamide
d Parsalamid

1 parsalamide
e parsalamida

2773 parsley
Biennial herbaceous plant (Petroselinum sativum) whose aromatic leaves are used for culinary purposes. The fleshy roots and the fruit are employed medicinally.
f persil
d Petersilie
i prezzemolo; petrosello
e persille; perejil

2774 parsley fruit
The ovoid fruits of the Petroselinum hortense, which are used medicinally for their diuretic and mildly cathartic properties.
f fruit de persil
d Petersilienfrucht; Petersiliensamen
i frutto di prezzemolo
e fruto de perejil

2775 parsley oil
Yellowish to greenish liquid obtained by distillation of the parsley fruit; it is chiefly constituted by apiole (parsley camphor) and pinene and is used in medicine as taenifugal agent and emmena gogue (menstruation inducing).
f essence de persil
d Petersilienöl
i olio essenziale di prezzemolo
e esencia de perejil

2776 parsley root
The root of the tuberosum petroselinum, which is used medicinally for its diuretic and mildly cathartic properties.
f racine de persil
d Petersilienwurzel
i radice di prezzemolo
e raíz de perejil

2777 parsnip
European biennial herb having large pinnate leaves and yellow flowers. The rather long fusiform roots, marked by a sweetish taste, are used for the table both as food and for their supposedly medicinal properties.
f panais
d Pastinak
i pastinaca
e chirivia

2778 pashydrazide
Hydrazide of the aminosalicylic acid. Therapeutic agent structurally related to PAS (para-aminosalicylic acid) and isoniazid, but considerably less toxic. It is used in all forms of pulmonary tuberculosis, in particular when chronic, whenever primary therapy with PAS and isoniazid have proved useless owing to bacterial resistance. The drug should be administered in association with other hydrazides.
f pashydrazide
d Pashydrazid

i pasidrazide
e pashidrazida

2779 pasiniazid
Chemotherapeutic antitubercular agent capable of associating the action of isoniazid with that of the p-aminosalicylic acid. It is mainly used, also in the case of long-lasting treatment, to combat pulmonary and extra-pulmonary tuberculosis. Gastro-intestinal disorders may occur as side effects.
f pasiniazide
d Pasiniazid
i pasiniazide
e pasiniazida

2780 PAS (sodium)
Para-aminosalicylic sodium. Chemotherapeutic agent displaying a bacteriostatic action against M. tuberculosis. It is particularly used when other antituberculous agents prove ineffective in the treatment of pulmonary tuberculosis.
f PAS sodique
d PAS (Natrium)
i PAS sodico
e PAS sódico

2781 pasteurization
Process aiming at checking the fermentation or putrefaction of liquid and semiliquid aliments of various nature and consisting in exposing them to temperatures of between 60 and 80 centigrades for a period of time lasting from 30 minutes to 15 seconds (the higher the temperature, the shorter the exposure). The technique aims at destroying all pathogenic organisms present in the aliment and part of the saprophitic flora which may cause alterations of a varying nature.
f pasteurisation
d Pasteurisierung
i pastorizzazione
e pasteurización

2782 patchouli
Pogostenum patchouli, a plant growing in Indochina, from which an essential oil is obtained, that is used for perfumes and in the pharmaceutical industry.
f patchouli
d Patschuli; Patchouli
i paciuli; patchouli
e patchuli

2783 patchouli oil
Essential oil obtained by the distillation with water vapour of the patchouli leaves which have been dried in the open air and partly fermented with or without their small twigs. It is a green to brown liquid used for soap and cosmetics and for some pharmaceutical products.
f essence de patchouli
d Patchouliöl

i olio essenziale di patchouli
e esencia de patchuli

2784 peach-kernel oil
Either of the two oils which can be obtained from the kernels of the peach and are very similar to almond oil in their properties and use: the colourless or pale-yellow non-drying oil obtained by expression (persic oil) and the aromatic toxic essential oil obtained by distillation.
f huile de noyaux de pêche
d Pfirsichkernöl
i olio essenziale di pesca; olio di noccioli di pesca
e aceite de semillas de melocotón

2785 pearl barley
Barley, q.v. ground into small pellets.
f orge perlé; orge mondé
d Gerstengraupen; geschälte Gerste
i orzo perlato; orzo brillato
e cebada perlada; cebada mondada

* **Pearson's solution** s. **sodium arsenate solution**

2786 pecazine
Phenothiazine-related tranquillizer with a central (subcortical) mechanism of action. It is indicated in the treatment of anxiety and hyperemotional states accompanied by insomnia, obsessive and phobic reactions and erethistic conditions. It does not affect motorial activity and strengthens the action of analgesic and narcotic drugs. Accommodation disorders and hypotension may occur as side effects.
f pécazine
d Pecazin
i pecazina
e pecazina

2787 pecilocin
Antibiotic obtained from cultures of Paecillomyces varioti banier or the same substance produced by some other means. An antimycotic agent used in the treatment of fungous infections of the skin, nails and scalp caused by Trichophyton, Epidermophyton, Microsporum and cryptococcus.
f pécilocine
d Pecilocin
i pecilocina
e pecilocina

2788 pectic acid
Polygalacturonic acid formed by the re-bonding of molecules of galacturonic acid in such number as to attain colloidal property. Pectic acids are found in a free state or as salts in vegetable tissues but their extraction is rather difficult and it is therefore preferred to obtain them from the pectines by de-methylation with acids, alkalis or enzymes.

f acide pectique
d Pektinsäure
i acido pectico
e ácido péctico

2789 pectine
Generic term denoting the pectic acids with a varying content of methoxylic groups and varying degree of neutralization capable of forming gel with sugar and with acids in certain conditions. An ideal pectine should have all the carboxylics esterified, something that has not so far been found in nature. As a therapeutic agent, pectin may be used topically to heal skin lesions and ulcerations, as a haemostatic and as a vehicle to retard the absorption of medicinal drugs.
f pectine
d Pektin
i pectina
e pectina

2790 pectinic acid
The same as pectine; pectinic acids differ from pectic acids because the latter have all or almost all free carboxylic groups. Pectinic acids are found in nature with a greatly variable percentage of methoxylic groups, a percentage, however, which is always below the theoretical value (16.3%) corresponding to a complete methylation. The salts of the pectinic acids constitute the pectinates.
f acide pectinique
d Pektininsäure
i acido pectinico
e ácido pectínico

2791 pelargonic acid
Aliphatic acid with nine carbon atoms. Colourless crystals obtained by oxidation of oleic acid and present in the oil of Pelargonium reseum, in lavender oil and in Japan wax. It is used in several organic syntheses and as an insect repellent.
f acide pélargonique
d Pelargonsäure
i acido pelargonico
e ácido pelargónico

2792 pelargonic aldehyde
Colourless, rose- and orange-smelling liquid, present in several essential oils and used in perfumery and in pharmaceutical industry. It can be prepared synthetically.
f aldéhyde pélargonique
d Pelargonaldehyd
i aldeide pelargonica
e aldehído pelargónico

2793 pelletierine
Or punicine. Alkaloid closely related to coniine and present as a racemic isomer in the root bark of the pomegranate. It is mostly used as a tannate,

almost always consisting of the mixture of the tannates of the many pomegranate alkaloids. It is used in medicine as a vermifuge. In fact, the taenias living as parasites in the intestine are paralysed by the drug after a short period of excitation and subsequently expelled by peristalsis intensified by a cathartic agent.
f pelletiérine
d Pelletierin; Punicin
i pelletierina
e peletierina; punicina

2794 **pellitory**
Name of two herbs of the Germinaceae family: Anacyclus pyrethrum growing in Mediterranean countries and Anacyclus officinalis growing in Germany, having roots respectively known as Roman and German roots which are used medicinally as antineuralgic and odontalgic agents. Also called Spanish camomile.
f pyrèthre romain; salivaire
d Römischer Bertram; Speichelwurz
i piretro; pilatro
e pelitre

2795 **pellitory root**
The root of the pellitory, from which a principle is obtained for medicinal use. See pellitory.
f racine de pyrèthre
d römische Bertramwurzel; Franzosenwurzel; Zahnwurzel
i radice di piretro
e raíz de pelitre romano

2796 **pemoline**
Central-nervous-system stimulating agent having minimal sympathomimetic effects, which, though chemically different from amphetamine and caffein, displays an action which is between the two drugs. The exact mechanism of its action is not completely known, but it appears that it increases the rate of synthesis of dopamine in the brain. It is chiefly indicated in the treatment of the minimal brain dysfunction (hyperkinetic syndrome in adolescents), fatigue, reactive depression and epilepsy accompanied by lethargy. The drug should be used with extreme caution.
f pémoline
d Pemolin
i pemolina
e pemolina

2797 **pempidine**
1,2,2,6,6-pentamethylpiperidine. Therapeutic agent having a hypotensive action by a sympatholytic mechanism, i.e. a ganglionic blocking agent. It may be used by itself or in association with a diuretic for the treatment of severe essential hypertension.
f pempidine
d Pempidin

i pempidina
e pempidina

2798 **penamecillin**
Semisynthetic penicillin used in the treatment of throat infections, e.g. tonsillitis and pharyngitis, as well as sinusitis, bronchitis, pneumonia, erysipelas and otitis media caused by penicillin-sensitive organisms. The drug should not be administered to patients who are hypersensitive to penicillin. Urticaria and allergic reactions may occur as side effects.
f pénamécilline
d Penamecillin
i penamecillina
e penamecilina

2799 **penbutolol**
Non-selective beta-adrenoreceptor blocking drug marked by actions, uses and side effects similar to those of propanolol, q.v. Its long half-life makes it possible to administer it in daily doses.
f penbutolol
d Penbutolol
i penbutololo
e penbutololo

2800 **penfluridol**
Long-acting neuroleptic agent of the diphenylbutylpiperidine group marked by a retarded action. It is indicated in the treatment of acute and chronic psychoses, in schizophrenic and paranoid symptomatology. Alcoholic drinks should be avoided in the course of the therapy. Extrapyramidal disorders may occur as side effect.
f penfluridole
d Penfluridol
i penfluridolo
e penfluridolo

2801 **pengitoxin**
Semisynthetic glycoside, a cardiotonic agent used in the treatment of all forms of cardiac insufficiency. High doses could prove dangerous. Excessive salivation, nausea, vomiting, diarrhoea, headache, general malaise and arrhythmias may occur as side effects.
f pengitoxine
d Pengitoxin
i pengitossina
e pengitoxina

2802 **penicillamine**
Chelating agent capable of binding certain toxic metals, e.g. copper, and increasing their excretion. It is thus indicated in the treatment of metal poisoning and rheumatoid arthritis. The product of chelation is rapidly eliminated. Allergic reactions, headache, fever, loss of taste, gastro-intestinal disorders, leukopenia and agranulocytosis may occur as side effects.
f pénicillamine

d Penicillamin
i penicillamina
e penicilamina

2803 penicillin
The antibiotic produced from Penicillium
notatum, originally described by Fleming
in 1929 and later purified and adapted
for therapeutic uses. Its main characteristic is its powerful bacteriostatic effect
against several bacteria which it destroys by inhibiting cell-wall synthesis
which in turn is blocked by transpep-
tidation.
f pénicilline
d Penizillin
i penicillina
e penicilina

2804 penicillinase
Enzyme produced by Bacillus cereus, capable of splitting penicillin into penicillanic acid and dimethylcysteine. It is
used to neutralize the effects of penicillin overdosage or of penicillin administered to hypersensitive subjects.
f pénicillinase
d Penicillinasis
i penicillinasi
e penicilinasa

2805 penicilloic acid
Acid produced by alkaline hydrolysis
of various penicillins. It is inactive as
antibiotic.
f acide pénicilloïque
d penizilloische Säure
i acido penicilloico
e ácido peniciloico

2806 penicycline
Broad-spectrum antibiotic formed by a
salt of tetracycline and phenoxymethylpenicillin, acid-stable penicillin which
is absorbed orally. The drug is slowly
excreted and the blood concentrations
persist for a long time. It is indicated
in the treatment of mixed infections
caused by bacteria which are sensitive
to the two antibiotics with a different
process, including the penicillin-resistant forms. The drug is not indicated
in the presence of gastric ulcer and
severe nephropathies, as well as in the
course of pregnancy. Decalcification of
the teeth has been observed as side effect.
f pénicycline
d Penicyclin
i peniciclina
e peniciclina

2807 penimepicycline
Active tetracycline-penicillin compound
(bacteriostatic and bactericide) on Gram-
-positive and Gram-negative microorganisms. It is indicated in the treatment
of infectious diseases caused by susceptible organisms, in particular bacteria
which prove resistant to penicillin. The

effect of the drug is rapid and lasts
for a long time. The drug should not
be prescribed to patients who are allergic to penicillin and tetracycline.
Intestinal secondary infections may occur
as an adverse effect.
f pénimépicycline
d Penimepicyclin
i penimepiciclina
e penimepiciclina

2808 penimocycline
Compound of penicillin and tetracycline,
a broad-spectrum antibiotic active a-
gainst Gram-positive and Gram-negative
microorganisms. It is chiefly used in the
treatment of acute, chronic and recurrent infections of the respiratory and
gastro-intestinal tracts, of infections
affecting ears, nose and mouth and in
the treatment of various other infections
which are caused by susceptible organisms. Allergic reactions, nausea and
diarrhoea may occur as side effects.
f pénimocycline
d Penimocyclin
i penimociclina
e penimociclina

2809 penmesterol
Androstane-derived androgenic hormone
acting by a physiological mechanism and
used in the treatment of disorders caused by deficiency of male sex hormones.
It is also indicated in the treatment of
uterine fibroma. The drug should not
be used in pregnancy. Slight virilism,
headaches and palpitations may occur
as side effects.
f penmestérol
d Penmesterol
i penmesterolo
e penmesterolo

2810 pentaerythritol tetranitrate
Long acting vasodilator and cardiac
sedative. It is used in the prophylaxis
of angina pectoris.
f tétranitrate de pentaérithritole
d Pentaerythritoltetranitrat
i pentaeritritolo tetranitrato
e tetranitrato de pentaeritritolo

2811 pentagastrin
Gastrin-like polypeptide used to stimulate gastric secretion especially for
diagnostic purposes. It should not be
used in the presence of active peptic
ulcer. Nausea, abdominal cramps, vertigo, a feeling of torpor in the limbs
and tachycardia may occur as side effects.
f pentagastrine
d Pentagastrin
i pentagastrina
e pentagastrina

2812 pentagestrone
Progesterone derivative with progestin-
-related action. It is used in threatened

abortion and premature labour. It has a trophic action on the endometrium. Menstrual disorders may occur as side effect.
f pentagestrone
d Pentagestron
i pentagestrone
e pentagestrona

2813 pental
Trimethyl- ethylene, ß-iso-amylene. Colourless oily liquid used in the past as a general anaesthetic. It was discontinued because of its toxicity.
f pental
d Pental
i pental
e pental

2814 pentamethonium
Organic radical derived by substitution from ammonium. Its simple salts act like the homologous hexamethonium, q.v. by inhibiting transmission across the ganglionic synapse and are used to lower blood pressure.
f pentaméthonium
d Pentamethonium
i pentametonio
e pentametonio

2815 pentamethonium bromide
Salt of pentamethonium. Its action resembles that of pentamethonium iodide, q.v.
f bromure de pentaméthonium
d Pentamethoniumbromid
i pentametonio bromuro
e bromuro de pentametonio

2816 pentamethonium iodide
Salt of metamethonium used to block the sympathetic and parasympathetic ganglia in the treatment of hypertension.
f iodure de pentaméthonium
d Pentamethoniumjodid
i pentametonio ioduro
e yoduro de pentametonio

2817 pentamidine
Chemotherapeutic and chemoprophylactic agent especially active on infections caused by Trypanosoma and Leishmania. Also used in the treatment of multiple myeloma, particularly in the septicaemic phase. Hypotension, tachycardia, gastro--intestinal disorders and transient diabetes may occur as side effects.
f pentamidine
d Pentamidin
i pentamidina
e pentamidina

2818 pentanobornamide
Antispasmodic agent marked by an elective action on the smooth muscles and chiefly used for the relief of pain affecting the digestive, biliary and genito urinary tracts.
f pentabornamide
d Pentabornamid
i pentabornamide
e pentabornamida

2819 pentapiperide
Anticholinergic agent marked by spasmolytic and antisecretive action on the gastro-intestinal and biliary tracts. It is indicated in the treatment of all spastic, hyperkinetic and hypersecretive states of the gastro-enteric tract, e.g. gastritis, duodenitis, peptic ulcers, pylorospasms and spastic colon. Xerostomia, sight disorders, tachycardia and arrhythmia may be observed as side effects.
f pentapipéride
d Pentapiperid
i pentapiperide
e pentapiperido

2820 pentaquine
Antimalarial drug which has generally been superseded by primaquine, a drug given in succession to chloroquine or quinine so as to eliminate the persistent exo-erythrocyte phase of the parasite in infections with Plasmodium vivax, P. malariae or P. ovale.
f pentaquine
d Pentachin
i pentachina
e pentaquina

2821 pentazocine
Narcotic analgesic with action resembling that of morphine and pethidine. It is used to relieve very acute and/or chronic pains. It is also indicated as coadjuvant in surgery if anaesthesia is required. Dependence and addiction are less common than with morphine. Mild euphoria, sometimes haunting dreams, hallucinations, withdrawal symptoms, as well as nausea, vomiting and drowsiness may occur as side effects. The drug should not be used in the presence of severe respiratory depression.
f pentazocine
d Pentazocin
i pentazocina
e pentazocina

2822 pentetrazol
Cardiokinetic drug acting on the central nervous system and capable of stimulating respiration. It is chiefly indicated in the treatment of circulatory collapse, asphyxia and intoxications caused by cardiotoxic and neurotoxic agents. It is administered either by itself or in association with ephedrine, glucose etc.
f pentétrazol
d Pentetrazol
i pentetrazolo
e pentetrazolo

2823 penthienate
Anticholinergic agent whose action and side effects are similar to those of atro-

pine. Mostly indicated to relieve acute gastric pains.
f penthiénate
d Penthienat
i pentienato
e pentienato

2824 pentobarbital
Barbiturate hypnotic, sedative and anticonvulsant agent which has proved very useful whenever sedation is recommended. It is used in psychiatrics as an antipsychopathic drug, in essential insomnia and, especially as a sedative, in very painful abdominal and cardiac syndromes. It is an antagonist of convulsant poison. Also used in obstetrics and in the post-operative treatment.
f pentobarbital
d Pentobarbital
i pentobarbitale
e pentobarbital

2825 pentolinium tartrate
Hypotensive agent indicated in the treatment of malignant hypertension, arteriosclerosis accompanied by hypertension etc. Also used to induce hypotension during general anaesthesia. The drug should not be used in the presence of severe cardiopathies.
f tartrate de pentolinium
d Pentoliniumtartrat
i pentolinio tartrato
e tartrato de pentolinio

2826 pentorex
Sympathomimetic amine marked by anorexia-inducing and lipolytic activity used in the treatment of all forms of obesity. The drug should be administered with great caution in the course of pregnancy and for children under the age of 12. Palpitations and tachycardia sometimes insomnia and restlessness may occur as side effects.
f pentorex
d Pentorex
i pentorex
e pentorex

2827 pentoxyphylline
Theobromine derivative marked by an elective activity on peripheral blood vessels; the action is of the metabolic type with intracellular increase of the cyclic AMP (adenosine monophosphate), reduction of blood viscosity and relaxation effect on the smooth vasal muscles. The drug is indicated in the treatment of arterial or venous circulatory disorders of a diabetic, arteriosclerotic or inflammatory nature, angioneuropathies and cerebral or ocular circulatory disorders.
f pentoxyphylline
d Pentoxyphyllin
i pentossifillina
e pentoxifilina

2828 pentoxyverine
Antitussive agent with a central mechanism of action. It is mainly indicated in the treatment of whooping cough and other persistent and violent coughs. Also used for cough caused by poison gas. The drug has no effect on expectoration. Some depression and drowsiness may be observed as side effects.
f pentoxyvérine
d Pentoxyverin
i pentossiverina
e pentoxiverina

2829 peonin
The amino derivative of aurin, formed by a diglycoside of the peonidine and as such widely present in many plants. It is a pararosaniline dye used as a pH indicator.
f péonine
d Päonin
i peonina
e peonina

2830 peony
Or piney. Herbaceous or shrubby plant; the petals of its flowers were used in the past as a sedative.
f pivoine; herbe chaste
d Päonia; Gichtrose
i peonia; erba casta; rosa della Madonna
e peonia hembra

2831 pepo
The pumpkin seed used when dried and ripe as a non-toxic vermifuge.
f graine de courge; semence de courge
d Kürbiskern; Kürbissamen
i seme di zucca
e semilla de calabaza; semilla de cucúrbita

2832 pepper
The dried fruit of several plants of the genus Piper, e.g. black pepper consisting mostly of terpene, phellandrene and dipenthene, piperine (an alkaloid) and chavicine (a resin); red pepper; Cayenne pepper etc. Some of them find employment in medicine, particularly externally, in the form of plaster, to relieve irritation.
f poivre
d Pfeffer
i pepe
e pimienta

2833 peppermint
Genus of herbs mainly growing in temperate climates. The most cultivated species is the Mentha piperita whose leaves and blossomed tops, marked by an aromatic odour, yield an essential oil that finds wide use in the pharmaceutical industry. It contains menthol, menthone, pinene and limonene.
f menthe poivrée
d Pfefferminze
i menta; menta piperita; piperita; menta

inglese
e menta piperita; hierbabuena

2834 peppermint leaves
The dried leaves of peppermint from
which peppermint oil (q.v.) is obtained.
f feuilles de menthe poivrée
d Pfefferminzblätter; Edelminzblätter
i foglie di menta piperita
e hojas de menta piperita

2835 peppermint oil
By distillation in vapour stream of the
various species of peppermint, essential
oils are obtained marked by particular
chemical and commercial characteristics.
The most common is the oil of mentha
piperita; it is a yellow-green, or colourless liquid, with a tendency to darken,
the usual odour of peppermint and a
penetrating taste which leaves a very
pleasant freshness on the tongue. Chemically, it consists of menthol, either in
the free state or in the form of ethers,
menthone, pinene, cadinene, limonene,
cineole etc. It finds wide employment in
medicine.
f essence de menthe poivrée
d Pfefferminzöl
i olio essenziale di menta
e aceite de menta piperita; esencia de
menta piperita

2836 peppermint spirit
Alcoholic solution of peppermint oil
obtained from the leaves and the flowering tops of peppermint. It is used in
the preparation of various medicinal
drugs.
f esprit de menthe; teinture d'essence de
menthe
d Pfefferminzspiritus
i acqua spiritosa di menta; spirito di
menta
e espíritu de menta piperita; alcohol de
menta piperita

2837 pepper oil
The essential oil found in black pepper,
containing an acrid resin and piperine.
f huile de poivre noir
d Schwarzpfefferöl
i essenza di pepe nero
e esencia de pimienta negra

2838 pepsin
Proteolytic enzyme of the gastric juice.
Active pepsin derives from the peptino-
gen secreted by the pyloric glands and
by the principal cells of the glands of
the fundus of the stomach. It is used
medicinally to aid digestion. In fact,
lack or insufficiency of pepsin charac-
terize most chronic gastritides, gastric
neoformations, pernicious anaemia and
congenital acheilia.
f pepsine
d Pepsin
i pepsina
e pepsina

2839 peptide
Compound resulting from two or more
molecules of amino acids linked by the
respective carboxyl and amino groups
into a molecular chain. Peptides are
distinguished into di-, tri-, tetra-,
octa-, deca- or poly- peptides according
to the number of amino acids involved.
Proteins, as they result from a very
large number of amino-acid molecules,
may be considered polypeptides.
f peptide
d Peptid
i peptide
e péptido

2840 peptone
The product of the partial hydrolysis
of proteins, e.g. by the action of pepsin,
or cf the artificial digestion of albu-
minoids. It is distinguished according
to the origin of the latter (casein, albumin etc.). It is soluble in water and
generally used to prepare culture beds,
but also in proteinotherapy and as an
easily assimilable food when mixed with
milk, soup etc.
f peptone
d Pepton
i peptone
e peptona

2841 peptonization
The transformation of proteins into pep-
tones occurring by a process of incomplete enzymatic splitting of the proteins
themselves by proteolytic ferments.
f peptonisation
d Peptonisierung
i peptonizzazione
e peptonización

2842 peracetic acid
Colourless liquid marked by a penetrating odour, solidifying into easily explosive crystals. It is generally found in
aqueous solution characterized by oxidizing properties and used as a polymerization catalyzer and as oxidant.
f acide peracétique
d Peressigsäure
i acido peracetico
e ácido peracético

2843 peracid
Oxyacid containing two atoms of oxygen
in the form of peroxide, i.e. directly
linked (-O-O-) and which tend to liberate oxygen, when in solution, thus acting as oxidants. In organic chemistry,
any acid containing the radical -COOH.
Solutions of peracid, treated with sulphuric acid, develop hydrogen peroxide.
f peracide
d Persäure
i peracido
e perácido

2844 perborates
Salts of perboric acid, all strong oxidiz

ing agents used as bleaches and disinfectants. Sodium perborate, for example, finds employment in the treatment of tonsillitis and Vincent's angina (tonsillitis accompanied by ulcerations).

f perborates
d Perborate
i perborati
e perboratos

2845 perboric acid
Hypothetic acid of the boron (HBO_3). While the acid is not known in its free state, its perborates are well known and important in medicine (see perborate).

f acide perborique
d Perboronsäure
i acido perborico
e ácido perbórico

2846 perbromate
Salt of perbromic acid.

f perbromate
d Perbromat
i perbromato
e perbromato

2847 perbromic acid
Acid of the bromine, $HBrO_4$. Perbrom is a prefix used in chemistry to indicate that all atoms of hydrogen in the molecule of a compound have been substituted with atoms of bromine.

f acide perbromique
d Perbromsäure
i acido perbromico
e ácido perbrómico

2848 percaine
Term applied to the hydrochloride of butyloxycinchonate of diethylethylendiamide. Colourless, tasteless, water-soluble crystals marked by topical prolonged anaesthetic properties.

f percaïne
d Percain
i percaina
e percaina

2849 percarbonate
The name given to the salts of the hypothetical precarbonic acid, its general formula being $Me_2C_2O_6$ in which Me indicates a monovalent metal. Percarbonates contain the characteristic groupings of the peroxides and this allows them to decompose when in solution, yielding hydrogen peroxide and bicarbonate; by action of an acid, they yield hydrogen peroxide, development of carbon dioxide and the salt of the added acid. They are mainly used for bleaching, in photography and radiography.

f percarbonate
d Percarbonat; überkohlensaures Salz
i percarbonato
e percarbonato

2850 percarbonic acid
Hypothetical acid ($H_2C_2O_6$) of which the

salts, i.e. the percarbonates are known.

f acide percarbonique
d Percarbonsäure
i acido percarbonico
e ácido percarbónico

2851 perchlor
Prefix used in organic chemistry to indicate those compounds in which all hydrogen atoms have been substituted with chlorine.

f perchloro-
d perchlor-
i percloro-
e percloro-

2852 perchlorate
Term denoting the salts of the perchloric acid whose formula is $Me\ ClO_4$ where Me indicates a monovalent metal. Perchlorates are water-soluble compounds and decompose under strong heat yielding oxygen. Sodium perchlorate, very easily soluble in water and thus easy to prepare, ammonium, and magnesium perchlorates are the best known.

f perchlorate
d Perchlorat; überchlorsaures Salz
i perclorato
e perclorato

2853 perchloric acid
Powerful oxidizing agent used in organic syntheses and a quantitative and qualitative reagent for potassium. It can be prepared by treating potassium perchlorate with a mixture of nitric and hydrochloric acid.

f acide perchlorique
d Perchlorsäure
i acido perclorico
e ácido perclórico

2854 perchloride
Chloride which contains more chlorine than that which is found in the normal chloride of the element.

f perchlorure
d Perchlorid
i percloruro
e percloruro

2855 perezone
Pipitzahoic acid or pipitzahoinic acid. Yellow substance obtained from the roots of the Mexican Perezia adnata and used as a vegetable indicator in volumetric analysis. In medicine, it is used as a cathartic agent.

f pérézone; acide pipitzahoïque
d Perezon; Pipitzahoinsäure
i perezone
e perezón

2856 perflavon
Coronary vasodilator used in the treatment of angina pectoris, incompetence of the coronary arteries, myocardial infarction and sclerosis of the coronary arteries.

f perflavon
d Perflavon
ı perflavone
e perflavona

2857 performic acid
Acid used to disrupt disulphide bridges
in proteins by oxidation thus separating
peptide chains linked through disulphide
bridges. The structural analysis is thus
facilitated.
f acide performique
d Perameisensäure
ı acido performico
e ácido perfórmico

2858 perhexilene
Antianginal agent indicated in the prophylaxis and treatment of angina pectoris. It has not much use in the case
of acute attacks. It reduces the heart
frequency under stress but not at rest
and this without causing myocardial
depression and without appreciably altering the systemic arterial pressure.
The drug also reduces the myocardial
consumption of oxygen and, though
moderately, the pulmonary resistance in
asthmatic patients. Nerve and liver
damage, nausea, vomiting, dizziness,
flushing and general weakness may occur as side effects.
f perhéxilène
d Perhexilen
ı peresilina
e peresilina

2859 pericyazine
Tranquillizer of the phenothiazinic series
indicated in the treatment of symptoms
generally associated with obsessive and
phobic reactions, manic-depressive
states, melancholia and hysteria. Also
used to control states of anxiety.
f péricyazine
d Pericyazin
ı periciazina
e periciazina

2860 perilla oil
Or Zegoma oil. Oil obtained by expression from the seeds of perilla; a pale-
-yellow liquid with the highest iodine
number (206). It is characterized by important siccative properties. Also used
as an edible oil.
f huile de perilla
d Perillaöl
ı olio di perilla
e aceite de albahaca; esencia de albahaca

2861 perimetazine
Phenothiazine-related neuroleptic agent
used in the treatment of acute paranoid
psychosis, schizophrenia, chronic delirium, anxiety, neurosis, behaviour disorders in both adults and children,
insomnia and psychosomatic diseases.
Drowsiness and hypotension may occur
as side effects.

f périmétazine
d Perimetazin
ı perimetazina
e perimetazina

2862 periodate
Term denoting the salts of periodic acid.
There are two series of periodates, those
with a general formula of $MeIO_4$ in
which Me indicates a monovalent metal
and one with formula M_5IO_6. The former
correspond to the metaperiodic acid and
are stable, the latter to the paraperi-
odic acid (also simply called periodic)
and are unstable. They have oxidizing
properties.
f périodate
d Perjodat; überjodsaures Salz
ı periodato
e peryodato

2863 periodic acid
Oxygenated acid of the heptavalent
iodine (H_5IO_6) which can be obtained
by the action of iodine on the perchloric
acid. It is a strong oxidizing agent.
f acide periodique
d Perjodsäure; Uberjodsäure
ı acido periodico
e ácido peryódico

2864 peristaltin
Glycoside derived from cascara bark and
used in the treatment of gastro-intestinal disorders.
f péristaltine
d Peristaltin
ı peristaltina
e peristaltina

2865 perlapine
Non-barbiturate hypnotic used in the
treatment of insomnia and states of anxiety. Drowsiness and vertigo may occur
as side effects.
f perlapine
d Perlapin
ı perlapina
e perlapina

2866 permanganate
Term denoting the salts of permanganic
acid, obtained through oxidation of the
manganates by carbon dioxide or chlorine or electrolytically. For their properties, see permanganic acid.
f permanganate
d Permanganat; übermangansaures Salz
ı permanganato
e permanganato

2867 permanganic acid
Unstable acid, readily decomposing into
manganese dioxide and free oxygen. Its
salts (permanganates) are important
oxidizing agents as well as antiseptic
and disinfectant.
f acide permanganique
d Permangansäure
ı acido permanganico

e ácido permangánico

2868 peroxide
In a series of oxides formed by an element, the term is commonly applied to the oxide which has the largest proportion of oxygen, i.e. an oxide containing the group -O-O- which yields hydrogen peroxide with diluted acids. Also known as superoxide.
f hyperoxyde; peroxyde
d Hyperoxyd; Superoxyd
i perossido
e peróxido

2869 perphenazine
Tranquillizer of the phenothiazine group acting at subcortical level and indicated for the treatment of acute and chronic psychoses (both in adults and the aged) characterized by states of anxiety. Also used as antiemetic. The side effects are the same as those of chloropromazine. Extrapyramidal effects, particularly a marked dystonia, have also been observed.
f perphénazine
d Perphenazin; Ethaperazin
i perfenazina
e perfenacina

2870 persulphate
Any salt of the persulphuric acid. The most important persulphates are those of ammonium and potassium, both powerful oxidizing agents.
f persulfate
d Persulfat; überschwefelsaures Salz
i persolfato
e persulfato

2871 persulphuric acid
The acid which is formed by the electrolysis of sulphuric acid. It is a powerful oxidizing agent.
f acide persulfurique
d Persulfursäure
i acido persolforico
e ácido persulfórico

2872 Peru balsam tree
Tropical American tree from which a balsam can be obtained, which is used in medicine to dress wounds and to treat some skin diseases.
f baumier du Perou
d Perubalsambaum
i balsamino del Perù
e balsamero peruviano

2873 peruvoside
Cardiotonic glycoside having the same action and uses of digitalis and strophantus preparation. The drug, taken orally, is rapidly absorbed and equally rapidly eliminated. It is chiefly used in the treatment of heart disorders, particularly incompetence, when accompanied by a low frequency of ventricle, chronic pulmonary heart and senile

heart. Some gastro-intestinal disorders may occur as side effect.
f péruvoside
d Peruvosid
i peruvoside
e peruvosido

2874 pesticide
Generic term denoting products capable of destroying harmful animal organisms and plants. Particular attention must be paid to the pollution danger presented by some pesticides which are not easily biodegradable and may cause undesirable effects of ecological importance.
f pesticide
d Pestizid; Schädlingsmittel
i pesticida
e pesticido

2875 pethidine
Carbetoxymethylphenyl derivative of the piperidine. Synthetic narcotic analgesic whose potent effects are essentially similar to those of morphine. The drug is indicated for the relief of severe pains, e.g. those that are caused by neoplastic diseases. Nausea, vomiting, euphoria and respiratory depression may occur as side effects. Coma with real danger of death may follow an overdosage.
f péthidine
d Pethidin
i petidina
e petidina

2876 petit-grain oil
Essential oil which is obtained by the distillation with steam of the leaves and twigs or bitter orange; a yellow liquid formed by limonene, dipentene, pinene, linalyl acetate, linalol etc.
f huile de petit-grain
d Petitgrainöl
i olio essenziale di petit-grain
e esencia de petit-grain; esencia de naranja verde

2877 petrolated gauze
Loose-meshed muslin which has been rendered antiseptic by impregnation with petrolatum, a soft mixture of semi-solid hydrocarbons produced during the distillation of petroleum. It is used for surgical operations and for dressings. Also known as petrolatum gauze and vaseline dressing.
f gaze vaselinée
d Vaselingaze; Vaselinmull
i garza vaselinata
e gasa vaselinada

2878 petroleum ether
Light petroleum, a solvent consisting of mixtures of liquid paraffin hydrocarbons and prepared in various boiling ranges.
f éther de pétrole
d Petroläther

i etere di petrolio
e éter de petróleo

2879 peucedanin
Chemical compound isolated from the roots of Peucedanum ostruthium: colourless, tasteless, alkali-and ether-soluble crystals which are poisonous to marine fauna.
f peucédanine
d Peucedanin
i peucedanina; imperatorina
e peucedanina

2880 phanquinone
Chemotherapeutic amoebicide agent indicated in the treatment of acute and chronic amoebiasis, lambliasis (infestation with the intestinal flagellate Giardia intestinalis) and, more generally, enteric bacterial infections. Inflammation of the gastro-intestinal tract may occur as a side effect.
f phanquinone
d Phanquinon
i fanquinone
e fanquinona

2881 phenacemide
Anticonvulsant agent with a central mechanism of action indicated in the treatment of epileptic seizures of the psychomotor type. It is not as active in the presence of grand mal and petit mal. Extreme caution should be taken in its administration and only used when other anticonvulsants have proved useless. Liver and kidney damage, bone--marrow depression, skin rashes, nausea and vomiting may occur as side effects.
f phénacémide
d Phenacemid
i fenacemide
e fenacemida

2882 phenacetin
Analgesic antipyretic agent marked by a central mechanism of action and indicated in the symptomatic treatment of states of fever. The drug should be used with caution, as several side effects, e.g. respiratory depression, methaemoglobinaemia, hypothermia, convulsions etc. may occur.
f phénacétine
d Phenacetin
i fenacetina
e fenacetina

2883 phenadoxone
Powerful analgesic agent so far considered free from addiction danger. The drug has a very slight hypnotic action. It is used as a substitute of morphine in the treatment of chronic pain. It does not cause constipation.
f phénadoxone
d Phenadoxon
i fenadossone
e fenadoxona

2884 phenamidine
One of several diamidines used in leishmaniasis and trypanosomiasis but not with great success. The drug has also been used as an antiseptic agent.
f phénamidine
d Phenamidin
i fenamidina
e fenamidina

2885 phenampromide
Analgesic agent used for the relief of rheumatic and arthritic pains, as well as headaches. The drug has also a hypnotic action.
f phénampromide
d Phenampromid
i fenampromide
e fenampromida

2886 phenanthrene
Crystalline hydrocarbon isomeric with anthracene and, like the latter, present in coal tar. A blue fluorescence appears when dissolved in alcohol. Its structure of three benzene rings forms the basis of the cyclopentenophenanthrene skeleton from which steroids, bile acids and sex hormones are built up.
f phénanthrène
d Phenanthren
i fenantrene
e fenantreno

2887 phenanthroline
Any of three crystalline nitrogen bases related to phenanthrene and derivable from ortho-, meta- and para-phenilenediamine. It is a strong oxidizing agent.
f phénanthroline
d Phenanthrolin
i fenantrolina
e fenantrolina

2888 phenarsone sulphoxylate
The sodium formaldehyde sulphoxylate derivative of 3-amino-4-hydroxybenzene arsonic acid. It is an organic arsenical indicated in the treatment of infections of the cervix and vagina by Trichomonas vaginalis.
f sulfoxylate de phénarsone
d Phenarsonsulfoxylat
i fenarsone sulfossilato
e sulfoxilato de fenarsona

2889 phenazocine
Narcotic analgesic agent as powerful as morphine and, like the latter, marked by stupefacient action. It is a drug of addiction. It is used against acute and chronic painful syndromes, e.g. of a neoplastic nature, and in the course of surgical operations.
f phénazocine
d Phenazocin
i fenazocina
e fenazocina

2890 phenazoglycodol

Minor tranquillizer acting directly on the cerebral cortex and indicated in the treatment of patients suffering from emotional instability, states of anxiety and tension and insomnia. The drug should not be prescribed in association with thiazine-related preparations. Drowsiness, vertigo, nausea, ataxia and allergic skin reactions may occur as side effects.
f phénazoglycodol
d Phenazoglycodol
i fenazoglicodol
e fenazoglicodol

2891 phenazone
Colourless and odourless crystalline solid marked by a slightly bitter taste and soluble in water and alcohol. It is a rapidly acting antipyretic as it reduces the body temperature by depressing the heat centres in the hypothalamus. Its action being quite safe, it is also indicated for the relief of neuralgia. The drug has a slight haemostatic action. Its salicylate which is obtained by heating phenazone and salicylic acid in equimolecular proportions, is used for the relief of rheumatic pains and sciatica.
f phénazone
d Phenazon
i fenazone
e fenazona

2892 phenazopyridine
Analgesic and antiseptic agent mainly used in the symptomatic treatment of acute and chronic cystitis, pyelonephritis, prostatitis and urethritis. The drug is particularly active on the mucosae of the urinary tract. Methaemoglobinaemia may occur as side effect.
f phénazopyridine
d Phenazopyridin
i fenazopiridina
e fenazopiridina

2893 phenbenicillin
Acid-resistant penicillin to be taken orally. It is an alpha-phenoxybenzyl penicillin. Antibiotic active against most Gram-positive and some Gram-negative organisms. Hypersensitivity reactions, which may be immediate or delayed, may occur as side effect.
f phenbénicilline
d Phenbenicillin
i fenbenicillina
e fenbenicilina

2894 phencyclidine
Analgesic and antipyretic drug used in the treatment of rheumatism and arthrosis.
f phencyclidine
d Phencyclidin
i fenciclidina
e fenciclidina

2895 phendimetrazine
Phenyl-ethyl-morpholine derivative whose mechanism of action and effects are similar to those of amphetamine. Though a sympathomimetic, the drug does not alter cardiovascular conditions. It is chiefly indicated in the treatment of obesity, particularly of a psychic origin, and seasickness. It should not be administered in the course of pregnancy. Tachycardia and insomnia may occur as side effects.
f phendimétrazine
d Phendimetrazin
i fendimetrazina
e fendimetrazina

* **phenecyclamine** s. phenetamine

2896 phenelzine
Antidepressant agent capable of inhibiting monoamine oxidase thus increasing tissue concentrations of noradrenaline, dopamine and 5-hydroxytryptamine. Some of these amines are involved, as neurohumours (the chemicals set free by stimulation of nerve-endings), in the functioning of the brain and nervous system. The drug has proved effective in the treatment of depressed patients who show symptoms of anxiety and depression of a phobic or hypochondrial nature.
f phénelzine
d Phenelzin
i fenelzina
e fenelzina

2897 phenelzine (sulphate)
Phenethylhydrazine hydrogen sulphate, a monoamine oxidase inhibitor and an effective antidepressant indicated in the treatment of neurotic depressive states. It has a pungent odour and a particular taste.
f phénelzine
d Phenäthylhydrazin
i fenelzina; feniletilidrazina
e fenelzina; feniletilhidrazina

2898 phenetamine
Fecleminum. A drug used to arrest spasms and convulsive attacks.
f phénétamine
d Phenetamin
i fenetamina
e fenetamina

* **phenethicillin** s. phenbenicillin

* **phenethylhydrazine** s. phenelzine

2899 phenetidine
Any of three isomeric liquid basic amino derivatives of phenetole, the ortho- and para- isomers of which are used for the production of pharmaceutical products.
f phénétidine
d Phenetidin
i fenetidina

e   fenetidina

2900 phenformin
N-phenethyldiguanide. Synthetic hypoglycaemic agent effective when administered orally and mainly used in the
treatment of diabetes mellitus associated
with obesity and senile diabetes. As it
increases fibrilytic activity, it has been
frequently used, together with ethyoestre-
nol, to relieve the consequences of vascular occlusions. It is also used in the
treatment of rheumatoid arthritis.
f   phenformine
d   Phenformin
i   fenformina
e   fenformina

2901 phenglutarimide
Anticholinergic agent whose hydrochloride, a synthetic preparation having
atropine-like actions, was used in the
past for the treatment of Parkinsonism.
Dryness of the mouth, difficulties of
accommodation, confusional psychosis and
tachycardia in the foetus following the
intravenous administration in the mother
have been observed as side effects.
f   phenglutarimide
d   Phenglutarimid
i   fenglutarimide
e   fenglutarimida

2902 phenindamine
Antihistaminic agent, usually administered orally as the hydrogen tartrate. It
differ from most antihistamines as it
causes stimulant side effects and may
be used when sedation proves a problem.
Insomnia, convulsions and gastro-intestinal disorders may occur as adverse effects.
f   phénindamine
d   Phenindamin
i   fenindamina
e   fenindamina

2903 phenindione
Anticoagulant which acts by neutralizing vitamin K in the organism and inhibiting the synthesis of prothrombin
and factor VII in the liver. It is completely eliminated within a maximum of
48 hours and therefore does not produce
accumulation in body tissues. It is indicated in the treatment of thromboembolic
states, phlebitis, arteritis, arteriascle-
rosis and coronaritis. Overdosage may
cause haemorrhages and a prolonged use
may cause some hypersensitivity reactions.
f   phénindione
d   Phenindion
i   fenindione
e   fenindiona

2904 pheniodol
Opaque contrast medium used in cholecystography. Given only by mouth. It
has largely superseded iodophthalein

as it is more palatable and less likely
to cause nausea, vomiting and diarrhoea.
f   phéniodol
d   Pheniodol
i   feniodolo
e   feniodolo

2905 pheniramine
Rapidly-acting powerful antihistaminic
having prolonged effect in its retard
form. It is mainly used for the symptomatic relief of several allergic conditions, e.g. rhinitis, hay fever, asthma,
urticaria etc. Drowsiness and palpitations may occur as side effects.
f   phéniramine
d   Pheniramin
i   feniramina
e   feniramina

2906 phenmetrazine
Anorectic-sympathomimetic with a central
mechanism of action, indicated in the
treatment of obesity, particularly if
psychogenous. It alters only very slightly psychic activities and metabolic processes at the tissue level. The drug
should not be used in pregnancy.
f   phenmétrazine
d   Phenmetrazin
i   fenmetrazina
e   fenmetrazina

2907 phenobarbital
Phenobarbitone. Slow-acting, long lasting
barbiturate, i.e. a sedative and hypnotic agent with a particularly strong action on the motor cortex. It is mainly
used to combat insomnia and as a pre-
narcotic, but is widely indicated in the
treatment of epilepsy, especially the
grand mal, although the development
of a tolerance to the drug is bound to
determine some limitations. It also must
be kept in mind that intoxication by
alcohol, narcotics, hypnotics and psycho
drugs, as well as myocardial, hepatic
and renal disorders, represent important
contraindications and a great deal of
precaution is required in the administration of the drug. Cardiovascular disorders may occur as a side effect.
f   phénobarbitale
d   Phenobarbital
i   fenobarbitale
e   fenobarbital

2908 phenobarbital sodium
Sedative, hypnotic and anticonvulsant
agent of the barbiturate group. It is
also a prenarcotic and its action is
rapid and lasts for a fairly long time.
It potentiates the effects of other central
-nervous-system-depressing drugs. By
inducing enzymes of the steroidal catabolism, the drug appears capable of
neutralizing the contraceptive activity
of the oral oestro-progestetional combinations.

f sodium de phénobarbitale
d Phenobarbitalnatrium
i fenobarbitale sodico
e fenobarbitona sódica

2909 phenobutiodil
Opaque contrast medium used for the examination of the biliary tract. Nausea and diarrhoea may occur as side effects.
f phénobutiodil
d Phenobutiodil
i fenobutiodil
e fenobutiodil

2910 phenocoll
Aminoacetophenitidin. Analgesic agent usually administered in the form of hydrochloride. Colourless, crystalline, slightly saline-tasting solid easily soluble in water. It is used in the treatment of rheumatism and to soothe neuralgia and headache.
f phénocolle
d Phenocoll
i fenocollo
e fenocola

2911 phenol
Carbolic acid. Colourless crystalline compound, soluble in water but more easily in alcohol or ether. Phenol is obtained from coal tar and is widely used in the manufacture of drugs. It was one of best known and greatly used disinfectants in the past, but its use has been very considerably restricted since it has proved toxic to tissue cells in the concentration that is demanded to kill bacteria. Phenol may be used topically as a cauterizing agent.
f phénol; acide phénylique
d Phenol; Carbolsäure
i fenolo; acido carbolico
e fenolo; ácido carbólico

2912 phenolated water
Water treated with phenol for disinfecting purposes. Also called carbolic acid water.
f eau phéniquée
d Phenolwasser
i acqua fenica; acqua carbolica
e agua carbólica; agua fenicada

2913 phenol bismuth
Or bismuth phenolate. Insoluble white powder containing phenol in a varying amount with bismuth oxide. It is used in medicine as an internal disinfectant. In fact, it is capable of liberating phenol in the intestine.
f phénol bismuth
d Phenolwismut
i fenolo bismuto
e fenolo bismuto

2914 phenol camphor
Fungistatic and topical anaesthetic agent medicinally used to relieve cutaneous itching. It is also used in the treatment of epidermophytosis, though with caution as it may cause some tissue damage.
f camphre phéniqué
d Carbolkampfer
i canfora fenicata
e alcanfor fenicato

2915 phenol oil
Phenolated oil, oleum phenolatum, carbolic oil. Solution of phenol in arachis oil mainly used for its emollient action and some anaesthetic effect on the skin. Its bactericidal effectiveness is rather limited.
f huile phéniqué
d Phenolöl
i olio fenicato; olio carbolico
e aceite fenicato

2916 phenolphthalein
Triphenylmethane derivative obtained by heating phenol with phthalic anhydride in the presence of concentrated sulphuric acid. Colourless, slightly water-soluble compound used as an indicator. Phenolphthalein is a potent purgative agent which acts directly on the walls of the intestine causing an abundant and watery secretion usually within eight hours from the administration.
f phénolphtaléine
d Phenolphthalein
i fenolftaleina
e fenolftaleina

2917 phenolphthalol
Contact purgative capable of increasing the mucous secretion of the large intestine and thus softening the faeces. It is not absorbed and is indicated in all forms of constipation and intestinal atony.
f phénolphtalol
d Phenolphthalol
i fenolftalolo
i fenolftalolo

2918 phenol salicylate
Compound of phenol and salicylic acid which, when taken orally, is hydrolyzed in the intestine liberating phenol. It was used in the past for the treatment of enteritis, but the practice was abandoned owing to the impossibility of administering an adequate dose without toxic effects.
f salicylate phénolique
d Phenolsalicylat
i salicilato di fenolo
e salicilato fenólico

2919 phenolsulphonate
Salt or ester of phenolsulphonic acid. Also called sulphocarbolate. Phenolsulphonates are widely used as antiseptics.
f phénolsulfonate
d Phenolsulfonat
i fenolsolfonato
e fenolsulfonato

2920 **phenolsulphonic acid**
Or sulphocarbolic acid. Sulphonic acid obtained by treating phenol with concentrated sulphuric acid. Three isomeric forms exist, all of them characterized by antiseptic properties. Also, the salts of the phenolsulphonic acid are marked by antiseptic properties; in medicine, the salts of aluminium, calcium, magnesium, zinc etc. are used. The salts of potassium and magnesium are used as intestinal antiseptics.
f acide phénolsulfonique; acide sozolique
d Phenolsulfonsäure; Sozolsäure
i acido fenolsolfonico; acido solfofenico
e ácido fenolsulfónico; ácido sulfofénico ácido sozólico

2921 **phenolsulphophthalein**
Phenolphthalein derivative only used to assess the level of the kidney functions, i.e. glomerular filtration and tubular secretion. To this purpose, the drug is injected intravenously and its concentration is measured in the urine.
f phénolsulfophtaléine
d Phenolsulfophthalein
i fenolsolfoftaleina
e fenolsulfoftaleina

2922 **phenomorphan**
Potent hypnotic and sedative agent, potentially a drug of addiction.
f phénomorphane
d Phenomorphan
i fenomorfano
e fenomorfano

2923 **phenoperidine**
Narcotic analgesic with action and side effects similar to those of morphine. It is used for the relief of severe pains, in particular the pain that is associated with the metastatic lesions of malignant tumours. Also used with other drugs, especially droperidol, to produce neuroleptanalgesia, a state of consciousness characterized by almost total indifference, thus allowing the patient to cooperate with the surgeon.
f phénopéridine
d Phenoperidin
i fenoperidina
e fenoperidina

2924 **phenothiazine**
Anthelmintic agent used in the treatment of conditions related to infestation of Nematodes. The absorbed component of the drug may act as a mild diuretic.
f phénothiazine
d Phenothiazin
i fenotiazina
e fenotiazina

  * **phenoxazole** s. pemoline

2925 **phenoxazoline**
Vasoconstrictor and nasal decongestant having a prolonged effect without a

secondary vasodilation. It is a topical anaesthetic and antihistaminic used in the treatment of subacute coryza, chronic vasomotor rhinitis, common rhinitis, sinusitis (as an adjuvant) and those conditions which follow nasal surgery. The drug should not be used for children under six years of age.
f phénoxazoline
d Phenoxazolin
i fenossazolina
e fenoxazolina

2926 **phenoxyacetic acid**
Compound obtained by reaction of the phenol with monochloroacetic acid. It is characterized by fungicide properties.
f acide phénoxyacétique
d Phenoxyessigsäure
i acido fenossiacetico
e ácido fenoxiacético

2927 **phenoxybenzamine**
Alkylating agent characterized by alpha-adrenoceptor blocking and antihistamine effects. It is indicated in the treatment of phaeochromocytoma (chromaffinoma mainly occurring in the adrenal medulla). It is also used as a peripheral vasodilator and in the treatment of essential hypertension. Large doses may cause postural hypotension for up to two hours. Myosis and malaise may occur as side effects.
f phénoxybenzamine
d Phenoxybenzamin
i fenossibenzammina
e fenoxibenzamina

  * **phenoxybenzylpenicillin** s. phenoxymethylpenicillin

2928 **phenoxyethanol**
Oily, colourless, water-soluble liquid marked by antiseptic properties. Also a bactericide very effective against Pseudomonas aeruginosa even in the presence of serum.
f phénoxyéthanol
d Phenoxyäthanol
i fenossietanolo
e fenoxietanolo

  * **phenoxyethylpenicillin** s. phenoxymethylpenicillin

2929 **phenoxymethylpenicillin**
Penicillin-G derivative which is resistant to the action of gastric juices and can therefore be administered orally. The peak plasma level appears after about 30 minutes and lasts for 4 to 6 hours. The drug is effective against all infections caused by micro-organisms susceptible to penicillin G i.e. both septicaemic and topical infections.
f phénoxyméthylpénicilline
d Phenoxymethylpenicillin
i fenossimetilpenicillina
e fenoximetilpenicilina

2930 phenprobamate
Minor tranquillizer, analgesic, antipyret‑
ic and antiphlogistic agent, as well as
muscle relaxant. It is indicated in the
treatment of pains of various nature,
i.e. arthritis, gout etc. and also to
soothe neuroses marked by anxiety and
insomnia. Headaches and general malaise
may occur as side effects.
f phenprobamate
d Phenprobamat
i fenprobamato
e fenprobamato

2931 phenprocoumon
Coumarin-related synthetic anticoagulant
acting by depressing prothrombin concentration in the blood. Its effect appears after a latency period of 12 to 24
hours. The drug is indicated in the prophylaxis and treatment of intravascular
clotting, thrombophlebitis, pulmonary
embolia and coronary occlusion. A tendency to haemorrhages has been observed as a side effect.
f phenprocoumon
d Phenprocoumon
i fenprocumone
e fenprocumona

2932 phensuximide
Anticonvulsant and antiepileptic agent
with a central mechanism of action,
chiefly used in the prophylaxis and
treatment of epilepsy (petit mal), vertigo and convulsions. Nausea and vomiting, asthenia and general weakness
may occur as side effects.
f phensuximide
d Phensuximid
i fensuccimide
e fensuximido

2933 phentermine
Anorectic-sympathomimetic amine with
a central mechanism of action chiefly
indicated in the treatment of obesity,
in particular obesity of a psychogen
nature. Insomnia, tachycardia, sight
disorders and hypertension may occur as
side effects.
f phentermine
d Phentermin
i fentermina
e fentermina

2934 phentolamine
Alpha-adrenoreceptor blocking drug with
partial agonist and smooth-muscle relaxant activity. It is indicated in the
treatment of hypertension and peripheral
vasculopathies. It does not alter the
cardiac activity. Sight disorders may
occur as a side effect.
f phentolamine
d Phentolamin
i fentolamina
e fentolamina

2935 phenylacetic acid

Acid marked by an acidity which is
equal to that of benzoic acid and by a
high reactivity of the methylene group.
It therefore finds employment in several
organic syntheses. In its hydroxy form
it occurs in the organism as the result
of tyrosine decomposition in the metabolism.
f acide phénylacétique
d Phenylessigsäure
i acido fenilacetico
e ácido fenilacético

2936 phenylalanine
Amino acid found in many proteins but
not in the protamines. It belongs to
the category of essential amino acids
and therefore its presence in the ordinary diet is indispensable as the organism cannot synthesize it. Phenylalanine
provides the benzene nucleus for the
metabolism.
f phénylalanine
d Phenylalanin
i fenilalanina
e fenilalanina

* phenylamine   s. aniline

* phenylaminobenzoic   acid   s. phenylanthranilic acid

* phenylaminopropionic   acid   s. phenylalanine

2937 phenylaminosalicylic acid
Phenyl derivative of the aminosalicylic
acid, the isomer with the aminic group
in the para-position. A chemotherapeutic
preparation indicated in the treatment
of tuberculosis.
f acide phénylaminosalicylique
d Phenylaminosalicylsäure
i acido fenilamminosalicilico
e ácido fenilaminosalicílico

2938 phenylanthranilic acid
Acid which is used as an indicator in
the assessment of ferrous iron in organic
mixtures.
f acide phénylanthranilique
d Phenylanthranilsäure
i acido fenilantranilico
e ácido fenilantranílico

2939 phenylbutazone
Colourless bitterish powder, soluble in
alkaline solutions. It has a considerable
pharmaceutical application owing to its
analgesic and antiphlogistic properties
and is therefore successfully used in the
treatment of rheumatoid arthritis, gout,
ankylosing spondylarthritis etc. The
drug must be administered with great
caution owing to its many counterindica-
tions and side effects.
f phénylbutazone
d Phenylbutazon
i fenilbutazone
e fenilbutazona

2940 phenylcinchoninic acid
Pale yellow powder soluble in alkalis
and used as an antipyretic and analgesic agent. It increases the excretion
of uric acid and is therefore indicated
in the treatment of gout.
f acide phénylcinchoninique
d Phenylcinchoninsäure
i acido fenilcinconinico
e ácido fenilcinconínico

2941 phenylcyclopentanecarboxylic acid
Diethylaminoethyl derivative, a parasympatholytic agent which has proved
useful in the treatment of muscular
spasms and tremor in parkinsonism.
f acide phénylcyclopentanecarboxylique
d Phenylcyclopentankarboxylsäure
i acido fenilciclopentancarbossilico
e ácido fenilciclopentanocarboxílico

2942 phenylephrine
White crystalline powder, almost insoluble in water and not easily soluble in
alcohol. Its hydrochloride is marked by
non-toxic vasoconstricting properties. It
is widely used against nasal congestion
in sinusitis, hay fever, rhinitis and a
mydriatic if applied topically in the
eye. Also indicated in the treatment of
hypertension.
f phényléphrine
d Phenylephrin
i fenilefrina
e fenilefrina

2943 phenylethanolamine
Ephedrine-related chemical compound
marked by a range of actions which are
similar to those of ephedrine, q.v.
f phényléthanolamine
d Phenylethanolamin
i feniletanolammina
e feniletanolamina

2944 phenylethylbarbituric acid
Compound of the barbituric class. Hypnotic and sedative agent particularly
used in the treatment of epilepsy. Known
as phenobarbitone.
f acide phényléthylbarbiturique
d Phenylethylbarbitursäure
i acido feniletilbarbiturico
e ácido feniletilbarbitúrico

2945 phenylglycinamide-p-arsonic acid
Acid the sodium salt of which, known as
tryparsamide, is indicated in the treatment of trypanosomes and spirochaetal
infections.
f acide phénylglycinamide-p-arsonique
d Phenylglycinamid-p-arsonsäure
i acido fenilglicinammide-p-arsonico
e ácido fenilglicinamido-p-arsónico

2946 phenylglycine
Intermediate compound formed in the
course of one of the synthesis processes
of the indigo and derived from the reaction between chloroacetic acid and aniline. From this product, by treatment
with alkalis, indoxyl can be obtained.
f phénylglycine
d Phenylglycin
i fenilglicina
e fenilglicina

2947 phenylhydracrylic acid
Tropic acid; phenylhydroxypropionic
acid; hydroxysaturated acid prepared
from atropine, in itself a tropine ester,
by hydrolysis. Tropine is a substance
constituting the nucleus of two important
series of alkaloids; those of the coca
marked by a topical anaesthetic activity
and those of the solanaceae (atropine,
q.v.).
f acide phénylhydracrylique; acide tro-
   pique
d Phenylhydracrylsäure; Tropasäure
i acido fenilidracrilico; acido tropico
e ácido fenilhidracrílico; ácido trópico

2948 phenyllactic acid
Or atrolactinic acid. White crystalline
acid derived from acetophenone. Like
phenylacetic acid, it may account for
hereditary metabolic errors correlated
with tyrosine.
f acide atrolactinique; acide phényllac-
   tique
d Atrolaktinsäure
i acido fenillattico; acido atrolattinico
e ácido atrolactínico

2949 phenylmercuric
Relating to a compound of mercury containing the phenyl radical (C_6H_5). Several of these compounds are important
and widely used bacteriostatic agents.
f phénylmercurique
d Phenylquecksilber-; Phenylmerkuri-
i fenilmercurico
e fenilmercúrico

2950 phenylmercuric acetate
Organic mercury compound having an
antibacterial and antimycotic action.
A spermicide at pH below 7.2. It is
used as vaginal antiseptic in the prophylaxis and therapy of leucorrhoea,
vulvitis, vaginitis, cervicitis, for hygienic purposes and as a chemical contraceptive. Also used as sterilizer and
preservative for eye lotions. Skin inflammation accompanied by erythema and
vesicles may sometimes occur as a side
effect.
f acétate phénylmercurique
d Phenylmerkuracetat
i acetossifenilmercurio
e acetato fenilmercúrico

2951 phenylmercuric borate
Topical mercury-related antiseptic and
disinfectant very active against bacteria, spores, viruses, mycetes and protozoa. It is not an irritant and does
not produce resistant strains. It is
mainly used for surgical disinfection in

gynecology, ophthalmology and otorhinolaryngology.
f bcrate phénylmercurique
d Phenylmerkurborat
i fenilmercurio borato
e borato fenilmercúrico

2952 **phenylmercuric nitrate**
Colourless crystalline compound not easily soluble in water, yet a strong bacteriostatic agent even when greatly diluted. It is used for the direct medication of wounds and, more generally, for
the disinfection of the skin. At a concentration of 0.1%, it is used as a bacteriostatic agent when preparing a solution for injections.
f nitrate phénylmercurique
d Phenylmerkurinitrat
i nitrato fenilmercurico
e nitrato fenilmercúrico

2953 **phenylphenol**
Chemical compound used in the state of
sodium salt; its chloride derivatives are
used as germicides and disinfectants.
f phénylphénol
d Phenylphenol
i fenilfenolo
e fenilfenol

2954 **phenylphosphine**
Aromatic compound of the phosphorus
with a composition that is analogous to
that of aniline but considerably less
stable. It is a bad-smelling, colourless,
water-soluble liquid which, by a reaction with dichlorophenylphosphine yields
phosphobenzene, a compound corresponding to azobenzene. In its turn, the dichlorophenylphosphine yields the phenyl-
phosphorous acid by hydrolysis, some
of whose derivatives are used in medicine as stimulating agents for the metabolism.
f phénylphosphine
d Phenylphosphin
i fenilfosfina
e fenilfosfina

* **phenylpropanolamine   s. norephedrine**

2955 **phenylpyruvic acid**
Crystalline keto acid found in urine as
a metabolic product of phenylalanine,
particularly in phenylketonuria, a
hereditary metabolic error which may
cause some lesions in the central nervous system, mental deficiency, convulsions, especially in children, and partial albinism.
f acide phénylpyruvique
d Phenylbrenztraubensäure
i acido fenilpiruvico
e ácido fenilpirúvico

2956 **phenylthiourea**
Compound containing the group =N-C=S;
the ability to taste it is inherited and
dependent on a single gene pair. It is

bitter for some people and tasteless for
others. Those who can taste it are either homozygous or heterozygous for the
dominant allele.
f phénylthiourée
d Phenylthioharnstoff; Phenylthiocarbamid
i feniltiourea
e feniltiourea

2957 **phenyltoloxamine**
Antihistamine, antispasmodic, analgesic
and antipyretic agent with central mechanism of action. It is indicated in the
prophylaxis and therapy of all allergic
conditions. It can be administered orally
or for local use. Some drowsiness and
dizziness may occur as side effects.
f phényltoloxamine
d Phenyltoloxamin
i feniltolossamina
e feniltoloxamina

2958 **phenyracillin**
Benzylpenicillin salt of 2.5 diphenyl-
piperazine. Benzylpenicillin is the only
type of penicillin that is permitted in
some countries.
f phényracilline
d Phenyracillin
i feniracillina
e feniracilina

2959 **phenyramidol**
Analgesic, antispasmodic and psychotherapeutic agent especially indicated
in orthopaedics. The drug is also used
in functional reeducation, for the relief
of skeletal-muscle spasms, arthritis and
lumbago, as well as of painful menstruations, premenstrual syndrome and post-
-partum pains. Gastro-intestinal disorders and general malaise may occur as
side effects.
f phényramidol
d Phenyramidol
i feniramidolo
e feniramidolo

2960 **phenytoin**
Anticonvulsant agent capable of suppressing epileptic discharge in the brain.
The drug is used orally to prevent convulsions and by injection to suppress
irregular heart rhythms. A long-term
treatment with the drug requires an
ingestion of at least 100 micrograms of
vitamin D weekly. Nausea accompanied
by vomiting, ataxia, sight disorders,
tremor, insomnia and tenderness with
hyperplasia of gums may occur as side
effects.
f phénytoïne
d Phenytoin
i fenitoina
e fenitoina

2961 **phloridzin**
Or phlorizin. Glycoside found in the root
and bark of some fruit trees (apple,
pear, cherry etc.). Silky, colourless,

bitter-sweet-tasting needles, not easily soluble in water. Phloridzin, by treatment with diluted acids, yields glucose and phloretin which in its turn yields floretic acid by oxidation. It has the property of inhibiting alcoholic fermentation and modifying intestinal and renal permeability to glucose. Experimentally administered to animals, it causes a pseudodiabetes (phloridzinic diabetes) characterized by glycosuria without alteration of the concentration of glucose in the blood.

f phloridzine; phlorizine
d Phloridzin; Phlorizin
i florizina
e floricina; floridzina

2962 phloroglucinol
Spasmolytic agent acting electively on the smooth-muscle cells of the urinary and biliary tract. Also active on the uterine and intestinal muscles. It is indicated for the treatment of renal colics and painful disorders of the urinary tract, of hepatic colics, spastic conditions of the female genital apparatus and intestine. The drug cannot be administered together with morphine.

f phloroglucinol
d Phloroglucinol
i floroglucina
e floroglucinolo

2963 pholcodine
Codeine-related cough sedative. It has a very mild analgesic action and an equally mild sedative action. It is sometimes preferred to codeine owing to an appreciably diminished possibility of drug dependence. It is particularly indicated in the treatment of dry coughs.

f pholcodine
d Pholcodin
i folcodina
e folcodina

2964 pholedrine
Sympathomimetic agent marked by a vasoconstricting action and used to restore and maintain normal blood pressure in all states of acute hypotension and shock. Tachycardia may occur as side effect.

f pholédrine
d Pholedrin
i foledrina
e foledrina

2965 phosphagen
Creatine phosphate in vertebrate tissues and arginine phosphate in invertebrate tissues, acting as energy store therein. Phosphagens may be metabolized in muscles during contraction and in nerves during conduction of a nerve impulse. Creatine phosphate can therefore react with ADP to yield creatine and with ATP which supplies the energy for muscle

contraction and nerve conduction. In cardiology, phosphocreatine is indicated in the treatment of cardiopathies, particularly myocardiopathies, yet cannot be considered a substitute of cardiokinetic therapy.

f phosphagène; phosphocréatine
d Phosphagen; Phosphocreatin
i fosfageno; fosfocreatina
e fosfageno; fosfocreatina

2966 phosphate
Salt or ester of phosphoric acid, i.e. formed by orthophosphoric acid. Phosphates are found naturally in minerals and in deposits of sea-birds' excreta.

f phosphate
d Phosphat; phosphorsaures Salz
i fosfato
e fosfato

2967 phosphatidic acid
Ester of glycerol in which two of the three alcohol groups are esterified with fatty acids and the third one with phosphoric acid. Phosphatidic acids have a considerable importance as intermediates in the biosynthesis of lipids.

f acide phosphatidique
d Phosphatidsäure
i acido fosfatidico
e ácido fosfatídico

2968 phosphatidum cerebri
Cerebral phosphatide. Neurotrophic agent capable of stimulating the endogenous biosynthesis of the cerebral phospholipids by activating the functional metabolism of the neuron in cases of membrane alterations caused by anoxia or trauma. It is chiefly indicated in states of depression and anxiety even when due to arteriosclerotic encephalopathies, cephalalgic syndromes, secondary symptomatology of electroshock therapy and the administration of lytic cocktails (a mixture of analgesic and phenothiazine derivatives used to potentiate anaesthesia).

f phosphatide cérébrale
d Phosphatidum cerebri
i fosfatide cerebrale
e fosfatida cerebral

2969 phosphide
Binary compound of phosphorus with a more electropositive element or radical. Phosphides of alkaline metals (and aluminium) may form by substitution of one or more hydrogen atoms of phosphine with the relative elements and be hydrolyzed by treatment with diluted acids or water. Zinc phosphide is used in medicine.

f phosphure
d Phosphid
i fosfuro
e fosfuro

2970 phosphite

Salt or ester of phosphorous acid. The salts are obtained by substitution of one, two or all three hydrogen atoms with metals. It is thus possible to obtain neutral or acid phosphites. Acid phosphites are mostly water-soluble and, under strong heat, develop phosphoretted hydrogen. The most important phosphites are those of ammonium, calcium, sodium and potassium.

f phosphite
d Phosphit
i fosfito
e fosfito

* **phosphocreatine   s. phosphagen**

2971 phosphocreatinine
Cyclic form of phosphocreatine marked by a high-energy nitrogen phosphate bond. It affects chemical and physiological processes leading to the rapid reconstitution of ATP. It constitutes the energy reserve of voluntary and involun̲tary striated muscles. The drug is there̲fore indicated in the treatment of all pathological conditions deriving from an altered energy balance. In cardiology, phosphocreatinine is used for the treatment of myocardial diseases caused by defective metabolism, ischaemia of the coronary arteries, intoxication caused by cardio-active drugs etc. The drug is also used in neurology (primitive and secondary myodystrophies) and traumatology.

f phosphocréatinine
d Phosphocreatinin
i fosfocreatinina
e fosfocreatinina

2972 phosphoglyceric acid
Compound existing in two isomeric forms according to whether the phosphoric group is esterified with a primary or secondary alcoholic group. It is an intermediate product of the metabolism of carbohydrates.

f acide phosphoglycéride
d Phosphoglyzerinsäure
i acido fosfoglicerico
e ácido fosfoglicérico

2973 phosphomolybdate
Salt of the phosphomolybdic acid; ammonium phosphomolybdate is used as reagent in chemical analysis.

f phosphomolybdate
d Phosphomolybdat
i fosfomolibdato
e fosfomolibdato

2974 phosphomolybdic acid
Yellowish, water-soluble crystals used as reagent and for the preparation of pigments. Its sodium salt, dissolved in nitric acid yields a yellow precipitate with the alkaloids.

f acide phosphomolybdique
d Phosphomolybdänsäure

i acido fosfomolibdico
e ácido fosfomolíbdico

2975 phosphomycin
Antibiotic active against Gram-positive and Gram-negative micro-organisms, including strains that are resistant to other antibiotics, in particular staphylococci resistant to penicillin. The drug is rapidly but not completely absorbed when taken orally and reaches peak blood level in 30 to 60 minutes. Its bond with plasma proteins is rather poor. It is easily excreted in the urine and is used against all infections caused by susceptible organisms.

f phosphomycine
d Phosphomycin
i fosfomicina
e fosfomicina

2976 phosphorated cod liver oil
Cod liver oil, q.v. combined with phosphorus or some substance containing phosphorus.

f huile de foie de morue phosphoré
d Phosphorlebertran
i olio di fegato di merluzzo fosforato
e aceite de hígado de bacalao fosforado

2977 phosphorated oil
One percent solution of phosphorus in almond oil used in the past as a nerve stimulant and a tonic.

f huile phosphoré
d Phosphoröl
i olio fosforato
e aceite fosforado

2978 phosphoric acid
Tribasic acid, a white crystalline solid in the pure state, which melts at 42° yielding a sirupy liquid. It is found in nature in the form of phosphates and phosphatides and prepared by treating animal bones with sulphuric acid. It is used therapeutically as a tonic for the nervous system and for its acidifying action on the urine.

f acide phosphorique
d Phosphorsäure
i acido fosforico
e ácido fosfórico

2979 phosphorodithioic acid
Phosphoric acid, q.v. characterized by the presence of two sulphur atoms in its molecule.

f acide dithiophosphorique
d Dithiophosphorsäure
i acido ditiofosforico
e ácido ditiofosfórico

2980 phosphorothioic acid
Thiophosphoric acid $H_2=PO_3S$. It is deriv̲ed from phosphoric acid by substitution of one or more oxygen atoms with the same number of atoms of sulphur. The corresponding salts are called thiophosphates.

f acide thiophosphorique
d Thiophosphorsäure
i acido tiofosforico
e ácido tiofosfórico

2981 **phosphorotrithioic acid**
Acid derived from phosphoric acid by
the substitution of the thionic acid's
oxygen atoms with the same number of
sulphur atoms.
f acide trithiophosphorique
d Trithiophosphorsäure
i acido tritiofosforico
e ácido tritiofosfórico

2982 **phosphorous acid**
Acid of the trivalent phosphorus. White
or yellowish crystalline masses which
melt at 70° and decompose at 200°. It
is obtained by treating calcium phosphate with sulphuric acid or by hydro-
lyzation of the phosphorus trichloride.
It is used as reducing agent and in
making phosphites.
f acide phosphoré
d phosphorige Säure
i acido fosforoso
e ácido fosforoso

2983 **phosphorus**
Non-metallic multivalent element of the
nitrogen family widely occurring in
combined forms, particularly as inorganic phosphates in minerals, soil, natural
waters, bones and teeth and as organic
phosphates in all living cells. Phosphorus plays a very important part in
metabolism (carbohydrate, lipid, calcium
etc.). Phosphorus has several allotropic
forms. It is tri- and pentavalent and
yields a wide range of oxides, acids
and halogen derivatives.
f phosphore
d Phosphor
i fosforo
e fósforo

2984 **phosphorus pentoxide**
Chemical compound known in various
polymeric forms usually obtained by
burning phosphorus in an excess of dry
air and found as a white powder which
reacts very strongly and sometimes ex-
plosively with water to form phosphoric
acids irritating to the skin and mucous
membranes. It is generally used as a
condensing agent in organic synthesis
and in the production of phosphoric
acids and their derivatives.
f pentoxyde de phosphore
d Phosphorpentoxyd
i pentossido di fosforo
e pentóxido de fósforo

2985 **phosphorus trichloride**
Compound obtained by direct combination
of the elements in controlled temperature
conditions. Fuming, colourless, pungent
poisonous liquid rapidly decomposing in
humid air. It is an effective solvent of

phosphorus and miscible with benzene,
ether, chloroform etc., thus finding
employment in many organic syntheses
(saccharine etc.).
f trichlorure de phosphore
d Phosphortrichlorid
i tricloruro di fosforo
e tricloruro de fósforo

2986 **phosphorylation**
Chemical reaction between the phosphoric
acid and another compound, usually organic, with the elimination of a molecule
of water. Specifically, the enzymatic
conversion of carbohydrates into their
phosphoric esters in metabolic processes.
Phosphorylation is therefore important
for the highly energetic compounds which
are formed when it is associated with
biological oxidation. The energy released in the oxidation of lipids and gly-
cids is stored and subsequently made
available in the organism through phosphorylation of ADP and ATP.
f phosphorylation
d Phosphorylierung
i fosforilazione
e fosforilación

2987 **phosphorylcholine**
Lipotropic and hepatoprotective agent
indicated in the treatment of hepatobiliary insufficiency. It is also used
in association with other lipotropic
amino acids. In the form of calcium
salt, it is used to balance a condition
of calcium and phosphorus deficiency,
e.g. in pregnancy, convalescence etc.
It is possible that it promotes the biosynthesis of phosphatides and acetylcholine.
f phosphorylcholine
d Phosphorylcholin
i fosforilcolina
e fosforilcolina

2988 **phosphorylcolamine**
Synthetic amino acid characterized by
a high phosphorus content and apparently capable of improving metabolism. It
is generally used in debilitated patients
as a tonic.
f phosphorylcolamine
d Phosphorylcolamin
i fosforilcolamina
e fosforilcolamina

2989 **phosphotungstate**
The salt of phosphotungstic acid.
f phosphotungstate
d Phosphotungstat; Phosphowolframat
i fosfotungstato
e fosfotungstato; fosfowolframato

2990 **phosphotungstic acid**
Yellow crystals which are obtained by
the evaporation of a solution containing
phosphoric and tungstic acids. Its
salts are used as reagents in analytical
chemistry. Also known as tungstophos-

phoric acid.
f acide phosphotungstique
d Phosphorwolframsäure
i acido fosfotungstico
e ácido fosfotúngstico

2991 phthalazine
Chemical compound. Yellow water-soluble
needles. Various chlorinated derivatives
are known. It is the azine of the dial-
dehyde related to phthalic acid.
f phtalazine
d Phthalazin
i ftalazina
e ftalazina

2992 phthalein
Any of dyes prepared from phthalic anhydride and a phenol. Some of them
are used as indicators and find employment in medicine for their purgative action.
f phtaléine
d Phthalein
i ftaleina
e ftaleina

2993 phthalein amide
Heart and respiratory analeptic marked
by a general euphoristic (euphoria-induc
ing) and stimulating action. It is chiefly indicated in the treatment of infectious diseases, states of collapse, drug
or carbon-dioxide intoxications and
post-operative complications.
f amide de phtaléine
d Phthaleinamid
i ftaletamide
e ftaletamida

2994 phthalic acid
Any of three isomeric dicarboxylic acids
that are obtained by oxidation of various benzene derivatives. A crystalline
acid usually obtained by hydration of
phthalic anhydride. It is used in making various solvents, dyes, benzoic acid
and intermediates.
f acide phtalique
d Phthalsäure
i acido ftalico
e ácido ftálico

2995 phthalylsulphathiazole
Sulphonamide which is not absorbed by
the intestinal tract and remains in the
intestine in a high concentration, acting on the bacterial flora that causes
non-specific enteritis, i.e. enteritis due
to E.coli, Proteus etc. Owing to the
action of the drug on the saprophytic
intestinal flora, hypovitaminosis may
occur as a side effect.
f phtalylsulfathiazole
d Phthalylsulfathiazol
i ftalilsulfatiazolo
e ftalilsulfatiazolo

2996 phthiocol
Chemical compound of quinone nature;
antibiotic isolated from cultures of Mycobacterium tuberculosis (the bacillus of
human tuberculosis). It is related to
vitamin K and has antihaemolytic properties.
f phtiocol
d Phthiocol
i ftiocolo
e ftiocol

2997 phylloquinone
Vitamin K. A vitamin that is appreciably
important for normal blood coagulation
and thus essential for the formation of
prothrombin. It is present in fresh
vegetables and more generally in all
green plants. It is also known as antihaemorrhagic vitamin.
f phylloquinone; vitamine K; vitamine anti
hémorrhagique
d Phyllochinon; antihämorrhagisches Vitamin;
Koagulationsvitamin; Vitamin K
i fillochinone; vitamina K
e filoquinona; vitamina antihemorrágica;
vitamina K

2998 physostigmine
Alkaloid obtained from Calabar beans,
i.e. the dried seeds of Physostigma vene
nosum. It inhibits the action of cholines
terase, the enzyme which destroys acetyl
choline in the organism. Physostigmine
potentiates all the muscarinic effects
of acetylcholine but it has a more complex influence on the nicotinic properties. The drug increases the cholinergic
response at neuromuscular junctions in
skeletal muscles thus causing a tetanic
response instead of the normal single
twitch of the muscle. Also, in case the
muscle has been curarized, it tends to
restore conduction. The drug is therefore
used to reduce intra-ocular pressure and
to reverse the effects of homatropine
when administered for diagnostic purposes. It has also been tried in myasthe
nia gravis.
f physostigmine
d Physostigmin
i fisostigmina
e fisostigmina

2999 physostigmine salicylate
Natural cholesterasic agent resembling
neostigmine methylsulphate and used as
a miotic agent as it counteracts the
dilatation effect of atropine in the
pupil. It is also indicated in the treatment of hypertension in glaucoma thus
differing from its methylsulphate, it
crosses the blood-brain barrier and can
reverse the peripheral effects of anticholinergic drugs. Side effects are similar to those of neostigmine methylsulphate, q.v.
f salicylate de physostigmine
d Physostigminsalicylat
i fisostigmina salicilato
e salicilato de fisostigmina

3000 physovenine
Alkaloid occurring together with physio-
stigmine in the Calabar bean. Its myotic
properties are similar to those of physio
stigmine, q.v.
f physiovénine
d Physiovenin
i fisiovenina
e fisiovenina

3001 phytic acid
Acid obtained on acidification of phytin,
the calcium-magnesium salt of inositol
phosphoric acid, i.e. the phytic acid,
which is found in several plants and
particularly in their seeds. It is indicat
ed in the treatment of anaemia and
neuroasthenia.
f acide phytinique; acide phytique
d Phytinsäure
i acido fitico
e ácido fítico

3002 phytin
Calcium magnesium salt of phytic acid
occurring as reserve material, particular
ly in seeds. It is used as a source of
inositol, q.v.
f phytine
d Phytin
i fitina
e fitina

3003 phytolacca
Or poke root. The root of a North American plant (Phytolacca americana) which
contains a bitter resin, a saponin, and
has cathartic and emetic properties. See
also poke berry.
f phytolaque
d Kermes; Nachtschatten
i fitolacca
e fitolaca

3004 phytomenadione
Synthetic vitamin K used in the prophylaxis or treatment of hypothrombo-
naemia tending to produce haemorrhages
(insufficient production of factor II, VII,
IX and X). The drug may be used for
babies.
f phytoménadione
d Phytomenadion
i fitomenadione
e fitomenadiona

3005 picolamine
Analgesic agent characterized by a deep
haemodynamic action and indicated in
the treatment of osteoarticular or muscular pains. The compound is sprayed
on the affected surface.
f picolamine
d Pikolamin
i picolamina
e picolamina

3006 picoline
Methylpyridine. Compound which is homologous with pyridine and occurs in three

isomeric forms, all three in the form of
pungent liquids extracted from bone oil
and coal tar. The alpha-compound finds
employment in medicine as a sedative.
f picoline
d Pikolin
i picolina
e picolina

3007 picolinic acid
Carboxylic acid, isomer of the nicotinic
acid, which is obtained by the oxidation
of picoline, q.v. Colourless crystals,
soluble in water and in alcohol. In the
form of a derivative it constitutes an
organic substance corresponding to the
methylbetaine of the pyruvic acid which
has been found in the muscles of the
lobster and other Invertebrates. The
hydrazide of picolinic acid has tuberculostatic properties.
f acide picolinique
d Pikolinsäure
i acido picolinico
e ácido picolínico

3008 picoperine
Non-narcotic antitussive agent acting on
the cough centre and displaying a
spasmolytic activity in the bronchi.
The drug is chiefly indicated for the
treatment of cough independently of its
origin. Drowsiness, anorexia and palpitations are possible side effects.
f picopérine
d Pikoperin
i picoperina
e picoperina

3009 picramic acid
Acid obtained by partial reduction of
the picric acid, mainly with sodium
hydrosulphite, stannous chloride and
hydrochloric acid. Red, alcohol-soluble
crystals constituting an important intermediate agent for the synthesis of dyes.
f acide picramique
d Pikraminsäure
i acido picrammico
e ácido picrámico

3010 picrate
Salt of the picric acid obtained by substitution with a metal of the hydroxylic
groups of the picric acid.
f picrate
d Pikrat; pikrinsaures Salz
i picrato
e picrato

3011 picric acid
Or picronitric acid. Trinitrophenol, an
organic compound so called because of
its very strong acid properties and its
bitter (picric) taste. It is a yellow,
toxic, crystalline solid formed by the
action of nitric acid on many organic
compounds, e.g. phenol, aniline etc. It
is used to precipitate albumen and to
detect glucose in urine. It is a good

antiseptic and was used in the past for the dressing of burns; a practice which has been discontinued owing to the danger of absorption.
f acide picrique
d Pikrinsäure; Trinitrophenol
i acido picrico; trinitrofenolo; acido picrinico
e ácido pícrico; ácido picronítrico; trinitro fenol

3012 **picrolonic acid**
Pyrazolone derivative used in analytical chemistry as a calcium reagent with which it yields a crystalline salt and as a precipitant for alkaloids.
f acide picrolonique
d Pikrolonsäure
i acido picrolonico
e ácido picrolónico

3013 **picrosclerotine**
Ergotinine, a crystalline alkaloid of ergot, originally obtained in 1875 (it was later called secaline) and considered inert. It is, in fact, readily transformed into ergotoxine which contains an extra molecule of water.
f picrosclérotine
d Pikrosklerotin
i picrosclerotina
e picrosclerotina

3014 **picrotoxin**
Active principle isolated from the seeds of Anamirta cocculus, an equimolecular compound of picrotoxinine and picrotin. It is a potent respiratory stimulant and convulsive agent and acts by stimulating the cerebrospinal axis from the top downward, affecting chiefly the cerebral cortex, mid-brain and medulla. In the unanaesthesized patient it causes convulsions first tonic and subsequently clonic. The stimulating of the central nervous system is followed by depression and respiratory failure. Its principal use in antagonizing the central depression following overdoses of narcotic drugs, in particular barbiturates. Hypertension, tachycardia and increase of the respiration volume may occur as side effects.
f picrotoxine
d Picrotoxin
i picrotossina
e picrotoxina

3015 **pifarnine**
Therapeutic agent, efficacious in checking the progress of ulcers and marked by an effective gastric antisecretory action without anticholinergic or antihistaminic activity at the H receptors. The drug has a good, papaverine-like spasmolytic activity and is indicated in the treatment of peptic ulcer, even in cases of recidivation, gastritis, duodenitis, gastric hyperacidity and hypermotility and any gastric disorders that are caus

ed by drugs originating toxic effects in the stomach. Neurotoxic reactions may result from high doses administered to patients suffering from kidney or central-nervous-system disorders.
f pifarnine
d Pifarnin
i pifarnina
e pifarnina

3016 **pifoxime**
Antiinflammatory, antirheumatic and analgesic agent indicated in acute conditions of inflammatory, degenerative, extra-articular rheumatism, gout and sciatica. The drug may affect digestion. Headache, vertigo, itching, insomnia and asthenia may occur as side effects.
f pifoxime
d Pifoxim
i pifossima
e pifoxima

3017 **pill**
Any medicine in the form of a small rounded mass, sometimes coated, to be swallowed whole.
f pilule
d Pille
i pillola
e pildora

3018 **pilocarpine**
Alkaloid derived from imidazole and butyrolactone which is obtained from jaborandi as an oily syrup crystallizing when pure, that is a strong sialogue and diaphoretic. Pilocarpine is mostly used in the form of its hydrochloride or nitrate as a miotic in glaucoma and to combat mydriasis caused by atropine.
f pilocarpine
d Pilokarpin
i pilocarpina
e pilocarpina

3019 **pilocarpine chloride**
Hydrochloride of the alkaloid obtained from the dried leaves of Pilocarpus microphyllus and other species of the Rutaceae family. Parasympathomimetic agent chiefly used to antagonize the effects of atropine. It increases secretions (e.g. milk, biliary etc.) and acts as cardiac depressant and vasodilator. It also promotes peristalsis and decreases intraocular pressure in glaucoma. Some colic pains and occasionally diarrhoea may occur as side effects.
f chlorure de pilocarpine
d Pilokarpinchlorid
i pilocarpina cloruro
e cloruro de pilocarpina

3020 **pilocarpine nitrate**
Nitrate of the alkaloid obtained from the dried leaves of Pilocarpus microphyllus and other species of the Rutaceae family. Parasympathomimetic agent capable of increasing salivary, sudoriparous, gas-

tric, pancreatic and bronchial secretion. Albeit in a minor way, it also affects the secretion of milk, bile and urine. Diarrhoea, bradycardia, hypotension and myosis marked by diminished intraocular pressure generally accompany the administration of the drug.

f nitrate de pilocarpine
d Pilokarpinnitrat
i pilocarpina nitrato
e nitrato de pilocarpina

3021 pimaric acid
Chemical compound, a constituent of the resinic acids of the oleoresins of various species of pine. The acid is optically active and occurs in the two forms, i.e. dextro- and laevo-. Three benzene rings are present in its structure.

f acide pimarique
d Pimarsäure
i acido pimarico
e ácido pimárico

*** pimaricine   s. natamycin**

3022 pimeclone
Respiratory analeptic acting directly on the respiration centres and chiefly used in asphyxia neonatorum, respiratory depression or paralysis caused by anaesthesia, intoxications due to morphine, various barbiturates and other hypnotics, as well as respiratory disorders in the course of pneumonia.

f piméclone
d Pimeclon
i pimeclone
e pimeclona

3023 pimefylline
Haemokinetic vasodilator tending to diminish peripheral resistance thus increasing arterial flow. It is used in the form of nicotinate (salt of nicotinic acid). It is chiefly indicated in the treatment of arterial circulatory disorders, both acute and chronic, peripheral or cerebral vasculopathies, vascular ophthalmic affections, sclerosis of the coronary arteries and pulmonary circulatory disorders. Some itching and headaches are possible side effects.

f pimefilline
d Pimephyllin
i pimefillina
e pimefilina

3024 pimelic acid
Aliphatic bicarboxylic acid occurring in the oxidation products of fats; it can be obtained by oxidation of suberone (cycloheptanone) and by several syntheses, starting from cyclohexanone, salicylic acid etc. It is used in organic syntheses, for the preparation of esters etc. It is also capable of stimulating the growth of the diphtheria bacillus.

f acide pimélique
d Pimelinsäure
i acido pimelico
e ácido pimélico

3025 pimethixene
Amine-antagonist agent marked by sedative properties. It inhibits the action of histamine, serotonin, acetylcholine, bradykinin and epinephrine. It is mainly used in paediatrics against allergic cutaneous diseases, sleep disorders as well as disorders of the respiratory tract. Drowsiness, nausea and dizziness are possible side effects.

f piméthixène
d Pimethyxen
i pimetixene
e pimetixeno

3026 piminodine
Analgesic stupefacient agent generally used in preoperative care and to soothe the intense pains which accompany a myocardial infarction, angina pectoris, cholecystitis, pleurisy and neoplasms. Nausea, anorexia, confusional states, urinary bladder disorders, vertigo and restlessness may occur as side effects.

f piminodine
d Piminodin
i piminodina
e piminodina

3027 pimozide
Tranquillizer causing selective depression of the brain structures that are responsible for the control of behaviour and wakefulness. It is indicated in the treatment of psychotic disorders. For side effects, see Chlorpromazine.

f pimozide
d Pimozid
i pimozide
e pimozido

3028 pimpernel oil
Essential oil contained in the root of the pimpernel; a yellow-golden liquid marked by a pleasant smell and a pungent bitter taste. It is used in preparations of several medicinal drugs.

f huile de boucage
d Pimpinellöl
i olio essenziale di pimpinella
e esencia de pimpinela

3029 pimpernel root
The root of the Pimpinella anisum, Pimpinella major and Pimpinella saxifraga, which yields a bitter principle used in the preparation of several medicinal drugs.

f racine de boucage
d Pimpinellwurzel;   Bibernellenwurzel; Pfefferwurzel
i radice di pimpinella; radice di tragoselino
e raíz de pimpinela

3030 pimpinella
Pimpernel. Genus of herbs of the family

Umbrelliferae with 200 species growing in temperate regions. The most important are the Pimpinella anisum, the Pimpinella saxifraga and the Pimpinella major which have roots yielding active medicinal principles.
f boucage; anagallide
d Bibernelle; Pimpinelle
i pimpinella; primula
e pimpinela

3031 pinaverium bromide
Papaverine-type spasmolytic agent and contact analgesic. It is indicated in the treatment of oesophageal dyskinesia, gastroduodenal ulcer, colitis, colopathies and biliary dyskinesia. The drug is also used for gastro-intestinal radiological examination. Constipation may occur as side effect.
f bromure de pinavérium
d Pinaveriumbromid
i pinaverio bromuro
e bromuro de pinaverio

3032 pinazepam
Benzodiazepine-related tranquillizer and efficient muscle relaxant, sedative and anticonvulsant agent indicated in the treatment of states of anxiety, neuroses, depressive syndromes and various, partly psychosomatic, disorders connected with old age involution. It also finds employment in the treatment of alcoholism and as an adjuvant for epileptic crises. Drowsiness and nausea are the most frequent side effects.
f pinazépam
d Pinazepam
i pinazepam
e pinazepam

3033 pindolol
Beta-adrenergic receptor blocking agent marked by a mild action on the bronchial beta-adrenergic receptors and an equally mild negatively inotropic action. The drug is indicated in the treatment of coronary insufficiency, arrhythmias and cardiac disorders due to sympathetic hypertonia. Bradycardia, hypotension, congestive heart failure, potentiation of A-V block are, among many others, the principal side effects of the drug.
f pindolol
d Pindolol
i pindololo
e pindololo

3034 pineapple
The fruit of the Ananas sativus of the Bromeliaceae family. It contains bromelin, a proteolytic enzyme and is used medicinally as a diuretic agent. The unripe fruit has an irritant action.
f ananas
d Ananas
i ananas
e ananas

3035 pineapple weed
Annual aromatic herb native to the Pacific coast of North America but widely naturalized. It produces yellow flowers which are very similar to, and have the same effects, of chamomile.
f matricaire discoïde
d Kopfkamille; Zigeunerkamille
i matricaria discoidea
e matricaria discoidea

3036 pine-needle oil
Essential oil obtained from the needles and twigs of several conifers. A colourless or yellowish bitter aromatic oil used in medicine as an expectorant and inhalant agent. Also, the oil obtained from the Siberian fir which is used as inhalant.
f huile de bourgeons de pin
d Fichtensprossenöl
i olio di gemme di pino
e aceite de yemas de pino

3037 pine-needle syrup
Syrupy liquid obtained from the needles and twigs of several conifers and used as an expectorant.
f sirop de bourgeons de pin
d Fichtensprossensirup
i sciroppo di gemme di pino
e jarabe de pino silvestre

3038 pine oil
Liquid obtained by distillation of the trunks and branches of the pine and other conifers and containing tar, methyl alcohol and acetic acid. It is mostly colourless or light-amber with an aroma of pine. It is used as a disinfectant, deodorant and insecticide.
f huile de pin; essence de térébenthine de bois
d Holzterpentinöl; Kienöl
i olio di pino; olio di legno di pino
e aceite de pino; esencia de trementina de madera

3039 pine sprout
The buds of the trees of the Pinaceae family used medicinally for their balsamic action, e.g. in the treatment of some catarrhal affection of the respiratory tract (tracheitis, bronchitis etc.) and for their antiseptic and diuretic action in the treatment of some forms of cystitis.
f bourgeon de pin
d Fichtensprosse; Fichtenspitze; Tannenspitze
i gemma di pino
e yema de pino

3040 pine tar
Tar obtained by destructive distillation of pine wood. A viscid dark-brown phenolic liquid used as a softener and, in medicine, in the treatment of cutaneous diseases. Also known as wood tar, Stockholmtar.

f goudron végétal; goudron de bois
d Holzteer; Nadelholzteer
i catrame di legna; catrame vegetale;
  pece liquida
e alquitrán de pino; alquitrán de madera;
  brea vegetal

3041 pine-tar oil
Dark-brown phenolic liquid obtained by distillation of pine tar and mainly used as an antiseptic, a deodorant and insecticide. Also known as wood-tar oil.
f huile de goudron
d Teeröl
i olio di catrame
e aceite de brea

3042 pinic acid
Bicarboxylic acid obtained from the oxidation of pinonic acid, q.v.
f acide pinique
d Pininsäure; Fichtensäure
i acido pinico
e ácido pínico

3043 pinkroot
Perennial woodland herb originally cultivated in North America and used for its anthelminthic properties.
f racine de spigélie du Maryland
d Pinkwurzel; marylandische Spiegelwurzel
i radice di spigella; rizoma di spigella
e raíz de espigelia

3044 pinonic acid
Crystalline keto acid derived from dimethyl-cyclobutane and obtained by oxidation of alpha-pinene.
f acide pinonique
d Pinonsäure
i acido pinonico
e ácido pinónico

3045 pipamperone
Tranquillizer of the butyrophenone series indicated in the treatment of paranoic, depressive and melancholic states. Also used to check involutional psychoses and nervous disorders caused by alcoholism.
f pipampérone
d Pipamperon
i pipamperone
e pipamperona

3046 pipazetate
Antitussive agent marked by a central and peripheral mechanism of action and thus acting as spasmolytic as well. It is indicated in the treatment of cough due to smoking, acute and chronic infections of the respiratory tract, silicosis, tuberculosis and other forms of upper--respiratory organ irritation. It does not affect breathing and does not seem to cause habituation.
f pipazétate
d Pipazetat
i pipazetato
e pipazetato

3047 pipebuzone
Non-steroid antiinflammatory analgesic used in the treatment of affections of throat, ear and nose, as well as in ophthalmology, phlebology, gynecology, surgery and traumatology.
f pipébuzone
d Pipebuzon
i pipebuzone
e pipebuzona

3048 pipenzolate
N-ethyl-3-piperidyl benzilate, used in the treatment of peptic ulcer.
f pipenzolate
d Pipenzolat
i pipenzolato
e pipenzolato

3049 pipenzolate bromide
Parasympatholytic and antispasmodic agent characterized by a mechanism of action which is similar to that of atropine. It is used in the treatment of spasm-causing gastro-intestinal and biliary tract disorders and has proved helpful in cases of gastric ulcer. Constipation and mydriasis may occur as side effects.
f bromure de pipenzolate
d Pipenzolatbromid
i pipenzolato bromuro
e bromuro de pipenzolato

3050 piperacillin
Broad-spectrum injectable penicillin particularly effective against Gram-negative micro-organisms, e.g. Pseudomonas, Proteus, anaerobes. The drug is mainly indicated in severe, potentially lethal infections. Hypersensitivity reaction may occur as adverse effect.
f pipéracilline
d Piperacillin
i piperacillina
e piperacilina

3051 piperazine
Uric-acid-solvent heterocyclic compound also useful as an anthelminthic for the treatment of nematodiasis. It has also found employment as an antirheumatic agent and a urinary antiseptic.
f pipérazine
d Piperazin
i piperazina
e piperacina

3052 piperazine adipate
Compound indicated in the treatment of threadworm and roundworm infestation.
f adipate de pipérazine
d Piperazinadipat
i piperazina adipato
e adipato de piperacina

3053 piperazine calcium edetate
Chelate (organic complex in which an atom of hydrogen or of a metal, e.g. iron, nickel, cobalt platinum etc., is

firmly held in a molecular ring-structure by residual valencies) produced by treating ethylenediamine-N,N,N',N'-tetra-acetic acid with calcium carbonate and piperazine. The compound is used as an anthelminthic agent.
f édétate calcique de pipérazine
d Calciumedetatpiperazin
i piperazina calcio-edetato
e edetato cálcico de piperacina

3054 **piperazine camsilate**
Cardiac and circulatory analeptic agent having the same actions and uses as camphor. The drug may be administered intravenously and is rapidly absorbed, but its effect is less prolonged than that of camphor. It is used in the treatment of heart failure, circulation disorders, lipothymia (transient and sudden loss of sensibility and movement) and in all those cases in which it is advisable to stimulate circulation and respiration, particularly in the course of infectious diseases, as well as convalescence and senescence.
f camsilate de pipérazine
d Piperazinkampfersulfonat
i piperazina camsilato; canfosulfonato di dietilendiammina
e camsilato de piperacina

3055 **piperazine citrate**
Compound having the same properties and uses as piperazine hydrate.
f citrate de pipérazine
d Piperazincitrat
i piperazina citrato
e citrato de piperacina

3056 **piperazine hydrate**
Hexahydrate of piperazine used in the treatment of threadworm and roundworm infestation.
f hydrate de pipérazine
d Piperazinhydrat
i piperazina idrato
e hidrato de piperacina

3057 **piperazine oestrone sulphate**
Synthetic oestrogen conjugate with therapeutic properties.
f sulfate de pipérazine-oestrone
d Piperazinöstronsulfat
i solfato di estrone e piperazina
e sulfato de piperacina estrona

3058 **piperidine**
Chemical compound, hexahydroderivative of the pyridine. Colourless, ammonia--smelling liquid obtained by electrolytic or catalytic hydrogenation of pyridine. Treated with oxidants, it decomposes yielding amino acids. It has solvent properties for uric acid. In fact, it was used in the past in the treatment of gout prior to the introduction of uricosuric drugs. The nucleus of piperidine participates in the constitution of several alkaloids.
f pipéridine
d Piperidin
i piperidina
e piperidina

3059 **piperidione**
Antitussive agent with central mechanism of action and a mildly sedative effect. It is indicated in the cure and relief of various types of cough, e.g. cough caused by laryngitis, bronchitis, tobacco smoke etc.
f pipéridione
d Piperidion
i piperidione
e piperidiona

3060 **piperidolate**
Parasympatholytic, anticholinergic agent marked by a peripheral mechanism of action. The antispasmodic properties of the drug are enhanced by its topical anaesthetic effects. It is therefore indicated against biliary spasms, pylorospasm and post-operative spasm. Dryness of the mouth, some tachycardia and mydriasis may occur as side effects.
f pipéridolate
d Piperidolat
i piperidolato
e piperidolato

3061 **piperine**
Compound of a basic nature extracted from the common pepper, of which it contains up to 8%. It forms colourless or yellowish crystals, not easily soluble in water, with a burning taste similar to that of pepper. Treated with alcoholic potash it decomposes into piperidine. It is used in medicine for its stomachic and febrifuge properties.
f pipérine
d Piperin
i piperina
e piperina

3062 **piperocaine**
Compound used in the form of the hydrochloride as synthetic local analgesic agent.
f pipérocaïne
d Piperocain
i piperocaina
e piperocaina

3063 **piperoxan**
Alpha-adrenergic blocking factor marked by a vasodilator action and used for the diagnosis of pheochromocytoma and the control of hypertension in the course of the surgical removal of the tumour. Tachycardia, dyspnoea, anginal pains and general malaise may occur as side effects.
f pipéroxane
d Piperoxan
i piperossano
e piperoxano

3064 pipestrone
Piperazine oestrone sulphate. Synthetic
oestrogen indicated in the treatment of
psychic and vasomotor disorders of the
menopause, vulvar pruritus and atrophic
vaginitis. Depression, headache, liver
disorders, urticaria etc. may occur as
side effects.
f pipestrone
d Pipestron
i pipestrone
e pipestrona

3065 pipethanate
Tranquillizer with a parasympatholytic
mechanism of action used in the treatment of gastroenteric neuroses. Also
used in the treatment of persistent
states of tension and anxiety. Nausea
and some gastro-intestinal disorders may
occur as side effects.
f pipéthanate
d Pipethanat
i pipetanato
e pipetanato

* **pipitzahoic acid s. perezone**

3066 pipobroman
Antineoplastic agent acting like a poly-
functional alkylating agent and indicated in the treatment of polycythaemia vera
(polycythaemia of unknown origin and
relative chronic duration, marked by a
considerable increases in the red cell
count) and chronic granulocytic leukaemia. Hypochromic anaemia, leukopenia
and granulocytic leukaemia may occur
as side effects. The drug should not be
administered in pregnancy and in the
presence of bone-marrow depression resulting from x-ray therapy.
f pipobroman
d Pipobroman
i pipobromano
e pipobromano

3067 pipothiazine
Phenothiazine tranquillizer mainly used
in the treatment of chronic psychoses,
in particular schizophrenia, chronic and
paranoid delirium and any acute psychosis. Drowsiness, asthenia and extrapyramidal symptoms may occur as side
effects.
f pipothiazine
d Pipothiazin
i pipotiazina
e pipotiazina

3068 pipoxolan
Myotropic spasmolytic marked by the
absence of atropine-like effects. Its action is more effective than that
of papaverine. It is chiefly indicated
in the treatment of abdominal spasms,
visceral spasms, biliary colic and irritable colon. Dysmenorrhoea and bronchial spasms may draw some benefit from
the drug. Tachycardia and mydriasis

may occur as side effects.
f pipoxolan
d Pipoxolan
i pipossolano
e pipoxolano

3069 pipradol
Piperidine derivative. Central nervous
system stimulating agent which causes
less anorexia, insomnia and euphoria
than amphetamine. It is indicated in the
treatment of depression due to anxiety,
obsessive states and fatigue. Also used
in narcolepsy (the uncontrollable tendency to fall asleep). Nausea, vomiting,
anorexia and insomnia may occur as
side effects.
f pipradol
d Pipradol
i pipradolo
e pipradolo

3070 pipratecol
Peripheral vasodilator with an alpha-
adrenolytic-type mechanism of action. By
lowering the resistance of the arterioles,
the peripheral blood flows without affecting the heart output. It is therefore
indicated in the treatment of peripheral
vascular disorders and, in association
with raubasine (q.v.), of cerebral and
neurosensory circulatory insufficiency,
senile vascular diseases, arteriopathies
and circulatory disorders of the lower
limbs.
f pipratécol
d Pipratekol
i pipratecolo; piratecolo
e pipratecolo

3071 piprinhydrinate
Antihistamine characterized by antipruriginous, secreto-inhibitory and antihaemat-
ic properties indicated in the treatment
of pruritus, urticaria, sunburn, insect
bites, hay fever, bronchial asthma,
allergic conjunctivitis, kinetosis, rhinitis etc. Drowsiness may occur as side
effect.
f piprinhydrinate
d Piprinhydrinat
i piprinidrinato
e piprinhidrinato

3072 piprozolin
Choleretic and cholagogue agent indicated in the treatment of biliary secretory
insufficiency or cholestasis in cases of
epatopathy, cholangitis, functional and
organic cholecystopathy and dispeptic
syndrome.
f piprozoline
d Piprozolin
i piprozolina
e piprozolina

3073 piracetam
Psychostimulant drug acting on the cerebral cortex and chiefly indicated in the
treatment of confusional states, irrita-

bility, behaviour disorders and pathological desire to be left alone, particularly noticeable in some aged patients. It
is also used for comatose and subcoma-
tose states. Mild psychomotor agitation
and, though rarely, aggressivity may
be observed as side effects.

f piracétam
d Pirazetam
i piracetam
e piracetam

3074 pirbuterol
Broncho-selective beta-adrenoreceptor
used in bronchial asthma by inhalation,
orally or intravenous infusion. Tachycardia, arrhythmias and muscle cramps
may occur as side effects.

f pirbutérol
d Pirbuterol
i pirbuterol
e pirbuterol

3075 pirenzepine
Inhibitor of the gastric secretion without anticholinergic activity and without
any influence on the gastric mobility
and the pancreatic secretion. The drug
is indicated in the treatment of acute
and chronic gastric and duodenal ulcer
and gastric hyperacidity. A notable increase of the appetite may occur as side
effect.

f pirenzépine
d Pirenzepin
i pirenzepina
e pirenzepina

3076 piretamide
Diuretic agent causing greater reduction
of sodium reabsorption by the kidney
than is the case with the thiazide diuretics. The drug may be used in emergency treatment of fluid overload.

f pirétamide
d Piretamid
i piretamide
e piretamido

3077 piribedil
Circulation-regulating drug acting by
mobilizing blood supplies from splanchnic
and venous areas towards ischaemic
peripheral areas by the action of a
sympatholytic mechanism with ensuing
reduction of peripheral resistance. The
increase of the tone of the mesenteric
vasal walls and the metabolic action
reducing CO in the ischaemic area make
the drug effective in the treatment of
cerebral and peripheral vascular disorders, as well as of circulation disorders
of vascular origin. Gastric intolerance
sometimes occur as side effect.

f piribédil
d Piribedil
i piribedile
e piribedilo

3078 piridoxilate

Vitamin B6 precursor capable of promoting the cellular utilization of oxygen
and in particular the anabolism of the
myocardium in states of coronary incompetence and inadequate oxygenation of
the tissues caused by peripheral angiopathies.

f piridoxilate
d Piridoxilat
i piridossilato
e piridoxilato

3079 piritramide
Potent analgesic agent used to relieve
post-operative pains. The drug should
be administered with caution in pregnancy and in the presence of hepatobiliary disorders.

f piritramide
d Piritramid
i piritramide
e piritramida

3080 piroheptine
Parasympatholytic agent chiefly indicated in the treatment of the Parkinson's
disease. Administration of the drug is
not advisable in the presence of glaucoma, hypertrophy of the prostate gland,
coronary and cardiac incompetence. Accommodation difficulty, tachycardia and
constipation are the principal side effects.

f piroheptine
d Piroheptin
i piroeptina
e piroheptina

3081 piroxicam
Non-steroid antiphlogistic and analgesic
agent marked by a long duration of action. The drug reduces inflammation by
inhibition of prostaglandin synthesis
which forms part of the inflammatory
process. It also inhibits the same enzymes in the gastric mucosa. It is used
in the treatment of arthritis, in particular rheumatoid arthritis. Gastro-intestinal disorders may occur as side effects.

f piroxicam
d Piroxicam
i piroxicam
e piroxicam

3082 piscidic acid
Bicarboxylic acid obtained from the
Piscidia, a plant of the Leguminosae
family, the best known variety of which
is the Piscidia erythrina whose bark is
used in medicine for its sedative and
slightly hypnotic action. It has also
the property of depressing uterine contractions. It is thus used therapeutically to relieve utero-ovarian pains in
dysmenorrhoea and as haemostatic in
menorrhagies and metrorrhagies.

f acide piscidique
d Piscidinsäure
i acido piscidico
e ácido piscídico

3083 piscidine
Bitter substance having a neutral reaction, extracted from the Piscidia and used medicinally for its antispasmodic and hypnotic properties.
f piscidine
d Piscidin
i piscidina
e piscidina

3084 pituitary extract
Extract of the posterior lobe of the pituitary mainly used in midwifery to promote uterine contractions. It is also indicated as an antidiuretic and in postoperative ileus.
f extrait de l'hypophyse
d Hypophysenhinterlappenextrakt
i estratto della pituitaria
e extracto de la pituitaria

3085 pivampicillin
Ampicillin-derived broad-spectrum penicillin with bactericide action against ampicillin-sensitive Gram-positive and Gram-negative microorganisms. When administered orally, the drug is three times as potent as ampicillin. It is chiefly indicated in the treatment of infections affecting the respiratory, genitourinary and gastro-intestinal systems. Pruritus, skin rushes, urticaria, diarrhoea and gastro-intestinal disorders may occur as side effects.
f pivampicilline
d Pivampicillin
i pivampicillina
e pivampicilina

3086 pivcephalexine
Semisynthetic cephalosporin orally active and marked by a broad antibacterial spectrum including staphylococci, streptococci, diplococci, gonococci, E.coli, Haemophilia influenzae, Salmonella, Klebsiella pneumoniae. Also active against penicillase-producing strains of microorganisms. Oral absorption is well-nigh total and peak serum concentrations are generally attained within one hour. It is indicated in the treatment of all infections caused by Gram-positive and Gram-negative microorganisms. Allergic reactions, gastro-intestinal disorders, high bilirubinaemia and azotaemia, leukopenia etc. may occur as side effects.
f pivcéphaléxine
d Pivcephalexin
i pivcefalessina
e pivcefalexina

3087 pivmecillinam
Mecillinam derivative, an antibiotic used in the treatment of infections of the urinary tract. The drug is readily absorbed from the gastro-intestinal tract and metabolized without delay to the active mecillinam whose actions, uses and side effects (allergic reactions, diar-

rhoea) it shares.
f pivmécillinam
d Pivmecillinam
i pivmecillinam
e pivmecillinam

3088 pizotifen
Or pizotyline. Polyvalent inhibitor of biogenous amines, especially serotonin and histamine. The drug is indicated in the treatment and prophylaxis of typical or atypical migraines, as well as headaches caused by vascular disorders.
f pizotifène; pizotyline
d Pizotifen; Pizotylin
i pizotifene; pizotilina
e pizotifeno; pizotilina

3089 placentae extractum
Non-specific opotherapeutic agent capable of activating cellular metabolism and the physiological functions of the organism. Topically applied it promotes the formation of granulation tissue and the reepithelization. In general medicine, it is indicated for the treatment of asthenia, degenerative osteopathies and skin diseases. It is also used to accelerate the process of cicatrization and to treat varicose ulcers, psoriasis and burns. In ophthalmology, the drug finds employment in the treatment of retinal degeneration, severe myopia, keratitis and optic atrophies.
f extrait de placenta
d Plazentaextrakt
i estratto di placenta
e extracto de placenta

3090 placental extract
Immune globulin, an extract of human placenta containing globulins and used to immunize children against measles.
f extrait de placenta
d Plazentaextrakt
i estratto di placenta
e extracto de placenta

3091 plasma humanum congelatum
Liquid fraction of total human blood separated from the corpuscular part and subjected to a freezing process. It is used as a blood volume replenisher in the treatment of haemophilia and haemophilioid states.
f plasma humanum congelatum
d Plasma humanum congelatum
i plasma humanum congelatum
e plasma humanum congelatum

3092 plasma humanum liquidum
Liquid fraction obtained by sedimentation or centrifugation of total human blood, made uncoagulable by a suitable anticoagulant agent. It is used intravenously for the treatment of haemophilic and haemophiloid states and of hypoprothrombinaemia.
f plasma humanum liquidum

d Plasma humanum liquidum
i plasma humanum liquidum
e plasma humanum liquidum

3093 platinocyanide
Salt of the platinohydrocyanic acid. Platinocyanides form coloured crystals which are fluorescent both in the solid state and in solution. Among the best known is the barium platinocyanide, which is widely used for the preparation of fluorescent screens for X-rays.
f platinocyanure
d Platincyanid
i platinocianuro
e platinocianuro

3094 platinum
Very heavy precious metallic element, non corroding, ductile and malleable. It has a very high melting point and is chemically resistant. It is used in dentistry and in surgery where other metals would be subject to corrosion by body fluids.
f platine
d Platin
i platino
e platino

3095 platinum black
Soft black powder of finely divided metallic platinum obtained by reduction and precipitation from solutions of its salts, which is capable of occluding large volumes of oxygen, hydrogen or other gases. It is used as a catalyst for hydrogenation or oxidation.
f noir de platine
d Platinschwarz; Platinmohr
i nero di platino
e negro de platino

3096 platinum sponge
Metallic platinum in a grey spongy form which is obtained by reducing ammonium chloroplatinate. It is capable of occluding large volumes of oxygen, hydrogen and other gases and is used as a catalyst.
f mousse de platine
d Platinschwamm
i spugna di platino
e esponja de platino

3097 pleurisy root
Or swallow root. Common name for the root of the butterfly weed and of Asclepias tuberosa, used medicinally for its diaphoretic and expectorant properties.
f racine d'asclépiade; racine de domptepoison
d Sanktlorenzwurzel; Schwalbenwurzel
i radice di vincetossico
e raíz de vincetóxico

3098 podophyllic acid
Monobasic acid obtained by hydrolysis from podophyllin. It exists in two isomeric forms.

f acide podophyllique
d Podophyllinsäure
i acido podofillico
e ácido podofílico

3099 podophyllin
Resinous substance obtained by extraction with alcohol from the roots of podophyllum. A bitter, light-brown to greenish yellow resin, hardly soluble in water and containing the active principles of podophyllum, i.e. podophyllic acid, podophyllotoxin, picropodophyllin. It has some therapeutical use as it is capable, when taken in small doses, to increase secretion and strengthen intestinal peristalsis thus exerting a laxative action. Higher doses may affect the intestinal mucosa and provoke general toxic effects, e.g. the alteration of the mitotic processes.
f podophylline
d Podophyllin
i podofillina
e podofilina

3100 podophyllotoxin
Active principle of the roots of Podophyllum; needle-shaped crystals, hardly soluble in water, which are split into picropodophyllin and podophyllic acid in the presence of alcohol. Podophyllotoxin has purgative properties.
f podophyllotoxine
d Podophyllotoxin
i podofillotossina
e podofilotoxina

3101 poison
Any substance which, in given quantities, proves harmful or lethal to an organism when brought into contact or absorbed.
f poison
d Gift
i veleno
e veneno

*** poison nut   s. nux vomica**

3102 poke berry
The fruits of a North American plant (Phytolacca americana) which have the same properties as those of the phytolacca roots (q.v.) but are milder in their action.
f fruit de phytolaque
d Kermesbeere; Scharlachbeere
i frutto di fitolacca
e fruto de fitolaca

3103 poldine methylsulphate
Parasympatholytic and anticholinergic agent mainly used for its capacity to reduce gastric secretion and thus treat gastrosuccorrhoea, peptic ulcer and, more generally, dyspeptic conditions marked by hyperchlorhydria. Reactions are like those of atropine. The drug should not be administered to small

children.
f méthylsulfate de poldine
d Poldinmethylsulfat
i poldina metilsolfato
e metilsulfato de poldina

3104 polidexide
Hypocholesterolaemic agent acting by inhibiting the reabsorption of biliary acids in the intestines. A new synthesis of the latter is thus initiated from cholesterol. The drug should not be administered to children. Constipation and tympanitis may occur as side effects.
f polidéxide
d Polidexid
i polidesside
e polidexida

3105 polidocanol
Surface anaesthetic used in the sclerosing therapy of varices and teleangiectasia. A transient oedema accompanied by slight pains may occur as side effect when the drug is administered by perivenous injection.
f polidocanol
d Polidocanol
i polidocanolo
e polidocanolo

3106 poligeenan
Therapeutic agent capable of protecting the gastric mucosa and thus indicated in the treatment of gastralgia, gastro-
-duodenal ulcer, peptic ulcer etc.
f poligéenane
d Poligeenan
i poligeenano
e poligeenano

3107 polipeptidum articulare
Lysate (the product of hydrolysis) of animal organs constituted by low-molecular weight (not more than 2000:36%) poly peptides. Biological preparation marked by antiphlogistic action. The drug is also capable of counteracting prostaglandin, antiserotonin, antihistamine and antibradykinin action and is indicated in the treatment of chronic articular rheumatism.
f polypeptide articulaire
d Gelenkpolypeptid; Polipeptidum articulare
i polipeptide articolare
e polipéptido articular

3108 polipeptidum cerebrale
Lysate (the product of hydrolysis) of bovine brain constituted by low-molecular weight (not more than 2000:32%) poly peptides. Biological preparation marked by a thymoleptic and anticephalalgic action. It also has an antiserotoninic and antibradykininic, as well as an adrenaline-sensitizing action. The preparation is indicated in the treatment of endogenous and reactive depression and headaches caused by vasomotor, post-traumatic and psychogenic disorders.

f polypeptide cérébrale
d Hirnpolypeptid
i polipeptide cerebrale
e polipéptido cerebral

3109 pollen extract
Pollinis extractum. A vaccine used for the desensitization from hay fever and asthma caused by pollen, i.e. the very fine dust which is left off from the anthers of seed plants.
f extrait de pollen
d Blütenstaubextrakt; Pollenextrakt
i estratto di polline
e extracto de polen

3110 poloxalcol
Synthetic purgative agent marked by surface-active properties whereby the faeces are kept soft for an easy evacuation. Used in the treatment of acute and chronic constipation.
f poloxalcol
d Poloxalkol
i polossalcolo
e poloxalcolo

3111 polyamine
Synthetic resin acting by anionic exchange as an antacid (if taken orally) temporarily binding together in the stomach hydrochloric acid and pepsin subsequently released in the intestine. It is then eliminated unaltered from the guts without any change in the electrolytic balance of the organic liquids. The preparation is indicated in the treatment of gastric and duodenal ulcer and hyper hydrochloric gastritis. Nausea and vomiting may occur as side effects when large doses are required.
f polyamine
d Polyamin
i poliamina
e poliamina

3112 polycresolsulphonate
Creosulphonic acid-formaldehyde condensation product used topically as an antiseptic, haemostatic and re-epithelizing agent. It is indicated in the treatment of sores, varicose ulcers, chaps, chilblains etc. In gynecology, the drug is used to treat cervical ulcerations and cervicovaginitis.
f polycrésolsulfonate
d Polycresolsulfonat
i policresolsolfonato
e policresolsulfonato

3113 polyenacidine
Combination, in variable proportions, of unsaturated fatty acids: arachidonic acid, linoleic acid, 9.11+ and 9.12- linoleic acid or a correspondent ethyl ether. Polyenacidine is an essential growth factor for some animals but not for man. It has a trophic and skin-protective, as well as lipotropic action which is capable of lowering the level of choles-

terol. It is thus indicated in the treatment of burns, eczemas, several skin diseases, hypercholesterolaemia and arteriosclerosis.
f polyénacidine
d Polyenacidin
i polienacidina
e polienacidina

3114 polygeline
Plasma substitute having an average half-life of up to 4 hours. It is eliminated from the blood stream within 48 hours. Chiefly indicated in states of shock due to reduction of the blood volume, loss of plasma caused by burns or dehydration, prophylaxis prior to major surgery and hypotension provoked by anaesthesia shock.
f polygéline
d Polygelin
i poligelina
e poligelina

3115 polyhydric alcohol
Polyvalent alcohol. Alcohol containing two or more oxydrylic groups.
f acide polyvalent; polyalcool
d mehrwertiger Alkohol
i alcool poliidrico; alcool polivalente
e alcohol polivalente

3116 polymyxin B
Antibiotic active against Gram-negative micro-organisms. The drug is not absorbed when taken orally but is quite effective when administered locally as it acts by alteration of the cellular membrane. Injections may be painful and some neurological symptoms may occur.
f polymyxine B
d Polymyxin B
i polimixina B
e polimixina B

3117 polynoxylin
Polyoxymethylenurea. Antimicrobial preparation effective against Gram-positive and Gram-negative micro-organisms, several mycetes and staphylococci which have proved resistant to antibiotics. It is either used by itself or in association with corticosteroids.
f polynoxyline
d Polynoxylin
i polinossilina
e polinoxilina

3118 polypeptide
Substance constituted by a minimum of two to a very high number of amino acids which are united by a peptidic of carboaminic bond. Polypeptides represent the intermediate stage both of the degradation of protein into amino acids and of the synthesis of the proteic molecule starting from the amino acids. They are the substrate of particular enzymes called peptidases.
f polypeptide
d Polypeptid
i polipeptide
e polipéptido

3119 polysorbate
Polyoxyethylene derivative of cyclic anhydrides of the sorbitol partially esterified with a fatty acid. Polysorbates are non-ionic surface active agents used as emulsifying and dissolving or wetting substances. The type "60" is an emulsifying agent and aids water-in-oil mixtures and the solubilizing of fat-soluble substances. Type "80" is used therapeutically in the form of aerosol as a liquefacient of substrates for pharmaceutical substances in chronic rhinitis. It also finds employment in the preparation of ceruminal plugs and in nasal infections in association with antibiotics.
f polysorbate
d Polysorbat
i polisorbato
e polisorbato

3120 polythiazide
Benzothiadiazine-related diuretic acting by carbonic anhydrase inhibition. Water, sodium chloride, potassium and bicarbonate ions are more copiously excreted by the action of the drug on the distal tubules. It is thus used for the treatment of patients suffering from oedema borne of various disorders. Either by itself or in combination with reserpine, it is also employed in cases of mild to moderate hypertension. Dizziness, nausea and in some cases muscle spasms may occur as side effects.
f polythiazide
d Polythiazid
i politiazide
e politiazida

3121 polyvidone
Blood plasma substitute used to increase blood volume in states of shock causing a decrease of the blood mass, for haemorrhages, burns etc. Cutaneous reactions and the possibility of cumulative effects have been observed.
f polyvidone
d Polyvidon
i polividone
e polividona

3122 poppy
Common herb growing in various European countries. The most important species is the opium poppy (Papaver somniferum) which is widely grown in Bulgaria, Macedonia, Turkey, Iran and India. The latex obtained from the poppy heads, or capsules, is dried in order to extract opium that contains the medicinally very important alkaloids morphine, codeine, papaverine, narcotine and thebaine.
f pavot

d Mohn
i papavero
e adormidera

3123 poppy capsules
Or poppy heads. The dried fruit of the Papaver somniferum (s. poppy) marked by mild sedative action caused by the very small quantities of opium alkaloids, in particular morphine, contained therein. They are widely used for the preparation of poultices and the syrup of poppy which is added to cough mixtures to act as sedative.
f capsules de pavot; têtes de pavot
d Mohnkapseln
i capsule di oppio; capi di papavero
e cápsulas de adormidera; cabezas de ador̄ midera

3124 poppy seed
The seed of the poppy, especially of the opium poppy used as the main source of poppy-seed oil, q.v.
f graine de pavot
d Mohnsamen
i seme di papavero
e semilla de adormidera

3125 poppy-seed oil
Oil obtained by expression of seeds of opium poppy, a pale-yellow odourless liquid with a faint almond flavour and a very bland taste. It is used as a substitute of olive oil for medical purposes as it contains the glycerides of stearic and palmitic acid as well as those of oleic and linoleic acids. Also used for the preparation of iodized oil.
f huile d'oeillette
d Mohnöl
i olio di papavero
e aceite de adormidera

3126 populin
Glycoside found in the bark and leaves of poplar. A colourless, water-soluble powder which splits by hydrolysis into benzoic acid and salicylin and displays antipyretic properties.
f populine
d Populin
i populina
e populina

3127 potassium
An element. A sodium-related soft white metal, more highly reactive than sodium itself. It is present in sea water, as silicate in primary rocks and granites and as chloride in vast deposits in various regions of central Europe and in California. It is a constituent of all plants and animals. Although it is essential for tissue excitability, potassium has a depressant action on the central nervous system and the heart and on voluntary and unstriped muscles. When taken orally, the ion is rapidly excreted by the kidneys. It also has a value as

diuretic.
f potassium
d Kalium
i potassio
e potasio

* potassium aminosalicylate   s. kalii aminosalicylas

* potassium bitartrate   s. kalii hydrogeno̲ tartras

* potassium bromide   s. kalii bromidum

* potassium canreonate   s. kalii canrenoas

* potassium chlorate   s. kalii chloras

* potassium chloride   s. kalii chloridum

* potassium citrate   s. kalii citras

* potassium clorazepate s. kalii clorazepas

* potassium ferrocyanide   s. kalii ferrocyanidum

* potassium gluconate   s. kalii gluconas

* potassium glycerophosphate   s. kalii gly̲ cerophosphas

* potassium iodide   s. kalii jodidum

* potassium nitrate   s. kalii nitras

* potassium perchlorate s. kalii perchloras

* potassium permanganate   s. kalii permanganas

* potassium sorbate   s. kalii sorbas

* poultice   s. cataplasm

3128 practolol
Cardioselective beta-adrenoceptor blocking drug which manages proportionally to diminish undesirable beta-adrenergic stimulation of the myocardium, causing a reduction in hear rate and force of contraction. It is therefore indicated in the treatment of angina pectoris due to coronary insufficiency, arrhythmias and hyper- beta-adrenergic syndrome with or without hypertension. Cutaneous and gastro-intestinal disorders may occur as side effects.
f practolol
d Practolol
i practololo
e practololo

* pralidoxime   s. pralidoxime methylsulphate

3129 pralidoxime iodide
Antidote of paradion (dinitrothiophosphate) and other organophosphorous esters. It acts by reactivating cholines-

terase. It is advisable to follow up the therapy with the administration of atropine.

f iodure de pralidoxime
d Pralidoximjodid
i pralidossima ioduro
e yoduro de pralidoxima

3130 pralidoxime methylsulphate
Cholinesterase reactivator indicated in the treatment of organo-phosphorus cholinesterase poisoning. It is generally administered in combination with atropine.

f méthylsulfate de pralidoxime
d Pralidoximmethylsulfat
i pralidossima metilsolfato
e metilsulfato de pralidoxima

3131 pramiverine
Spasmolytic agent with a myolytic-papaverine-like and anticholinergic-atropine-like action. Its effect is rapid and prolonged. The drug is indicated in the treatment of spasms affecting the urinary tract, as well as the biliary and gastroenteric systems. It is also used in cases of threatened abortion caused by uterine hypertone.

f pramivérine
d Pramiverin
i pramiverina
e pramiverina

3132 pramocaine
Local anaesthetic of the morpholine series used in the treatment of haemorrhoids, rhagades, various dermatoses, burns and sunburn. It is also used for endotracheal and intragastric intubation, but not for broncoscopy and gastroscopy. The drug, which is administered in the form of ointment, has a prompt effect lasting 3 to 4 hours.

f pramocaïne
d Pramocain
i pramocaina
e pramocaina

3133 pranosal
Antiinflammatory, analgesic, topical anaesthetic and vasodilator agent with muscle relaxant action. It is used topically for the treatment of arthritis, acute articular rheumatism, pains and aches due to stress etc.

f pranosal
d Pranosal
i pranosal
e pranosal

3134 prasterone
Adrenocortical hormone marked by antidepressant and psychotonic action. It is indicated in the treatment of exhaustion due to physical or mental stress, psychoneuroses, neurasthenia and psychic disorders caused by the menstrual cycle or the menopause.

f prastérone

d Prasteron
i prasterone
e prasterona

3135 prazosin
Vasodilator and antihypertensive agent with a depressant action on the smooth muscles of the arterioles. It is indicated in the treatment of essential arterial hypertension mainly caused by a nephropathy. Asthenia, sudoresis, palpitations, hypotension, nausea etc. may occur as side effects. The drug should not be administered in pregnancy and to children.

f prazosine
d Prazosin
i prazosina
e prazosina

3136 precipitant
Or precipitating agent. Substance which, added to a compound in solution, transforms it into an insoluble one and subsequently causes its precipitation.

f précipitant
d Fällungsmittel
i precipitante
e precipitante

3137 precipitated bismuth
Purified and finely divided preparation of metallic bismuth from which various other medicinal preparations are made, e.g. injection of Bismuth.

f bismuth précipité
d Wismuthpräzipitat
i bismuto precipitato
e bismuto precipitado

3138 precipitated sulphur
Sulphur reduced to a pale-yellow or greyish amorphous or microcrystalline powder by precipitation. It is used medicinally in the treatment of cutaneous diseases.

f soufre précipité
d Schwefelpräzipitat
i precipitato di zolfo
e precipitado de azufre

3139 precipitin
Particular antibody capable of causing the coagulation of certain albumins that are present in organic fluids and subsequently their precipitation. The specific action of the precipitin is used in the biological tests which aim at the identification of animal proteins of unknown origin, in the classification of certain micro-organisms, in the dosage of some toxins etc.

f précipitine
d Präzipitin
i precipitina
e precipitina

3140 precursor
Any substance from which another substance is formed. Specifically, a sub-

stance which intervenes or has been
formed in a preliminary stage of a reaction or process and which later is trans
formed into another substance, e.g. a
provitamin is a compound which in itself
the precursor of vitamin.
f précurseur
d Vorstufe
i precursore
e precursor

3141 prednazoline
Association of a corticosteroid with antiphlogistic and antiallergic activity with
a vasoconstrictor having decongestant effect. It is indicated in the topical treatment of rhinitis, rhinopharyngitis and
inflammatory or allergic sinusitis. Dryness of the mucous membranes may occur
as side effect.
f prednazoline
d Prednazolin
i prednazolina
e prednazolina

3142 prednilidene
Corticosteroid hormone marked by chiefly
glucocorticoid action and indicated in
the treatment of rheumatic conditions
(rheumatoid arthritis), collagenosis,
hypopituitarism, inflammatory and allergic skin diseases, autoimmune haemolytic
anaemia, ulcerous colitis, sarcoidosis,
multiple sclerosis and, in association
with a mineralocorticoid, adrenocortical
insufficiency. Increased susceptibility
to infections, osteoporosis, hypertension,
hypercoagulability of the blood, myopathies etc. may occur as side effects.
f prednilidène
d Predniliden
i prednilidene
e prednilideno

3143 prednisolamate
Prednisolone 21-diethylaminoacetate: the
hydrochloride, a white odourless crystalline powder is indicated in the treatment
of inflammatory skin conditions.
f prednisolamate
d Prednisolamat
i prednisolamato
e prednisolamato

3144 prednisolone
Synthetic steroid marked by a formula
differing from that of 17-oxycorticoste-
rone or hydrocortisone from which it
derives owing to the presence of a double bond between positions 1 and 2.
Therapeutic indications are the same as
those of cortisone and hydrocortisone,
i.e. rheumatic arthritis, rheumatism of
the joint, allergic diseases, e.g. urticaria, asthma etc., mesenchymal disorders, e.g. lupus erythematosus, scleroderma, acute leukaemia, haemopathies
etc. It must be noted that prednisolone
has an activity three to five times that
of cortisone or hydrocortisone though the

unfavourable action on hydrologic and
saline balance is the same.
f prednisolone
d Prednisolon
i prednisolone
e prednisolona

3145 prednisolone acetate
The acetyl derivative of prednisolone
used for the same purposes of prednisolone, q.v.
f acétate de prednisolone
d Prednisolonacetat
i prednisolone acetato
e acetato de prednisolona

3146 prednisolone pivalate
Odourless white crystalline powder which
can be prepared by partial synthesis
and is indicated in the treatment of
rheumatic diseases and inflammatory conditions of the skin.
f pivalate de prednisolone
d Prednisolonpivalat
i prednisolone pivalato
e pivalato de prednisolona

3147 prednisolone sodium phosphate
Disodium salt of the 21-phosphate ester
of prednisolone, i.e. an adrenocortical
steroid which is administered intravenously.
f phosphate sodique de prednisolone
d Prednisolonnatriumphosphat
i prednisolone sodio fosfato
e fosfato sódico de prednisolona

3148 prednisolone steaglate
Antiphlogistic and antirheumatic agent,
an adrenocortical steroid indicated in
the treatment of bronchial asthma, aller-
gopathies, rheumatic fever, blepharitis
and several dermatoses.
f stéaglate de prednisolone
d Prednisolonsteaglat; Dehydro-Hydrokorti-
sonsteaglat
i prednisolone steaglato
e steaglato de prednisolona

3149 prednisone
Deltadehydrocortisone. Synthetic steroid
whose formula differs from that of cortisone, from which it derives, owing to
the presence of a double bond between
positions 1 and 2; crystalline white
powder decomposing at about 235°, poorly soluble in water. The therapeutic
advantages in comparison with cortisone
and hydrocortisone are the same as
those of prednisolone. Dyspepsia, ulcers
and gastroenteric haemorrhages, hyperglycaemia, hypertension, myopathies etc.
may occur as adverse effects of the
drug.
f prednisone
d Prednisone
i prednisone
e prednisona

3150 pregnandiol

One of the two biologically inactive isomers into which the progesterone is transformed in order to be eliminated with urine. It is present in considerable quantity in the urine of pregnant women.

f pregnandiol
d Pregnandiol
i pregnandiolo
e pregnandiolo

3151 pregneninolone
Synthetic preparation having an action that is similar to that of progesterone. Its advantage in comparison with the latter consists of the fact that it can be taken orally and by injection.

f pregnéninolone
d Pregneninolon
i pregneninolone
e pregneninolona

3152 prenalteron
Synthetic beta-adrenoreceptor agonist capable of increasing the force of cardiac contraction without appreciably altering the heart rate and peripheral vascular actions. It is indicated in the treatment of difficult-to-treat heart failure. It also appears to be useful as an antidote in cases of overdosage of beta-adrenoreceptors antagonists.

f prénaltéron
d Prenalteron
i prenalterone
e prenalterona

3153 prenoxdiazine
Antitussive agent with a peripheral mechanism of action chiefly indicated in the treatment of spasmodic and congestive cough independently of the age of the patient. Large doses may cause drow siness and fatigue.

f prénoxdiazine
d Prenoxdiazin
i prenoxdiazina
e prenoxdiazina

3154 prenylamine
Diphenylpropylamine-related vasodilator chiefly acting on the coronary arteries and favouring the oxygenation of the myocardium. The drug is thus indicated in the treatment of angina pectoris and all states of coronary insufficiency of both organic and functional origin. It also acts as sympatholytic and sedative. Atrioventricular conduction disorders may be noticed in the course of the therapy.

f prénylamine
d Prenylamin
i prenilamina
e prenilamina

3155 preproinsulin
The beginning polypeptide chain in the lumen of the rough endoplasmic reticulum containing an amino-terminal sequence of sixteen residues, which is absent from insulin. It may well be that this amino-terminal hydrophobic region has the function of a signal sequence directing the sequence which begins to form to the endoplasmic reticulum.

f préproïnsuline
d Präproinsulin
i preproinsulina
e preproinsulina

3156 preservative
Any substance or agent which is added to chemicals, natural products, food articles etc. to preserve them against decay or deterioration.

f agent conservateur; préservateur
d Konservierungsmittel; Vorbeugungsmittel
i conservante; agente conservante; preservativo
e preservativo; conservante

3157 pridinol
Skeletal-muscle relaxant agent acting by opposing the effects of impulses transmitted by the parasympathetic nerves. It is used in all parkinsonian syndromes, including the postencephalitic ones, muscular spasms and tremors. Collateral effects, which resemble those of atropine, are very mild.

f pridinol
d Pridinol
i pridinolo
e pridinolo

3158 prilocaine
Topical anaesthetic of the amide type characterized by a toxicity which is half that of the lidocaine, q.v. It has a tendency to produce methaemoglobin by its metabolites.

f prilocaïne
d Prilocain
i prilocaina
e prilocaina

3159 primaquine
Antimalarial agent indicated in the treatment and prophylaxis of Plasmodium vivax malaria. Nausea accompanied by vomiting, abdominal pains and cramps, haemolytic anaemia, methaemoglobinaemia and leukopenia may occur as side effects.

f primaquine
d Primaquin; Primachin
i primachina
e primaquina

3160 primidone
Pyrimidine-related anticonvulsant agent with a central mechanism of action. It does not involve hypnotic affects and is indicated in the treatment of grand-mal and petit-mal by suppressing epileptic discharges in the brain. It is used orally to prevent convulsions. The drug must be used with great caution in elderly patients. Transient and mild gastric and cutaneous disorders of a nervous nature, polyuria and sexual-

-function disorders may be observed as side effects.

f primidone
d Primidon
i primidone
e primidona

3161 pristinamycin

Antibiotic isolated from cultures of Streptomyces pristina spiralis and mainly active on Gram-positive microorganisms, particularly on Staphylococcus aureus, and strains which prove resistant to other antibiotics. Gastro-intestinal disorders may occur as side effects.

f pristinamycine
d Pristinamycin
i pristinamicina
e pristinamicina

3162 probenecid

Uricosuric capable of reducing the rate of uric acid in the blood by increasing by 30 to 50% its excretion. It is mainly indicated in the treatment of chronic uricaemia. It inhibits the excretion of penicillin and sulphonamides through the kidneys. Haematuria, anorexia, polyuria and anaemia may occur as adverse effects.

f probénécide
d Probenecid
i probenecid
e probenecid

3163 probucol

Compound capable of reducing elevated concentrations of serum cholesterol and indicated in the treatment of primary hypercholesterolaemia, also when the latter is accompanied by hypertriglyceridaemia. Abdominal disorders, in particular meteorism and diarrhoea, nausea, vertigo, palpitations and angioneurotic oedema may occur as side effects.

f probucol
d Probucol
i probucolo
e probucolo

3164 procainamide

Compound capable of depressing irritability of the thyro-epiglottic muscle and characterized by an effect which is more prolonged than that of procaine hydrochloride and considerably better tolerated. The drug is indicated in the treatment of ventricular and atrial arrhythmias and extrasystoles. Administered intravenously, it causes less hypotension than procaine hydrochloride. Granulocytopenia has been diagnosed following prolonged administration.

f procaïnamide
d Prokainamid
i procainamide
e procainamida

3165 procaine

Chemical compound, 2-diethylaminoethyl p-aminobenzoate, the principal local anaesthetic of the ester type (ester of p-aminobenzoic acid). Colourless crystalline synthetic base administered as hydrochloride. It is water-soluble and frequently used instead of cocaine owing to its considerably lower degree of toxicity. Its action is rapid in onset, particularly if injected, and can easily be prolonged when given with adrenaline which constricts the blood vessels around the site of the injection. It may also be given by intrathecal injection as well as orally especially to soothe gastro-intestinal pains and spasms.

f procaïne
d Prokain
i procaina
e procaina

3166 procaine benzylpenicillin

Retard-action penicillin with effect lasting 12 hours. It is indicated in the treatment of infections caused by penicillin G-sensitive bacteria. Heart hypokinesis may occur as a side effect.

f procaïne benzylpénicilline
d Penizillinbenzylprocaine
i benzilpenicillina procainica
e benzilpenicilina procaínica

3167 procaine borate

Salt which is considered less toxic and more active than procaine hydrochloride. It is used in surface analgesic in ophthalmic and dental surgery.

f borate de procaïne
d Prokainborat
i procaina borato
e borato de procaina

3168 procaine hydrochloride

See definition of procaine. Also known as ethocaine hydrochloride.

f chlorhydrate de procaïne
d Prokainhydrochlorid
i procaina idrocloridrato
e hidrocloruro de procaina

3169 procaine penicillin

Long-acting form of benzylpenicillin (see procaine benzylpenicillin) having a similar action and similar side effect.

f pénicilline procaïnique
d Prokainpenicillin
i procaina penicillinica
e procaina penicilínica

3170 procarbazine

Cytotoxic drug of the methylhydrazine group used in the treatment of neoplastic diseases, in particular malign lympho granulomatosis, lymphosarcoma and reticulosarcoma. It has practically no effect on solid tumours. Nausea, vomiting, diarrhoea, neurotoxicity, skin rashes and bone marrow depression may occur as side effects.

f procarbazine
d Procarbazin

ı procarbazina
e procarbazina

3171 prochlorperazine
Phenothiazine-related tranquillizer acting
on the reticular system and indicated in
the treatment of schizophrenia, obsessive
and phobic reactions and manic depressive states. Owing to its antiemetic action, which is more potent though less
sedative than that of chlorpromazine,
it is chiefly used as antiemetic. Extrapyramidal symptoms may be noticed as
a side effect.
f prochlorpérazine
d Prochlorperazin
ı proclorperazina
e proclorperazina

3172 procyclidine
Anticholinergic and parasympatholytic
agent with an atropine-like action and
a marked antispastic effect on the
smooth musculature. It is used in all
forms of parkinsonism, including the
drug-induced one, as it reduces rigidity
and improves muscle coordination. Side
effects are those of atropine sulphate
and, in addition, suppurative parotitis.
f procyclidine
d Procyclidin
ı prociclidina
e prociclidina

*** profenamide   s. ethopropazine**

3173 proflavine
Sulphate of the diamine-derivative of
acridine, a compound acting as fairly
potent germicide against Gram-positive
micro-organisms in alkaline media. It
is excreted in the urine and can act as
an effective urinary antiseptic. A solution (0.1%) is used as a germicide for
mucous membranes and to disinfect
wounds. It is also incorporated in ointments having an antiseptic action. Oleate,
hydrochlorate etc. of the same acridine-
-derived compound have analogous uses.
f proflavine
d Proflavin
ı proflavina
e proflavina

3174 progesterone
The steroid which is produced by the
corpus luteum and the placenta. The
secretion of progesterone which is connec
ted with the formation of the corpus
luteum (supported by the chorionic
gonadotropin) begins with the dehides-
cence of the follicle to end, unless fertilization has occurred, immediately
before menstruation, i.e. the luteinic
phase of the menstrual cycle. As a medi
cinal drug, progesterone is indicated
in the treatment of uterine haemorrhage,
functional dysmenorrhoea, endometritis,
premenstrual tension and habitual or
threatened abortion. Also used for breast

and uterine tumours, prostatic carcinoma
and senility. Acne, weight gain, enlarge
ment of the breasts and gastro-intestinal
disorders may occur as side effects.
f progestérone; hormone lutéale
d Progesteron; Luteohormon
ı progesterone; ormone luteale
e progesterona; hormona lutea

3175 proglumide
Chemical compound marked by an antisecretory non-anticholinergic action and
used in the treatment of gastric and
duodenal ulcers, gastritis and duodenitis
and, more generally, of gastro-intestinal
disorders due to the effect of drugs.
Some constipation may ensue the treatment as side effect.
f proglumide
d Proglumid
ı proglumide
e proglumido

3176 proguanil
Antimalarial agent acting as schizonticide and gametocide by inhibiting the
metabolism of the red corpuscles and
cells which shelter the malarial parasites. The drug is indicated in the prophylaxis and treatment of both acute
and chronic forms of malaria caused by
Plasmodium vivax and Plasmodium falciparum. Owing to the reduction of gastric
secretion, anorexia, accompanied by
nausea, may occur as side effect.
f proguanil
d Proguanil
ı proguanile
e proguanilo

3177 prolactin
Lactogenic hormone. Polypeptide hormone
produced by the anterior pituitary and
responsible for the stimulation of milk
production. It also carries out luteotro-
pic and antifolliculinic action and is
indicated in the treatment of hypogalactia and agalactia, menstrual disorders
marked by hyperfolliculin syndrome and
threatened abortion.
f prolactine; hormone lactogène; hormone
  lutéotrope
d Prolaktin; Laktotropin; Laktationshormon
ı prolattina; ormone lattogeno; galattina;
  ormone lattogeno ipofisario
e prolactina; hormona lactógena; hormona
  luteótropa

3178 proline
Imino-acid abundantly occurring in vege
table proteins, particularly in protamines. In animal organisms, it appears
to have some importance in the differen-
tiation, participating in the construction
of specific structures or of morphogenet-
ic hormones. It is also found in nature
in the L-form, i.e. colourless crystals
very easily soluble in water.
f proline
d Prolin

ı prolına
e prolına

3179 prolintane
Central nervous system stimulant which
may considered as an intermediate (with
reference to its effects) between caffeine
and amphetamine. As centrally acting
antidepressant, it stimulates metabolism,
blood pressure and the respiration. It
is used to treat lethargy, nervous exhaustion, hypotension and senile debility.
Insomnia and arrhythmia may occur as
side effects.
f prolintane
d Prolintan
ı prolintano
e prolintano

3180 prolonium iodide
Organic compound which contains injectable iodine. It is indicated in the treat‾
ment of arteriosclerosis, mild hypertension, rheumatism, neuralgias and neuritides. Acne caused by iodine and inflam‾
mation of the mucous membranes may occur as side effects.
f iodure de prolonium
d Proloniumjodid
ı prolonio ioduro
e yoduro de prolonio

3181 promazine
Phenothiazine-related tranquillizer and
hypnotic synergic with analgesic. It is
chiefly indicated in the treatment of
various psychoses, alcoholism, states of
anxiety and of hyperemotivity. Hypotensive effects appear to be greater than
those produced by chlorpromazine, q.v.
f promazine
d Promazin
ı promazina
e promazina

3182 promethazine
Phenothiazine antihistamine whose actions are very similar to those of chlorpromazine. It probably acts by occupying the receptors sites in the effector
cells to the exclusion of histamine. It
is used in internal medicine for its
sedative effects. It also finds employment in dermatology and pneumology, as
well as certain psychotic illnesses.
Extrapyramidal reactions and photosensitivity reactions may occur as adverse
effects.
f prométhazine
d Promethazin
ı prometazina
e prometazina

3183 prontosil
Sulphamidocrysoidine. Purely of histori-
cal interest. The first sulphonamide to
be prepared and replaced by the far
more effective and easily obtainable
sulphonamides, e.g. sulphanilamide, sulphathiazole, sulphacetamide etc.

f prontosil; sulfamidochrysoïdine
d Prontosil; Sulfamidocrysoidin
ı prontosil; solfammidocrisoidina
e prontosil; sulfamidocrisoidina

3184 propallylonal
Barbiturate hypnotic indicated in the
treatment of insomnia. The drug must
be used with extreme precaution in the
presence of intoxications due to alcohol,
analgesic and psychotropic drugs. Also
in heart diseases, bronchial asthma,
shock and malignant hypertension, a
very strict monitoring is required.
f propallylonal
d Propallylonal
ı propallilonal
e propallilonal

3185 propamidine
Antiseptic agent with antibacterial and
antifungal actions, indicated in the
topical treatment of skin and conjunctiva
infections. In case of prolonged application, i.e. over a week, the danger of
tissue damage may emerge.
f propamidine
d Propamidin
ı propamidina
e propamidina

3186 propane
Aliphatic saturated hydrocarbon. Colourless, odourless non-toxic gas burning
with a luminous flame. It is a constituent of natural gas, oil gas, i.e. any
gas obtained by cracking and dehydro-
genation of carbons, tar and mineral oil.
It also represents the raw material of
several chemical processes. Ethylene,
propylene, propylic alcohol, propionic
aldehyde, acetone, acetic acid etc. can
be obtained from it.
f propane
d Propan
ı propano
e propano

3187 propanidid
Very short-acting anaesthetic that is
injected intravenously. A non-barbituric
narcotic agent. The induction is very
rapid and the anaesthesia lasts for
about five minutes. Nausea and transient
hypotension may occur as side effects.
f propanidide
d Propanidid
ı propanidide
e propanidida

3188 propanolamine
Amino-derivative of the propylic alcohol.
Propanolamines may contain one, two
and three radicals that are derived from
propylic alcohol, both normal and isopropylic. Prepared by reaction of propylene oxide with ammonia, they are colourless, water-soluble solids widely used
as intermediates, germicides and several
pharmaceutical products.

f propanolamine
d Propanolamin
i propanolammina
e propanolamina

3189 propantheline
Parasympatholytic having peripheral and toxic effects similar to atropine and chiefly used to reduce gastric secretion in peptic ulcers. The drug is also used as an antispasmodic agent for gastro--intestinal and urinary disorders. It is generally administered in the form of its bromide. Dryness of the mouth, blurred vision and urinary retention may occur as adverse effects.
f propanthéline
d Propanthelin
i propantelina
e propantelina

3190 propathylnitrate
Coronary vasodilator marked by a rapid and prolonged action by relaxing smooth muscles, in particular the vasal ones. It is chiefly used for the prevention and therapy of anginoid pains and coronary sclerosis.
f propathylnitrate
d Propathylnitrat
i propatilnitrato
e propatilnitrato

3191 propene
Unsaturated olefine hydrocarbons gas, a by-product in the cracking of petroleum hydrocarbons (of high molecular weight) for petrol fractions. Propene was widely used as an inhalant anaesthetic before more potent and safer gaseous anaesthetics were developed.
f propène
d Propen
i propene
e propeno

3192 propetandrol
Hormonal anabolic agent which has proved useful in the treatment of denutrition, osteoporosis, calcification disorders and arthrosis. The drug is also used in the course of a prolonged corticosteroid treatment and when there is a delay in the healing of a fracture. Sodium retention and jaundice may occur as side effects.
f propétandrol
d Propetandrol
i propetandrolo
e propetandrolo

3193 prophylactic
Relating to the prevention of the development of disease. More specifically, relating to any agent and/or pharmaceutical preparation used to avoid or check an infection.
f prophylactique
d Prophylactisch
i profilattico
e profiláctico

3194 propicillin
Semisynthetic acid-stable orally active penicillin marked by a rapid absorption and diffusion. Its peak serum levels are reached without delay and the therapeutic concentration is generally prolonged. It is mainly active against Gram-positive and Gram-negative microorganisms which are sensitive to the drug, particularly streptococci. Propicillin has proved resistant to destruction by penicillase and is also effective against staphylococci showing some resistance. In addition to being indicated for the treatment of infections, the drug has proved quite active in the prophylaxis of rheumatic fever.
f propicilline
d Propicillin
i propicillina
e propicilina

3195 propiolactone
Antiseptic substance used as the inactivating agent in the preparation of certain bacterial and viral vaccines. It is active against Gram-positive and Gram--negative micro-organisms, fungi and viruses and less active against bacterial spores.
f propiolactone
d Propiolakton
i propiolattone
e propiolactona

3196 propiomazine
Phenothiazine-related tranquillizer and a hypnotic at high doses. It is used in the treatment of psychopathies mainly when marked by mania and/or delirium. Propiomazine enhances the action of barbiturates. Drowsiness and gastro-intestinal disorders may occur as side effects.
f propiomazine
d Propiomazin
i propiomazina
e propiomazina

3197 propiram
Analgesic agent characterized by action and uses which are intermediate between those of analgesic antipyretics and pure analgesic, e.g. opium alkaloids. The drug is indicated for the relief of acute pains of various origin.
f propiram
d Propiram
i propiram
e propiram

3198 propizepine
Tricyclic antidepressant indicated in the treatment of endogenous depression states in neurotic subjects. The drug is also used in cases of asthenia accompanied by symptoms of depression. It should not be administered to patients suffering from epilepsy, glaucoma, hyperthyroidism, prostatitis and heart diseases. Dryness of the mouth, accommodation disorders

and some insomnia may occur as side
effects.
f   propizépine
d   Propizepin
i   propizepina
e   propizepina

3199   propoxycaine
Local anaesthetic used for infiltration
and nerve block. Tenseness, vertigo,
tremor and blurred vision may occur
among other adverse effects.
f   propoxycaïne
d   Propoxycain
i   propossicaina
e   propoxicaina

3200   propranolol
Used in the form of its hydrochloride.
A specific blocker acting on beta-recep-
tors and indicated in the treatment of
cardiac arrhythmias caused by heart
disease, digitalis or anaesthetics. Brady
cardia, potentiation of A-V block, nervous depression, insomnia, nausea and
bronchospasms are among the adverse
effects which may occur in the course
of the treatment.
f   propranolol
d   Propranolol
i   propranololo
e   propranololo

3201   propylbenzene
Choleretic agent marked by a specifically hydrocholeretic action which is both
potent and prolonged. Being at the same
time a cholagogue, the drug is indicated
in the treatment of chronic and acute
cholecystitis, cholelythiasis, consequences of hepatic disorders and mild
hepatic insufficiency. The drug should
not be used in the presence of a severe
liver disease or in case of complete
obstruction of the biliary tract.
f   propylbenzène
d   Propylbenzol
i   propilbenzene
e   propilbenceno

3202   propyl docetrizoate
Diagnostic medium chiefly used in bronchography, sialography and fistulology.
It is also used in the radiographic examination of the nasal fossae.
f   docétrizoate de propyle
d   Propyldocetrizoat
i   propile docetrizoato
e   docetrizoato de propilo

*   propylene   s. propene

3203   propylene glycol
A solvent used in the extract of some
crude drugs and as a vehicle for some
injections and local applications. It is
generally less toxic than other glycols
as it is rapidly broken down and excreted. Propylene glycol, used by itself,
inhibits the growth of fungi and micro-

-organisms with the same effectiveness
as that of ethyl alcohol.
f   propylèneglycol
d   Propylenglykol
i   glicole propilenico; glicole di propilene
e   propilenglicol

3204   propylhexedrine
Methylaminopropane-derived anorexia-inducing agent capable of reducing appetite and at the same time the sense of
fatigue. It is chiefly used against various forms of obesity. The drug should
not be administered in the presence of
neurovegetative disorders and insomnia.
f   propylhexédrine
d   Propylhexedrin
i   propilesedrina
e   propilexedrina

3205   propyliodone
Iodine-containing contrast medium chiefly
used for bronchography, cystography
and for the radiographic examination of
fistulae. Patients suffering from idiosyncrasy for iodine may confidently be
treated with propyliodone as the drug
does not produce iodism.
f   propyliodone
d   Propyljodon
i   propiliodone
e   propilyodona

3206   propylthiouracil
Synthetic preparation capable of depressing the formation of thyroid hormone
and indicated in the treatment of hyperthyroidism. Allergic rashes, headache,
diarrhoea and blood dyscrasias may occur as side effects.
f   propylthiouracile
d   Propylthiourazil
i   benziltiouracile
e   propiltiouracilo

3207   propyphenazone
Antipyrine-related analgesic and antipyretic agent used by itself or in association with other analgesics in the treatment of painful conditions of various
origin. Drowsiness, allergic dermatoses
and nausea may occur as side effects.
f   propyphénazone
d   Propyphenazon
i   propifenazone
e   propifenazona

3208   propyromazine bromide
Phenothiazine-derived spasmolytic agent
with an atropinic-, papaverinic- and
ganglioplegic-type peripheral action.
The drug is indicated in the treatment
of spasms, hypermotility and hypersecretion in gastro-intestinal diseases, hepato
biliary disorders and pathological conditions of the urinary tract. It also finds
employment in gynecology, paediatrics
and pneumology. Dryness of the mouth
and accommodation difficulties may occur
as side effects.

f bromure de propyromazine
d Propyromazinbromid
ı propiromazina bromuro
e bromuro de propiromazina

3209 proquazone
Anti-inflammatory, antipyretic and analgesic agent indicated in the treatment
of articular degenerative diseases, e.g.
osteoarthritis affecting the ilium, knee
and other joints. It is also used in the
treatment of chronic inflammatory conditions borne of acute attaks of gout,
sciatica, lumbago, tendinitis, synovitis
etc. Abdominal pains, nausea, tachycardia and headache may occur as side
effects.
f proquazone
d Proquazon
ı proquazone
e proquazona

3210 proscillaridin
Cardiotonic agent marked by positive
inotropic action and used, either by itself or in association with theophylline,
in the treatment of cardiac incompetence
that is resistant to digitalis- and stro-
phatin-related drugs. No cumulative
effect. Nausea and diarrhoea may occur,
though rarely, as side effects.
f proscillaridine
d Proscillaridin
ı proscillaridina
e proscilaridina

3211 prospidium chloride
Cytostatic (capable of causing the blocking of the capillaries by the leukocytes)
agent used in the treatment of papillo-
mas in the upper respiratory tract, of
tumours in the larynx, in the lungs, in
the stomach, breast, ovaries and urinary bladder. It is also used in the treat
ment of lymphogranulomas and all cancer
ous metastases. It is sometimes associated with other cytostatics and corticosteroids. Vertigo, nausea, cutaneous swelling and hypersensitivity to cold may
occur as side effects.
f chlorure de prospidium
d Prospidiumchlorid
ı prospidio cloruro
e cloruro de prospidio

3212 prostaglandine
One of a group of fatty acids affecting
a large number of physiological processes. It is probable that prostaglan-
dins, rather than acting as hormones,
have the function of modulating the action of the latter. They frequently alter
the activity of the cells in which they
are synthesized. Among the most relevant physiological processes affected by
them, the most important are their car-
diological and gastric effects, the dilata
tion of the bronchi and some metabolic
actions. They also provoke uterine contractions and some local inflammation.

f prostaglandine
d Prostaglandin
ı prostaglandina
e prostaglandina

3213 prostate extract
Prostatae extractum. Deproteinized extract obtained from the prostatic gland
of healthy animals. Opotherapeutic agent
capable of restoring the functional balance of the prostatic gland by strengthening the vesical muscular tone and
improving micturition. It is indicated in
the treatment of the initial phase of
prostatic hypertrophy, urinary incontinence and pre- as well as post-operative
treatment in the surgery on the prostate.
f extrait de prostate
d Prostataextrakt
ı estratto di prostata
e extracto de próstata

3214 prosulthiamine
Thiamine (vitamin B1) derivative with
antineuritic action. Also useful in cases
of nervous exhaustion. It may be associated with high doses of vitamin B12
in the treatment of some neuritides.
f prosulthiamine
d Prosulthiamin
ı prosultiamina
e prosultiamina

3215 protamine
Each of a class of simple proteins which
are strongly basic, do not coagulate by
heat and are soluble in water and dilute ammonia. Protamines yield a considerable amount of amino acids on hydrolysis. They do not contain sulphur and
precipitate with phosphotungstic acid.
They have no antigenic or toxic power
and are not affected by pepsin but only
by a specific enzyme (protaminase).
Protamines easily combine with insulin
thus delaying its absorption.
f protamine
d Protamin
ı protamina
e protamina

3216 protamine sulphate
Chemical compound capable of antagonizing the anticoagulant effect of heparin
and other coagulation-preventing substances. It is indicated in the treatment
of haemorrhagic conditions. The therapy
must be carefully monitored owing to the
danger of embolism. Some allergic reactions have been noticed as side effects.
f sulfate de protamine
d Protaminsulfat
ı protamina solfato
e sulfato de protamina

3217 protamine zinc insulin
Buffered sterile suspension of insulin
with added zinc chloride and protamine.
It has a hypoglycaemic action which appears slowly but lasts for a long time.

It is used in the treatment of severe
diabetes accompanied by glycosuria and
a tendency to diabetic coma. Allergic
reactions may occur as side effect.
f préparation d'insuline zinc protamin
d Protamin-Zink-Insulin; PZ-insulin
i insulina zinco protaminato
e insulina-protamina-cinc

3218 protein
Natural, extremely complex substance
containing carbon, hydrogen, oxygen,
nitrogen and sometimes phosphorus, sulphur and metals, e.g. iron and copper.
Proteins invariably contain long chains
of amino acids obtained by condensation
between aminic and carboxylic groups
of the latter. They are present in every
living cell and form an essential constituent of it. They are classified on the
basis of three main groups, i.e. simple,
conjugated or compound and derived.
f protéine
d Protein
i proteina
e proteína

3219 protein A
Protein found on the surface of certain
strains of Staphylococcus aureus cells.
It binds very tightly to the Fc portion
of an antibody molecule when the antibody is bound to an antigen.
f protéine A
d A-Protein
i proteina A
e proteína A

3220 protein extract
Extract of the protein of a substance to
which sensitivity must be diagnosed. The
substance may be used for both diagnosis and therapeutic treatment.
f extrait de protéine
d Proteinextrakt
i estratto di proteina
e extracto de proteína

3221 protein hydrolysate
Complex of amino acids and short-chain
proteins obtained by enzymatic, acid or
some other type of hydrolysis from animal proteins, e.g. casein, lactalbumin
etc. Over 50% of the total nitrogen present must be in the form of alpha-aminic
nitrogen. The compound is thus a well
balanced mixture of free amino acids
which can be immediately and completely
assimilated. It is indicated as a dietary
supplement when the proteic requirement
is increased, e.g. in the course of
pregnancy, during lactation, convalescence etc. It is also used parenterally
in order to reestablish a positive nitrogen balance in cases of protein deficiency caused by inadequate nutrition, some
severe disease or surgical operations affecting the gastro-intestinal tract.
f hydrolysat de protéine; protéolysat
d Proteinhydrolysat; Eiweisshydrolysat

i proteolisato; idrolisato proteico
e hidrolizado de proteínas; proteolisato

3222 protheobromine
Theobromine-derivative with a cardio-
stimulating action. It is also effective
as a vasodilator and bronchial antispastic agent. It finds employment in peripheral circulatory disorders.
f prothéobromine
d Protheobromin
i proteobromina
e proteobromina

3223 prothionamide
Antituberculous agent indicated in the
treatment of pulmonary and extrapulmonary tuberculosis. It sometimes proves
better tolerated than ethionamide, q.v.
Anorexia, excessive flow of saliva, head
ache, hypotension, stomatitis, diarrhoea,
impotence and loss of hair may occur,
among others, as side effects.
f prothionamide
d Prothionamid
i protionamide
e protionamido

3224 prothipendyl
Neuroleptic agent indicated in the treatment of neurovegetative and psychomotor
disorders. It is also used to relieve
states of confusion, sleep disorders and
lack of mental concentration in children.
Drowsiness, photosensitivity, tachycardia
and hypotension are the most probable
side effects.
f prothipendyl
d Prothipendyl
i protipendile
e protipendilo

3225 protirelin
TRH (thyrotropin-releasing hormone).
Synthetic prohormone capable of stimulat
ing the release of thyrotropin and used
as a diagnostic agent in difficult cases
of hypothyroidism and to assess the
supply of thyrotropin contained in the
hypophysis. Nausea, dizziness and the
sudden urgent need of passing urine
may occur as side effects.
f protireline
d Protirelin; Thyroliberin; TRH
i protirelina
e protirelina

3226 protocatechuic acid
Acid found in the fruit of Illicium reli-
giosum, formed by alkaline fusion of
lignin, catechins etc. It is obtained by
various syntheses, e.g. by treating the
pyrocatechin with carbon dioxide. Vanillic acid and veratric acid constitute the
methyl- and dimethyl ether of the protocatechuic acid.
f acide protocatéchique
d Protocatechinsäure
i acido protocatechico
e ácido protocatéquico

3227 protokylol
Bronchodilator and sympathomimetic agent indicated in the treatment of bronchial asthma, asthmatic bronchitis, pulmonary emphysema and spastic bronchitis. Its action is more prolonged that that of epinephrine and resembles that of iso-prenaline. Tachycardia may occur as a side effect.
f protokylol
d Protokylol
i protochilolo
e protoquilolo

3228 protoveratrine
Alkaloid indicated in the treatment of hypertension. It may easily cause brady cardia and peripheral vasodilation.
f protovératrine
d Protoveratrin
i protoveratrina
e protoveratrina

3229 protriptyline
Tricyclic antidepressant with an action that is more immediate than that of imipramine and amitriptyline but has central stimulating effects. It is indicat ed in the treatment of withdrawal and lack of energy. It has no sedative or tranquillizing properties.
f protriptyline
d Protriptylin
i protriptilina
e protriptilina

3230 provalamide
Sedative used for the relief of psychic agitation. Chemically, ethyl butylethyl-malonylamide.
f provalamide
d Provalamid
i provalamide
e provalamida

3231 provitamin
Any of those substances which, owing to metabolic modifications or the interfer-ence of factors that are extraneous to the organisms, e.g. ultraviolet radia-tions, acquire a vitaminic action. Typi-cal examples are the carotenes which are contained in fruit and vegetables and transformed into vitamin A in the intestinal mucosa and in the liver, and the 7-dehydrocholesterol which is trans-formed into vitamin B_3 under the action of ultraviolet rays.
f provitamine
d Provitamin
i provitamina
e provitamina; previtamina

3232 proxamine
Vasoconstrictor chiefly used in the treat-ment of gastro-intestinal haemorrhages.
f proxamine
d Proxamin
i proxamina
e proxamina

3233 proxazole
Spasmolytic agent capable of causing relaxation of the smooth muscles by a mechanism that is analogous to that of papaverine. It is chiefly indicated in the treatment of congestive oedematous processes accompanied by spasms of the smooth muscles.
f proxazole
d Proxazol
i prossazolo
e proxazolo

3234 proxibarbal
Barbiturate derivative practically with-out hypnotic effect yet characterized in a singular manner by the inhibiting diencephalic action of the classic bar-biturates. It increases the tolerability of the nervous cell for the deficit of oxygen and the transport of glycogen to the brain thus displaying a specific therapeutic effect in vascular cephalal-gias, migraine and other painful syn-dromes due to hypertension, traumas etc.
f proxibarbal
d Proxibarbal
i prossibarbale
e proxibarbalo

3235 proxifezone
Antiphlogistic and analgesic agent indi-cated in the treatment of pains caused by arthritis, periarthritis and rheuma-tism, as well as traumas and a general phlogistic condition. Vertigo and stomach ache may occur as side effects.
f proxifézone
d Proxyfezon
i prossifezone
e proxifezona

3236 proxymetacaine
Ester-type surface anaesthetic used in ophthalmology for surgical operations on the eyeball, extraction of foreign bodies and painful medications. It is more po-tent than tetracaine when prepared in equal concentrations but does not cause irritation or mydriasis. Side effects are similar to those of lignocaine.
f proxymétacaïne
d Proxymetacain
i prossimetacaina
e proximetacaina

3237 proxyphilline
Vaso- and broncho-dilator without any effect on gastric motility. It is used in the treatment of bronchial asthma, chron ic bronchitis, emphysema and dyspnoea of cardiac origin or connected with pul-monary oedema. The drug is also used as a coronary dilator.
f proxyphylline
d Proxyphyllin
i prossifillina
e proxifilina

3238 prozapine

Relaxant of the smooth musculature mark
ed by a selective action on the digestive
tract. It is 20 times more active than
papaverine and is used in association
with sorbitol for the treatment of gastro-
-intestinal or hepatobiliary disorders
and inflammation of the biliary tract.
f prozapine
d Prozapin
i prozapina
e prozapina

3239 pseudocumene
Chemical compound, isomer of cumene;
together with the other two trimethylben-
zene isomer, it is found in the tar of
coal and used as intermediate in several
syntheses.
f pseudocumène
d Pseudocumen; Pseudocumol
i pseudocumene
e pseudocumeno

3240 pseudoephedrine
Sympathomimetic agent marked by a
nasal decongestant action. It is chiefly
indicated in the symptomatic treatment
of cold, influenza, rhinitis and acute
otitis media serosa (see ephedrine). The
drug should not be used, or used with
great precautions, in patients suffering
from hypertension, heart disease, hyperthyroidism and hypersensitivity to sympathomimetic preparations. Tachycardia,
insomnia and nervousness may occur as
side effects.
f pseudoéphédrine
d Pseudoephedrin
i pseudoefedrina
e pseudoefedrina

3241 pseudopelletierine
Alkaloid of the bark of the pomegranate
containing two nuclei of the piperidine
which have a nitrogen atom and two
carbon atoms in common. Colourless, opti
cally inactive crystals. Its derivatives
have some value as local anaesthetics.
f pseudopelletiérine
d Pseudopelletierin
i pseudopelletierina
e pseudopeletierina

3242 psoralen
Chemical compound, of the natural couma
rin group. It is a furane derivative of
the coumarin. Several compounds are
found in nature, which can be considered psoralen derivatives, e.g. bergaptene,
bergaptol etc.
f psoralène
d Psoralen
i psoralene
e psoraleno

3243 pterin
Any of a group of chemical compounds
derived from the pteridin (tetrazonaph-
thalene) and considered products of
degradation of nucleic acids. They have

some biological importance owing to their
correlation with substances characterized
by their antianaemic properties.
f ptérine
d Pterin
i pterina
e pterina

3244 pteroic acid
Chemical compound resulting from the
union of a pteridine derivative and
p-aminobenzoic acid. By condensing the
pteroic acid with glutaminic acid, the
pteroylglutaminic acid, i.e. the folic
acid, is obtained.
f acide ptéroïque
d Pteroinsäure
i acido pteroico
e ácido pteroico

* pteropterin   s. pteroic acid

3245 pteroylglutamic acids
Any of the acids which are a conjugate
of pteroic acid (acid occurring in the
wing pigments of butterflies, a source
of vitamin B) and glutamic acid (an
amino acid, a common constituent of protein), in particular the folic acid. Some
of them, especially the one found in
yeast, are conjugates of pteroic with
more than one molecule of glutamic acid.
f acides ptéroylglutamiques
d Pteroylglutaminsäure
i acidi pteroilglutamminici
e ácidos pteroilglutamínicos

3246 pulmonis extractum
Extract obtained from the lung tissue
of hog. It is a specific opotherapeutic
agent displaying a stimulating and
eutrophic action on the pulmonary tissues in order to strengthen defence processes. It is chiefly indicated in pretu-
bercular conditions, inflammatory pulmonary primary and postprimary states
and all exudative forms, as well as
those prevailing in the course of convalescence following pneumonia.
f extrait de poumon
d Lungenextrakt
i estratto di polmone
e extracto de pulmón

3247 puncture vine
Small caldrop. Tribulus lanuginosus, a
plant growing in warm regions, whose
fruit has diuretic and purgative properties. Also known as puncture weed.
f tribule lanugineux
d Bürzeldorn
i tribolo lanuginoso
e tríbulo lanuginoso

3248 puncture vine fruit
The fruit of the puncture vine containing an active principle which acts as
a diuretic and purgative agent.
f fruit de tribule
d Bürzeldornfrucht; Erdstachelnuss

i frutto di tribolo lanuginoso
e fruto de tríbulo lanuginoso

* punicine   s. pelletierine

3249 purging cassia
The dried pods of the drumstick tree which is native of warm regions. Its pulp may be used medicinally for its laxative action.
f casse officinale; casse en bâtons
d Purgierkassie; Röhrenkassie
i cassia in canna; cassia in bastoni
e cañafístula

3250 purging genista
Plant belonging to a large genus of spiny plants whose flowers and leaves yield a purgative agent.
f genêt purgatif
d Purgierginster
i ginestra purgativa
e retama purgante

3251 purging globe daisy
Plant of the genus Globularia whose leaves and flowers are used for the preparation of a purgative substance.
f globulaire purgative; séné de Provence; séné sauvage; globulaire turbith
d strauchartige Kugelblume
i alipio bianco
e alipio blanco; bocha; cardenilla; serventia

3252 purine
Complex cyclic diureide that does not appear free in nature but is widely distributed in the form of hydroxy-, amino- or methyl-derivatives. The amino purine which contains an amino group, e.g. adenine, appears in the decomposition products of the nucleic acids. Hydroxy--purine contains hydroxyl instead of hydrogen. Methyl purine is one in which one or more methyl groups have been substituted, e.g. caffeine, theobromine etc.
f purine
d Purin
i purina
e purina

3253 purpurin
Chemical compound, either orange or red in colour, occurring as a glycoside in the roots of Rubia tinctorum or prepared by oxidation of alizarin with pyrolusite and sulphuric acid. It is used, among other things, as a chemical reagent.
f purpurine
d Purpurin
i purpurina
e purpurina

3254 pygelinum
Pygeum africanum. Phytotherapeutic agent specifically indicated in the treatment of prostatic adenoma. It improves the histologic and functional tones of the glandular epithelium of the prostate with the result of normalizing micturition. It is also indicated in the treatment of the functional disorders caused by prostatic hypertrophy, inoperable forms of the condition, and in the preparations required for adenomectomy.
f Pygeum africanum; prunier d'Afrique
d Pygeum africanum
i Pygeum africanum; pigelina
e Pygeum africanum

3255 pyocyanin
Toxic blue crystalline pigment found in green pus and elaborated in the metabolism of a bacterium (Pseudomonas aeruginosa), which is a quinone imine related to phenazine and has an antibiotic action. Pyocyanin stimulates the cellular respiration; it is the only natural substance capable of substituting methylene blue as an intermediate transporter between a coupling of hydrogenases and thus making possible the contemporaneous oxidation of a system and the reduction of another.
f pyocyanine
d Pyocyanin
i piocianina
e piocianina

3256 pyrantel
Anthelminthic agent which acts by paralysing mature as well as immature forms thus making their excretion easier. It is only very moderately absorbed from the intestine so that its activity is concentrated at the site of the infection. It is used as a single dose (10 mg/kg body weight) for threadworm, hookworm, roundworm, whipworm and trichostrongyliasis. Gastro-intestinal disorders, diarrhoea, insomnia and dizziness may occur as side effects.
f pyrantel
d Pyrantel
i pirantele
e pirantel

3257 pyrazinamide
Minor antituberculous chemotherapeutic agent chiefly used as tuberculostatic in those cases in which major antituberculous drugs yield no result, have met with chemoresistance or idiosyncrasy. It has given some fairly good results in the treatment of chronic pulmonary tuberculosis. Anorexia, an abnormal increase in the number of eosinophil leukocytes in the blood and pains in the joints may occur as side effects.
f pyrazinamide
d Pyrazinamid
i pirazinamide
e pirazinamida

3258 pyrazinobutazone
Phenylbutazone piperazine. Non-steroid antiinflammatory agent marked by rapid analgesic effect. It is indicated in all

phlogistic syndromes, in rheumatology, phlebology, otorhinolaryngology, pneumology and gynecology. Abdominal pains, occasionally some nausea and dizziness may occur as side effects.

f pyrazinobutazone
d Pyrazinobutazon
i pirazinobutazone
e pirazinobutazona

3259 pyrazole
Heterocyclic pentatomic compound, an isomer of the imidazole. Pyrazole can be obtained in various ways, e.g. by condensation of epichlorhydrine and hydrazine in the presence of zinc chloride. Many derivatives of pyrazole are used in the pharmaceutical industry.

f pyrazole
d Pyrazol
i pirazolo
e pirazol

3260 pyrazoline
Heterocyclic compound containing two hydrogen atoms more than pyrazole from which it is obtained by reduction. A colourless liquid miscible with water and alcohol. It has basic properties.

f pyrazoline
d Pyrazolin
i pirazolina
e pirazolina

3261 pyrazolone
Keto derivative of pyrazoline. Several isomers occur which can be derived from the various pyrazolines. They are prepared on the basis of Knorr synthesis, i.e. condensation of the acetacetic ether with phenylhydrazine and reaction of the product with methyl iodide. Many of them are widely used in pharmaceutical products.

f pyrazolone
d Pyrazolon
i pirazolone
e pirazolona

3262 pyrethrum
Genus of composite plants with finely divided leaves which are generally included in the genus Chrysanthemum. It is cultivated in several regions of the earth and particularly in Japan. The tops, with flower not yet fertilized, contain the active principles (pyrethrine) that are highly toxic for insects.

f pyrèthre
d Pyrethrum
i piretro
e pelitre

3263 pyrethrum flowers
Also known as insect flowers. The flowers of the pyrethrum which yield a strong insecticide (pyrethrum powder).

f fleurs de pyrèthre
d Pyrethrumblüten; Insektenblumen; Chrysanthemumblüten
i fiori di piretro
e flores de pelitre

3264 pyrethrum powder
The powder that is obtained from the pyrethrum flowers and is used as insecticide.

f poudre de pyrèthre; poudre persane
d Insektenpulver; Kapuzinerpulver
i polvere di piretro; razzia
e polvo de pelitre

3265 pyridinolcarbamate
Therapeutic agent used in arteriosclerosis. The drug is capable of protecting the walls of the blood vessels. It has an antibradykinic and antithrombic action inhibiting capillary permeability. It also inhibits platelet aggregation and is indicated in the treatment of peripheral vascular circulatory disorders due to arteriosclerosis obliterans, diabetes, retinal vasculitis, Bürger's disease (thrombo-angiitis obliterans) etc.

f pyridinolcarbamate
d Pyridinolkarbamat
i piridinolcarbamato
e piridinolcarbamato

3266 pyridophylline
Vasodilator, diuretic agent and respiratory stimulant. It is used in the treatment of coronary incompetence and respiratory disorders. Also used in arteriosclerosis.

f pyridophylline
d Pyridophyllin
i piridofillina
e piridofilina

3267 pyridostigmine bromide
Anticholinesterase, i.e. an inhibitor of the cholinesterasic activity, principally indicated in the treatment of myasthenia gravis and occasionally in ophthalmology. It is sometimes used in association with neostigmine. The methylsulphate of the latter has a less sustained improvement if administered by itself. Gastro-intestinal pains and disorders, accommodation difficulties, myosis, ECG alterations etc. may occur as side effects.

f bromure de pyridostigmine
d Pyridostigminbromid
i piridostigmina bromuro
e bromuro de piridostigmina

3268 pyridoxine
Vitamin B_6. An important factor whose deficiency has shown in rats a form of florid symmetrical dermatitis of the extremities, epileptic fits and haematuria. In the organism, it is convertible into pyridoxal and pyridoxamine which especially occur in cereals. The drug protects the metabolism of protides, glucides and lipides.

f pyridoxine; adermine; vitamine B_6
d Pyridoxin; Adermin; Vitamin B_6
i piridossina; vitamina B_6; adermina

e piridoxina; vitamina B$_6$; adermina

3269 pyridoxine oxoglutarate
Antitoxic agent capable of counteracting ammonaemia. It acts in the liver where it intervenes in the metabolism of glucides and protides (alpha-keto-glutaric acid) and in the decarboxylation and transamination processes. It has a neutralizing effect on hyperammonaemia by the fixation of ammonia and the formation of glutamine. It is thus indicated in the treatment of hepatic incompetence and conditions of hyperammonaemia due to acute and chronic hepatopathies.
f oxoglutarate de pyridoxine
d Pyridoxinoxoglutarat
i piridossina ossoglutarato
e oxoglutarato de piridoxina

3270 pyrimethamine
Antimalarial agent indicated in the prophylaxis and treatment of falciparum malaria and vivax malaria. The drug has also proved useful in the treatment of toxoplasmosis, i.e. the infection caused by Toxoplasma gondii. Gastro-intestinal disorders, skin rashes and folate-deficient anaemia may occur as side effects.
f pyriméthamine
d Pyrimethamin
i pirimetamina
e pirimetamina

3271 pyrimidine
A cyclic compound which is not found in nature and whose derivatives are of great biological importance as they enter into the structure of vitamin B$_1$, vitamin B$_2$ and the nucleic acids. They also enter into the structure of barbituric and uric acid.
f pyrimidine
d Pyrimidin
i pirimidina
e pirimidina

3272 pyritinol
Pyrithioxine. Sedative and psychostimulant agent apparently capable of promoting the acceptance of glucose by the brain thus improving cerebral metabolism. The drug improves the psychomotor activity and relieves fatigue and nervous stress that are due to specific events, senile involution or arteriosclerosis. It is also effective for the recovery of normal cerebral conditions following a trauma, stroke, viral encephalitis etc., as well as migraine and neuralgias. The effects of the drug are slow to appear; a few weeks or even months are required to show some results. Insomnia is the principal side effect.
f pyritinol
d Pyritinol
i piritinolo
e piritinolo

3273 pyrocatechin
Or pyrocatechol. Bivalent phenol. Colourless, phenol-smelling crystals that are soluble in water. It is found in nature (the leaves of Ampelopsis hederacea, the juice of cascarilla etc.) or is obtained by dry distillation of various species of tannins. It can also be prepared by alkaline fusion of the o-chlorphenol or phenol-disulphonic acid. It is widely used as a fungicide and a bactericide.
f pyrocatéchine; pyrocatéchol
d Pyrocatechin; Brenzcatechin
i pirocatechina
e pirocatequina

3274 pyrogallic acid
Pyrogallol. Trivalent phenol obtained by heating gallic acid. Needle-shaped, light, colourless but easily darkened and water-soluble crystals. Pyrogallol is used medicinally as an antiseptic and keratolytic (capable to dissolve epidermis) agent.
f pyrogallol; acide pyrogallique
d Brenzgallussäure; Pyrogallol
i acido pirogallico; pirogallolo
e ácido pirogálico; pirogalol

3275 pyroligneous
Relating to a substance obtained by the distillation of wood.
f pyroligneux
d Holz-
i pirolegnoso
e piroleñoso

3276 pyroligneous vinegar
Wood vinegar, pyroligneous acid. Impure solution of acetic acid obtained by the dry distillation of wood. It is an acid, reddish-brown aqueous liquid mainly containing acetic acid, methanol, wood oils and tars.
f vinaigre de bois; vinaigre pyroligneux
d Holzessig
i acido acetico pirolegnoso; acido pirolegnoso
e ácido acético piroleñoso

3277 pyrophosphate
Salt of pyrophosphoric acid. Pyrophosphates may be acid or neutral. Acid pyrophosphates are soluble in water and the solution displays a weakly acid reaction; of the neutral pyrophosphates only those of alkaline metals are soluble in water and the solution displays a weakly basic reaction owing to the hydrolysis to which they are subjected.
f pyrophosphate
d Pyrophosphat
i pirofosfato
e pirofosfato

3278 pyrophosphoric acid
Oxygenated acid of the phosphorus formed from the common phosphoric acid by the elimination of water due to prolonged heating at 200 ÷ 300 centigrades. A

strong acid whose solutions are transformed, slowly at cold temperatures,
rapidly in hot temperature, into phosphoric acid.
f acide pyrophosphorique
d Pyrophosphorsäure
i acido pirofosforico
e ácido pirofosfórico

* **pyroracemic acid s. pyruvic acid**

3279 **pyrovalerone**
Psychostimulant indicated in the treatment of asthenia and states of depression of various origin. It is also used
to combat states of lethargy caused by
drugs. A general malaise and palpitations may occur as side effects.
f pyrovalérone
d Pyrovaleron
i pirovalerone
e pirovalerona

3280 **pyrocaine**
Chemical compound acting as local anaesthetic. Also spelt pirrhocaine.
f pirrocaïne
d Pyrrocain
i pirrocaina
e pirocaina

3281 **pyrrole**
Heterocyclic compound, the parent substance of a large number of important
natural compounds, e.g. blood, bile
pigments, chlorophyll, some alkaloids
such as nicotine and some of the amino
acids that are present in proteins.
f pyrrole
d Pyrrol
i pirrolo
e pirrol

3282 **pyrrolidine**
Chemical compound containing the heterocyclic ring of pyrrole, yet completely
saturated. It is fairly toxic, inflammable and a skin irritant. It gives origin
to important products: the ring opens by
oxidation and forms amino-butyrric acid,
while it yields pyrrole by hydrogenation.
Derivatives of some interests are obtained by the introduction of substituting
groups into the atom of nitrogen. Some
of these derivatives find employment in
the preparation of antioxidants, insecticides etc.
f pyrrolidine; tétrahydropyrrole
d Pyrrolidin; Tetrahydropyrrol
i pirrolidina; tetraidropirrolo
e pirolidina; tetrahidropirolo

3283 **pyrrolnitrin**
Topical fungicide, a broad-spectrum anti
biotic capable of acting against infections caused by various species of fungi.
It is chiefly used for the treatment of
superficial mycoses, e.g. dermatophytosis,
pityriasis, candidiasis and their bacterial complications.
f pyrrolnitrine
d Pyrrolnitrin
i pirrolnitrina
e pirrolnitrina

3284 **pyruvic acid**
Aliphatic keto acid; a liquid characterized by a penetrating odour and obtained by dry distillation of tartaric acid,
saponification of acetylcyanide, oxidation of acetone or calcium lactate and
by fermentation of sugar.
f acide pyruvique
d Brenztraubensäure
i acido piruvico
e ácido pirúvico

Q

3285 **quassic acid**
Product obtained by acid hydrolysis of quassin, q.v.; sericeous crystals, soluble in alkaline solution, marked by a yellow-reddish colour.
f acide quassique
d Quassinsäure
i acido quassico
e ácido cuásico

3286 **quassin**
The bitter principle found in the bark and wood of quassia. A solid, colourless crystalline substance, colourless yet brilliant, soluble in alcohol, used in medicine as a tonic and stimulant. Quassin is also prepared as an amorphous substance but in this form it appears to be less active.
f quassine; picrasmine
d Quassin
i quassina
e cuasina; picrasmina

3287 **quatacaine**
Topical anaesthetic used in surgery, obstetrics, dentistry, otorhinolaryngology and nephrology. Some local inflammation may occur as a side effect.
f quatacaïne
d Quatacain
i quatacaina
e quatacaina

3288 **quebracho**
Tree yielding quebracho bark, q.v., also used as a source of tannin. It mainly grows in Brazil and Argentina.
f quebracho
d Quebracho
i quebracho
e quebracho

3289 **quebracho bark**
The dried bark of the white quebracho used as a respiratory sedative in dyspnoea and asthma.
f écorce de quebracho
d Quebrachorinde
i corteccia di quebracho
e corteza de quebracho

3290 **queen bee jelly**
Royal jelly. Highly nutritious secretion of the pharyngeal gland of the honey bee, which may be used medicinally.
f gelée royale; lait des abeilles

d Bienenkönigin-Futtersaft
i gel reale; pappa reale
e gelo real

3291 **quercetagetin**
Chemical compound, hexahydroxyflavone, a yellow colouring substance extracted from the Tagetes patula. The glucoside of the quercetagetin, called quercetagitrine, is found in the flowers of some Compositae.
f quercétagine
d Quercetagetin
i quercetagina
e quercetagina

3292 **quercetin**
Chemical compound, pentahydroxyflavone; a yellow colouring substance occurring as a glucoside in the bark of Prunus serotina and Quercus tinctoria. Its dimethyl ester constitutes the gynotermone of the undifferentiated sexual cells of green algae.
f quercétine
d Quercetin
i quercetina
e quercetina

3293 **quercitol**
Chemical compound, pentaoxyderivative of cyclohexane; colourless, sweet-tasting, water soluble crystals that are extracted from the bark of the oak.
f quercitol
d Quercitol
i quercitolo
e quercitolo

3294 **quercitrin**
Glucoside occurring in the bark of Quercus tinctoria; yellow crystals, moderately soluble in warm water, yielding rhamnose and quercetin by hydrolysis.
f quercitrine
d Quercitrin
i quercitrina
e quercitrina

3295 **Quercus stenophylla**
Oak genus of the family Fagaceae. The drug obtained therefrom is used in the treatment of lithiasis. Owing to its antiphlogistic and diuretic action it is also indicated in the treatment of renal and ureteral calculosis to promote the elimination of calculi and check their growth.

Some gastric disorders, in particular diarrhoea, may occur as side effects.
f Quercus stenophylla
d Quercus stenophylla
i Quercus stenophylla
e Quercus stenophylla

3296 quillaia
Soap tree. Genus of trees of the family Rosaceae native to Brazil Peru and Chile and pharmaceutically useful for their saponaceous bark.
f quillaya
d Seifenbaum
i quillaia; legno saponario
e quilaya; jabonosa

3297 quillaia saponins
The saponins which are obtained from the bark of the Quillaia saponaria in which they are contained by approximately 10% and yield sapogenine (quillaic acid) and galactose by hydrolysis. An amorphous yellowish powder and aqueous solution displaying some therapeutic properties as expectorant and diuretic.
f quillayasaponines
d Quillayasaponinen
i quillaiasaponine; quillaina
e quilayasaponinas

3298 quinacilline
Sodium salt of a quinoxalinic derivative of penicillin; colourless, water-soluble crystals taking up a yellow colour in the light. It displays a good antimicrobial activity especially against Staphylo̲coccus aureus.
f quinacilline
d Quinacillin
i quinacillina
e quinacilina

* **quinacrinum chloride   s. mepacrine**

3299 quinalbarbitone
Barbiturate hypnotic having practically the same effects as secobarbital, q.v.
f quinalbarbitone
d Chinalbarbiton
i chinalbarbitale
e quinalbarbital

3300 quinaldic acid
Isomer of the cinchoninic acid, an oxida̲tion product of cinchonine.
f acide quinaldique
d Chinaldinsäure
i acido chinaldico
e ácido quináldico

3301 quinate
Any salt or ester of quinic acid, q.v.
f quinate
d Chinat; chinasaures Salz
i chinato
e quinato

3302 quinazoline
Yellow crystalline compound characteriz-

ed by an odour resembling that of naph̲thalene, derived from quinoline by substitution of a nitrogen atom for a methylidyne group in the 3-position.
f quinazoline
d Chinazolin
i chinazolina
e quinazolina

3303 quinbolone
Androstane-related synthetic steroid mark̲ed by an anabolic action. It supports the protein synthesis and diminishes nitrogen loss. It also acts on the nucleic acids and on the metabolism of the amino acids.
f quinbolone
d Chinbolon
i quinbolone; chinbolone
e quinbolona

3304 quince
The fruit of a widely cultivated central Asiatic tree, having many seeds in each carpel and hard acid flesh used for the production of mucilage.
f coing
d Quitten
i cotogna
e membrillo

3305 quince juice
The juice which is used for the preparation of mucilage and of several lotions. It is also used as a suspending agent.
f suc de coing
d Quittensaft
i succo di cotogna
e zumo de membrillo

3306 quince mucilage
The product of the quince's cell activity secreted in the cell. It is used as a water-miscible base for various pharmaceutical preparations, especially dermatological ones.
f mucilage de graines de coing; bandoline
d Quittenschleim
i mucillagine di semi di cotogna
e mucilago de membrillo

3307 quince seed
Each of the many seeds of the quince, containing approximately 20% of mucilage with a fixed oil and protein.
f pépin de coing
d Quittensamen; Küttenkern
i seme di cotogna
e pepita de membrillo; semilla de membrillo

3308 quinestradiol
Oestriol-derived oestrogen principally act̲ing on the lower segments of the feminine genital apparatus. The drug is indicated in the treatment of pruritus vulvae, senile dystrophic vaginitis and cervicovaginal ulcers, especially during menopause.
f quinestradiol

d Chinestradiol
i chinestradiolo
e quinestradiol

3309 quinestrol
Oestrogen capable of inhibiting prolactin production by a physiological action. It is used in order to stop milk secretion. Also indicated in the treatment of circulatory disorders and menopausal neuralgias. Gastric pains and nausea may occur as side effects.

f quinestrol
d Chinestrol
i chinestrolo
e quinestrolo

3310 quinethazone
Sulphamide-related diuretic agent capable of increasing the excretion of sodium, chloride ions and, to a lesser extent, potassium ions. It is especially indicated in the treatment of oedemata involving salt retention, e.g. the oedemata that are typical consequence of a heart congestion condition or a nephrotic syndrome. The drug is also used in the treatment of hypertension. Nausea, intestinal disorders accompanied by diarrhoea, dizziness and general debility may occur as side effects.

f quinéthazone
d Chinetazon
i chinetazone
e quinetazona

3311 quingestanol
Oral contraceptive (progestogen) characterized by a peripheral action on the cervical mucus whereby the passage of sperms is inhibited. Such action does not involve the endocrine balance: ovulation is thus normal and menstruation quite regular. Some intermenstrual bleeding and amenorrhoea may occur as side effects.

f quingestanol
d Chingestanol
i chingestanolo
e quingestanolo

3312 quinic acid
Or kinic acid. Acid occurring in the cinchona bark, combined with the alkaloids and in coffee beans. White crystalline water-soluble substance used in the treatment of rheumatism and particularly of gout.

f acide quinique
d Chinasäure
i acido chinico
e ácido quínico

3313 quinicine
Or quinotoxine. Bitter poisonous reddish-yellow amorphous alkaloid isomeric with quinine and obtained from cinchona bark. It can also be obtained by heating a salt of quinine.

f quinicine

d Chinicin
i chinicina
e quinicina

3314 quinidine
Stereo-isomer of quinine characterized by pharmaceutical action and effects similar to those of quinine in cases of malaria. The drug is usually administered to people who show idiosyncrasy to quinine. Quinidine has proved useful in the treatment of auricular flutter and paroxysmal tachycardia. In some cases, some idiosyncrasy has been observed.

f quinidine
d Chinidin
i chinidina
e quinidina

3315 quiniephytine
Quinine anhydroxymethylene diphosphate which is used in medicine as a tonic and antiperiodic, i.e. acting against the periodic recurrence of a disease.

f quiniéphytine
d Chiniephytin
i chiniefitina
e quiniefitina

3316 quinine
The principal alkaloid of the cinchona bark and the most active constituent of its therapeutic value. It is used in high concentrations as a general protoplasmic poison and bacteriostatic agent, depressing phagocytosis and ciliary action. It has a specific effect on the malaria parasite, especially the erythrocytic forms. In small doses, it has an analeptic action on the voluntary muscles thus imparting a greater agility to the organism. It also excites the salivary, gastric and biliary secretion thus increasing the appetite and favouring digestion.

f quinine
d Chinin
i chinina
e quinina

3317 quininic acid
Acid obtained from quinine and quinidine by oxidation and splitting of the ring system. It is used in organic syntheses.

f acide quininique
d Chininsäure
i acido chininico
e ácido quínico

3318 quininone
Ketone which is prepared from the quinine by oxidation in acid medium. In it the secondary alcoholic group of the quinine, =CHOH, is transformed into the ketonic group =CO.

f quininone
d Chininon
i chininone
e quininona

3319 quinisocaine
Topical anaesthetic with effects lasting
up to 6 hours and principally used for
infiltration anaesthesia. The drug may
also be used to soothe itching and burning caused by prolonged exposure to sun
and itching caused by urticaria.
f quinisocaïne
d Chinisocain
i chinisocaina
e quinisocaina

3320 quinoline
Heterocyclic compound occurring in small
quantities in coal tar and in Dippel
animal oil. Colourless liquid, miscible
with alcohol and ether, insoluble in
water. It is obtained from aniline, glycerine and sulphuric acid and is used
in organic syntheses and in the treatment of diphtheria.
f quinoléine
d Chinolin
i chinolina
e quinolina

3321 quinolinic acid
Heterocyclic bycarboxylic acid, an isomer of cinchomeronic acid formed during
the biodegradation of tryptophan. It is
an inhibitor of phosphoenolpyruvate
carboxykinase.
f acide quinolinique
d Chinolinsäure
i acido chinolinico
e ácido quinolínico

3322 quinone
Chemical compound, bioxyderivative of
an aromatic nucleus. Quinones are colour
ed substances, usually from yellow to
red, to which the coloration of fungi,
moulds and various vegetable pigments
is due. Also called benzoquinone.
f quinone; benzoquinone
d Chinon; Benzochinon
i chinone; benzoquinona

e quinona; benzoquinona

3323 quinotannic acid
Tannin isolated together with quinic
acid from the bark of various species
of cinchona; a yellow powder, alcohol-
and water-soluble which, by hydrolysis
with diluted acids splits into glucose
and Chinese red.
f acide quinotannique
d Quinotanninsäure
i acido quinotannico; acido chinotannico
e ácido quinotánico

3324 quinotoxin
Isomer base of quinine from which it is
obtained through the action of diluted
acetic acid. Amorphous, bitter mass of
a yellow-red colour insoluble in water.
It has the same reactions as quinine.
f quinotoxine
d Chinotoxin
i chinotossina
e quinotoxina

3325 quinoxaline
Heterocyclic compound; crystalline, colourless powder soluble in alcohol but
not easily soluble in water. It is obtained through condensation of glyoxal
with orthophenylendiamine. It is used
in organic syntheses.
f quinoxaline
d Chinoxalin
i chinossalina
e quinoxalina

3326 quinuclidine
Ethylic derivative of piperidine occurring in the molecule of alkaloids of the
Cinchona group, e.g. quinine and cincho
nin.
f quinuclidine
d Chinuclidin
i chinuclidina
e quinuclidina

R

3327 racefemine
Spasmolytic agent acting on the uterine muscles and indicated in the treatment of both essential and symptomatic dysmenorrhoea and afterpains.
f racéfémine
d Racefemin
i racefemina
e racefemina

3328 racemic acid
Tartaric acid, optically inactive, consisting of equal parts of dextro- and laevotartaric acid into which it can be separated. It is frequently present, with dextrotartaric acid in the juice of grapes, and is formed mainly by oxidation of mannitol or dulcitol.
f acide racémique
d Racemsäure
i acido racemico
e ácido racémico

3329 racemic compounds
Mixture in equal quantities of the two optical isomers of the same compound; as these optical antipodes present a specific rotatory power which is equal but of opposite sign, so that the mixtures themselves will result optically inactive toward polarized light.
f composés racémiques
d Razemverbindungen
i composti racemici
e compuestos racémicos

3330 racemoramide
Chemical compound, pyrrolidine. It is indicated in the treatment of hypertension.
f racémoramide
d Racemoramid
i racemoramide; racemorammide
e racemoramida

3331 racemorphan
Narcotic analgesic producing effects which are equal to those of morphine. It is generally used in the treatment of acute pains, e.g. renal colic, infarction, traumas and cancer. Also used as a preoperative medication. It acts in association with barbiturates without adverse effects.
f racémorphan
d Racemorphan
i racemorfano

e racemorfano

3332 racephedrine
Racemic synthetic ephedrine having an action which is similar to that of the laevorotatory ephedrine but less potent.
f racéphédrine
d razemisches Ephedrin; Ephedrinrazemat
i racefedrina; efedrina racemica
e efedrina racémica

3333 radioactive tracer
A material which is easily recognizable by its radioactivity and which is introduced into an organism so as to trace the condition and/or the behaviour of a specific part of it.
f traceur radioactif; radiotraceur
d Radioaktiver Tracer
i tracciante radioattivo
e trazador radiactivo

3334 radiostrontium
Radioactive strontium widely used in medicine for diagnostic purposes.
f radiostrontium; strontium radioactif
d Radiostrontium
i radiostronzio
e radioestroncio

3335 raffinose
Trisaccharide present in sugar beets, so called because it is found in the mother liquor of raffination of saccharose. It is also obtained by aqueous extraction of cottonseed, in eucalyptus manna and various cereals. It yields D-fructose and melibiose on hydrolysis.
f raffinose
d Raffinose
i raffinosio
e rafinosa

3336 ramson oil
Oil extracted from the bulbous root of the ramson, a broad-leaved garlic, used therapeutically to relieve gastric disorders.
f essence d'ail des ours
d Bärenlauchöl
i essenza d'aglio orsino
e esencia de ajo de oso

3337 ranitidine
Therapeutic drug which blocks selectively histamine receptors mediating gastric acid secretion. It is indicated in peptic

ulceration and hyperacidity, oesophageal reflux and other gastro-intestinal disorders. Diarrhoea, dizziness and breast enlargement in males may occur as side effects.
f ranitidine
d Ranitin
i ranitidina
e ranitidina

3338 raubasine
Tetrahydroserpentine. Peripheral and cerebral vasodilator capable of improving the blood supply through the dilatation of the arterial vessels and the increase of tone of the venous vessels with mobilization of the blood from the veins and increase of its flow. It is mainly indicated in the treatment of peripheral and cerebral vascular diseases, especially when caused by arteriosclerosis and diabetes. Dizziness and flush may occur following intravenous injection of the drug.
f raubasine; ajmalicine
d Raubasin
i raubasina
e raubasina

3339 rauwolfia
Pantropical genus of mildly poisonous trees and shrubs having small cymose flowers with a salver-shaped corolla and bicarpellary ovary and yileding emetive and cathartic substances. Rauwolfia growing in India is used in the treatment of hypertension. Cases of depression have been noted in this therapy.
f rauwolfia
d Rauwolfia; Chinlen
i rauwolfia
e rauwolfia

3340 Rauwolfia serpentina
Genus of rauwolfia whose root is very rich in active principles of an alkaloid nature, e.g. reserpine, rauwolfine, serpentine etc. Its total extracts have been used in the treatment of hypertension and some of its alkaloids as hypotensive and psychotropic agents.
f Rauwolfia serpentina
d Rauwolfia serpentina
i Rauwolfia serpentina
e Rauwolfia serpentina

3341 rauwolfine
Alkaloid found in the Rauwolfia serpentina. Exagonal crystals, soluble in warm water and in methyl alcohol. Rauwolfine lowers arterial pressure and appears to have some effect as a renal vasodilator.
f rauwolfine
d Rauwolfin
i rauwolfina
e rauwolfina

3342 razoxane
Antineoplastic agent used in the combined treatment with radiotherapy for sar-

comas, chondrosarcomas, osteosarcomas, acute leukaemia and lymphosarcomas. Some cases of leukopenia have been reported in the course of the treatment, in which case the therapy should be discontinued. Nausea, vomiting, diarrhoea, alopecia may occur as side effects, especially when the therapy is carried out in combination with radio-therapy.
f razoxane
d Razoxan
i razossano
e razoxano

3343 red baneberry
North American perennial herb with small white flowers, yielding a therapeutically useful solution from its bright -red oval berries.
f actée rouge
d rotes Christofskraut
i attea rossa
e actea roja

3344 red cedar
Or Virginia cedar; American juniper whose dark green closely imbricated needle-shaped leaves yield an active principle used for drugs indicated in the treatment of gastro-intestinal disorders.
f cèdre rouge; cèdre de Virginie
d virginische Zeder; Bleistiftzeder
i cedro rosso; ginepro della Virginia
e enebro virginiano

3345 red poppy flowers
The flowers of the wild poppy (Papaver rhoeas) characterized by four vivid-red petals, often with a black spot on their base. The flowers and the leaves of the young plant are used for their mild sedative properties.
f fleurs de coquelicot
d Klatschmohnblüten; Feldrosen
i rosolaccio; fiori di papavero selvaggio
e ababol; amapola

3346 red soap root
The root of the Saponaria officinalis, which contain saponin, a glycoside capable of producing haemolysis when solutions are injected into the bloodstream.
f racine de saponaire rouge
d rote Seifenwurze
i radice di saponaria; radice di saponaria rossa
e raíz de saponella roja

3347 reductinic acid
Reductic acid. Compound developing in carbohydrate treated with alkali, or dehydrated. It is produced for its strong reducing action.
f acide reductique
d Reduktinsäure
i acido reduttinico
e ácido reductico

3348 reductone
Chemical compound belonging to a group of reducing substances of an aromatic or aliphatic nature which have a considerable biological importance, e.g. in the cellular respiration. They include vitamin C, echinochrom, adrenalin etc.
f réductone
d Redukton
i reduttone
e reductona

3349 relaxin
Substance of a peptidic nature secreted from the corpora lutea in pregnancy. It relaxes the pelvic ligaments in the course of pregnancy.
f relaxine
d Relaxin
i relassina
e relaxina

3350 renis extractum
Pharmaceutical preparation obtained from fresh kidneys of healthy bovines. Renal opotherapeutic drug used as a trophic and protective agent in the treatment of kidney disease for its diuretic action. It is also indicated as an adjuvant in the treatment of renal insufficiency, acute and chronic nephritis, hypertension and metabolic diseases affecting the kidneys.
f renis extractum
d Renis Extractum
e renis extractum
e renis extractum

3351 rennin
Enzyme capable of coagulating milk and especially occurring with pepsin in the gastric juice of some animals. It is usually obtained, either in powder, grains or scales, by extracting mucous membrane of the fourth stomach of calves. It is also known as labferment or chymosin.
f rennine; chymosine; labferment
d Rennin; Labferment; Chymosin
i rennina; lab-fermento
e renina; quimosina; fermento lab

3352 reovirus
Group of spherical viruses mainly found in the intestinal canal and in the respiratory tract of several animals, including man. They contain a double-stranded RNA and form a distinct group with three antigenic types. Although isolated in children suffering from intestinal and respiratory diseases, it is as yet not quite certain that they are the cause of infections.
f réovirus
d Reovirus
i reovirus
e reovirus

3353 reproterol
Beta-stimulant bronchodilator marked by a selective action on the beta-bronchial receptors. It is principally indicated in the treatment of bronchial asthma, asthmatic and chronic bronchitis accompanied or not by emphysema, silicosis, bronchiectasis, tuberculosis and bronchial cancer. It is also used as an adjuvant to inhalation therapies with antibiotics, mucolytic agents, secretolytic drugs and corticosteroids.
f réprotérol
d Reproterol
i reproterolo
e reproterolo

3354 rescinnamine
Reserpine-derivative which acts by inhibiting the orthosympathetic nervous system. It is indicated in the treatment of hypertension and as a sedative. Drowsiness and Parkinsonian symptoms, as well as some depression and lipothimia may occur as adverse effects.
f rescinnamine
d Rescinnamin
i rescinammina
e rescinamina

3355 reserpic acid
Or reserpinolic acid. Polycyclic compound obtained by controlled hydrolysis of the reserpine in an alkaline medium.
f acide réserpique
d Reserpinsäure
i acido reserpico
e ácido resérpico

3356 reserpine
Crystalline alkaloid of Rauwolfia serpentina principally indicated in the treatment of neuropsychiatric disorders and hypertension. Its action is probably due to depletion of serotonin and other biogenous amines in the brain. The therapy may occasionally cause some depression when applied to hypertension. Abdominal disorders accompanied by pains, drowsiness and allergic rhinitis may occur as side effects.
f réserpine
d Reserpin
i reserpina
e reserpina

3357 resinate
Salt of a resin acid; any compound that is formed by the acids of resin with alkaline or metallic bases. Resinates have various properties according to the type of acid from which they derive.
f résinate
d Resinat; harzsaures Salz
i resinato
e resinato

3358 resin oil
Volatile oil contained in resins. They occur in the form of more or less viscous or limpid liquids which become dense when exposed to air.

f huile de résine
d Resinöl; Harzöl
i olio di resina
e aceite de resina

3359 resorcin
Antiseptic and antipruritic agent used
in dermatological treatment. Topically,
it is used in the treatment of acne and
dandruff, as well as ear drops. Suppres
sion of the thyroid gland has been observed if the drug has been absorbed
for a long time. It should not be ingest
ed as it is a corrosive agent and may
cause kidney damage, coma and convulsions.
f résorcine
d Resorcin
i resorcina
e resorcina

* **resorcinol   s. resorcin**

3360 resorcylic acid
Benzoic acid derivative occurring in the
three isomer forms alpha, beta and
gamma. It is obtained from many natural substances and it can be prepared
by introducing a carboxylic group into
the molecule of the resorcine. It is used
as intermediate for various pharmaceutical products.
f acide résorcylique
d Resorcylsäure
i acido resorcilico
e ácido resorcílico

3361 retinal extract
Ophthalmic opotherapeutic agent capable
of stimulating the metabolic and biochemical processes of the retinal tissue involved in the reception of sight stimuli.
It improves the trophism of the cellular
elements which have not yet been affected by some irreversible degenerative
process. It is therefore indicated in the
treatment of juvenile myopathies, chorio-
retinitis, initial stages of degenerative
conditions of the retinal tissues and as
a coadjuvant in the treatment of detachment of the retina and of retinitis pig-
mentosa (genetically determined bilateral,
primary degeneration of the retina).
f extrait rétinien; extrait de rétine
d Netzhautextrakt
i estratto di retina
e extracto de retina

3362 retinoic acid
Chemical compound obtained by oxidation
of the aldehyde of vitamin A; it represents the acid of this vitamin.
f acide rétinoïque
d Retinosäure
i acido retinoico
e ácido retinoico

3363 retinol
The principal form, biologically the most
active, of the group of substances that
are very similar to one another and are
present in animal tissues. Such substances are generically described as
"vitamin A". Retinol can be extracted
from animal organs or produced synthetically. It is indispensable for the normal development of epithelia and for reti
nal accommodation.
f rétinol
d Retinol
i retinolo
e retinolo

3364 rhamni frangulae cortex
Dried bark of the stem and branches of
Rhamnus frangula, whose active principles consist of emodin-derived anthra-
quinonic glucoside, e.g. frangulin and
glycofrangulin. Both drugs have a good
cathartic action by acting on the large
intestine. Some abdominal pains may
occur as side effects. Also known as
frangula bark, alder buckthorn bark.
f écorce de bourdaine
d Faulbaumrinde; Geibholzrinde
i corteccia di frangola; frangola
e frángula

3365 rhamnose
Monosaccharide. The most important of
the methylpentoses, which occurs free
in the leaves and flowers of· Rhus toxi-
codendrus and, combined, in many glyco
sides, e.g. quercetrin, rutin, xantho-
rhamnin, frangulin etc. from which it
can be obtained by hydrolysis with dilut
ed acids. Rhamnose reduces the
Fehling's reagent.
f ramnohexose
d Rhamnose
i ramnosio
e ramnosa

3366 rhaponticin
Glycoside occurring in rhapontic rhubarb,
which by hydrolysis with diluted sulphuric solutions yields glucose and rhapontigenin. White-yellowish crystals easi
ly soluble in warm alcohol, benzene and
acetone and decompose at a temperature
of 231°C.
f rhaponticine; rhapontine
d Rhaponticin
i raponticina; rapontina
e raponticina; rapontina

3367 rhapontic rhubarb
Variety of rhubarb cultivated in England
and in Siberia. It does not contain the
aloe-emodin derivatives that occur in
Chinese rhubarb, but rhaponticin, a
specific glycoside. See also rhaponticin.
f rhapontic; rhubarbe pontique
d Rhapontikarhabarber
i rapontico
e rapóntico

3368 rhapontigenine
Stilbene derivative obtained by hydrolysis of rhaponticin and forming colour-

less tubular crystals with a molecule of water of crystallization, i.e. the water which is retained in a molecular association by a compound that crystallizes from an aqueous solution. Such crystals are not easily soluble in water, but are soluble in ether, alcohol, acetone and acetic acid. They are insoluble in the ether of petroleum and in benzene.

f rhapontigénine
d Rhapontigenin
i rapontigenina
e rapontigenina

* **rhapontin** s. rhaponticin

* **rhatany** s. rhatany root

3369 rhatany root
The dried root of Krameria triandra, family Leguminosae. The root contains about 10% of tannin and is used as an astringent in lozenges with cocaine.

f racine de ratanhia
d Ratanhiawurzel
i radice di ratania
e raíz de ratania

3370 rhodanate
Sulphocyanate, thiocyanate. Salt of rhodanic acid. Potassium rhodanate was used in the treatment of malignant tumours but without much success.

f rhodanate; sulfocyanate
d Rhodanat; Sulfocyanat
i rodanato; solfocianato; tiocianato
e rodanato; sulfocianato

* **rhodanic acid** s. rhodanin

3371 rhodanin
Rhodanic acid; thiocyanic acid. Heterocyclic chemical compound obtained by the reaction of chloroacetate with ammonium dithiocarbamate. A volatile pungent liquid whose many salts and esters are important in medicine. Potassium and sodium rhodanins (thiocyanates) are indicated in the treatment of hypertension.

f acide rhodanique; rhodanine
d Rhodanin
i rodanina
e rodanina

3372 rhubarb
The dried rhizome and roots of any of various herbs of the genus Rheum characterized by large leaves with succulent petioles. It is widely used as a purgative and stomachic bitter.

f rhubarbe
d Rhabarber
i rabarbaro
e ruibarbo

3373 ribavirine
Antiviral agent administered by inhalation in the treatment of severe chest infections affecting infants and children and due to the respiratory syncytial virus.

f ribavirine
d Ribavirin
i ribavirina
e ribavirina

3374 riboflavine
Vitamin B_2, lactoflavine. Water-soluble and heat stabile vitamin. Yellow pigment present in milk, malt, some grasses and algae, as well as yeast and meat. It can also be prepared synthetically. Riboflavine is an essential part of the enzyme system in the carbohydrate metabolism of cells. Its absence or insufficiency causes arrest of growth, cataract, cutaneous eczema mainly round the nose and eyes, dystrophic lesions of the tongue and vision disorders.

f riboflavine
d Riboflavin
i riboflavina
e riboflavina

3375 ribonuclease
Crystalline enzyme, mainly present in the pancreas, acting on ribonucleic acid by catalyzing its hydrolysis only to nucleotides. It also catalyzes the transfer, but not the liberation, of phosphate groups from the nucleotides. It is further characterized by antiphlogistic and analgesic properties and is indicated in the treatment of pains caused by traumas and articular rheumatism by means of an ointment.

f ribonucléase
d Ribonuklease
i ribonucleasi
e ribonucleasa

3376 ribonucleic acid
Nucleic acid having a chemical structure that is similar to that of the deoxyribonucleic acid. It is a polynucleotide principally formed by 4 ribonucleotides, i.e. adenylic acid, guanylic acid, cytidylic acid and uridylic acid, that is polymers of the monophosphate of adenosine, guanosine, cytidine and uridine. The "messenger" RNA (ribonucleic acid) codes for the synthesis of protein in translation; the "ribosomal" RNA which is characterized by a high molecular weight is the slowest of these nucleic acids within the cell. The "soluble" or "transfer" RNA is the final amino acid acceptor before peptide bond synthesis in the course of translation.

f acide ribonucléique; RNA
d Ribonukleinsäure; RNS
i acido ribonucleico; RNA
e ácido ribonucleico; ARN

3377 ribose
Crystalline aldose sugar of the pentose class of which both the dextrorotatory and laevorotatory forms are known. The dextrorotatory form is a constituent of many nucleosides, e.g. adenosine, cyti-

dine etc. and is usually obtained from
ribonucleic-acid hydrolysis or from
altrose by degradation.
f ribose
d Ribose
i ribosio
e ribosa

3378 ribostamycin
Aminoglycosidic antibiotic indicated in
the treatment of infections due to staphylococci, streptococci, pneumococci,
gonococci, Escherichia coli and Klebsiella pneumoniae. Cutaneous inflammation and respiratory depression may
occur as side effects.
f ribostamycine
d Ribostamycin
i ribostamicina
e ribostamicina

3379 ricinoleic acid
One of the oxyacids which are constituent of natural fats. It is an oily
unsaturated hydroxy fatty acid occurring
in castor oil in the form of a glyceride.
It polymerizes rapidly on heating.
f acide ricinoléique
d Ricinolsäure
i acido ricinoleico
e ácido ricinoleico

*** rifamide   s. rifamycin**

3380 rifampicin
Bacteriostatic and bactericidal antibiotic
used in the treatment of all types of
tuberculosis, as well as in otorhinolaryngology and dermatology. It is characterized by a rapid and potent effect,
in particular against penicillase-producing staphylococci. Skin rushes are possible side effects, as well as liver toxicity and influenza-like symptoms. Owing
to the induction of liver enzymes, the
effectiveness of other drugs, e.g. oral
contraceptives and corticosteroids is
reduced.
f rifampicine
d Rifampicin
i rifampicina
e rifampicina

3381 rifamycin
Antibiotic agent particularly effective
against Gram-positive micro-organisms
and indicated in the treatment of septicaemic or topical infections caused by
staphylococci. Also used in the treatment of tuberculosis by intrapleural
introduction in association with PAS
(para-aminosalicylic acid) and isoniazid.
Spasms of the gallbladder have been observed as side effects.
f rifamycine
d Rifamycin
i rifamicina
e rifamicina

3382 rimazolium metisulphate

Omopyrinidazole-derived analgesic agent
marked by a central mechanism of action. It has no depressive action on
respiration and does not induce habit
or dependence. It is mainly used in the
treatment of acute or chronic pains of
various origin.
f métisulfate de rimazolium
d Rimazoliummetisulfat
i rimazolio metisolfato
e metisulfato de rimazolio

3383 rimiterol
Beta-adrenoceptor agonist, bronchoselective, principally indicated in the treatment of bronchial asthma by inhalation,
intravenous infusion or orally. Nervousness, tachycardia, occasionally tremors,
palpitations and headache may occur as
adverse effects.
f rimitérol
d Rimiterol
i rimiterolo
e rimiterolo

3384 ristocetin
Each of the antimicrobial substances
produced by a species of Nocardia. Used
as antibiotic. It is administered intravenously. Its composition and action are
similar to those of vancomycin chloride,
q.v.
f ristocétine
d Ristocetin
i ristocetina
e ristocetina

3385 ritodrine
Spasmolytic agent characterized by a
potent beta-adrenergic stimulant and a
weak alpha-adrenergic blocking action.
It is principally used as a myorelaxant
in premature birth and has proved useful in the treatment of acute fetal acid
osis caused by uterine hypermotility or
compression of the umbilical cord and
in all those cases in which it is necessary or advisable to obtain relaxation
of the uterine musculature. Flushes,
tremors and increase of the cardiac
rhythm have been observed as side effects.
f ritodrine
d Ritodrin
i ritodrina
e ritodrina

3386 ritrosulfan
Alkylating and cytostatic agent principally used in the treatment of lymphoid
leukaemia, Hodgkin's disease, reticulo-
-lymphosarcomatosis, pulmonary and
mammary neoplasms, intracavitary me-
tastasis-inducing tumours and in the
treatment of auto-immune myositis, recur
rent endogenous uveitis and myasthenia
gravis. A transient form of leukopenia
and thrombopenia may occur as side
effect.
f ritrosulfan

d Ritrosulfan
i ritrosulfan
e ritrosulfan

3387 rock cress
Any of various rock-loving cresses, characterized by small white flowers. The rock cress is very rich in vitamins E and C.
f cresson des roches
d Steinkresse
i crescione delle rocce
e berro de las rocas

* **rocky mountains spotted fever vaccine s. vaccinium rickettsiale**

3388 rolitetracycline
Tetracycline-derived broad-spectrum antibiotic. Owing to a solubility which is greater than that of tetracycline, the drug is absorbed without delay and rapidly attains a high concentration in the blood, tissues and urinary apparatus. It is quite effective against Gram-positive and Gram-negative micro-organisms, in particular spirochaetes, rickettsiae, and some large viruses. A yellowish or grey coloration of the teeth and a sensation of warmth, especially in the face, may be observed as side effects. The drug should not be administered in pregnancy.
f rolitétracycline
d Rolitetracyclin
i rolitetraciclina
e rolitetraciclina

3389 root
The part of the plant axis that is developed from the radicle of the embryo and grows downwards. Innumerable roots are very important as they contain active principles.
f racine
d Wurzel
i radice
e raíz

3390 rose
Wild or cultivated flowers of several species of Rosa, some of which, e.g. the rosa canina, produce fruits which contain malic and citric acid, sugar, tannin and some ascorbic acid, that is one of the principal sources of natural vitamin C.
f rose
d Rose
i rosa
e rosa

3391 rose bengal
Rose bengal sodium. A staining agent used for the detection of damage in the cornea.
f rose bengale sodique
d Bengalrosa
i rosa bengala sodico
e rosa bengal sódico

3392 rosemary
Or romero. The flowering tops of the evergreen shrub Rosmarinus officinalis, cultivated in all over the evergreen mediterranean plain and yielding a volatile oil which contains borneol, bornyl acetate and other acetates. It is used medicinally for its carminative properties.
f romarin; encensier
d Rosmarin; Kranzenkraut
i rosmarino; ramerino
e rosmarino; romero

3393 rosemary oil
See definition of rosemary.
f essence de romarin
d Rosmarinöl
i essenza di rosmarino
e esencia de romero

3394 rose root
Perennial fleshy herb having rose-smelling roots which are used in medicinal preparations.
f rhodiole; orpin odorant
d Rosenwurz; Rosswurz
i radice idea
e raíz rodia

3395 rotenone
Crystalline chemical substance which, together with amytal, specifically inhibits electron transfer within the NDAH-H reductase to prevent the generation of a proton gradient at site 1.
f roténone
d Rotenon
i rotenone
e rotenona

3396 round turmeric
Zedoary. Fragrant East Indian drug characterized by a bitter aromatic taste used in the East, particularly in India, as a stimulant. It is derived from the rhizome of various plants of the genus Curcuma.
f zedoaire
d Zitwer; Zepterwurz
i zedoaria
e cedoaria; zedoaria

3397 rubeanic acid
Chemical compound used as a reagent of several ions, e.g. copper ions, nickel ions etc.
f acide rubéanique
d Rubeansäure
i acido rubeanico
e ácido rubeánico

3398 rubella vaccine
Live attenuated rubella virus used for immunization against German measles (rubella). It is advisable always to use it in girls before the onset of puberty. It may be used in women during childbearing age provided they are not pregnant. Enlarged lymph glands, rash and

fever may occur as side effects.
f vaccin antirubéole
d Rubellavakzine; Rötelnvirus
i vaccino antirubeola
e vacuna antirubéola

3399 rubidium
Element. Soft white alkaline metal not unlike sodium and potassium and very reactive. Mineral and sea waters contain some rubidium. It also occurs in human tissues. Its salts are used medicinally like potassium salts but they do not seem to display any particular advantage, unless it be a better tolerance.
f rubidium
d Rubidium
i rubidio
e rubidio

* **rubidomycin   s. daunorubicin**

3400 rue oil
Colourless to yellow fluorescent essential oil marked by an intense smell yielded by the rue and other plants of the genus Ruta. It is used medicinally as a topical irritant.
f essence de rue
d Rautenöl
i essenza di ruta
e esencia de ruda

3401 rufocromomycin
Antibiotic produced by Streptomyces rufochromogenus, marked by antineoplastic activity. It is mainly indicated in the treatment of lymphatic-reticulocytic sarcomas, Hodgkin's disease, chronic myeloid leukaemia, breast and ovary tumours as well as solid tumours affecting the testicles. Leukopenia and thrombocytopenia may be observed as side effects.
f rufocromomycine
d Rufocromomycin

i rufocromomicina
e rufocromomicina

3402 ruscogenines
Sapogenines obtained from the rhizomes of Ruscus aculeatus. Antihaemorrhoidal and antiphlogistic agents used in the treatment of internal and external haemorrhoids and of lesions due to haemorrhoidectomy.
f ruscogénines
d Ruscogenine
i ruscogenine
e ruscogeninas

3403 rutin
Flavonic glycoside, rhamnoglycoside of the quercetrin, occurring in the leaves of ruta, in buckwheat etc. Needle-shaped, tasteless, yellowish crystals which darken when exposed to the air. They hydrolyze yielding quercetrin and rutinose. Rutin is used therapeutically owing to its capacity to reduce capillary fragility thus finding a wide application, e.g. in retinal haemorrhages, hisstaminic shock, hypertension etc.
f rutine
d Rutin
i rutina
e rutina

* **rutinic acid   s. rutin**

3404 rutinose
Chemical compound, disaccharide, formed by rhamnose and glucose, occurring in rutin, q.v. from which it is possible to obtain by enzymatic hydrolysis. It is a colourless hygroscopic powder.
f rutinose
d Rutinose
i rutinosio
e rutinosa

* **rutoside   s. rutin**

S

3405 sabal
Small genus of Central American dwarf fan palms characterized by creeping horizontal or subterranean stems and long petioled leaves. They are medicinally interesting owing to their fruit (see sabal berry).
f sabal
d Sabal; Sägepalme
i sabal
e sabal

3406 sabal berry
See sabal. Some species of the plant are cultivated in various European areas as well as in Central America owing to their diuretic properties.
f baie de sabal
d Sabalfrucht; Sägepalmenfrucht
i bacca di sabal
e baya de sabal

3407 saccharate
Terms denoting various types of chemical compounds. Calcium saccharate is used in medicine for its antacid and absorbent properties; also used in medicine are: iron saccharate, a red powder indicated in the treatment of anaemia; sodium saccharate, a white water-soluble powder used as a coadjuvant of sodium chloride in physiological solutions; copper saccharate, a blue crystalline water-soluble powder used for vaginal hygiene etc. More generally, any salt or ester of saccharic acid.
f saccharate
d Saccharat
i saccarato
e sacarato

3408 saccharated iron oxide
Ferric drug for oral and intravenous iron therapy, indicated in the treatment of hypochromic hypoferric anaemias and in all conditions caused by iron deficiency. It is very well tolerated, even at high doses.
f oxyde de fer saccharé
d Saccharateisenoxyd
i ossido di ferro saccarato
e óxido férrico sacarado

3409 saccharic acid
Solid dicarboxylic acid, deliquescent, known in 10 stereoisomeric forms, which is obtained by oxidation of its derivatives, e.g. sucrose, by nitric acid. It rapidly undergoes inner esterification to a lactone (glutaric acid). Saccharic acid can also be defined as either of two dicarboxylic acids from the other hexoses having the pair of central hydroxyl groups on opposite sides of the molecule like glucose. Saccharic acid has not yet been isolated in animal organisms.
f acide saccharique
d Zuckersäure
i acido saccarico
e ácido sacárico

3410 saccharide
Compound formed by the union of a sugar and a base. The term is also used as a synonym of carbohydrate.
f saccharide
d Saccharid
i saccaride
e sacarida

*** saccharin s. benzosulphinide**

3411 saccharose
Any of the compound sugars, including disaccharides and trisaccharides. Also, a synonym of sucrose.
f saccharose; sucre
d Saccharose; Sucrose
i saccarosio
e sacarosa

3412 saffron
Reddish substance consisting of the aromatic dried stigmas of saffron crocus, used in medicine as a stimulant antispasmodic emmenagogue.
f oxycrocéum; safran
d Safran
i zafferano
e azafrán

3413 safrazine
Therapeutic agent used in the treatment of depression and depressive psychosis. Hypotension, vertigo, cutaneous eruptions and dryness in the mouth may occur as side effects. The drug should be administered with great caution to aged patients.
f safrazine
d Safrazin
i safrazina
e safrazina

3414 safrol
Or safrole. Chemical compound contained in the essential oils of sassafras, of which it is the principal constituent, camphor, asarabacca (wild ginger), badiana (stellated or star anise), nutmeg etc. It is used in rubefacient mixtures and lice-destroying preparations.
f safrol
d Safrol
i safrolo
e safrol

3415 salazosulphamide
Sulphamidic agent obtained by denitrification of the sulphanylamides and coupling with salicylic acid. Yellowish crystals, both easily soluble in water, used as a germicide.
f salazosulfamide
d salazosulfamid
i salazosolfammide
e salazosulfamida

3416 salazosulphapyridine
A sulpha drug indicated in the treatment of chronic ulcerative colitis and marked by a prompt but short-lasting action. A cumulative effect is possible. Also known as sulphasalazine.
f salazosulfapyridine
d Salazosulfapyridin
i salazosolfapiridina; solfasalazina
e salazosulfapiridina

3417 salbutamol
Sympathomimetic agent with mainly ß--adrenergic action. A long-acting bronchodilator, it acts selectively on the ß1-receptors of the bronchial muscles without affecting the ß2-receptors of the myocardium. The drug is indicated in the treatment of bronchial asthma, bronchitis and pulmonary emphysema. Also used to inhibit premature labour. Tremors and muscle cramps may occur as side effects.
f salbutamol
d Salbutamol
i salbutamolo
e salbutamolo

3418 salcatonin
Hormone obtained from salmon or prepared synthetically. It acts by controlling calcium metabolism, i.e. it diminishes the level of calcium in the blood and increases the plasma phosphate levels by inhibiting bone reabsorption. It is therefore indicated in the treatment of osteoporosis, Paget's disease and principally hypercalcaemia.
f salcatonine
d Salcatonin
i salcatonina
e salcatonina

3419 salicin
Salicoside. Antipyretic and analgesic glycoside obtained from the bark of several species of willow and poplar. Colourless bitter crystals or powder, soluble in cold water or alcohol, hydrolyzed by acids or enzymes into glucose and salicyl alcohol.
f salicine
d Salicin
i salicina
e salicina

3420 salicylamide
Chemical compound; colourless and tasteless crystals, not easily soluble in cold water but readily soluble in warm water, alcohol and aqueous solution of sodium carbonate. It has good analgesic properties and is much less toxic than acetylsalicylic acid even when administered repeatedly. It is therefore used as an antipyretic and antirheumatic agent.
f salicylamide
d Salicylamid
i salicilammide
e salicilamida

3421 salicylamidophenazone
The product of the combination of a pyrazole radical with salicylamide. It is an antipyretic, antiinflammatory, antirheumatic and analgesic agent used against rheumatism and indicated in the symptomatic treatment of influenza, especially when accompanied by a cold and high temperature.
f salicylamidophénazone
d Salicylamidophenazon
i salicilamidofenazone
e salicilamidofenazona

3422 salicylanilide
Chemical compound; colourless or slightly rose-coloured and odourless crystals, soluble in alcohol and chloroform, less so in water. It has a considerable fungicide action and is used in the preparation of non-fatty ointment indicated in the treatment of infections of the scalp.
f salicylanilide
d Salicylanilid
i salicilanilide
e salicilanilida

3423 salicylate
Salt and ester of salicylic acid. Of particular importance for medicinal uses are: sodium, ammonium, aluminium, calcium, lithium and magnesium salicylates. Sodium salicylate is used in medicine as an antipyretic, analgesic and antirheumatic. Antipyretic action is due to a considerable increase of perspiration with consequent increase of thermodispersal.
f salicylate
d Salicylat
i salicilato
e salicilato

3424 salicylic acid
Colourless and odourless crystals, not

easily soluble in water but very easily in alcohol and ether. It is mainly used as a bactericide, a preservative and, in concentrated solutions, for the removal of warts and corns. It is also indicated in the treatment of various skin diseases.
f acide salicylique
d Salicylsäure
i acido salicilico
e ácido salicílico

3425 **salicylsulphonic acid**
Or sulphosalicylic acid. Aromatic acid which can be derived from salicylic acid by substitution of a hydrogen atom of the benzene ring with a sulphonic group. It is known in several isomeric forms that are distinguished by indicating with a number the position of the above mentioned group. The 5-s acid is obtained by heating salicylic and concentrated sulphuric acid. It is used in medicine as a reagent for albumin.
f acide sulfosalicylique
d Salicylsulfonsäure
i acido salicilsolfonico; acido solfosalicilico
e ácido sulfosalicílico

3426 **salicylsulphuric acid**
Sulphuric ester of the salicylic acid. Colourless, water-soluble crystals which are prepared by treating salicylic acid with chlorosulphuric acid. The sodium salt of this acid has analgesic and anti rheumatic properties.
f acide salicylsulfurique
d Salicylschwefelsäure
i acido salicilsolforico
e ácido salicilsulfúrico

3427 **saline solution**
Any solution which contains one or more salts. Also, a solution prepared for parenteral drip, which consists of 0.9% of sodium chloride or other salts.
f solution salée
d Salzlösung
i soluzione salina
e solución salina

3428 **salsalate**
Bimolecular ester of the salicylic acid. Antiinflammatory and analgesic agent, insoluble in gastric juices. It is rapidly absorbed and used in the treatment of rheumatoid arthritis, osteoarthritis, fibrositis and bursitis. Some allergic reactions may occur as side effects.
f salsalate
d Salsalat
i salsalato
e salsalato

3429 **salt**
The compound which is formed by a base with an acid, usually a metallic base with an inorganic acid. It is soluble in water and a good electrolyte.
f sel
d Salz
i sale
e sal

3430 **sandalwood**
Two species are known: the red sandalwood and the white sandalwood. The latter is characterized by a typical smell due to the essential oil contained therein, which is medicinal. The red sandalwood yields a glycoside, the santalin, a red colouring matter (santalic acid, or santalenic acid). Also known as sanders or saunders tree.
f santal
d Ambraholz; Sandelholz
i sandalo
e sándalo

3431 **sangreeroot**
Virginia serpentary root, Virginia snake root. Birthwort of the eastern United States known in medicine for its rhizoma.
f racine de serpentaire de Virginie; racine de vipérine
d virginische Osterluzeiwurzel; virginische Schlangenwurzel
i radice di serpentaria della Virginia
e raíz de serpentaria; raíz viperina

3432 **sanicle root**
The root of a plant of the genus Sanicula mainly used in popular medicine as an anodyne or astringent.
f racine de sanicle
d Sanikelwurzel
i radice di sannicola
e raíz de sanícula

3433 **santonic acid**
Oxyacid, isomer of the santoninic acid, q.v. Colourless crystals.
f acide santonique
d Santoninsäure
i acido santonico
e ácido santónico

3434 **santonica oil**
European wormwood. Also, an anthelmintic drug obtained from the unexpanded dried flower heads of the plant or a closely related one.
f essence de semen-contra
d Wurmsamenöl
i olio di cina
e esencia de santónico

3435 **santoninate**
Salt of the santoninic acid. The lithium santoninate is considered capable of dissolving vesical calculi. The sodium santoninate is marked by anthelmintic properties.
f santoninate
d Santoninat; santoninsaures Salz
i santoninato
e santoninato

3436 **santonine**

Lactone of the santoninic acid contained in the flower tops of santonica. Poisonous, slightly bitter crystalline compound used as an anthelmintic agent, particularly effective against roundworms. It acts by raising the motility of the parasites which can thus be expelled more easily.
f santonine
d Santonin
i santonina
e santonina

3437 santoninic acid
Organic oxyacid; colourless crystals easily soluble in water. Its lactone constitutes the santonin.
f acide santoninique
d Santonininsäure
i acido santoninico
e ácido santonínico

3438 santoninoxime
Santonin derivative. Needle-shaped crystals, insoluble in water, soluble in alcohol and ether. It has an anthelmintic action that is similar to that of santonin yet less toxic.
f santoninoxime
d Santoninoxim
i santoninossima
e santoninoxima

3439 saralasin
Antihypertensive agent used for the diagnosis of hypertension caused by angiopathy and for the treatment of severe hypertension.
f saralasine
d Saralasin
i saralasina
e saralasina

3440 sarsaparilla root
Sarsa. The dried roots and rhizome of species of Smilax, a group of plants growing in Central America. Its principal constituent is a crystalline glycoside known as sarsaponin and used medicinally for its emetic properties.
f racine de salsepareille
d Sarsaparillewurz; Sarsewurzel
i radice di salsapariglia
e raíz de zarzaparrilla

3441 sassafras bark
The bark of the root of the sassafras, a very tall tree growing in North America, marked by an aromatic smell resembling that of fennel, and used in medicine as a diuretic and diaphoretic.
f écorce de sassafras
d Sassafrasrinde
i corteccia di sassafrasso; corteccia di sassofrasso
e corteza de sasafrás

3442 sassafras oil
Yellowish aromatic essential oil obtained from the roots and stumps of sassafras

and mainly used as disinfectant.
f essence de sassafras
d Sassafrasöl
i essenza di sassofrasso
e esencia de sasafrás

3443 savin oil
The oil which is obtained from the bitter and acrid tops of the savin and is considered in popular medicine an abortifacient and a therapeutic agent for amenorrhoea.
f essence de sabine; huile essentielle de sabine (genévrier fetudi)
d Sadebaumöl
i essenza di sabina
e esencia de sabina

3444 scammony
Twining plant mainly growing in Asia having a thick root up to a meter long and white flowers from which a resin is obtained that is used as a potent purgative.
f scammonée
d Skammonia; Purgierwinde
i scammonea
e escamonea

3445 scopolamine
Alkaloid found in the seeds of henbane, in the leaves of Duboisia, in the roots of belladonna (deadly nightshade) and in various Solanaceae. It is obtained from the bittern of the preparations of henbane derivatives and atropine. Its hydrobromide is used as an antispastic agent (q.v.).
f scopolamine
d Scopolamin; Hyoscin
i scopolamina
e escopolamina

3446 scopolamine hydrobromid
Water-soluble salts of scopolamine (hyoscyamine) indicated in the treatment of neuralgia, mental excitement and mania. It has a mydriatic action which is greater than that of atropine. Sight disorders, dryness of the mouth and tachycardia may occur as side effects. The drug should not be administered to patients suffering from glaucoma or prostatic tumours accompanied by urinary retention.
f hydrobromure de scopolamine
d Scopolaminhydrobromid
i scopolamina bromidrato
e hidrobromuro de escopolamina

3447 scopolamine methylnitrate
Methylhyoscini nitras. Parasympatholytic marked by peripheral effects that are similar to those of atropine. It decreases gastric secretion and gastro-intestinal motility. The drug is indicated in the treatment of gastric and duodenal ulcer and, more generally, of diseases caused by hyperacidity or gastrointestinal spasm, renal and biliary colics,

ptyalism and hyperchlorhydria. Mydriasis, accommodation disorders, dryness of the skin and occasionally tachycardia may occur as side effects.

f méthylnitrate de scopolamine
d Hyoscinmethylnitrat
i metilnitrato di scopolamina; metilioscina nitrato
e metilnitrato de escopolamina

3448 scopolia
The dried rhizome of Scopolia carniolica, a herb of the family Solanaceae, used as an adulterant of and a substitute for belladonna root. It contains several alkaloids (hyoscine and hyosciamine, as well as atropine) and is therefore used for the treatment of various diseases.

f belladone de Hongrie
d Skopolia; Glockenbilse
i scopolia
e escopolia

3449 scurvy grass
Or spoonwort. One of various cresses which were used in popular medicine for the treatment or prevention of scurvy.

f cranson; herbe au scorbut
d Skorbutkraut; Bitterkresse
i coclearia; erba cocchiara
e coclearia; hierba de las cucharas

3450 sea salt
The salt which is obtained through the evaporation of seawater. It chiefly contains sodium chloride and very small quantities of magnesium chloride, magnesium sulphate and calcium sulphate.

f sel marin
d Meersalz; Seesalz
i sale marino
e sal marina

3451 sebacic acid
Aliphatic saturated bicarboxylic acid. Colourless tabular crystals, not easily soluble in water but easily soluble in alcohol, which are prepared by treating at high temperature castor oil or the fat acids contained therein with caustic soda. In these conditions, the ricinoleic acid decomposes forming 2-octanol and sebacic acid.

f acide sébacique
d Sebacinsäure
i acido sebacico
e ácido sebácico

3452 secbutobarbitone
Hypnotic agent having the same action and effects of butalbital (q.v.). Also used as a sedative.

f secbutobarbital
d Secbutobarbital
i secbutobarbitale
e secbutobarbital

3453 secobarbital
Barbituric agent used as hypnotic and sedative. The drug is mainly used in obstetrics, but also in neuropsychiatry and in surgery especially in after-operation pains. A fall in blood pressure and some vasodilatation may occur as side effects.

f sécobarbital
d Secobarbital
i secobarbitale
e secobarbital

3454 secretin
Hormone secreted by the upper small intestine in response to free fatty acids in the duodenum which provokes the secretion of dilute pancreatic juice rich in bicarbonate. It is indicated in the treatment of pancreatic exocrine insufficiency, peptic ulcer, duodenal disorders and gastrectomy-caused states of denutrition. Some allergic reactions may occur as side effects.

f secrétine
d Secretin
i secretina
e secretina

3455 sedative
Any substance which is capable of diminishing the activity of an organism, an organ or a tissue. More specifically, any substance capable of acting on the nervous system in order to obtain sedation.

f sédatif
d Sedativum; Beruhigungsmittel
i sedativo
e sedativo

3456 Seidlitz's powder
Effervescent powder used in the preparation of artificial mineral waters. It principally consists of sodium potassium tartrate and sodium bicarbonate.

f poudre de Seidlitz
d Seidlitz's Pulver; Brausepulver
i polvere di Seidlitz
e polvo de Seidlitz

3457 selegiline
Selective monoamine oxidase inhibitor. By inhibiting the breakdown of dopamine in the brain, the drug increases and prolongs the action of levodopa. In conjunction with the latter, it is indicated in the treatment of Parkinson's disease. Hypotension, nausea, mental confusion and some involuntary movements may occur as side effects.

f sélégiline
d Selegilin
i selegilina
e selegilina

3458 selenic acid
Oxyacid obtained in a pure state by electrolytic oxidation of the selenious acid (q.v.); crystalline, colourless, hygroscopic solid, water soluble and somewhat stronger than sulphuric acid. When heated to 210°C, it decomposes

liberating oxygen.
f acide sélénique
d Selensäure
i acido selenico
e ácido selénico

3459 **selenious acid**
Acid analogous with sulphurous acid, forming salts and selenites. It is derived from selenium dioxide. Its aqueous solutions are weakly acid.
f acide sélénieux
d Selenigsäure
i acido selenioso
e ácido selenioso

3460 **selenium**
Element closely related to sulphur in its chemical properties and existing in seve̱ral allotropic forms.
f sélénium
d Selen
i selenio
e selenio

3461 **selenium sulphide**
Topical antiseborrhoeic and antibacterial agent used in the treatment of seborrhoeic dermatitis of the scalp, dandruff and Tinea versicolor (a cutaneous infection).
f sulfure de sélénium
d Selensulfid
i solfuro di selenio
e sulfuro de selenio

3462 **senega**
Senega root. The dried root of Polygala senega, containing the saponins, senegin and polygalic acid, which are not dissimilar from those of quillaia (q.v.). A fixed oil and a very small quantity of methyl salicylate are also found in it. Senega is an expectorant and is indicated as such in the treatment of bronchitis. It should not be used on patients suffering from bronchial asthma.
f sénega; racine de sénega
d Senega; Senegawurzel; Klappenschlangenwurzel
i senega; radice di senega; radice di poligala senega; radice di poligala virginiana
e raíz de polígala de Virginia; raíz de senega

3463 **senna**
Term denoting some species of Cassia and the medicinal drugs yielded by them. The leaves and pods contain some active principles marked by a bland purgative action, principally due to the excitation of the peristalsis of the large intestine. Senna should be avoided in inflammatory conditions of the abdominal organs and in the course of lactation.
f séné
d Senne
i sena
e sen

3464 **senna leaves**
The leaves of the senna, which yield some active principles used for the preparation of infusions acting as a mild purgative.
f feuilles de séné
d Sennesblätter
i foglie di sena
e hojas de sen

3465 **senna pods**
Or senna fruit. The pods of the senna (q.v.) which are capable, together with the leaves of the plants, of yielding some active principles.
f fruit de séné; follicules de séné
d Sennesfrucht; Mutterblätter
i frutto di sena
e folículos de sen; fruto de sen

3466 **serine**
Oxyderivative of alanine, a constituent of many proteins, belonging to the non--essential amino acids.
f sérine
d Serin
i serina
e serina

3467 **serinphosphate**
Phosphorated derivative of serine generally used as a tonic in the treatment of nervous exhaustion, asthenia and convalescence. Gastroenteric disorders may occur as side effects.
f sérinphosphate
d Serinphosphat
i serinfosfato
e serinfosfato

3468 **serinphosphoric acid**
Amino acid contained in the caseinogen; it is the only example of esterified oxidrylic and not free aminoacid.
f acide sérinphosphorique
d Serinphosphorsäure
i acido serinfosforico
e ácido serinfosfórico

3469 **serpentine**
Alkaloid contained in the Rauwolfia serpentine; crystalline, needle-shaped solid having a strong hypotensive action.
f serpentine
d Serpentin
i serpetina
e serpentina

3470 **serpentinine**
Alkaloid whose chemical constitution is very similar to that of serpentine. Its pharmacological and its origin are the same as those of the latter.
f serpentinine
d Serpentinin
i serpentinina
e serpentinina

3471 **serum albumin**
Crystallizable albumin or mixture of

albumins normally constituting more than half of the protein of blood serum, blood plasma and other serous fluids. Serum albumin can be isolated after precipitation of the globulins. It is synthesized in the liver and serves to maintain the osmotic pressure of the blood. It is also used in transfusion for the treatment of shock and other conditions.

f sérum-albumine; séro-albumine
d Serumalbumin
i sieroalbumina; plasmalbumina
e seroalbúmina; albúmina sérica

3472 serum haemopoieticum
Serum obtained from the blood of a healthy horse when there is a maximum content of haematoforming substances. It has an antianaemic and tonic effect and is indicated in the treatment of anaemia, convalescence accompanied by debility, pregnancy and lactation.

f serum haemopoieticum
d Serum haemopoieticum
i serum haemopoieticum
e serum haemopoieticum

* **sesame seed   s. oily grain**

3473 shark liver oil
Yellowish fatty oil obtained from the livers of various sharks and used princi̲ pally as a source of vitamin A.

f huile de foie de requin
d Haifischleberöl
i olio di fegato di squalo
e aceite de hígado de tiburón

3474 sialic acid
Neuraminic acid, first isolated in 1936 from the salive of oxen. It is an important constituent of mucopolysaccharides, gangliosides and blood group substances. As neuraminic acid may present some structural differences, several sialic acids exist, e.g. N-acetyl neuraminic acid, characteristic of the gangliosides, which makes it possible to distinguish it from the other sphingolipids.

f acide sialique
d Acetylneuraminsäure
i acido sialico
e ácido siálico

3475 siderophilin
Iron-carrying pseudo-globulin molecule. Ferrous iron is oxidized to ferric acid and a complex one CD_2 molecule for each ferric atom is attached to the protein molecule. Also known as transferrin.

f sidérophiline; transferrine
d Transferrin
i siderofillina; transferrina
e siderofilina; transferina

3476 silbenate
Chemotherapeutic agent acting as anti--atherosclerotic by its preventive and stabilizing action on the sclerotic plaques of the connective tissue. It is

indicated in the treatment of circulatory ischaemia, in particular cerebral, coronary and mesenteric ischaemias. Also used in peripheral arteriopathies, primary and secondary osteoporosis, microcystic and macrocystic mastopathies. The drug may also be prescribed to combat advanced states of senescence and senile cataract.

f silbenate
d Silbenat
i silbenato
e silbenato

3477 silica
Silicon dioxide. The anhydride of silicic acid widely occurring in nature (quartz, flintsilicate rocks, rock crystals etc.). Hard, generally colourless compound, whose gel, i.e. a coagulated hydrated form, is used as an absorbing and coagulating agent.

f silice; anhydride silicique
d Siliciumdioxid; Kieselsäureanhydrid
i silice; anidride silicica
e anhídrido silícico; óxido de silicio

3478 silicate
Any salt of silicic acid, in which silicon and oxygen are always present.

f silicate
d Silicat; kieselsäures Salz
i silicato
e silicato

3479 silicic acid
Term denoting two hypothetical acids to which reference is made when describing soluble silicates. They are known as orthosilicic acid and metasilicic acid, from which the orthosilicates and the metasilicates are respectively derived. Many others may be formed (polysilicic acids), though not in a free state.

f acide silicique
d Kieselsäure
i acido silicico
e ácido silícico

3480 silicotungstic acid
Or tungstosilicic acid. Yellow crystalline solid substance, soluble in water and alcohol, used in chemical analysis for the qualitative and quantitative determination of alkaloids.

f acide silicotungstique
d Siliciumwolframsäure; Kieselwolframsäure
i acido silicotungstico
e ácido silicotúngstico

3481 silver arsphenamine
A sodium salt of silver, a dark brown powder which is soluble in water and was used in the past in the treatment of syphilis.

f arsphénamine d'argent
d Silberarsphenamin
i arsfenammina d'argento
e arsfenamina de plata

3482 silver nitrate
Colourless crystalline substance, astringent and caustic and therefore used to destroy warts. Its solutions are indicated in the treatment of lesions involving lacerated skin and as an eye antiseptic.
f nitrate d'argent; pierre infernale
d Silbernitrat; Höllenstein
i nitrato d'argento
e nitrato de plata; piedra infernal

3483 silver picrate
Yellow crystals used for their astringent and germicide actions. It is principally indicated in the treatment of vaginal trichomonas infection.
f picrate d'argent
d Silberpikrat
i picrato d'argento
e picrato de plata

3484 silver protein
Compound of silver with a protein, e.g. albulmin or casein, prepared in the form of a brown powder or as a dark granular substance. In solution, it has a mild antiseptic property. It is weaker than silver nitrate but is not caustic. It is generally used in nasal and conjunctival infections.
f protéine argentée
d Proteinsilber
i argento proteinato
e proteína argéntica

3485 silver sulphadiazine
Sulphonamide derivative characterized by a broad spectrum of activity, mainly used topically in the treatment of severe burns to avoid infection.
f sulfadiazine d'argent
d Silbersulfadiazin
i solfadiazina d'argento
e sulfadiacina de plata

* **silver trinitrophenate   s. silver picrate**

3486 silymarine
Physiological antidote capable of rendering normal the detoxicant, metabolic and synthetic function of the liver by a regenerating, stabilizing and protective action of the hepatocellular biomembranes. It is also capable of modifying hepatocellular impairment especially in relation to the detoxicant function. The drug is thus indicated in the treatment of toxic-metabolic liver disorders caused by exogenous or endogenous poisons, chronic, persistent and severe hepatitis, as well as acute hepatitis and cirrhosis of the liver.
f silymarine
d Silymarin
i silimarina
e silimarina

3487 simaldrate
Antacid agent indicated in the treatment of gastric disorders due to inordinate use of alcohol, tobacco, coffee, excessive eructation and gastrointestinal inflammation and fermentation.
f simaldrate
d Simaldrat
i simaldrato
e simaldrato

3488 simaruma
Or simarouba. Genus of plants of the family Simarubaceae, with 10 species growing in Central America and in Mediterranean countries. The bitter bark of the root or stem is frequently used against diarrhoea and fever. Also used as a tonic.
f simarouba
d Simarouba
i simaruba
e simaruba

3489 simethicone
Or dimethycone. Silicone marked by water-repellent and skin-adherent properties. It is used for topical application for protection against irritant substances, the prevention decubitus ulcers or lesions caused urinary or faecal incontinence. Also employed in combination with antacids for the elimination of meteorism as it reduces surface tension of small gas bubbles, thus allowing them to coalesce into larger pockets of gas which can be expelled without difficulty.
f diméthicone
d Dimethicon
i simeticone; dimeticone
e dimeticona

3490 sinapis nigrae semen
The seed of the black mustard. Therapeutic agent characterized by a revulsive action due to sinigrin (potassium mironate) which splits by the action of myrosin into glucose, acid potassium sulphate and allyl iso-sulphocyanate. The drug is used as a counterirritant and is capable of causing the decongestion of deep tissues.
f graines de moutarde grise; graines de moutarde noire
d Senfkohlsamen
i semi di senape nera; semi di senape
e semillas de mostaza negra

3491 sincalide
Choleretic principally used as a diagnostic agent of the biliary tract. It is administered intravenously.
f sincalide
d Sincalid
i sincalide
e sincalida

3492 sitosterol
Sterol found in wheat embryo and indicated in the treatment of hyperlipoproteinaemia. Anorexia and diarrhoea accompanied by intestinal pains may occur as

side effects.
f sitostérole
d Sitosterol
i sitosterolo
e sitoesterina

3493 sneezing powder
Or sternutatory agent. Any powder or
substance which causes sneezing, i.e.
a spasmodic and explosive expiration of
air through the mouth and nose.
f sternutatoire
d Niesmittel; Sternutatorium
i starnutatorio
e estarnutatorio

3494 soapberry
Or soapberry tree. Tree of the genus
Sapindus. Its fruit contains up to 37%
saponin.
f saponnière
d Seifennussbaum
i albero saponario
e jabonero

3495 sobrerole
Crystalline alcohol formed by oxidation
of alpha-pinene.
f sobrérole
d Sobrerol
i sobrerolo
e sobrerolo

3496 soda
Generic term denoting alkaline sodium
salts. Unless qualified, soda used by
itself means sodium carbonate.
f carbonate de soude
d Soda
i soda; carbonato di sodio
e soda; carbonato sódico

3497 soda lime
Granular mixture of calcium hydroxide
with sodium hydroxide or potassium hydroxide or some other substances used
to absorb moisture and acid gases, in
particular carbon dioxide. Soda lime is
therefore used in the rebreathing technique of inhalation anaesthesia, in gas
masks and in oxygen therapy.
f chaux sodée
d Natronkalk
i calce sodata
e cal sodada

3498 sodium acetate
Chemical compound used in medicine as
a diuretic agent and to increase the
alkalinity of urine.
f acétate de sodium
d Natriumacetat
i acetato di sodio
e acetato sódico

3499 sodium acetrizoate
X-ray contrast medium used sometimes
instead of diodone (intravenous contrast
medium used to visualize the urinary
tract), as its side effects are less ad-

verse.
f acétrizoate de sodium
d Natriumacetrizoat
i acetrizoato di sodio
e acetrizoato sódico

3500 sodium acetylarsanilate
Arsenical derivative used in medicine to
treat protozoal infections and syphilis.
Sodium aminoarsonate is used for the
same purposes.
f acétylarsanilate de sodium
d Natriumacetylarsanilat
i acetilarsanilato di sodio
e acetilarsanilato sódico

3501 sodium acetylsalicylate
Soluble compound used as antipyretic
and antirheumatic agent. It is obtained
by treating a suspension of acetylsalicylic acid in methyl alcohol with anhydrous sodium carbonate.
f acétylsalicylate de sodium
d Natriumacetylsalicylat
i acetilsalicilato di sodio
e acetilsalicilato sódico

3502 sodium alginate
The sodium salt of alginic acid, which
is extracted from species of seaweeds
and is used as an emulsifying agent.
f alginate de sodium
d Natriumalginat
i alginato di sodio
e alginato sódico

3503 sodium (para)-aminosalicylate
The sodium salt of para-amino salicylic
acid used in the treatment of pulmonary
tuberculosis. Optimum results are generally obtained if the drug is administered in conjunction with streptomycin.
f para-aminosalicylate de sodium
d Natrium-para-aminosalicylat
i para-aminosalicilato di sodio
e para-aminosalicilato sódico

3504 sodium and potassium tartrate
Or potassium sodium tartrate. Cathartic
agent prepared in a concentrate saline
solution. It is a constituent of the
Seidlitz powder.
f tartrate de sodium et potassium
d Kalium-Natriumtartrat
i sodio e potassio tartrato
e tartrato sódico y potásico

3505 sodium anoxynaphthonate
A diagnostic agent, i.e. a low-toxicity
dye used in the measurement of cardiac
output and in particular to determine
the position of cardiac shunts.
f anoxynaphtonate de sodium
d Natrium-Anoxynaphthonat
i anossinaftonato di sodio
e anoxinaftonato sódico

3506 sodium arsenate solution
Anhydrous sodium arsenate. Poisonous
secondary orthoarsenate or its hydrates

used in medicine as spirochaeticide and
trypanocide.
f soluté d'arséniate de soude
d Natriumarsenatlösung
i soluzione di arseniato di sodio
e solución de arsenato sódico

3507 sodium arsenotartrate
Arsenic salt sometimes used in dermatology instead of potassium arsenite.
f arsénite de sodium
d Natriumarsenit
i arsenito sodico
e arsenito sódico

3508 sodium ascorbate
Salt of ascorbic acid for parenteral use.
Its properties are analogous to those of
the acid. Sodium ascorbate is indicated
for the treatment of pathological conditions caused by inadequacy of vitamin
C supply, in particular when the gastro
-intestinal absorption is affected. The
drug has proved very useful against
scurvy, hypovitaminosis or avitaminosis,
haemorrhagic diathesis, intoxications,
allergic diseases etc. Side effects are
mostly irrelevant unless excessive doses
are administered.
f ascorbate de sodium
d Natriumascorbat
i ascorbato di sodio; sodio ascorbato;
  ascorbato sodico
e ascorbato sódico

3509 sodium aurothiomalate
Water-soluble organic derivative of gold,
a long-acting agent indicated in the
treatment of chronic polyarthritis, chron
ic juvenile arthritis and psoriatic arthritis. Dermatitis, stomatitis, vaginitis,
gastro-intestinal disorders, general
weariness and a metallic taste in the
mouth may occur as side effects.
f aurothiomalate de sodium
d Natriumthiomalat
i sodio aurotiomalato
e aurotiomalato sódico

3510 sodium aurothiopropanolsulphonat
Antirheumatic agent used in the chrysotherapy of rheumatoid arthritis. Itching,
erythema, stomatitis, diarrhoea, albuminuria and bilirubinuria may occur as
side effects.
f aurothiopropanolsulfonate de sodium
d Natriumaurothiopropanolsulfonat
i aurotiopropanolsolfonato di sodio
e aurotiopropanolsulfonato sódico

3511 sodium benzalbutyrate
Antithrombotic agent capable of normaliz
ing platelet aggregation and of inhibiting the aggregation of red corpuscles.
The drug is indicated in the treatment
of cerebral thrombosis, peripheral vascular and cerebrovascular disorders,
diabetic retinopathies, coronary vasculopathies and the prophylaxis of postopera
tive venous thrombosis.

f benzalbutyrate de sodium
d Natriumbenzalbutyrat
i benzalbutirrato di sodio
e benzalbutirato sódico

3512 sodium benzoate
Intermediate agent used as an antiseptic.
f benzoate de sodium
d Natriumbenzoat
i benzoato di sodio; benzoato sodico
e benzoato sódico

3513 sodium bicarbonate
Chemical compound obtained by scrub-
bing carbon anhydride in a solution
saturated with sodium carbonate. In medicine it is principally used to reduce
acidity in the stomach and, in the form
of tablets or lozenges, against dyspepsia
and flatulence. It is extensively used
in effervescent powders generally mixed
with citric and tartaric acids.
f bicarbonate de sodium
d Natriumbicarbonat
i bicarbonato di sodio
e bicarbonato sódico

3514 sodium bifluoride
Chemical compound principally used in
medicine as a preservative for histological specimens.
f bifluorure de sodium
d Natriumbifluorid
i bifluoruro di sodio
e bifluoruro sódico

3515 sodium biphosphate
Sodium acid phosphate. Sodium dihydrogen phosphate, a compound used to make
the urine acid during the administration
of hexamine. The drug is also used as
a diuretic agent.
f biphosphate de sodium
d Natriumbiphosphat
i bifosfato di sodio
e bifosfato sódico

3516 sodium bisulphate
Chemical compound widely used to eliminate typhoid bacilli from drinking water.
It is also used internally as an anti-
ferment.
f bisulfate de sodium
d Natriumbisulfat
i bisolfato di sodio
e bisulfato sódico

3517 sodium bisulphite
Analytical reagent and starting material
for a large number of compounds. Used
medicinally as an antiseptic in gastric
fermentation and in cutaneous diseases
caused by parasites.
f bisulfite de sodium
d Natriumbisulfit
i bisolfito di sodio
e bisulfito sódico

3518 sodium bromide
Sedative acting on the cortex and used

in the treatment of psychic excitability and insomnia. The drug shows a tendency to accumulate in the organism as it substitutes chlorine in sodium chloride.

f bromure de sodium
d Natriumbromid
i bromuro di sodio
e bromuro sódico

3519 sodium cacodylate
Organic compound of arsenic acid, in general better tolerated than inorganic compounds, mainly used as a tonic. Large doses may cause some harm to the liver and kidneys (fatty degeneration).

f cacodylate de sodium
d Natriumcacodylat
i cacodilato di sodio
e cacodilato sódico

3520 sodium calcium edetate
Chemical compound used as antidote in intoxications from heavy metals, e.g. lead and mercury, and in the elimination of radioactive substances, e.g. plutonium.

f calcium édétate de sodium
d Natrium-Calcium-Edetat
i edetato sodico calcico
e edetato sódico cálcico

*** sodium camsilate s. piperazine camsilate**

3521 sodium carbonate
Salt of carbonic acid, commonly known as soda. The compound is principally used externally; for internal treatment bicarbonate, being not irritative, is preferred. It is indicated in the treatment of cutaneous diseases in the form of lotion. Also used as a mouthwash. It is also used for the pH control of water.

f carbonate de sodium
d Natriumcarbonat
i carbonato di sodio
e carbonato sódico

3522 sodium carboxymethylcellulose
Chemical compound prepared from chloroacetic acid, cellulose and sodium hydroxide. It is used in pharmaceutics to suspend insoluble powders in aqueous preparations. Some grades of the substance may be used as a bulk purgative.

f carboxyméthylcellulose de sodium
d Natriumcarboxymethylzellstoff
i carbossimetilcellulosa di sodio
e carboximetilcelulosa sódica

3523 sodium choleate
Dried and purified ox gall used as a purgative in chronic constipation.

f choléate de sodium
d Natriumcholeat
i coleato di sodio
e coleato sódico

3524 sodium chloride
Chemical compound used for the preparation of physiological sodium chloride solution as well as other solutions, e.g. the Ringer's solution (used for intravenous injections).

f chlorure de sodium
d Natriumchlorid
i cloruro di sodio
e cloruro sódico

3525 sodium chondroitinsulphate
Non-hormonal agent used for demineralization. Indicated in the treatment of osteoporosis, decalcification, Paget's disease, delayed fracture healing and arthritis.

f chondroitinsulfate de sodium
d Natriumchondroitinsulfat
i condroitinsolfato di sodio
e condroitinsulfato sódico

3526 sodium citrate
Antacid and antiacetonaemic agent indicated in the treatment of hyperchlorhydria, acidosis and gastralgias.

f citrate de sodium
d Natriumzitrat
i sodio citrato; citrato sodico
e citrato sódico

3527 sodium citrobenzoate
Compound indicated in the treatment of asthma and bronchitis. Also used as a diuretic and antilithic agent.

f citrobenzoate de sodium
d Natriumzitrobenzoat
i citrobenzoato di sodio
e citrobenzoato sódico

3528 sodium cyclamate
Synthetic sweetening agent (30 times the action of sugar) principally used in the treatment of obesity, diabetes, dyspepsia and diarrhoea.

f cyclamate de sodium
d Natriumcyclamat
i ciclamato di sodio
e ciclamato sódico

3529 sodium dehydrocholate
The most active of the acids that are derived from natural bile acids. It has a choleretic, more precisely a hydrocholeretic action and is used as a stimulant of the flow of bile. It is also a diuretic agent.

f déhydrocholate de sodium
d Natriumdehydrocholat
i deidrocolato di sodio
e dehidrocolato sódico

3530 sodium dibunate
Antitussive agent indicated in the treatment of cough independently of its origin, also when accompanied by a phlogistic condition. Some gastro-intestinal disorders may occur as side effects.

f dibunate de sodium
d Natriumdibunat
i dibunato di sodio
e dibunato sódico

3531 sodium dimetocinnamate

Synthetic choleretic marked by a fairly potent hypocholesteraemic action by greatly increasing biliary secretion and by facilitating the elimination of cholesterol through the bile. It is used in association with sorbitol in pathological conditions of the biliary tract, e.g. cholangitis, incompetent biligenetic function and hypercholesterolaemia.

f dimétocinnamate de sodium
d Natriumdimetocinnamat
i dimetocinnamato di sodio
e dimetocinamato sódico

3532 sodium diphenylhydantoinate

Anticonvulsant agent marked by a variable or hypnotic action, indicated in the treatment of epilepsy which does not respond to phenobarbitone or bromides.

f diphénylhydantoïnate de sodium
d Natriumdiphenylhydantoinat
i difenilidantoinato di sodio
e difenilhidantoinato sódico

3533 sodium fluoride

Anticarious and osteogenetic agent capable of strengthening dentine and dental enamel, and of stimulating osteosynthesis and bone recalcification. In small doses, it is indicated in the prophylaxis of dental caries; in larger doses, it is frequently used in primary osteoporosis, in the therapy and prophylaxis of osteoporosis due to glucocorticoids, plasmacytoma and osteolytic bone metastases.

f fluorure de sodium
d Natriumfluorid
i fluoruro di sodio
e fluoruro sódico

3534 sodium formate

Chemical compound marked by a potent diuretic action. Also used for the treatment of lumbago and rheumatism.

f formiate de sodium
d Natriumformiat
i formiato di sodio
e formiato sódico

3535 sodium galactopolysulphate

Anticoagulant and antiphlogistic agent for topical use. It has a very slow absorption and is indicated in the prophylaxis and therapy of both superficial and deep thrombophlebitis and of ulcers.

f galactopolysulfate de sodium
d Natriumgalactopolysulfat
i galattopolisolfato di sodio
e galactopolisulfato sódico

3536 sodium gentisate

Antiphlogistic and antirheumatic, as well as antipyretic and analgesic agent. Nausea and some gastro-intestinal disorders may occur as side effects.

f gentisate de sodium
d Natriumgentisat
i gentisato di sodio
e gentisato sódico

3537 sodium glutamate

Compound indicated in various disorders of the nervous system and some psychopathies, e.g. mongolism, oligophrenia etc.

f glutamate de sodium
d Natriumglutamat
i glutammato di sodio
e glutamato sódico

3538 sodium hexacyclonate

Analeptic and psychostimulant agent used in the treatment of asthenia and depression, in particular when characterized by senile psychic involution. The drug is also used to increase the intellective potential and to improve mental concentration.

f hexacyclonate de sodium
d Natriumhexacyclonat
i esaciclonato di sodio
e hexaciclonato sódico

3539 sodium iodomethamate

Iodine contrast medium principally used for intravenous pyelography (the X-ray examination of the kidneys and ureters). Nausea and vomiting, general weakness, hypotension and some allergic reactions may occur as side effects.

f iodométhamate de sodium
d Natriumjodomethamat
i iodometamato sodico
e yodometamato sódico

3540 sodium iopodate

Orally-administered radiopaque agent mainly used in angiography. Opacity sets in after about ten hours.

f iopodate de sodium
d Natriumiopodat
i iopodato di sodio
e iopodato sódico

3541 sodium lauryl sulphate

Anionic surface active agent marked by detergent properties in acid and alkaline medium and in hard waters. It has a fairly potent bacteriostatic power against Gram-positive micro-organisms. It is indicated for external use in the preparation of detergent and antiseptic solutions and of emulsifier for creams.

f laurylsulfate de sodium
d Natriumlaurylsulfat
i solfato di laurile e sodio
e laurilsulfato sódico

3542 sodium methylarsinate

Arsenic-containing organic compound marked by a haemopoietic (blood-cell forming) and calcifying action.

f méthylarsinate de sodium
d Natriummethylarsinat
i metilarsinato di sodio
e metilarsinato sódico

3543 sodium metrizoate

Contrast medium mainly used in angiography, benography, pyelography,

lumbar discography (the demonstration of a disc of a joint) and hysterosalpingography. Nausea, general weakness, a sensation of warmth and thirst, itching, urticaria and tachycardia may occur as side effects.
f métrizoate de sodium
d Natriummetrizoat
i metrizoato di sodio
e metrizoato sódico

3544 sodium nitrite
Nitrite capable of slowly liberating the nitrite ion and used in angina pectoris and asthma to diminish arterial tension. More generally, the drug is used as a vasodilator.
f nitrite de sodium
d Natriumnitrit
i nitrito di sodio
e nitrito sódico

3545 sodium nitroferricyanide
Sodium nitroprusside. Poisonous chemical reagent used to identify acetone bodies and aldehydes. It is also used to reduce blood pressure in hypotensive anaesthesia owing to its immediate effects of the vessel walls. Nausea, vomiting, copious perspiration, anorexia and general weakness may occur as side effects.
f nitroferricyanure de sodium
d Natriumnitroferricyanid
i nitroferricianuro di sodio
e nitroferrocianuro sódico

3546 sodium nitroprusside
Poisonous chemical reagent used to detect acetone bodies and aldehydes. It is a red crystalline salt obtained from reaction of a ferricyanide and nitrous acid. It is employed in medicine to reduce blood pressure in the course of hypotensive anaesthesia.
f nitroprusside de sodium
d Natriumnitroprussid
i nitroprusside di sodio
e nitroprusido sódico

3547 sodium oxybate
Intravenous anaesthetic. Hypnotic agent of the hydroxybutyric series. It has a rapid, though short-duration, action. Also used in psychiatry in the treatment of sleep disorders. Bradycardia and respiratory depression have been observed as side effects.
f oxybate de sodium
d Natriumoxybat
i ossibato di sodio
e oxibato sódico

3548 sodium phosphate
Disodium hydrogen phosphate. Mild, tasteless saline cathartic.
f phosphate de sodium
d Natriumphosphat
i fosfato di sodio
e fosfato sódico

3549 sodium picosulphate
Hydrosoluble contact laxative capable of stimulating peristalsis by direct action on the membrane of the large intestine. Abdominal pains and some diarrhoea may follow large doses of the drug.
f picosulfate de sodium
d Natriumpicosulfat
i picosolfato di sodio
e picosulfato sódico

3550 sodium polystirene sulphonate
Sulphonic cation-exchange resin capable of removing from the organism potassium ions by exchange with sodium ions in the intestine and particularly the large intestine. Elimination occurs in the faeces. It is indicated in the treatment of hyperkalemia associated with renal incompetence. Anorexia, nausea, vomiting and constipation may occur as side effects.
f polystyrolsulfonate de sodium
d Natriumpolystyrolsulfonat
i polistirolsolfonato di sodio
e polistirolsulfonato sódico

3551 sodium potassium tartrate
Saline cathartic agent producing a watery evacuation without irritation. Small doses are also used as diuretic and to render urine less alkaline.
f potassium-tartrate de sodium
d Natrium-Kaliumtartrat
i potassio tartrato di sodio
e tartrato de potasio sódico

3552 sodium propionate
Chemical compound used as a fungicide. It is mostly used in the form of ointment or spraying powder together with other antimycotic agents in cutaneous mycosis, pytiriasis etc. Also used in solutions for external infections.
f propionate de sodium
d Natriumpropionat
i propionato di sodio
e propionato sódico

3553 sodium ricinoleate
Mixture of the sodium salts of the fatty acids from castor oil, capable of reducing the surface tension of emulsifying agents. Solutions are used in the treatment of infections of the mouth and as a sclerosing agent in the treatment of varicose veins.
f ricinoléate de sodium
d Natriumricinoleat
i ricinoleato di sodio
e ricinoleato sódico

3554 sodium salicylate
Antipyretic and antirheumatic agent which is also used as a mild gastric antiseptic and equally mild cholagogue action. It is also indicated in the treatment of varicose veins as a sclerosing agent. Gastro-intestinal disorders may occur as side effects.
f salicylate de sodium

d Natriumsalicylat
i salicilato di sodio
e salicilato sódico

3555 sodium stibocaptate
Antihelmintic agent marked by a specific action against schizostomes, indicated in the treatment of vesical, intestinal and arteriovenous bilharziosis. Nausea, abdominal and muscular pains as well as anorexia may occur as side effects.
f stibocaptate sodique
d Natriumstibocaptat
i stibocaptato di sodio
e stibocaptato sódico

3556 sodium stibogluconate
Pentavalent antimony derivative of gluconic acid indicated for the treatment of kala-azar. Being stable in solution, it can be prepared in ampules ready to be injected. The drug may be injected intravenously or intramuscularly.
f stibogluconate de sodium
d Natriumstibogluconat
i stibogluconato di sodio
e stibogluconato sódico

3557 sodium sulphate
Saline purgative. Orally administered, it reduces water absorption from the intestine and stimulates the enteric secretion thus increasing the bulk of the faeces, distending the loops and stimulating the peristalsis. The drug is also administered by intravenous injection in diluted solution so as to stimulate diuresis in severe forms of hypercalcaemia and in some forms of crystalluria due to sulphonamidic therapy. Also used as a lymphagogue for the treatment of septic wounds. Cardiac insufficiency, hypertension, pulmonary oedema and gestosis may occur as side effects.
f sulfate de sodium
d Natriumsulfat
i solfato di sodio
e sulfato sódico

3558 sodium sulphite
Antioxidant and reducing agent, as well as antiseptic and antizymotic, used internally in the treatment of fermentative dyspepsia and as a mouth wash in stomatitis. Also used in the form of lotion in parasitic skin diseases.
f sulfite de sodium
d Natriumsulfit
i solfito di sodio
e sulfito sódico

3559 sodium tetraborate
Mild disinfectant agent mainly used as a collutory.
f tétraborate de sodium
d Natriumtetraborat
i tetraborato di sodio; borato di sodio
e tetraborato sódico

3560 sodium tetrachloriodide

Antiseptic and disinfectant for external use. It develops nascent oxygen, iodine and chlorine thus exerting a very rapid and potent germicide action. Its broad spectrum makes it effective against both Gram-positive and Gram-negative micro-organisms. It is indicated as a prophylactic and adjuvant agent in the treatment of various diseases affecting the female genitalia, e.g. vulvitis, vulvovaginitis, leukorrhoea etc.
f tétrachloriodure de sodium
d Natriumtetrachlorjodid
i tetracloroioduro di sodio
e tetracloroyoduro sódico

3561 sodium tetradecyl sulphate
Anionic surface active agent marked by sclerosing properties and thus indicated in the treatment of varicose veins by intravenous injections. Also used in the treatment of haemorrhoids.
f tétradécyl-sulfate de sodium
d Natriumtetradecylsulfat
i sodio tetradecile solfato
e tetradecilo sulfato sódico

3562 sodium thiosulphate
Antianaphylactic and antiallergic agent, also used as an antidote for cyanide poisoning.
f thiosulfate de sodium
d Natriumthiosulfat
i tiosolfato di sodio
e tiosulfato sódico

3563 sodium tyropanoate
Contrast medium mainly used in radiography of the biliary tract. Nausea and diarrhoea may occur as adverse effects. The drug should not be used in patients suffering from acute nephritis, uraemia and gastro-intestinal disorders.
f tyropanoate de sodium
d Natriumtyropanoat
i tiropanoato di sodio
e tiropanoato sódico

3564 sodium valproate
Anticonvulsant agent probably acting by increasing the brain levels of gamma-aminobutyric acid. The drug is used in the treatment of all forms of epilepsy. Gastro-intestinal disorders, liver necrosis and prolonged bleeding times with thrombocytopaenia may occur as side effects.
f valproate de sodium
d Natriumvalproat
i valproato di sodio
e valproato sódico

3565 solapsone
Water-soluble derivative of diaminodiphenyl sulphone, principally indicated in the treatment of leprosy. It has also been used against the Mycobacterium tuberculosis but then discontinued as it was proved less effective than the combined treatment with PAS (para-aminosali

cyclic acid) and streptomycin. It is a
bacteriostatic agent of the sulphonamide
type as it is competitively inhibited by
para-aminobenzoic acid. Also called sola
sulphone.
f solapsone
d Solapson; Solasulfon
i solapsone
e solapsona; solasulfona

3566 somatotrophin
Human growth hormone secreted by the
eosinophil cells of the anterior lobe of
the pituitary gland. It stimulates the
skeletal, visceral and, more generally,
somatic growth, also promoting the forma
tion of new cartilages and bones in rela
tion to the epiphyses. It further stimulates the metabolism of carbohydrates
and lipids, glomerular filtration and the
excretion of water. Principally indicated
in the treatment of nanism, stress-induc-
ed ulcer and extended burns. Some anorexia and oliguria may be observed as
side effects at the beginning of the
therapy.
f somatotrophine; hormone somatotrope
d STH; somatotropes Hormon; Somatotrophin;
  Wachstumshormon
i somatotrofina; somatotropina
e somatropina

3567 somatrem
Synthetic hormone marked by an action
that is similar to that of somatotrophin
(q.v.). It replaces the growth hormones
extracted from the human pituitary
gland thus avoiding the danger of viral
infections from that source. It is used
in the treatment of nanism caused by
growth-hormone deficiency.
f somatrem
d Somatrem
i somatrem
e somatrem

3568 sorbic acid
Aliphatic acid containing two double
bonds, occurring in the unripe sorb-
-apples; crystalline powder, not easily
soluble in water, used as intermediate
in several syntheses and as a fungicide.
Also used to inhibit the growth of
moulds.
f acide sorbique
d Sorbinsäure
i acido sorbico
e ácido sórbico

3569 sorbinicate
Chemical compound capable of reducing
cholesterol content and used as an anti-
lipidaemic and peripheral vasodilator
agent. It stimulates biliary secretion
and is therefore indicated in the treatment of hypercholesterolaemia, arteriosclerosis, both generalized and localized,
and in the prophylaxis of hypertension
and cardiac infarction.
f sorbinicate

d Sorbinicat
i sorbinicato
e sorbinicato

3570 sorbitan
Chemical compound obtained by heating
very briefly sorbitol in the presence of
diluted sulphuric acid. Some of its
esters are used as emulsifiers and sur-
face-tension agents.
f sorbitan
d Sorbitan
i sorbitano
e sorbitano

3571 sorbitol
Hexahydric alcohol isomeric with mannitol, a sweetening agent which may be
used by diabetics instead of sugar. It
is a cholecystokinetic agent capable of
regulating biliary and intestinal functions. It stimulates duodenal secretions
of three digestive hormones, i.e. chole-
cystokinin which induces gall-bladder
contraction with the opening of Oddi's
sphincter, villikinin which accelerates
intestinal passage and absorption, and
pancreozymin which stimulates a pancreatic secretion that is rich in enzymes.
The drug is principally indicated in the
treatment of biliary, pancreatic and,
more generally, intestinal disorders.
f sorbitol
d Sorbitol
i sorbitolo
e sorbitolo

3572 sotalol
Beta-adrenoreceptor blocking drug used
in the treatment of hypertension, angina
pectoris and thyrotoxicosis.
f sotalol
d Sotalol
i sotalolo
e sotalol

3573 soya
The seeds of several varieties of Glycine
growing in America and the Far East.
Its principal constituents are protein
and a fixed oil.
f soja
d Soja
i soia; soja
e soja

3574 soziodolic acid
Or sozo-iodolic acid. Antiseptic having
the same uses as iodoform.
f acide sozoïodolique
d Sozojodolsäure
i acido sozojodolico
e ácido sozoyodólico

3575 sparteine
Tetracyclic alkaloid found in the broom
(Irish broom or genista) from which it
is obtained by extraction with alcohol.
It is used therapeutically owing to its
action on the heart which is roughly

similar to that of digitalis. It improves
the conduction of the stimuli and
strengthens the contractility of the myocardium. It also increases diuresis.
f spartéine
d Spartein
i sparteina
e esparteina

3576 **spectinomycin**
Antibiotic isolated from cultures of
Streptomyces spectabilis. A broad-spectrum antibiotic effective against Gram-
-positive and Gram-negative micro-organisms and in particular against most
strains of Neisseria gonorrhoeae. It is
therefore mainly used in the treatment
of gonococcal infections of the genito-
-urinary tract both in men and women.
f spectinomycine
d Spektinomycin
i spectinomicina
e espectinomicina; actinospectacina

3577 **spermine**
Deliquescent crystalline aliphatic tetra-
mine found in the semen in combination
with phosphoric acid in blood serum and
body tissues. It is also found in yeast
and can be prepared synthetically.
f spermine
d Spermin
i spermina
e espermina

3578 **sphingomyelin**
Phospholipid which splits by hydrolysis
yielding d-sphingoline, phosphoric acid,
choline and a fatty acid. Sphingomyelins
are especially found in the nervous system and accumulate, in pathological conditions, in the spleen and liver. They
probably are, like the cerebrosides,
stabilized intermediate products of the
transformation of carbohydrates into
fats.
f sphingomyéline
d Sphingomyelin
i sfingomielina
e esfingomielina

3579 **sphingosine**
Unsaturated amino glycol obtained by
hydrolysis of various sphingomyelins,
cerebrosides and gangliosides. A crystalline, needle-shaped, alcohol-soluble
solid.
f sphingosine
d Sphingosin
i sfingosina
e esfingosina

3580 **spike lavender oil**
The oil that is obtained by distillation
from the flowering herb Lavandula latifolia. Its properties are similar to those
of lavender oil. It is used, among other
things as an insect repellent.
f essence de spic; huile d'aspic
d Spiköl

i essenza di spigo
e esencia de espliego macho

* **spinacene** s. squalene

3581 **spindle tree**
Shrub or tree of the genus Euonymus,
especially a small erect shrubby tree
growing in Europe and western Asia. It
produces a pink capsule enclosing the
orange fruit. It is a potent cathartic
principle in the bark and the fruit.
f bois à lardoires; fusain d'Europe
d europäischer Spindelbaum
i evonimo europeo; fusaggine europea
e bonetero europeo

3582 **spiperone**
Butyrophenone-related neuroleptic principally used for the treatment of schizophrenia.
f spipérone
d Spiperon
i spiperone
e espiperona

3583 **spiramycin**
Antibiotic obtained from Streptomyces
ambofaciens, active against Gram-positive cocci. As it does not act on the
normal Gram-negative micro-organisms of
the intestinal canal, it hardly ever produces intestinal infections with Candida.
General malaise and some allergic reactions may occur as side effects.
f spiramycine
d Spiramycin
i spiramicina
e espiramicina

3584 **spironolactone**
Propionic acid lactone, a diuretic antagonist of aldosterone, used in the treatment of hypertension, resistant oedemata,
including idiopathic oedemata, cardiac
incompetence, cirrhosis of the liver and
nephrotic syndrome. Electrolyte imbalance due to the blocking of hormones
secreted by the cortex of the suprarenal
glands may occur as side effect.
f spironolactone
d Spironolakton
i spironolattone
e espironolactona

3585 **spleen extract**
Any preparation obtained from the splenic tissue of healthy animals. It is indicated for the treatment of neurovegetative disorders, gastritis, gastro-duodenal ulcer, vasomotor rhinitis, polycythaemia and as a coadjuvant in the
treatment of acute and chronic infections.
f extrait de rate
d Milzextrakt
i estratto di milza
e extracto de baso

3586 **sponge**
The dried skeleton of some marine ani-

mals formed by a tangled mass of extremely fine fibres of a horny and very elastic substance. Because of their porosity, sponges can absorb a large quantity of liquid and expel it when squeezed.
f éponge
d Schwamm
i spugna
e esponja

3587 squalene
Liquid acyclic triterpene hydrocarbon especially found in the liver oil of various sharks but also in several plant oils, in yeast and human sebum. It may be considered a precursor of cholesterol and related steroids.
f squalène
d Squalen
i squalene
e esqualeno

3588 squill
The dried fleshy inner scales of the bulb of the white-rooted form of the squill (a bulbous herb), which contain some physiologically active glucosides and are used as an expectorant, cardiac stimulant and diuretic.
f scille
d Scilla
i scilla
e escila

3589 squill bulb
The bulb of the white-rooted form of the squill, from whose scales one or more principles are extracted for the preparation of expectorant, cardiac stimulant and diuretic agents.
f bulbe de scille
d Scillazwiebel
i bulbo di scilla
e bulbo de escila

3590 stallimycin
Topical antiviral antibiotic particularly effective against herpes virus and vaccine virus. Its antibacterial action is rather weak. The drug is mainly indicated in the treatment of herpes labialis, diffuse and traumatic herpes, genitalis herpes, chickenpox and herpes zoster. Sensitization reaction may occur as side effect.
f stallimycine
d Stallimyzin
i stallimicina
e estalimicina

3591 stanozolol
Anabolic steroid. It displays a potent anabolic action owing to greatly increased protein biosynthesis. It is indicated in the treatment of debilitated patients, in particular when suffering from decalcification of the bones. A prolonged use may cause jaundice.
f stanozolol

d Stanozolol
i stanozololo
e estanozololo

3592 star anise oil
Or star aniseed oil. Essential oil obtained from star aniseed (the dried fruit of the Chinese anise) and is used as an expectorant and carminative agent.
f essence d'anis étoilé
d Sternanisöl
i essenza di badiana; essenza di anice stellato
e esencia de anis estrellado; esencia de badiana

3593 starch
White, odourless and tasteless granular or powdery carbohydrate which is the main storage form of carbohydrate in plants. Chemically, a polysaccharide used as an absorbant in dusting powder for skin lesions and as a disintegrating agent in tablets. Starch is also used as a mucilage for baby food and as an antidote in iodine poisoning.
f amidon
d Stärke
i amido
e almidón

3594 stearate
Salt or ester of stearic acid.
f stéarate
d Stearat
i stearato
e estearato

3595 stearic acid
Acid occurring widely, together with palmitic and oleic acids; as a glyceride in vegetable and animal fats. It is widely used as the basis of soaps.
f acide stéarique
d Stearinsäure
i acido stearico
e ácido esteárico

3596 stearine
A mixture of fatty acids, particularly stearic acid; obtained by hydrolysis with superheated steam of various fats. The resulting hard and semicrystalline mass is insoluble in water and soluble in alcohol, ether, chloroform and other fats. It is used either by itself or partly neutralized, in the preparation of various ointment bases.
f stéarine
d Stearin
i stearina
e estearina

3597 stearyl alcohol
Unctuous solid alcohol especially found in whale, dolphin and porpoise oils or obtained by hydrogenation of stearic acid. It is used in ointments and creams as its solubility promotes the incorporation of water.

f alcool stéarylique
d Stearylalkohol
i alcool stearilico
e alcohol estearílico

3598 stibamine
The sodium salt of stibanilic acid mostly injected in the form of glycoside in the treatment of several tropical diseases.
f stibamine
d Stibamin
i stibammina
e estibamina

3599 stibaminic acid
An unstable acid containing antimony that is analogous to arsanilic acid. It is principally used for the preparation of organic antimonial drugs.
f acide stibaminique
d Stibaminsäure
i acido stibamminico
e ácido estibamínico

3600 stibophen
Antimonial therapeutic agent effective against protozoa. The compound reacts with sulphahydryl groups and inhibits breathing. It is used in the treatment of schistosomiasis by intramuscular injection. Thrombocytopenia may occur as side effect.
f stibophène
d Stibophen
i stibofene
e estibofeno

*** stilbestrol   s. diethylstilbestrol**

3601 stilonium iodide
Anticholinergic spasmolytic agent capable of an elective ganglion-blocking action in relation to the parasympathetic system. It has no atropine-like effects for the arterial pressure and the cardiac frequency. It is principally indicated in the treatment of spasms of the smooth musculature: renal and hepatic colics, dyskinesia of the urinary tract and spastic colitis. Great caution must accompany the administration of the drug in the presence of paralytic ileus and glaucoma.
f iodure de stilonium
d Stiloniumjodid
i ioduro di stilonio
e yoduro de estilonio

3602 stimulant
Any substance or agent capable of acting as a stimulus upon a nerve centre, an organic tissue or the entire organism. Typical substances are the analeptics of circulation and respiration, hormones and various growth factors.
f stimulant
d Stimulans; Anregungsmittel
i stimolante
e estimulante

3603 stramonium
Short for stramonium leaves (thornapple leaves). The dried leaves and flowering tops of Datura stramonium, which contains the alkaloid hyoscyamine. It has a parasympatholytic action and its sedative activity on the central nervous system differs from that of atropine and hyoscyamine as it causes no excitation. Brachycardia and a blurred vision may occur as side effects.
f stramoine; chasse-taupe
d Stramonium; Stechapfel
i stramonio; mazzetone
e estramonio; datura

3604 streptodornase
Enzyme obtained from diverse strains of Streptococcus haemolyticus, capable of hydrolyzing the desoxyribonucleins. It is used, together with streptokinase (q. v.) for the removal of clotted blood or purulent fibrinous clots. It is generally injected intramuscularly or instilled into body cavities. Some pains, fever and nausea may occur as side effects.
f streptodornase
d Streptodornase
i streptodornasi
e estreptodornasa

3605 streptokinase
Coenzyme obtained from cultures of various strains of Streptococcus haemolyticus and capable of transforming the plasminogen into plasmin. It is used to dissolve blood clots and thus in the treatment of embolic or thrombotic diseases. It is frequently administered in conjunction with heparin. It also finds employment in arteriosclerosis. Some allergic reactions and mildly feverish conditions may occur as side effects.
f streptokinase
d Streptokinase
i streptochinasi
e estreptoquinasa

3606 streptomycin
A group of antibiotics produced by cultures of Streptomyces griseus, very effective in the treatment of a large number of infections, in particular tuberculosis. It is an antibiotic inhibitor of protein synthesis. It inhibits initiation and causes misreading of messenger RNA. Its tendency to produce resistant strains of organisms and the toxic effects which are caused in the course of a protracted administration present serious disadvantages. If it is administered orally it is not absorbed and acts on the intestinal flora.
f streptomycine
d Streptomyzin
i streptomicina
e estreptomicina

*** streptoniazid   s. streptonicozid**

3607 **streptonicozid**
Synthetic product of isonicotinylhydrazone and streptomycin used in the treatment of pulmonary tuberculosis and miliary tuberculosis. Ototoxicity and some psychic excitation may occur as side effects.
f streptonicozide; streptoniazide
d Streptoniazid
i streptoniazide
e estreptoniazida

3608 **streptonigrin**
Antibiotic obtained from cultures of Streptomyces flocculus; brown crystals, not easily soluble in water, having anti tumoral properties.
f streptonigrine
d Streptonigrin
i streptonigrina
e estreptonigrina

3609 **streptothricin**
Antibiotic substance obtained from culture filtrates of Actinomyces lavendulae. It is a bacteriostatic and bacteriocidal agent and is inactivated by concentrated mineral acids, though unaffected by blood, peptone or vitamin B. It is active against Gram-positive and Gram-negative micro-organisms. Also effective against fungi.
f streptothricine
d Streptothrizin
i streptotricina
e estreptotricina

3610 **strichnine nitrate**
Alkaloid capable of stimulating the posterior horns of the spinal cord and acting on the cerebral cortex. Owing to its bitter taste, it is also used as a eupeptic agent. It is administered orally and, rarely, intramuscularly. Muscular spasms and very rarely convulsions may occur as side effects.
f nitrate de strychnine
d Strychninnitrat
i nitrato di stricnina
e nitrato de estricnina

3611 **strontium**
A metal of the alkaline earths, similar to calcium in its chemical properties. It occurs in sea water and thus in the ash of marine plants; it has also been observed, although in very minute quantities, in human tissues. Its salts were used in medicine, but are now obsolete.
f strontium
d Strontium
i stronzio
e estróncio

3612 **strontium bromide**
Colourless deliquescent crystalline substance having the same general action of the bromides.
f bromure de strontium
d Strontiumbromid

i bromuro di stronzio
e bromuro de estróncio

3613 **strophanthus**
Genus of tropical Asiatic and African trees and shrubs with flowers that have a glandular calyx and tubular corolla. The seeds therein and the bark are very poisonous and were used in the past to prepare the poison for the arrows. The various types of seeds of strophanthus contain glycosides characterized by an elective action on the heart (strophanthin). The therapeutic indications correspond, in fact, to those of strophanthin, q.v.
f strophantus
d Strophanthus
i strofanto
e estrofanto

3614 **strophanthidin**
Steroid anglycone of the various strophandins from which it is obtained by hydrolysis.
f strophantidine
d Strophanthidin
i strofantidina
e estrofantidina

3615 **strophanthin**
Any of the glycosides which are contained in the seeds of the various species of strophanthus. Strophantins have an action that is analogous to that of digitalis but their absorption and their elimination are faster, with a reduced cumulative effect. They increase the contraction strength of the myocardium and depress in the latter the stimuli-conduction faculty as well as the capacity to promote stimuli in the atrial myocardium thus stimulating its excitability. They are used therapeutically in the treatment of cardio-circulatory decompensation due to acute pulmonary oedema, of cardiac asthma etc.
f strophantine
d Strophanthin
i strofantina
e estrofantina

3616 **strophanthin K**
Cardiac tonic acting more rapidly than digitoxin. It prolongs diastole, strengthens the systole and normalizes the heart rhythm. It has a diuretic action but no cumulative effect. The drug should not be used in the presence of degenerative lesions of the myocardium.
f strophantine K
d K-Strophanthin
i strofantina K
e estrofantina K

3617 **strychnaminoxid**
Strychnine derivative which is less toxic and more lasting than strychnine itself. It has a stimulating action on the bulbar centres and the spinal cord. It also

promotes gastro-intestinal secretions and motility. The drug is chiefly indicated in the treatment of neuro-muscular asthenia, gastro-intestinal atony, spinal and peripheral paresis and paralysis, circulatory collapse and amaurosis. Great caution must be used in the administration of the drug in the presence of cerebral paralysis.
f strychnine oxyde hydrochloride; strychna minoxyde
d Strychnaminoxid
i stricnaminossido
e estricnaminóxido

3618 strychnine
Alkaloid of the various species of Strychnos, in which it is found together with brucine, vomicine etc. It is extract ed from the plants with organic solvents in the presence of lime. It is highly poisonous and is marked by a convulsive action at large doses. Therapeutic doses are used in the treatment of peripheral paralysis. It may be used in granules or galenical preparations. Chloroform and chloral hydrate, as well as bromides are effective antidotes in case of strichnine poisoning.
f strychnine
d Strychnin
i stricnina
e estricnina

3619 Strychnos seed
The ripe dried seed of Strychnos nux vomica. Its principal alkaloids are strychnine and brucine. The strychnine, which constitutes 30 to 35% of the alkaloids, acts as a central nervous system stimulant by acting on the posterior horns of the spinal cord. It is also active on the cerebral cortex. Hypertension, apnoea, convulsions and cardiac disorders may occur as side effects.
f semence de Strychnos
d Strychnossamen
i seme di Strychnos
e semilla de Strychnos

3620 styphnic acid
Explosive yellow astringent acid, less powerful than picric acid to which it is very similar. It is obtained by direct nitration of resorcinol which is sulphonated with concentrated sulphuric acid. Also known as trinitroresorcinol.
f acide styphnique; trinitrorésorcinol
d Trinitroresorcinol
i acido stifnico; trinitroresorcinolo
e ácido estífnico; trinitroresorcinol

3621 styptic cotton
Cotton prepared with a styptic (having an astringent effect) agent and applied to small wounds to stop bleeding.
f coton hémostatique; coton au perchlorure de fer
d Eisenchloridwatte
i cotone emostatico

e algodón hemostático

3622 styramate
Skeletal-muscle relaxant which acts by depressing the transmission of nerve impulses in the spinal cord and brain stem. It is principally indicated to relieve spasms in several skeletal-muscle disorders, including post-traumatic conditions.
f styramate
d Styramat
i stiramato
e estiramato

3623 suberic acid
Octanedioic acid of the oxalic series, produced by the oxidation of cork.
f acide subérique
d Suberinsäure; Korksäure
i acido suberico
e ácido subérico

3624 succinamic acid
Crystalline compound which is the half amide of succinic acid.
f acide succinamique
d Succinamidsäure
i acido succinamico
e ácido succinámico

3625 succinamide
Crystalline compound: the amide of succinic acid, q.v. A phenyl succinamide has proved useful in the treatment of the petit mal.
f succinamide
d Succinamid
i succinammide
e succinamida

3626 succinate
The salt or ester of the succinic acid.
f succinate
d Succinat; bernsteinsaures Salz
i succinato
e succinato

3627 succinic acid
Bicarboxylic acid, found in nature both in vegetable and animal tissues or prepared by catalytic, electrolytic etc. hydrogenation of the maleic anhydride or the fumaric acid. It is widely used as an intermediate in the synthesis of pharmaceutical preparations, e.g. antispasmodic, expectorant and diuretic drugs.
f acide succinique
d Bernsteinsäure
i acido succinico
e ácido succínico

3628 succinimide
Succinic-acid derivative readily reduced to pyrrole. It is frequently used in the manufacture of medicinal drugs. Its chloralimide is used for the sterilization of drinking water. Medicinally, the drug is indicated in the treatment of nephrolithiasis by oxalic acid and hyper-

oxaluria.
f succinimide
d Succinimid
i succinimide
e succinimida

3629 **succinonitrile**
Compound derived from ethylene bromide
and hydrolyzed to succinic acid. Neurotrophic antidepressant agent participating in the ribonucleic metabolism of the
nerve cell and renewing its nucleoproteic
property. It is indicated in the treatment of depression, melancholia, mental
and physical exhaustion and hypochon-
dria. Intravenous injection of the drug
may cause some vasomotor disorders.
f succinonitrile
d Succinonitril
i succinonitrile
e succinonitrilo

3630 **succinylcholine**
Basic compound marked by an action
which is similar to that of curare. It
is a skeletal relaxant and is administer
ed intravenously in surgery mainly in
the form of its crystalline dihydrochloride.
f succinylcholine
d Succinylcholin
i succinilcolina
e succinilcolina

3631 **succinylsalicylic acid**
Chemical compound used as an antipyretic agent.
f acide succinylsalicylique
d Succinylsalicylsäure
i acido succinilsalicilico
e ácido succinilsalicílico

3632 **succinylsulphathiazole**
Sulphanilamide derivative in which both
the amino and amido groups are substituted. The drug is not easily absorbed;
well over 90% is lost in the alimentary
canal and therefore it is principally
used in the treatment of intestinal bacterial infections. Also used in bacillary
dysentery and prior to extensive surgery
of the gastro-intestinal tract to diminish
the risk of peritonitis. Nausea and vomit
ing may occur as a side effect.
f succinylsulfathiazole
d Succinylsulfathiazol
i succinilsolfatiazolo
e succinilsulfatiazolo

3633 **succisulphone**
Sulphamidic antibacterial agent used
topically in the treatment of infections.
f succisulfone
d Succisulfon
i succisolfone
e succisulfone

3634 **sucralfate**
Aluminium-sucrose complex indicated for
the treatment of peptic ulcers. It pro-

tects the gastro-duodenal mucosa by
forming with pepsin a substance which
adheres to still active ulcers. Constipation may set in as a side effect.
f sucralfate
d Sucralfat
i sucralfato
e sucralfato

3635 **sucrose**
Saccharose, cane sugar, beet sugar. Disaccharide occurring in sugar cane,
sugar beet, sugar maple and various
other plants. Colourless, crystalline sub
stance, very easily soluble in water and
yielding a dextrorotatory non-reducing
solution. It can be hydrolyzed by diluted acids and invertase to a laevorotatory mixture of glucose and fructose.
Both the enzymes acting on glucose and
those acting on fructose are active on
sucrose.
f sucrose
d Sucrose
i saccarosio; sucrosa
e sucrosa

* sugar s. sucrose

3636 **sulbenicillin**
Semisynthetic penicillin which is effective against Gram-positive and Gram-negative micro-organisms, including
Proteus and Pseudomonas, and is indicat
ed for the treatment of infections caused
by surgery. It is also used to treat
septicaemia and infections of the respira
tory and urinary tracts. Nausea, some
fever, gastro-intestinal disorders and
allergic reactions may occur as side effects.
f sulbénicilline
d Sulbenicillin
i sulbenicillina
e sulbenicilina

3637 **sulbentine**
Therapeutic agent used in the treatment
of fungal infections of the skin or infections that are caused by a dermatophyte,
e.g. erythrasma, genital mycosis, oto-
mycosis, tinea and anal and genital mycosis. The drug is also used in the ther
apy of mixed infections of the skin.
f sulbentine
d Sulbentin
i sulbentina
e sulbentina

3638 **sulglycotide**
Therapeutic agent capable of reducing
gastric secretion and of soothing inflammation. It is chiefly indicated in the
prophylaxis and the treatment of peptic
and duodenal ulcers. The drug is charac
terized by a physiological mechanism of
action.
f sulglycotide
d Sulglycotid
i sulglicotide

e sulglicotido

3639 sulisatin
Synthetic contact laxative mainly used in the treatment of chronic and acute constipation. It is also used prior to X-ray examination and abdominal surgery. Great caution is necessary in the presence of acute abdominal morbid conditions.
f sulisatine
d Sulisatin
i sulisatina
e sulisatina

3640 sulisobenzone
Sunscreen agent used for protection against ultraviolet radiation. It is mainly used for the prevention of burns and spots caused by prolonged exposure.
f sulisobenzone
d Sulisobenzon
i sulisobenzone
e sulisobenzona

3641 sulmarin
Water-soluble vitamin P factor capable of protecting the blood capillaries by diminishing their permeability and increasing their resistance. The drug is therefore indicated in the treatment of primary and secondary alterations of capillary permeability and for the prevention of haemorrhages due to vasculopathies. Also used in the treatment of serum effusions in the joints due to the impaired capillary permeability.
f sulmarine
d Sulmarin
i sulmarina
e sulmarina

3642 suloctidil
Peripheral vasodilator indicated in the treatment of occlusive arteriopathies, trophic disorders, vasospasms and angio spasms. It has also proved useful in the treatment of disorders which may be ascribed to cerebral arteriosclerosis. Not to be used in pregnancy.
f suloctidile
d Suloctidil
i suloctidile
e suloctidilo

3643 sulodexid
Natural heparin substitute which may be administered intravenously. It has an anti-hyperlipidaemic action and is capable of reducing the level of cholesterol in the blood. The drug inactivates thrombin, thus preventing the conversion of fibrinogen to fibrin. Its activity is powerful and constant. It is therefore indicated in the treatment of hyperlipaemia and vascular, myocardial and cerebral atherosclerosis, by reducing the complement consumption.
f sulodéxide
d Sulodexid

i sulodexide
e sulodexida

3644 sulphacarbamide
Sulphonamide antibacterial agent with specific action on the germs affecting the urinary tract (see also sulphadimidine). The drug is rapidly excreted with urine.
f sulfacarbamide
d Sulfacarbamid
i solfacarbamide
e sulfacarbamido

3645 sulphacetamide
Sulphonamide compound which is very soluble, in particular the sodium salt, i.e. sulphacetamide sodium, which promptly penetrates the cornea and the conjunctiva, as well as the skin. It is chiefly indicated for local application in case of surface lesions of the cornea or conjunctiva. It has also been used in infections of the urinary tract and also in meningitis as it penetrates the cerebrospinal fluid.
f sulfacétamide
d Sulfazetamid
i solfacetammide
e sulfacetamida

3646 sulphachlorpyridazine
Sulpha drug of the pyridazine series which is rapidly absorbed when taken orally but slowly eliminated in urine, thus assuring high and prolonged plasma levels.
f sulfachlorpyridazine
d Sulfachlorpyridazin
i solfaclorpiridazina
e sulfaclorpiridacina

3647 sulphachrysoidine
Antibacterial sulphonamide capable of releasing para-aminobenzenesulphonamide which acts effectively against susceptible micro-organisms. A reddish coloration of the skin may appear as side effect.
f sulfachrysoïdine
d Sulfachrysoidin
i solfacrisoidina
e sulfacrisoidina

3648 sulphacitine
Antibacterial sulphonamide especially effective against infections of the urinary tract due to E.coli bacteria, Klebsiella, Staphylococcus aureus, Proteus vulgaris and mirabilis. The drug is indicated for the treatment of pyelonephritis, pyelitis and cystitis. Gastrointestinal disorders, cutaneous rashes and blood dyscrasia, as well as some depression and toxic neurosis causing oliguria or anuria may occur as side effects.
f sulfacitine
d Sulfazitin
i solfacitina

e sulfacitina

3649 sulphadiazine
Or sulphapyrimidine. Sulphanimide compound characterized by a slow yet complete absorption and equally slow excretion. It is more active than sulphathiazole but less toxic, in fact one of the
least toxic sulphonamides. It is not easi
ly soluble in urine and tends to crystal
lize, thus occasionally causing haematuria and anuria. It promptly penetrates
the cerebrospinal fluid and into the eye
and has proved very useful in the treat
ment of meningitis when administered in
conjunction with sulphathiazole and sulphamerazine.
f sulfadiazine
d Sulfadiazin; Sulfapyrimidin
i solfadiazina; solfapirimidina
e sulfadiacina

3650 sulphadicramide
Antifungal sulphoamide used in the treat
ment of cutaneous infections, in surgery,
gynecology and obstetrics, as well as
in ophthalmology.
f sulfadicramide
d Sulfadikramid
i solfadicramide
e sulfadicramida

3651 sulphadimethoxine
Sulpha drug acting as metabolite on the
germ and indicated in the treatment of
septicaemic and localized infectious diseases due to susceptible organisms, in
particular cocci. The action is prompt
and prolonged. Nausea and anaemia may
occur as side effects.
f sulfadimethoxine
d Sulfadimethoxin
i solfadimetossina
e sulfadimetoxina

3652 sulphadimidine
Dimethyl derivative of sulphadiazine (q.
v.) marked by a much higher degree of
solubility. The absorption is thus more
rapid but the urinary excretion equally
slow. It is frequently used for its characteristic of combining an excellent
therapeutic activity with low toxicity
and an almost total absence of adverse
effects. The drug has proved very useful
to combat pneumococcal infections. Owing
to its rather slow diffusion into the
cerebrospinal fluid, it is not indicated
for the treatment of meningitis. However,
being very easily soluble in urine and
because of its high concentration, it is
indicated for the treatment of urinary
infections. Also known as sulphametha-
zine.
f sulfaméthazine
d Sulfadimidin
i solfametazina
e sulfadimidina

3653 sulphadimidine sodium

Water-soluble derivative of sulphadimidine administered by injection as a 33%
solution when oral administration of
sulphadimidine fails to produce a sufficiently high blood level.
f sulfadimidine sodium
d Natriumsulfadimidin
i solfadimidina sodica
e sulfadimidina sódica

3654 sulphadoxine
Antibacterial sulphanamide marked by
a prolonged action. The drug, which is
rapidly absorbed, has, even with one
administration, a concentration lasting
almost a week, i.e. up to 200 hours half
-life. It acts against Gram-positive and
Gram-negative micro-organisms which are
sensitive to sulpha drugs, M. leprae
and certain Mycetes. It is mainly indicated in the treatment of respiratory,
gastro-intestinal and genito-urinary infections. It is also used in cases of
spinal meningitis and some diseases caus
ed by a fungus.
f sulfadoxine
d Sulfadoxin
i solfadossina
e sulfadoxina

3655 sulpha drugs
Sulphanilamide and its derivatives used
widely in the chemotherapy of bacterial
infections. Although they are effective
against Gram-positive cocci and, to some
degree, against staphylococci, their
greatest value is shown against Gram-
-negative meningococci and gonococci.
f médicaments sulfamidiques
d Sulfonamidpräparate
i farmaci sulfamidici
e medicamentos sulfamídicos

3656 sulphaethidole
Antimicrobial agent of the sulphamide
group, rapidly absorbed and thus rapidly reaching high plasma levels. It is
mainly indicated for the treatment of
infections of the respiratory tract and
in the prophylaxis of tonsillitis, otitis,
pharyngitis etc.
f sulfaéthidole
d Sulfäthidol
i solfetidolo
e sulfaetidolo

3657 sulphafurazole
Highly soluble sulpha drug particularly
effective against infections of the urinary tract. Its absorption is very rapid
and plasma as well as urinary concentration is high. Gram-positive and Gram-
-negative, especially Escherichia coli
and Proteus vulgaris, are immediately
sensitive.
f sulfafurazole
d Sulfafurazol
i solfafurazolo
e sulfafurazolo

3658 sulphaguanidine
Chemical compound of the sulphamidic group, a derivative of sulphanilamide and guanidine. It crystallizes with a molecule of water. It is indicated in the treatment of intestinal infections caused by bacteria. One of the least potent antibiotics as its absorption, though rapid, is incomplete.
f sulfaguanidine
d Sulfaguanidin
i solfaguanidina
e sulfaguanidina

3659 sulphaguanole
Sulpha drug marked by bacteriostatic action against intestinal pathogenic germs, though Streptococcus faecalis, Streptococcus pyogenes and Salmonella are excepted. It is indicated for the treatment of intestinal infections caused by susceptible micro-organisms, of diarrhoea, gastroenteritis and dysentery.
f sulfaguanole
d Sulfaguanol
i solfaguanolo
e sulfaguanolo

3660 sulphalene
Sulphamethoxypyrazine. Prolonged-action sulphonamide active against Gram-positive and Gram-negative bacteria, Plasmodium falciparum and Toxoplasma gondii. It is indicated in the treatment of infections of the respiratory and urinary tracts, in malaria, leprosy, cholera and in the prophylaxis of meningitis. Nausea, photosensitivity, blood dyscrasia and liver toxicity may occur as side effects.
f sulfalène
d Sulfalen
i solfametopirazina
e sulfametopiracina

3661 sulphamerazine
Sulphamethyldiazine. Sulfanilamide- and methyldiazine-derived sulpha drug. It is effective against infections caused by susceptible Gram-positive and Gram-negative bacteria. Nausea and epigastric pains, as well as agranulocytosis, may occur in patients suffering from hypersensitivity.
f sulfamérazine
d Sulfamerazin
i solfamerazina
e sulfameracina

3662 sulphamethizole
Antibacterial sulphonamide, effective both against Gram-positive and Gram-negative micro-organisms. It is principally indicated for the treatment of infections of the urinary tract.
f sulfaméthizole
d Sulfamethizol
i solfametizolo
e sulfametizolo

3663 sulphamethomidine
Sulpha drug marked by a prolonged action on Gram-positive and Gram-negative micro-organisms and mainly indicated in the treatment of infections of the respiratory and urinary tracts. Also used against infections affecting ear, nose and throat, as well as in the course of surgical operations. Nausea, vomiting, some fever and allergic reactions may occur as side effects.
f sulfaméthomidine
d Sulfamethomidin
i solfametomidina
e sulfametomidina

3664 sulphamethoxazole
Sulpha drug characterized by a specific affinity for kidneys and very active against colibacilli, Proteus and Bacillus pyocyaneous. It is principally indicated for the treatment of infections of the urinary tract.
f sulfaméthoxazole
d Sulphamethoxazol
i solfametossazolo
e sulfametoxazolo

3665 sulphamethoxydiazine
Sulphonamide of the methoxy pyrimidine series marked by a rapid absorption and a slow excretion in the urine which account for the high plasma and tissue levels reached with one or two daily doses. The drug is indicated in the treatment of infections due to susceptible organisms. Allergic dermatoses may occur in very sensitive patients.
f sulfaméthoxydiazine
d Sulfamethoxydiazin
i solfametossidiazina
e sulfametoxidiacina

3666 sulphamethoxypyridazine
Sulphonamide drug marked by a rapid and prolonged-effect absorption. It is active against all Gram-positive and Gram-negative micro-organisms which are susceptible to sulphamidic agents. Also used in the treatment of epidemic meningitis. Some anaemia may occur as side effect.
f sulfaméthoxypyridazine
d Sulfamethoxypyridazin
i solfametossipiridazina
e sulfametoxipiridacina

3667 sulphamethylthiazole
Sulpha drug as active as sulphathiazole (q.v.). Its use has been discontinued owing to its tendency to cause peripheral neuritis.
f sulfaméthyldiazole
d Sulphamethyldiazol
i solfametildiazolo; metilsolfadiazolo
e sulfametildiazolo

* sulphamide s. sulphanilamide

3668 sulphamonomethoxine

Antimalarial sulpha drug marked by
long-lasting effects, especially on Plasmodium falciparum. It is mainly used to
combat chloroquinine-resistant malaria.
f sulfamonométhoxine
d Sulfamonomethoxin
i solfamonometossina
e sulfamonometoxina

3669 sulphamoxole
Sulphonamide chemotherapeutic agent
marked by rapid absorption and prolong
ed high plasma and tissue levels. It is
generally used to treat infections due
to susceptible micro-organisms.
f sulfamoxole
d Sulfamoxol
i solfamossolo
e sulfamoxolo

3670 sulphanilamide
The parent compound of most sulpha
drugs. It is the first of the simple sulphonamides used in medicine and the
first systemic bactericide. Being highly
soluble, it is rapidly and totally absorb
ed and promptly penetrates the tissues
including the cerebrospinal fluid. Its
therapeutic properties are similar to
those of the sulphonamides, but its use
is now limited against streptococci.
f sulfanilamide
d Sulfanilamid; p-Aminobenzolsulfonamid
i solfanilamide
e sulfanilamida

3671 sulphanilic acid
Parasulphonic acid prepared from aniline
by substitution of one or more hydrogen
atoms of the benzenic ring with $-SO_3$ H
groups. It is used in the manufacture of
organic chemicals.
f acide sulfanilique
d Sulfanilsäure
i acido solfanilico; acido amminobenzensol-
fonico
e ácido sulfanílico

3672 sulphaperin
Sulpha drug of the pyrimidine group,
characterized by a rapid intestinal absorption and a slow excretion in the
urine. The blood levels attained persist
for a fairly long time. It is principally
used in the treatment of septicaemic
infections or localized infections caused
by susceptible micro-organisms. Some
toxic reactions have been observed as
side effects in highly sensitive patients.
f sulfapérine
d Sulfaperin
i solfaperina
e sulfaperina

3673 sulphaphenazole
Prolonged-action therapeutic agent of the
sulphonamide series used in the treatment of all infectious diseases caused
by germs which are sensitive to sulpha
drugs, i.e. septicaemic infections and

infections localized in the respiratory
and urinary tracts as well as in the
skin.
f sulfaphénazole
d Sulfaphenazol
i solfafenazolo
e sulfafenazolo

3674 sulphapyridine
Slow-solubility sulphonamide compound
marked by a slow and varying absorption. Though of moderate potency and
low renal clearance, its toxicity is considerable and its tendency to crystallize
in the urine rather strong. The drug is
principally used in the treatment of
infections caused by streptococci,
pneumococci, Salmonella, Proteus vulgaris and Bacterium coli. Nausea, some
fever and drowsiness may occur as side
effects.
f sulfapyridine
d Sulfapyridin
i solfapiridina
e sulfapiridina

3675 sulpharsphenamine
Compound which, like neoarsphenamine,
is used in the treatment of syphilis. It
is less toxic and less irritant than neoarsphenamine and may be given by intra
muscular injection which makes it easier
to administer to children and adults
with inadequate veins. The drug has
also proved useful in the treatment of
trypanosomiasis. Some allergic reactions
and a tendency to drug-dependence have
been observed as side effects.
f sulfarsphénamine
d Sulfarsphenamin
i solfarsfenamina
e sulfarsfenamina

3676 sulphasomidine
Chemical compound acting as a sulphona
mide drug and very effective in gonococcal, meningococcal and pneumococcal
infections. It is also used in the treatment of rheumatic fever, as well as kidney infections, as it does not damage
the kidney. Its degree of toxicity is
very low.
f sulfasomidine
d Sulfasomidin
i solfasomidina
e sulfasomidina

* **sulphasuxidine s. succinylsulphathiazole**

3677 sulphate
Salt or ester of sulphuric acid.
f sulfate
d Sulfat; schwefelsaures Salz
i solfato
e sulfato

3678 sulphathiazole
The most effective of the sulphonamides.
Its absorption is very rapid and ac-
counts for its quick action. It is also

rapidly excreted so that blood concentration is difficult to control but generally
low, as are the concentrations in the
tissues. Sulphathiazole is eliminated in
the urine in the acetylated form which
is highly insoluble with risk of haematuria and oliguria. Fever, rashes and
the risk of sensitization dermatitis may
easily occur. Owing to all this, the
drug is now only used in combination
with sulphanilamide and penicillin for
application to wounds.
f sulfathiazole
d Sulfathiazol
i solfatiazolo
e sulfatiazolo

3679 sulphatolamide
Sulphathiourea salt of mafenide used
chiefly in the treatment of leucorrhoea.
f sulfatolamide
d Sulfatolamid
i solfatolamide
e sulfatolamida

3680 sulphenazone
Combination of a pyrazole radical with
a long-acting sulphonamide. It is an
antiinflammatory, antipyretic and antibacterial agent principally used in the
treatment of acute respiratory diseases,
influenza and other viral infections of
the upper respiratory tract. Anaemia
and some allergic reactions have been
observed as side effects.
f sulfénazone
d Sulfenazon
i solfenazone
e sulfenazona

3681 sulphide
Term denoting any compound of sulphur
with an element or base. Some sulphides
have a surface-active and detergent
action, others a bactericide action.
f sulfure
d Sulfid
i solfuro
e sulfuro

3682 sulphinpyrazone
Pyrazolone derivative capable of prolonging in man the surviving time of thrombocytes and of inhibiting their aggregation. It reduces platelet stickiness and
is therefore indicated for the prevention
of thrombosis in the coronary and cerebral circulation. Also used for a prophylactic treatment of gout, though of
no value in the treatment of the acute
form of this condition. The drug promotes the renal excretion of urates by
partly inhibiting reabsorption in the
renal tubules. It lowers uric acid levels
and gradually depletes urate deposits
in the tissues. A long-term use of the
drug may suppress bone marrow activity.
Caution is necessary with patients with
sensitivity to phenylbutazone and other
pyrazoles.

f sulfinpyrazone
d Sulfinpyrazon
i solfinpirazone
e sulfinpirazona

3683 sulphite
The salt of the sulphurous acid marked
by antifermentative action. Also, the
salt of the sulphurous acid. Some of the
latter have a germicide action.
f sulfite
d Sulfit
i solfito
e sulfito

3684 sulphoguaiacol
Potassium guaiacolsulphonate. Guaiacol-
-related expectorant and antiseptic agent
used to relieve cough and facilitate
bronchial drainage in acute and chronic
bronchitis independently of their origin.
The drug should not be used in the
presence of pneumonia. Some gastro-intestinal inflammation has been observed as
side effect.
f sulfoguaiacol
d Sulfoguaiacol
i solfoguaiacolo
e sulfoguaiacolo

3685 sulphonate
Relating to a compound containing the
sulphonic group, e.g. sulphonate amide,
sulphonate ester etc. Also, the salt of
a sulphonic acid.
f sulfonate
d Sulfonat
i solfonato
e sulfonato

3686 sulphonethylmethane
Sulphonal-related hypnotic drug. Like
other preparations of this group, its use
has been discontinued and replaced by
barbiturates.
f sulfonéthylméthane
d Sulfonethylmethan
i solfonetilmetano
e sulfonetilmetano

3687 sulphoniazide
Tetrahydrate sodium salt. Antituberculous therapeutic agent marked by a
specific bacteriostatic and bactericide
action as well as a eutrophic, appetite-
-stimulating, antiphlogistic and exudation-inhibiting action. The drug is indicated in the treatment of tuberculous
infections, paraspecific pathological conditions, e.g. erythema induratum, and
certain diatheses, e.g. a diathese of a
limphatic or exudative nature.
f sulfoniazide
d Sulfoniazid
i solfoniazida
e sulfoniazida

3688 sulphoridazine
Phenothiazine-related neuroleptic agent
principally used in states of restlessness,

psychomotor excitation, involutionary psy
choses, schizophrenia and depression
accompanied by anxiety. Drowsiness, dry
ness of the mouth, tachycardia, hypotension, blood dyscrasia and, though rarely, epileptoid crises may occur as side
effects.
f sulforidazine
d Sulforidazin
i solforidazina
e sulforidacina

3689 sulphur
Element which exists in several allotropic forms. It is found free in Sicily
(Italy) and the USA. Sulphur is mainly
used for the manufacture of sulphuric
acid. In medicine, it is used internally
as a mild intestinal antiseptic and stimu
lant; externally as an antiseptic and
parasiticide.
f soufre
d Schwefel
i zolfo; solfo
e azufre

3690 sulphur dioxide
Sulphurous anhydride. Gas prepared by
the combustion of sulphur or by the reac
tion of copper and sulphuric acid. It is
used as a preservative for foodstuffs
and some medicinal products.
f bioxyde de soufre; anhydride sulfureux
d Schwefeldioxid
i biossido di zolfo; anidride solforosa
e bióxido de azufre; anhidrido sulfuroso

3691 sulphur flowers
Sublimed sulphur. Commercial purified
sulphur in fine powder form, obtained
by sublimation.
f fleurs de soufre
d Schwefelblüte
i fiori di zolfo
e flores de azufre

3692 sulphuric acid
Highly corrosive acid widely used in
industry and in chemical manufacture.
Dilute sulphuric acid is indicated as an
astringent and antiseptic in diarrhoea.
Also used in staphylococcal infections.
f acide sulfurique
d Schwefelsäure
i acido solforico
e ácido sulfúrico

3693 sulphur iodide
Compound of sulphur and iodine indicated in the treatment of parasitic skin
diseases.
f iodure de soufre
d Jodschwefel
i ioduro di zolfo
e yoduro de azufre

3694 sulphur milk
Milk of sulphur, precipitated sulphur.
Impalpable yellowish powder prepared
by the precipitation of sulphur from

solution. It is mainly used in the treatment of skin diseases.
f soufre précipité; magistère de soufre
d Schwefelmilch; gefällter Schwefel
i zolfo precipitato; latte verginale
e azufre precipitado; leche de azufre

3695 sulphurous acid
Unstable weak dibasic acid unknown in
the free state but known in solutions of
sulphur dioxide in water and in the
form of salts. It is a good reducing
agent.
f acide sulfureux
d schweflige Säure
i acido solforoso
e ácido sulfuroso

3696 sulphur trioxide
Sulphuric anhydride. White crystalline
solid forming sulphuric acid with water.
f anhydride sulfurique
d Schwefelsäureanhydrid; Schwefeltrioxyd
i anidride solforica
e anhidrido sulfúrico

3697 sulpiride
Antipsychotic agent without sedative
action. The drug causes selective depres
sion of the brain structures which are
responsible for the control of behaviour.
It has anticholinergic alpha-adrenergic
blocking and dopaminergic effects together with other pharmacological effects.
It is used in several psychotic disorders,
in particular schizophrenia and agitated
depression. Also indicated in terminal
illnesses to enhance anaesthesia and to
control nausea and hiccups. Dry mouth,
blurred vision, photosensitivity and
cholestatic jaundice may occur as side
effects.
f sulpiride
d Sulpirid
i sulpiride
e sulpirida

3698 sulthiame
Anticonvulsant agent of the sulphonamidic series, a carbonic anhydrase inhibitor used for the prevention of epilepsy,
generally in association with other
drugs. It does not cause drowsiness, yet
mental confusion, psychotic reactions,
headache, nausea, insomnia and leucopenia may occur as side effects.
f sulthiame
d Sulthiam
i sultiame
e sultiamo

3699 sultroponium
Parasympatholytic agent marked by properties that are similar to those of atropine and indicated in the treatment of
spasms of the smooth muscles, particular
ly in morbid conditions due to oesophageal disorders, hiatal hernia, hiccup,
violent vomiting and gastroduodenal
ulcer. The drug has also proved useful

for endoscopies, pre- and post-operative treatment in case of abdominal surgical operations, kidney and biliary colic, spastic colitis, dysmenorrhoea, all pains caused by cancerous growths and after-pains. Caution is required in the presence of glaucoma and prostatic adenoma.

f sultroponium
d Sultroponium
i sultroponio
e sultroponio

3700 suppository

Generally small medicament, shaped like an elongated cone, having gelatin or cocoa butter as basis and usually introduced into the rectum for the treatment of various disorders. The basis itself is sufficiently solid to be handled at room temperature, but dissolves rather rapidly at body temperature.

f suppositoire
d Suppositorium; Darmzäpfchen
i supposta
e supositorio

3701 suramin

Suramin sodium. Antiworm agent active on African trypanosomiasis, particularly in its initial stages, and leishmaniasis. Block of the reticulo-endothelial system and impairment of kidney function may occur as side effects.

f suramine; suramine sodique
d Suramin; Natriumsuramin
i suramina; suramina sodica
e suramina sódica

3702 suxamethonium

Succinylcholine. The bischoline ester of succinic acid used in anaesthesia in the form of iodide, bromide or chloride. It is a muscle relaxant and acts by depolarization of the muscle end plate rendering the tissue totally unresponding to the neurotransmitter. It is used as an adjuvant to anaesthesia in surgery and, although short-acting, displays effects which are prolonged in patients with reduced psychocholinesterase levels.

f suxaméthonium
d Suxamethonium
i sussametonio
e suxametonio

3703 suxamethonium chloride

See suxamethonium.

f chlorure de suxaméthonium
d Suxamethoniumchlorid
i succinilcolina cloruro
e cloruro de suxametonio

3704 suxamethonium iodide

Synthetic quaternary ammonium compound marked by a neuromuscular blocking action similar to decamethonium but of short duration. Although the compound is chemically related to acetylcholine it has no muscarinic effect as the latter and does not cause bradycardia. In cases of high doses, the nicotinic effect on blood pressure is clearly shown. Repeated injections or constant infusion are practiced for long operations. A narcotic always precedes the administration of suxamethonium iodide and artificial respiration proves generally necessary during the paralysis.

f iodure de suxaméthonium
d Suxamethoniumjodid
i ioduro di sussametonio; sussametonio ioduro
e yoduro de suxametonio

3705 suxamidophylline

Theophylline derivative, highly soluble and capable of bringing together the eupnoic, cardiotonic and diuretic properties of theophylline and the analeptic and cardio-respiratory ones of the diethylaminic and succinic radicals. It is principally indicated for the symptomatic relief of acute and chronic asthma, paroxysmal dyspnoea accompanied by congestive heart failure and decompensation as well as vascular spasms. The drug is also used to check the danger of asphyxia and to treat infectious diseases. It is also employed in preoperative and post-operative treatment.

f suxamidophylline
d Suxamidophyllin
i sussamidofillina
e suxamidofilina

3706 syrosingopine

Semisynthetic reserpine derivative indicated in the treatment of hypertension. The drug is often used in association with diuretics. Some gastro-intestinal disorders may occur as side effects.

f syrosingopine
d Syrosingopin
i sirosingopina
e sirosingopina

3707 syrup

Liquid medicament for oral administration, usually containing a rather large percentage of sugar either to flavour it or to preserve it.

f sirop
d Sirup
i sciroppo
e jarabe

T

3708 tachysterol

One of the organic substances which are formed through the effects of ultraviolet irradiation of ergosterol which is transformed into lumisterol, subsequently in tachysterol and finally into vitamin D_2. Tachysterol, though displaying structural affinity with vitamin D_2, does not share its antirachitic action and, in fact, has a decalcifying action on the skeleton. It is therefore not used for therapeutic purposes, but its hydroderivative, marked by a potent and lasting hypercalcemia-promoting action is widely used in order to substitute parathormone in hypoparathyroid syndromes.

f tachystérol
d Tachysterol
i tachisterolo
e taquisterol

3709 tacrine

Cholinesterase inhibitor used as antagonist in the decurarization by non-depolarizing muscle relaxants, so as to prolong the myorelaxant action of succinylcholine. It is also used as a heart and respiratory stimulant.

f tacrine
d Tacrin
i tacrina
e tacrina

3710 talampicillin

Antimicrobial ester of ampicillin, q.v. to which it is rapidly metabolized after being absorbed. Its uses and effects are similar to those of ampicillin.

f talampicilline
d Talampizillin
i talampicillina
e talampicilina

3711 talastine

Benzylphthalazone. Antiallergic agent used in the treatment of pruritus, urticaria, exanthema, allergic eczema and allergic asthma. As the drug causes some drowsiness, caution is necessary when driving.

f talastine
d Talastin
i talastina
e talastina

3712 talbutal

Hypnotic and sedative agent indicated in the treatment of insomnia. The drug should be administered with caution to patients who are hypersensitive to barbiturates. Drowsiness, skin rashes and some excitement in elderly patients are among the most frequent side effects.

f talbutal
d Talbutal
i talbutale
e talbutalo

3713 talcum

Or talc. Mineral consisting of a basic magnesium tetrasilicate which is usually whitish, grey or greenish with a soapy feel. It acts as an adsorbent, water repellent and skin-protective agent. Mainly used to prevent inflammation and, in pharmaceutical technique, as a lubricant agent in the production of tablets and in the clarification of liquids.

f talc
d Talk
i talco
e talco

3714 tamarind

The fruit of the Tamarindus indica without its outer pericard and preserved by being mixed with sugar. The resulting mass, which contains tartaric acid, cream of tartar, glucose and fructose, as well as the sugar in which it must be preserved, is a cooling mild laxative and an ingredient of the confection of senna.

f tamarin
d Tamarinde
i tamarindo
e tamarindo

3715 tamoxifen

Antioestrogen which competes with oestrogen for tissue receptor sites. It is indicated for the palliative treatment of mammary carcinoma, as well of infertility caused by failure of ovulation. Some gastro-intestinal disorders and fluid retention may occur as side effects.

f tamoxifène
d Tamoxifen
i tamossifene
e tamoxifeno

3716 tannate

The salt of the tannic acid or its ester.

f tannate
d Tannat
i tannato
e tanato

3717 tannic acid
Colourless to yellowish powder of an astringent taste, soluble in water, alcohol but not in ether. Also soluble in alkaline solutions which on exposure to air take up a brownish tint. In medicine, the acid is employed as an astringent and coagulant, in throat diseases to check superficial bleeding and to provide a coagulum in the treatment of burns. When used as an astringent internally, tannic acid is generally associated with opium in diarrhoeas. It is also used as an antidote of the alkaloids and of some metallic salts with which insoluble tannates are formed.
f acide tannique
d Tanninsäure
i acido tannico
e ácido tánico

3718 tansy
Perennial herb containing an essential oil and used as anthelminthic for cattle and as emmenagogue. Also used as a tonic.
f tanaisie
d Wurmkraut; Rainfarn
i tanaceto
e tanaceto; hierba lombriguera

3719 tansy oil
Essential oil obtained by distillation of tansy flowers; a yellowish liquid turning to brown on exposure, marked by an unpleasant taste and an unpleasant odour, containing tanacetone, borneol and terpenes. It displays an epilectogen and tetanizing action.
f essence de tanaisie
d Rainfarnöl
i olio essenziale di tanaceto
e esencia de tanaceto

3720 tar ointment
Brownish ointment obtained by the redistillation of certain wood tars. It is mainly used for veterinary purposes.
f pommade au goudron
d Teersalbe
i unguento di catrame
e pomada de brea

3721 tartar
The mass of calcium and magnesium salts which forms a deposit around the teeth and on artificial dentures.
f tartre
d Zahnstein
i tartaro
e tártaro

3722 tartaric acid
The dextrorotatory acid that is widely distributed in plants, especially in fruits, either free or combined as salts. It is usually obtained from tartar and is frequently used in pharmaceutical preparations. Also the laevorotatory acid that is obtained by resolving the racemic acid.
f acide tartrique
d Weinsäure
i acido tartarico
e ácido tartárico

3723 tartrate
Salt or ester of the tartaric acid.
f tartrate
d Tartrat
i tartrato
e tartrato

3724 tartrazine
Orange-coloured dye mainly used to give colour to medicinal preparations. Some cross-sensitivity with acetylsalicylic acid is sometimes shown.
f tartrazine
d Tartrazin
i tartrazina
e tartracina

3725 taurine
Aminoethansulphonic acid. Colourless, water-soluble crystals formed by hydrolysis of the taurocholic acid (one of the bile acids). It is capable of activating tissular nutrition as it promotes the exchange of oxygen between blood and tissues. It is thus indicated in the treatment of dystrophic diseases or disorders caused by inadequate oxygenation, e.g. acute and chronic coronary arteriopathies, angina pectoris, peripheral arteriopathies, cerebral arteriopathies and, in some cases, epilepsy.
f taurine
d Taurin
i taurina
e taurina

3726 taurocholate
Generic term denoting the salts of the taurobiliary acids which are found in the bile usually salified with sodium.
f taurocholate
d Taurocholat
i taurocolato
e taurocolato

3727 taurocholic acid
Chemical compound resulting from the union of a molecule of taurine (aminoethansulphonic acid) and one of cholic acid. It is found in the bile. Its sodium salt represents the form in which normally, together with the combination with glycocholic acid, cholic acid is present in the bile. It is one of the sources of the neutral urinary sulphur.
f acide taurocholique
d Taurocholsäure
i acido taurocolico
e ácido taurocólico

3728 tea-seed oil
Olive-oil resembling fatty oil, obtained from the seeds of the sasanqua and main ly used as edible oil and for some medicinal preparations.
f huile de camellia; huile de thé
d Teesamenöl
i olio di semi di the
e aceite de té

3729 teclothiazide
Benzothiadiazine-related diuretic agent capable of augmenting the urinary excretion of water, sodium and chloride, yet without modifying kalaemia. It is mostly indicated in the treatment of patients suffering from oedemata of various origin, e.g. cardiac, hepatic and renal. Also useful in the treatment of obesity, premenstrual syndrome and some forms of hypertension. Gastro-intestinal disorders may occur as side effect.
f téclothiazide
d Teclothiazid
i teclotiazide
e teclotiazida

3730 temazepam
Benzodiazepinic neurotonic capable of restoring the psychosomatic balance and thus indicated in the treatment of states of anxiety and emotive disorders characterized by cardiac, circulatory and intestinal anomalous conditions. The drug must be avoided during the first three months of pregnancy.
f témazépam
d Temazepam
i temazepam
e temazepam

3731 teniposide
Cytostatic agent used in the treatment of reticulosarcoma, lymphosarcoma, malignant cerebral tumours, vesical and mammary tumours and tumoral serous shedding. Leukopenia, thrombocytopenia, alopecia, nausea, vomiting and other gastro-intestinal disorders, possible collapse after the first administration and sometimes collapse in the course of treatment may occur as side effects.
f téniposide
d Teniposid
i teniposide
e teniposido

3732 tephrosin
Chemical compound obtained from Tephrosia vogelii, having insecticide properties.
f téphrosine
d Tephrosin
i tefrosina
e tefrosina

3733 terbutaline
Beta-adrenoreceptor agonist. Bronchodilator with a negligible action on the heart and blood pressure. It is indicated in the treatment of bronchial asthma

and of reversible bronchospasm occurring in some cases of bronchitis and emphysema. A mild tachycardia and some tremors may be observed as side effects.
f terbutaline
d Terbutalin
i terbutalina
e terbutalina

3734 terebene
Oil marked by a pleasant smell and used to offset unpleasant odours or tastes. Also used as a vapour to relieve nasal congestion. Gastro-intestinal irritation may occur with large doses.
f térébène
d Tereben
i terebene
e terebeno

3735 terfenadine
Antihistamine marked by action and effects similar to those of promethazine, q.v. It appears to produce a lower degree of sedation. It is mainly used in the treatment of hay fever and some allergic cutaneous disorders.
f terfénadine
d Terfenadin
i terfenadina
e terfenadina

3736 terizidone
A minor chemotherapeutic antituberculous agent marked by bacteriostatic action which may be potentiated if associated with other antituberculous drugs. In this case the emergence of resistant strains is delayed.
f térizidone
d Terizidon
i terizidone
e terizidona

3737 terodiline
Therapeutic agent capable of relaxing smooth muscles through various mechanisms, in particular anticholinergic and calcium antagonist action. It is especially indicated to reduce the bladder tone in case of excessive urinary frequency and incontinence. Tachycardia, vision disorders and constipation may occur as side effects.
f térodiline
d Terodilin
i terodilina
e terodilina

3738 terpin
Diterpene alcohol obtained from turpentine oil and occurring in two stereoisomeric forms (cis- and trans-). The cis-compound is a solid which promptly absorbs water to form terpin hydrate, used as an expectorant and disinfectant. It acts on the non-specific flora of the bronchi and the lungs.
f terpine
d Terpin

i terpina
e terpina

3739 terpineol
Mixture of isomeric alcohols obtained
from terpin. The alpha form is found in
nature and is marked by lilac-like
odour. It is an aromatic solvent antiseptic.
f terpinéol
d Terpineol
i terpineolo
e terpineol

3740 testis extractum
Or orchitic extract. Preparations (powders, total extracts, enzymatic lysates)
obtained from the testicular tissue of
healthy animals, which contain the
complex of active principles produced by
the gland itself. It is an opotherapeutic
drug capable of stimulating selectively
the function of the testes and of regulat
ing those phenomena that are directly
related to the male genital system. It
is thus indicated in the treatment of
male hypogonadism, precocious senescence
and puberty disorders. Also useful for
the treatment of general debility, mental
and physical stress-related fatigue, sexual psychasthenia and hypertrophy of
the prostate.
f extrait orchidique; extrait total de
glande orchidique
d Testisextrakt
i estratto di testicolo
e extracto de testículo

3741 testolactone
Testosterone derivative without androgenic activity and indicated for the palliative treatment of advanced or disseminat
ed mammary cancer in case a hormonal
therapy is advisable. Some pains and
oedemata of the extremities, nausea,
vomiting may occur as side effects.
f testolactone
d Testolacton
i testolattone
e testolactona

3742 testosterone
The most potent natural androgenic hormone, secreted by the testicles. Anabolic
steroid. It is indicated in the treatment
of testicular insufficiency, the climacterium, hypogonadism, menorrhagia, dysmenorrhoea, chronic mastitis and gynecomastia. It is also used for the palliative treatment of malignant tumours of
the breast. Small doses of the drug may
counteract menopausal symptoms.
f testostérone
d Testosteron
i testosterone
e testosterona

3743 tetrabenazine
Non-phenothiazine tranquillizer marked
by a central mechanism of action. It is
principally used to suppress abnormal
movements (e.g. Huntington's chorea)
and, more generally, as an antipsychotic agent in the treatment of psychoneurotic states characterized by manic conditions and hallucinations, as well as
of disorders of the extrapyramidal system.
f tétrabénazine
d Tetrabenazin
i tetrabenazina
e tetrabenacina

3744 tetracaine
Powerful ester-type local anaesthetic
with greater potency and toxicity than
lidocaine (q.v.). The drug displays a
vasodilating action and is mainly used
as a surface anaesthetic and as topical
anaesthetic agent in ear, throat and
nose disorders, particularly before pro-
ceeding to an endoscopy, in the urethra,
prior to cytoscopy, in ophthalmology, in
spinal, caudal and infiltration anaesthesia. It is up to 20 times more potent
than cocaine.
f tétracaïne
d Tetracain
i tetracaina
e tetracaina

3745 tetrachloroethylene
Colourless mobile liquid insoluble in
water and miscible with most organic sol
vents. It is an anthelmintic used in the
treatment of worm infections, especially
hookworm in man and roundworm and
hookworm in animals. Central nervous
system depression, nausea, drowsiness,
respiratory depression, sometimes coma,
cyanosis and jaundice may occur as side
effects.
f tétrachloroéthylène
d Tetrachloroäthylen
i tetracloroetilene
e tetracloroetileno

3746 tetracosactrin
Synthetic corticotrophin whose action,
uses and side effects are similar to
those of corticotrophin, q.v. The drug
may be used intravenously as a test of
adrenal function or by depot injection
for the treatment of inflammatory or
degenerative disorders.
f tétracosactrine
d Tetracosactrin
i tetracosattrina
e tetracosactrina

3747 tetracycline
Bacteriostatic antibiotic effective against
several Gram-positive and Gram-negative
micro-organisms, some viruses and
chlamidia, Protozoa etc. Particular care
should be taken when the drug is admin
istered to pregnant women and young
children. Yellowish discoloration of the
teeth, diarrhoea, Candida bowel super-
infection, possible formation of stable

calcium complex which may interfere with bone growth in children and exacerbation of renal failure may occur as adverse effects.
f tétracycline
d Tetracyclin
i tetraciclina
e tetraciclina

3748 tetramethylammonium formate
Strychnine-like neurotonic acting on the heart and capable of improving metabolic tone. It is chiefly indicated in the treatment of organic debility, convalescence, senility, neurasthenia and toxic/infective neuritis.
f formiate de tétraméthylammonium
d Tetramethylammoniumformiat
i formiato di tetramletilammonio
e formiato de tetrametilamonio

3749 tetramethylammonium hydroxide
Compound acting on the sympathetic and parasympathetic nervous system causing, as it occurs with nicotine, first stimulation and subsequently paralysis. The drug provokes a peripheral muscarine--like inhibition of the heart causing a fall of blood pressure and then a nicotine-like stimulation of sympathetic ganglion cells with consequent rise in blood pressure. Finally there occurs a nicotine-like paralysis of ganglion cells and a consequent fall in blood pressure. The various actions being so complex, the tetraethylammonium compound is preferred for blocking ganglia.
f hydroxyde de tétraméthylammonium
d Tetramethylammoniumhydroxid
i idrossido di tetrametilammonio
e hidróxido de tetrametilamonio

3750 tetrammonium iodide
Therapeutic agent capable of modifying the thyrohyoid function and indicated in the treatment of arthritis, sciatica, lumbago, epigastric pains and visceral spasms, cholecystic and vesical atony. Vertigo and vision disorders may occur as side effects.
f iodure de tétrammonium
d Tetrammoniumjodid
i ioduro di tetrammonio
e yoduro de tetramonio

3751 tetranicotinoylfructose
Peripheral vasodilator marked by a direct relaxant effect on muscles in peripheral blood vessels and indicated in the treatment of reduced peripheral circulation. Flushing, rashes and excessively copious menstruations may occur as side effects.
f tétranicotinoylfructose
d Tetranicotinoylfruktose
i tetranicotinoilfruttosio
e tetranicotinoilfructosa

3752 tetrazepam
Antispasmodic, analgesic and skeletal-

-muscle relaxant mainly indicated in the treatment of muscular spasms in the hemiplegias of cortical, bulbar, medullary and vascular origin. The drug is also used in spastic paraplegias caused by medullary compression and multiple sclerosis, in infantile spastic paralysis, in sciatica, lumbago, arthritis etc. General tiredness, drowsiness and vertigo may occur as side effects.
f tétrazépam
d Tetrazepam
i tetrazepam
e tetrazepam

3753 tetrylammonium bromide
Ganglionic blocking agent mainly used as a vasodilator and a hypotensive agent which is indicated in the treatment of peripheral vascular diseases. The drug acts by inhibiting the transmission of nerve impulses through the sympathetic and parasympathetic ganglionic synapses of the nervous system. Diuresis and peristalsis are somewhat reduced. Some vertigo and mental confusion, as well as dyspnoea and constipation may occur as side effects.
f bromure de tétrylammonium
d Tetrylammoniumbromid
i bromuro di tetrilammonio
e bromuro de tetrilamonio

3754 tetryzoline
Sympathomimetic agent used as a vasoconstrictor in the treatment of inflammatory hyperaemia and oedema of the nasal and ocular mucosa, e.g. allergic conjunctivitis, rhinitis and congestive disorders of the upper respiratory tract. Systemic disorders on a rather reduced scale may be caused by overdosage.
f tétryzoline; tétrahydrozoline
d Tetryzolin; Tetrahydrozolin
i tetrizolina; tetraidrozolina
e tetrizolina; tetrahidrozolina

3755 thalidomide
Commercial name of a chemical compound, alpha-phthalimidoglutarimide, which is obtained by reaction of phthalylglutamic acid or anhydride with ammonium or urea. It was originally introduced as a hypnotic and sedative agent, but its use was later outlawed owing to its high teratogenic effect, amply proved by the considerable frequency of congenital malformations in children of mothers who had used it in the course of pregnancy.
f thalidomide
d Thalidomid
i talidomide
e talidomida

3756 thapsia bark
The bark of the lower part of thapsia, a perennial herb, from which a resin is obtained that is used, in particular in North Africa, as a revulsive agent.
f écorce de racine de thapsia

d Thapsiarinde
i corteccia di radice di tapsia
e corteza de raíz de tapsia

3757 thebaine
Minor alkaloid of opium in which it is contained in the percentage of 0,2÷0,5. It constitutes the methyl derivative of morphine. Colourless shining crystals not easily soluble in water and marked by a tetanizing action that is analogous to that of strychnine. It has a slight narcotic action.
f thébaïne
d Thebain
i tebaina
e tebaina

3758 thenalidine
Antihistaminic agent of the piperidine series, capable of blocking the histamine effects by occupying the receptor sites in the effector cells to the exclusion of histamine. The drug is principally used as an antiallergic in skin diseases, persistent pruritus and bronchial asthma. It has also proved useful for the immediate treatment of accidents in the course of a blood transfusion. Drowsiness and a general malaise may occur as side effects.
f thénalidine
d Thenalidin
i tenalidina
e tenalidina

3759 thenitrazole
Trichomonicide (chemotherapeutic agent used against parasites of the genus Trichomonas) used in the treatment of vaginal trichomoniasis and candidiasis. Alcohol should be avoided in the course of the treatment.
f thénitrazole
d Thenitrazol
i tenitrazolo
e tenitrazolo

3760 thenyldiamine
Antihistaminic agent mainly used as an antiemetic and in the treatment of allergic reactions. The drug is also useful for its marked sedative effects and is therefore indicated as a hypnotic in children and in pre-operative medication. Its action is rather short. It must be used with caution in liver diseases and epilepsy.
f thényldiamine
d Thenyldiamin
i tenildiammina
e tenildiamina

3761 theobromine
Dimethylxanthine marked by a diuretic action by extrarenal and renal mechanism at glomerular level. It is used as a myocardial stimulant and vasodilator.
f théobromine; diméthylxantine
d Theobromin

i teobromina; dimetilxantina
e teobromina; dimetilxantina

3762 theophylline
Dimethylxanthine marked by a diuretic action and acting by an extra-renal mechanism, i.e. at tissue level and above all at the level of glomerular ultrafiltration. It inhibits the reabsorption of the electrolytes in the proximal renal tubules. The drug has a myolytic action on the smooth muscles, causes hypotension and improves cardiac activity. Some nausea and palpitations may occur as side effects.
f théophylline
d Theophyllin
i teofillina
e teofilina

3763 theosalicine
Salt of calcium theobromine and calcium salicylate. See theobromine for action and effects. It does not cause any stomach irritation and is therefore principally indicated in a prolonged therapy. It is sometimes used in conjunction with some digitalis preparations in the treatment of cardiac insufficiency, cardiac asthma, vascular spasm, hypertension and angina pectoris.
f théosalicine
d Theosalicin
i teosalicina
e teosalicina

3764 thevetin
Cardiac glycoside with effects like those of strophantus and digitalis. It hydrolyzes to digitoxigenin, glucose and thevetose. The drug acts directly on the cardiac muscle by increasing the force of systolic contraction without affecting the rate of the pulse, unless it be for a slowing down. It is mainly indicated in the treatment of congestive heart failure. Some degree of bradycardia may be observed as side effect.
f thévétine
d Thevetin
i tevetina
e tevetina

3765 thiabendazole
Or tiabendazole. Chemical compound acting as anthelmintic against parasites and trichinosis.
f thiabendazol
d Thiabendazol
i tiabendazolo
e tiabendazol

3766 thiambutosine
Thiourea-related chemotherapeutic agent very effective against acido-resistant bacteria. The drug is used in the treatment of leprosy and against Mycobacterium tuberculosis. Gastro-intestinal disorders may occur as side effects.
f thiambutosine

d Thiambutosin
i tiambutosina
e tiambutosina

3767 thiamine
Or aneurine, vitamin B_1. Antineuritic vitamin, the result of the union of a pyrimidinic nucleus with a thiazolic nucleus. Widely diffused in plants, thiazine is contained in alimentary substances, particularly in yeast, the outer cuticle of cereals, vegetables, liver, fish eggs, pork etc. Free thiamine, once introduced with vegetable aliments, is transformed, especially in the liver, into thiaminpyrophosphoric acid or carboxylase by the action of the adenyl pyrophosphoric acid. This process of phosphorylation conditions the biological activity of the thiamine which is indispensable for a normal process of the metabolism of the carbohydrates.
f thiamine
d Thiamin
i tiamina
e tiamina

3768 thiamine bisulphide
A derivative of vitamin B_1 with equal vitaminic activity, but more intense and prolonged and better tolerated analgesic effect both locally and generally. It also has a good antiphlogistic action. In practice, it displays all indications relating to vitamin B_1, especially in the neuritic and analgesic field.
f bisulfure de thiamine
d Thiaminbisulfid
i bisolfuro di tiamina; disolfuro di neurina
e bisulfuro de tiamina

3769 thiamphenicol
Chloramphenicol analogue but less toxic and effective. It is generally used against Gram-positive and Gram-negative micro-organisms affecting the respiratory, genito-urinary, hepatobiliary and gastro-intestinal systems. Also known as thiophenicol.
f thioamphénicol; thiophénicol
d Thiamphenicol
i tiamfenicolo; tiofenicolo
e tiamfenicol; tiofenicol

3770 thiamylal sodium
Barbituric anaesthetic marked by a rapid but very short action and used in the induction of anaesthesia by itself or as a coadjuvant of various other anaesthetic agents. It is administered by intravenous injection for short surgical operations, mainly to produce unconsciousness. Respiratory depression to the point of apnoea and circulatory depression are the two most frequent and severe side effects.
f thiamylal sodique
d Thiamylalnatrium
i sodio di tiamilale

e tiamilal sódico

3771 thiazides
A group of related compounds marked by diuretic effects. They act at the distant convoluted tube of the kidney for the purpose of checking or reducing the reabsorption of salt and water. Thiazides are fairly potent, yet not as potent as the so-called loop diuretics, e.g. frusemide. Their action occurs within a couple of hours, their duration does not exceed 24 hours. Hypokalaemia, hyperuricaemia and hyperglycaemia may occur as side effects.
f thiazides
d Thiazides
i tiazidi
e tiazidas

3772 thiazinaminum methylsulphate
Antihistaminic and anticholinergic agent indicated in the treatment of asthma, acute eczema, chronic bronchitis and allergic gastritis. Drowsiness may occur as a side effect.
f méthylsulfate de thiazinaminium
d Thiazinaminiummethylsulfat
i metilsolfato di tiazinaminio
e metilsulfato de tiazinaminio

3773 thiazine
Heterocyclic chemical compound existing in several isomeric forms and used as chemotherapeutic agents.
f thiazine
d Thiazin
i tiazina
e tiazina

3774 thibenzazoline
Therapeutic agent indicated for the treatment of hyperthyroidism and consequent disorders. Some allergic haematological reactions may occur in highly sensitive patients.
f thibenzazoline
d Thibenzazolin
i tibenzazolina
e tibenzazolina

3775 thiethylperazine
Phenothiazine-related antiemetic with a central mechanism of action. It is used to treat nausea and vomiting of an intestinal, hepatobiliary and iatrogenic (relating to any disorder which may be attributed to a medical procedure) origin. Also indicated for the prophylaxis and the treatment of vertigo (in particular the Ménière's syndrome). Extrapyramidal symptoms may be observed as side effects in hypersensitive patients.
f thiéthylpérazine
d Thiäthylperazin
i tietilperazina
e tietilperazina

3776 thioacetazone
Antituberculous agent of the thiosemi-

carbazone series, an alternative -appreciably cheaper- to para-aminosalicylic acid. Gastro-intestinal discomfort, blurred vision and some allergic reactions may occur as side effects.

f thioacétazone
d Thioacetazon
i tioacetazone
e tioacetazona

3777 thioacetic acid
Colourless liquid marked by a pungent odour of acetic acid and sulphuretted hydrogen. It is chiefly used as a reagent.

f acide thioacétique
d Thioessigsäure
i acido tioacetico
e ácido tioacético

3778 thiobarbital
Chemical compound, diethyl-derivative of the thiobarbituric acid; light yellow, water-soluble crystals having thyroid-activity depressing properties.

f thiobarbital
d Thiobarbital
i tiobarbitale
e tiobarbital

3779 thiobarbituric acid
Barbituric acid with an atom of oxygen substituted by an atom of sulphur. Some of its derivatives have short-lasting anaesthetic properties and may thus be used without any danger of secondary effects.

f acide thiobarbiturique
d Thiobarbitursäure
i acido tiobarbiturico
e ácido tiobarbitúrico

3780 thiobutabarbital
Barbituric narcotic used for short anaesthesias or for pre-anaesthesia. Great caution is advisable in cases of acute intoxication by alcohol, severe nephropathies, myocardial disorders, asthma etc.

f thiobutabarbital
d Thiobutabarbital
i tiobutabarbitale
e tiobutabarbitalo

* **thiocarbamide   s. thiourea**

3781 thiocarbonate
Salt or ester of thiocarbonic acid.

f thiocarbonate
d Thiocarbonat
i tiocarbonato
e tiocarbonado

3782 thiocarbonic acid
Chemical compound corresponding to the carbonic acid in which one, two or all three oxygen atoms are substituted by the same number of sulphur atoms. Several acids may thus be obtained, some of which find employment in pharmaceu-

tical preparations.

f acide thiocarbonique
d Thiokohlensäure
i acido tiocarbonico
e ácido tiocarbónico

3783 thiocarlide
Drug used in the treatment of tuberculosis either in conjunction with PAS (para-aminosalicylic acid) or cycloserine. It also finds employment in the treatment of leprosy. Gastro-intestinal disorders may occur as side effect.

f thiocarlide; tiocarlide
d Thiocarlid
i tiocarlide
e tiocarlido

3784 thiocolchicoside
Relaxant of the skeletal muscles acting on the central nervous system and marked by analgesic action. The drug is indicated in the treatment of contractures due to hemiplegias and traumas and of contractures of an extrapyramidal nature. It may also be useful to combat sciatica and lumbago.

f thiocolchicoside
d Thiocolchicosid
i tiocolcicoside
e tiocolcicosido

3785 thioglycerol
Glycerin-derived sulphuretted chemical compound. Various thioglycerols may be obtained by substituting with the same number of sulphur one, two or three of the oxygen atoms. Monothioglycerin is a sirupy liquid, smelling of sulphuretted hydrogen, characterized by keratoplastic and cicatrizing properties.

f thioglycérol
d Thioglycerin
i tioglicerina
e tioglicerol

3786 thioguanine
Antineoplastic agent and antimetabolite capable of blocking the metabolism of purine. It is principally indicated for the treatment of acute leukaemia and chronic myelocytic leukaemia. Leukopenia, thrombocytopenia, both accompanied by bleeding, stomatitis and gastro-intestinal disorders are possible side effects.

f thioguanine
d Thioguanin
i tioguanina
e tioguanina

3787 thiomersal
Mercurial disinfectant marked by antifungal and antibacterial actions. It is generally used to prepare the skin for a surgical operation. Also used for urethral irrigations and as eye drops. Rashes have sometimes been observed as side effect.

f thiomersal
d Thiomersal

i tiomersale
e tiomersal

3788 thiomucase
Enzyme isolated from the testicles of mammals. Diffusion factor causing the depolymerization of chondrosulphuric acid thus reducing the viscosity and the hydrophilia of the fundamental substance of connective tissues; it relieves the ten sion of the capillaries and eases the drainage of oedemas and the softening of tissues. The drug is principally indicated for the treatment of obesity, cellu litis, water retention and premenstrual tension.
f thiomucase
d Thiomukase
i tiomucasi
e tiomucasa

3789 thionine
Stain used in microscopy owing to its affinity for acidic constituents of cells and intercellular matrices, e.g. nucleoproteins and mucopolysaccharides.
f thionine
d Thionin
i tionina
e tionina

3790 thiopentone sodium
Very short-acting barbiturate used intra venously for anaesthesia of very short duration or for the induction of anaesthesia before using other anaesthetics. Also used as a coadjuvant in crises of convulsions. Respiratory and cardiovascular depression may occur as side effect.
f thiopentone sodique; thiopental sodique
d Thiopentonnatrium
i tiopentone sodico; sodio tiopentale
e tiopentona sódica

3791 thiophene
Or thiofuran, thiole. Heterocyclic chemical compound. It is found in raw benzene obtained by distillation of tar. Colourless, water-insoluble liquid used for the preparation of several therapeutic compounds, in particular antihistamine agents.
f thiophène; thiofuran
d Thiophen; Thiofuran
i tiofene; tiofurano
e tiofeno; tiofurano

3792 thioporan
Anti-vitamin K anticoagulant mainly indicated in the treatment of venous or arterial thromboembolia and myocardial infarction. Also used prophylactically against post-operation thrombophilia.
f thioporan
d Thioporan
i tioporano
e tioporano

3793 thiopronin

Liver-protecting agent acting through the presence of the −SH radicals which interfere in the process of oxidation-reduction of the liver detoxication. Its hepatoprotective action occurs in correspondence of membranes. Its particular stability accounts for its constant and long-lasting effect. The drug is indicated in the treatment of liver dysfunction, acute and chronic hepatitis, pathological conditions caused by intoxication, allergies, cystinuria and some dermatides. It is also used for the treat ment of rheumatoid arthritis and diabetes.
f thiopronine
d Thiopronin
i tiopronina
e tiopronina

3794 thiopropamine
Antacid agent capable of stimulating the secretion of pancreatic bicarbonates which neutralize the hydrochloric acid, and of promoting the defensive activity of the gastric membranes. It is principally used for the treatment of duodenal ulcer, as well as gastric ulcer and gastroduodenitis.
f thiopropamine
d Thiopropamin
i tiopropamina
e tiopropamina

3795 thiopropazate
Phenothiazine-related tranquillizer indicated in the treatment of symptoms mostly associated with psychomotor disorders and states of anxiety. Also useful in the treatment of insomnia and psychosomatic syndromes. Blurred vision and extrapyramidal disorders may occur as side effects.
f thiopropazate
d Thiopropazat
i tiopropazato
e tiopropazato

3796 thioproperazine
Phenothiazine-related tranquillizer marked by a central mechanism of action and principally used in the treatment of acute and chronic schizophrenia, manic-depressive conditions, epilepsy, hallucinations and senile psychosis. Urinary retention and vision disorders are among the most frequent side effects.
f thiopropérazine
d Thioproperazin
i tioproperazina
e tioproperacina

3797 thioridazine
Phenothiazine-related tranquillizer acting on the subcortex and indicated in the treatment of insomnia, anxiety and psychomotor agitation. The drug does not alter psychic functions and proves synergic with analgesic and hypnotic agents. Side effects are very similar to

those of chloropromazine, q.v.
f thioridazine
d Thioridazin
i tioridazina
e tioridacina

3798 thiosalicylic acid
Yellow crystals, not easily soluble in water, obtained by treating halogenated benzoic acid with alkaline sulf-hydrates. It is used as a reagent in analytical chemistry.
f acide thiosalicylique
d Thiosalicylsäure
i acido tiosalicilico
e ácido tiosalicílico

3799 thiosulphuric acid
Acid which does not occur free but in the form of salts which are known as thiosulphates. It is used medicinally in the treatment of eczema and scabies and also in case of metallic poisoning.
f acide thiosulfurique
d Thioschwefelsäure
i acido tiosolforico
e ácido tiosulfúrico

3800 thio-TEPA
Alkylating cytostatic agent used in neoplastic disease. It acts by inhibiting deoxyribonucleic acid. It has a prompt absorption and is mainly indicated for the treatment of myeloid and lymphatic leukaemia, lymphosarcoma and adenosarcoma and, more generally, neoplastic diseases affecting the ovary and breast. General malaise, vomiting and anorexia may occur as side effects. Also bone marrow depression.
f thio-TEPA
d Thio-TEPA
i tiotepa; tio-TEPA
e tiotepa; tio-TEPA

3801 thiothixene
Psychotomimetic agent, i.e. a psychotropic drug of the thioxanthene series, e.g. chlorprothixen. It is characterized by certain chemical and pharmacological properties which are similar to those of the piperazine phenothiazines and at the same time different from the aliphatic group of phenothiazine. The drug is indicated in the treatment of psychotic disorders, e.g. paranoia and schizophrenia. The side effects are those of chlorpromazine. In addition, some extra-pyramidal symptoms are common.
f thiotixène
d Thiothixen
i tiotissene
e tiotixeno

3802 thiouracil
Chemical compound, also known as cyclic thiourea. Colourless, odourless bitter-tasting crystals, not easily soluble in water, characterized by an inhibiting action on the thyroid function.

f thiouracil
d Thiouracil
i tiouracile
e tiouracilo

3803 thiourea
Colourless crystalline substance which was used in the past as an antithyroid drug. It has now been replaced by its derivatives, i.e. the thiouracils, q.v.
f thiourée
d Thiocarbamid; Thioharnstoff
i tiocarbammide; tiourea
e tiocarbamida; tiourea

3804 thioxolone
Therapeutic agent used in the treatment of psoriasis owing to its capacity to promote hyperemization. It also has a disinfectant and antiphlogistic action. Mucous membranes should not be treated with this drug.
f thioxolone
d Thioxolon
i tiossolone
e tioxolona

3805 thonzylamine
Antihistamine which acts by antagonizing histamine in the receptor sites in the effector cells. It is mainly indicated for the treatment of allergic cutaneous diseases, hay fever, rhinitis and various allergic conditions caused by drugs. Drowsiness, vertigo and some gastro-intestinal disorders may occur as side effects.
f tonzylamine
d Tonzylamin
i tonzilamina
e tonzilamina

3806 threitol dimethane sulphonate
Cytotoxic drug indicated for the treatment of ovarian cancer. Bone-marrow suppression, gastro-intestinal disorders and rashes may occur as side effects.
f diméthane sulfonate de threitol
d Threitoldimethansulfonat
i dimetano solfonato di treitolo
e dimetano sulfonato de treitolo

3807 threonine
Amino acid that is essential for the animal organisms and which cannot be synthesized by them. It is present in milk, eggs, meat, oats etc., as well as in the hydrolysates of certain proteins, particularly of blood fibrin.
f thréonine
d Threonin
i treonina
e treonina

3808 thrombin
Substance which is formed in the initial phase of the coagulation of the blood by the action of various factors, e.g. thromboplastin, calcium ions etc. It is an albumin which, once inactivated at

60°C, determines, almost certainly by enzymatic action, the transformation of fibrinogen into fibrin. It further determines the labilization and disintegration of the platelets, the activation of the acceleration factors and catalyzes the prothrombin-thrombin transformation.

f thrombine
d Thrombin
i trombina
e trombina

3809 thrombine (pharm)
Physiological haemostatic agent mainly used by local application in surgery, gynecology, obstetrics, otorhinolaryngology etc. as well as haemorrhages caused by anticoagulants and haemophilia. Also used internally in cases of intestinal haemorrhages or of bleeding oesophageal varices. Thrombosis or embolism may occur when the drug is injected intravenously.

f thrombine
d Thrombin
i trombina
e trombina

* **thrombokinase   s. thromboplastin**

3810 thromboplastin
Lipoprotein present in animal tissues and organs, mainly brain, lungs, thymus, testes and placenta, and obtain ed from recently slaughtered mammals, preferably rabbits, oxen and horses. Thromboplastin performs an important catalyzing action in the first phase of the process of clotting (the transformation of prothrombin into thrombin). It is formed in the plasma further to the reaction of particular substances which belong to the plasma itself and of others which are liberated from the platelets. Particularly important are the first ones which are essentially represented by the antihaemophilic globulin and other plasmatic and platelet-connected factors which react according to strictly quantitative ratios.

f thromboplastine
d Thromboplastin
i tromboplastina
e tromboplastina

3811 thujone
Bridged terpene ketone obtained from the oils of panacetum, thuja, tansy, wormwood and sage. It has a menthol odour and is used in the preparation of various drugs.

f thujone
d Thuyon
i tujone
e tujona

3812 thyme
Common garden herb from which an essential oil is obtained that finds employment in several therapeutic prepara-

tions.

f thym; farigoule
d Thymian; Gartenthymian
i timo; pepolino
e tomillo

3813 thyme camphor
Thymol. Crystalline phenol marked by a pleasant aromatic odour and antiseptic properties. It may be obtained from thyme oil and other essential oils or prepared synthetically. It is principally used as a fungicide and in pharmacological and dental preparations. Also known as thymic acid.

f acide thymique; camphre de thym
d Thymiansäure; Thymol; Thymolkampfer
i canfora di timo; acido timico
e ácido tímico; alcanfor de tomillo

3814 thyme oil
Fragrant essential oil containing thymol (see thyme camphor) and carvacrol which is obtained from various thymes and is used mainly as an antiseptic in pharmaceutical and dental preparations.

f essence de thym
d Thymianöl
i essenza di timo; olio di timo
e esencia de tomillo

3815 thymic acid
Term denoting the mixture of nucleotides which is obtained by a cautious hydrolysis of the thymonucleic acid, q.v.

f acide thymique
d Thymussäure
i acido timico
e ácido tímico

3816 thymol
Disinfectant similar but less toxic than phenol. It acts on the oral, vaginal, urethral mucosa, as well as on the mucosa of the respiratory and digestive tracts. It is principally used in mouth washes and as an anthelmintic.

f thymole
d Thymol
i timolo
e timolo

3817 thymonucleic acid
Acid present in the thymus, spleen and other glandular tissues. It is a polynucleotide and each nucleotide consists of phosphoric acid, the sugar deoxyribose and one of the bases thymine, adenine, cytosine and guanine.

f acide thymonucléique
d Thymonukleinsäure
i acido timonucleico
e ácido timonucleico

3818 thymus extract
Preparation obtained from the thymus of a calf by partial lysis, subsequent deproteinization and final dilution with glycerine. It has a stimulant action on leukopoiesis and cell-mediated immuni-

tary response. It is especially indicated
for the treatment of primary or secondary leukopenic syndromes, the prophylaxis of leukopenias due to myelotoxic
agents and infectious and immuno-defi-
ciency-type dermatologic pathology.
f extrait de thymus
d Thymusextrakt; Thymusdrüsenextrakt
i estratto di timo
e extracto de timo

3819 thyroglobulin
Iodized protein synthesized and stored
in the thyroid gland in which it constitutes the principal constituent of the
colloid of the vesicle. Pharmaceutically,
a thyroid preparation indicated in the
treatment of all forms of hypothyroidism
also when resulting from surgical removal or radiation. In some cases, it
is used for thyroid-suppression therapy
in toxic goiter.
f thyroglobuline
d Thyroglobulin
i tireoglobulina; tiroglobulina
e tiroglobulina

3820 thyroid extrakt
Powder obtained from the thyroid gland
of oxen, sheep etc., fully defatted and
dried at a temperature not exceeding
60°C. The extract contains thyreoglobulin,
thyroxine and the various hormones produced by the thyroid gland. It is indicated in the treatment of hypothyroidism,
obesity, hypogalactia (inadequate secretion of milk) and goiter accompanied
by cretinism.
f extrait de thyroïde
d Schilddrüsenextrakt
i estratto di tiroide
e extracto de tiroides

3821 thyrotrophin
Thyrotrophic hormone, thyroid-stimulat-
ing hormone, TSH. Hormone produced by
the basophilic cells of the anterior lobe
of the hypophysis, which has a stimulat
ing action on the thyroid tissue. In
fact, in its absence, the gland atrophies
and its colloid substance increases considerably. Its production is stimulated
by inorganic iodine but is inhibited by
thyroxin and 3.5-diiodothyrosine. As a
drug, thyrotrophin is therefore used in
the treatment of hypothyroidism, obesity,
goiter, cretinism etc. Some allergic reactions may occur as side effect.
f thyréotropine; thyréotrophine; TSH; hormone thyréotrope
d Thyreotropin; thyreotropes Hormon; TSH
i tirotropina; tireotropina; ormone tireo-
tropo; TSH
e tireotropina; tirotropina; hormona tiro-
tropica; TSH

3822 thyroxine
Tetraiodo-p-oxyphenyl derivative of the
thyrosine. White or white-yellowish powder, insoluble in water, odourless and
tasteless. It represents the most active
hormone of the thyroid gland and can
be prepared synthetically. Its formation
is conditioned by the presence in the
blood of iodine and is regulated by the
thyreotropohypophysary hormone and by
the sympathetic nervous system. The
hormonal action of the drug is carried
out in correspondence with the cells and
consists of accelerating the intracellular
oxidations of the proteins, of the carbohydrates and of the lipids. It also in-
fluences the blood circulation, the water
balance, the nerve centres and the
gonads, all secondary to the fundamental
action which consists of the stimulation
of the metabolism.
f thyroxine
d Thyroxin
i tiroxina
e tiroxina

3823 tiadenol
Therapeutic agent capable of reducing
the content of cholesterol in the blood
and therefore used in the treatment of
hypercholesterolaemia, hyperglyceri-
daemia and mixed hyperlipidaemia. Care
ful monitoring is required when the drug
is administered together with anticoagulants.
f tiadénol
d Tiadenol
i tiadenolo
e tiadenolo

3824 tiapride
Antiemetic and sedative agent used in
the treatment of emesis, states of agitation caused by acute or chronic alcoholism, delirium tremens, Huntington's
chorea and cramps. Amenorrhoea and
galactorrhoea may occur as side effects.
f tiapride
d Tiaprid
i tiapride
e tiaprida

3825 tiaprofenic acid
Non-steroid antiphlogistic and analgesic
agent capable of reducing inflammation
by the inhibition of prostaglandin synthesis which is part of the inflammatory
process. Also, it inhibits the same enzymes in the gastric mucosa. Headache
and some other central nervous system
symptoms have been observed.
f acide tiaprofénique
d Tiaprofensäure
i acido tiaprofenico
e ácido tiaprofénico

3826 tiaramide
Antiinflammatory, analgesic and antipyretic agent marked by a potent antioedematous activity and indicated in the
treatment of oedemas of the respiratory,
digestive and urogenital tracts. A slight
loss of weight has been observed as a
side effect.

f tiaramide
d Tiaramid
i tiaramide
e tiaramida

3827 tibenzonium iodide
Topical antibacterial agent of the quater
nary ammonium salts series marked by
a bacteriostatic and bactericide activity,
particularly against infections affecting
the mouth and pharynx. It is also mark
ed by a good anaesthetic and analgesic
action. It is therefore frequently prescribed to combat inflammations and
infections of the mouth, gums and
throat, for the necessary pre- and post-
-operative treatment of tonsillectomy and
tooth extraction.
f iodure de tibenzonium
d Tibenzoniumjodid
i ioduro di tibenzonio
e yoduro de tibenzonio

3828 ticarcillin
Broad-spectrum penicillin principally
used to combat severe and/or potentially
lethal infections. Caution is required in
cases of severe renal impairment. Pains
in the site of injection and a slight,
reversible increase of transaminases may
occur as side effect.
f ticarcilline
d Ticarcillin
i ticarcillina
e ticarcilina

3829 ticlatone
Antimycotic agent used for the topical
treatment of dermatomycoses and bacterial superinfections. Contact with eyes
and mucous membranes should be avoid
ed.
f ticlatone
d Ticlaton
i ticlatone
e ticlatona

3830 tidiacic
Liver-protecting and detoxicant agent
indicated in the treatment of hepatic
insufficiency, precirrhosis and chronic
hepatitis.
f tidiacic
d Tidiacic
i tidiacic
e tidiacic

3831 tiemonium iodide
Antispasmodic and anticholinergic agent
marked by an action which appears to
be more potent than papaverine and
atropine and usually administered in
association with barbiturates for the
treatment of painful and extremely painful conditions due to contractions of the
abdominal organs, i.e. liver, kidneys
and uterus. Some slight cardiac disorders have been observed as side effects.
f iodure de tiémonium
d Tiemoniumjodid

i ioduro di tiemonio
e yoduro de tiemonio

3832 tiemonium methylsulphate
Anticholinergic spasmolytic agent as well
as myolytic and antihistaminic, acting
by inhibiting the cholinergic receptors
and by direct action on the smooth muscles. It is mainly indicated in the treat
ment of painful spastic conditions of the
gastro-enteric, urinary and biliary
tracts.
f sulfate de tiémonium
d Tiemoniumsulfat
i solfato di tiemonio
e sulfato de tiemonio

3833 tigloidine
Tiglylpseudotropeine. Therapeutic drug
chemically similar to atropine but with
mild anticholinergic effects, indicated
for the treatment of spasticity and muscle spasms, in particular when caused
by a lesion of the upper motor neurone.
f tigloïdine
d Tigloidin
i tigloidina
e tigloidina; tiglilseudotropina

3834 tilidine
Powerful analgesic agent mainly used in
pre-operative and post-operative medication and for the relief of pains due to
wounds, contusions, scalds, pathological
conditions of the biliary and urinary
tracts and osteoarthropathies. It is also
used in ophthalmology in the treatment
of endo-ocular hypertension, chronic
open-angle glaucoma and glaucoma-ac-
companied aphacia (the absence of the
lens), also when afflicting contact-lenses
bearers. Vertigo, drowsiness and nausea
may occur as side effects.
f tilidine
d Tilidin
i tilidina
e tilidina; tilidato

3835 timepidium bromide
Spasmolytic and anticholinergic agent
used for the treatment of smooth-muscle
spasms in cases of gastritis, enteritis,
duodenal ulcer and severe pains caused
by pancreatitis. Dryness of the mouth
and dysuria may occur as side effect.
f bromure de timépidium
d Timepidiumbromid
i bromuro di timepidio
e bromuro de timepidio

3836 timolol
Beta adrenoreceptor blocking drug mainly used in the treatment of angina pectoris caused by ischaemic cardiopathy
and essential hypertension. Also used in
ophthalmology for the treatment of endo-
ocular hypertension, chronic open-angle
glaucoma and aphacia. Gastro-intestinal
disorders, vertigo, general debility,
dyspnoea, insomnia and bradycardia may

occur as side effects.
f timolol
d Timolol
i timololo
e timololo

3837 tinidazole
Antiprotozoal agent particularly active
against Trichomonas vaginalis and
Trichomonas foetus and indicated in the
treatment of infections of the urogenital
tract caused by Trichomonas both in men
and women. The drug is rapidly absorbed in the mouth and excreted with urine
within 24 hours, by 20% in a biologically active form. Transient leukaemia and,
though rarely, some allergic reactions
may occur as side effects.
f tinidazole
d Tinidazol
i tinidazolo
e tinidazolo

3838 tinoridine
Or tienoridine. Analgesic antiphlogistic
agent mainly used to relieve pains, in
particular when caused by rheumatic
conditions. Nausea, diarrhoea, gastric
and liver disorders may occur as side
effects.
f tinoridine
d Tinoridin
i tinoridina
e tinoridina

3839 tiphenamil
Spasmolytic anticholinergic agent marked
by topical vasodilator and anaesthetic
properties. It is mainly used against
abdominal spasms, peptic ulcer-related
pains, angina pectoris and bronchial
asthma. Drowsiness may occur as side
effect.
f tiphénamil
d Tiphenamil
i tifenamile
e tifenamilo

3840 tiratricol
Thyroid hormone marked by a lipolytic
action and indicated in the treatment of
obesity, cellulitis and lipodystrophies.
Some allergic cutaneous reactions have
been observed as side effect.
f tiratricol
d Tiratricol
i tiratricolo
e tiratricolo

3841 tisopurine
A derivative of allopurinol capable of
inhibiting the synthesis of uric acid and
indicated in the treatment of hyperuricaemia. It may be used for patients suf
fering from renal incompetence. Fever
and some allergic skin rashes may occur
as side effects.
f tisopurine
d Tisopurin
i tisopurina

e tisopurina

3842 titanium dioxide
Astringent, absorbent and skin-protect-
ing agent mainly used for dermatological
treatment. The drug reduces itching and
absorbs ultraviolet rays. It is used
topically to protect against sunburn and
to heal some forms of eczema.
f bioxyde de titane
d Titandioxid
i biossido di titanio
e dióxido de titanio

3843 tobramycin
Amino-glycoside antimicrobial mainly
active against stafilococci and Gram-negative micro-organisms, particularly
enterobacteria and Pseudomonas aeruginosa. The drug is indicated in the treat
ment of infections of the genito-urinary,
urinary and gastro-intestinal tracts,
soft tissues and the central nervous system. Dizziness, nystagmus, some degree
of deafness and oliguria may occur as
side effects.
f tobramycine
d Tobramycin
i tobramicina
e tobramicina

3844 tocainide
Cardiac antiarrhythmic agent and a
local anaesthetic. The drug stabilizes
the nerve cell membranes so as to prevent the conduction of impulses. Thus
differing from lignocaine (q.v.), it is
also active after oral administration. A
prolonged use may cause blood dyscrasias.
f tocaïnide
d Tocainid
i tocainide
e tocainida

3845 tocamphyl
Choleretic, mild cholagogue and disinfectant agent acting directly on the hepatic
cells to promote the production of bile.
Its effects are fairly prolonged.
f tocamphyl
i Tocamphyl
i tocamfile
e tocamfilo

3846 tocopherols
The pale yellow fat-soluble oily liquid
phenolic compounds which derive from
chroman and differ in the number and
location of the methyl group in the benzene ring. They have antioxidant properties and vitamin E activity in various degrees. They occur, particularly
in their dextrorotatory form, in oils from
various seeds, e.g. wheat-germ oil and
cottonseed oil, in leaves and in fish-
-liver oil. They can be produced synthetically in their racemic form and are
used in a mixture mainly as antioxidants (e.g. for stabilizing vitamin A in

fats and oil and in nutrition and veterinary medicine. See also tocopheryl
acetate).
f tocophérols
d Tocopherole
i tocoferoli
e tocoferoles

3847 tocopheryl acetate
The acetate of natural alpha-tocopherol
which can be obtained from wheat-germ
oil or of synthetic ($\pm$)-alpha-tocopherol.
Tocopheryl acetate is one of the forms
of vitamin E, deficiency of which grave-
ly affects normal pregnancy. For this
reason, the drug is indicated in the
treatment of habitual abortion, although
the recorded results allow for some per-
plexity.
f acétate de tocophéryl
d Tocopherylacetat
i acetato di tocoferile
e acetato de tocoferilo

3848 tocopherylquinone
Hypotensive agent marked by a slow
and gradual but not very long-lasting
action. It is especially indicated in the
treatment of hypertension that is associated with abnormal adrenal function
and as a coadjuvant in the treatment of
arteriosclerosis. Tocopherylquinone being
a homeostatic (relating to the tendency
of a comparatively stable internal envi-
ronment) factor of the connective tissue,
it also has an antisclerotic and antiphlogistic action and is used in cutaneous, muscular and neurogenic diseases
caused by changes of connective tissue.
It is also used in inflammatory and degenerative rheumatism.
f tocophérylquinone
d Tocopherylchinon
i tocoferilchinone
e tocoferilquinona

3849 todralazine
Hypotensive agent marked by a spasmolytic action on the walls of the vessels
thus promoting glomerular filtration. It
is mainly indicated for the treatment of
arterial hypertension, especially when
caused by arteriosclerosis or endocrino-
pathy.
f todralazine
d Todralazin
i todralazina
e todralacina

3850 tofenacin
Therapeutic agent used in the treatment
of Parkinson's syndrome. Also used, as
antidepressant, to treat states of depression affecting elderly people. A marked
dryness of the throat, accommodation
difficulty, drowsiness, gastro-intestinal
disorders and a difficult miction may
occur as side effects.
f tofénacine
d Tofenacin

i tofenacina
e tofenacina

3851 tofisopam
Benzodiazepinic tranquillizer mainly
used in the treatment of psychic dystonia and consequent somatic phenomena.
Also indicated in the treatment of alcohol, states of anxiety and depression
both in adults and children. Drowsiness
and insomnia may occur as side effects.
f tofisopam
d Tofisopam
i tofisopam
e tofisopam

3852 tolazamide
Antidiabetic used orally in the treatment
of selected cases of diabetes mellitus.
It is a sulfonylurea compound capable
of promoting the secretion of insulin and
the biosynthesis of glycogen. Drug dependence may develop as side effect.
f tolazamide
d Tolazamid
i tolazamide
e tolazamida

3853 tolazoline
Alpha-adrenoreceptor blocking drug marked by partial agonist activity and relaxant properties for smooth muscles. It
is indicated in the treatment of peripheral vascular diseases. Vertigo, general malaise and tachycardia may occur
as side effects.
f tolazoline
d Tolazolin
i tolazolina
e tolazolina

3854 tolbutamide
Orally administered antidiabetic drug.
It is a sulphonyl urea compound capable
of stimulating the secretion of insulin
and of favouring the biosynthesis of
glycogen. The drug has the same effects
as chlorpropamide and, in addition,
some mild accumulation.
f tolbutamide
d Tolbutamid
i tolbutamide
e tolbutamida

3855 tolmetin
Non-steroid antiphlogistic, analgesic
agent used in the treatment of inflammatory arthropathies, extra-articular rheumatism and contusions. Some gastric
pains, pyrosis, nausea and dizziness
may occur as side effects.
f tolmétine
d Tolmetin
i tolmetina
e tolmetina

3856 tolnaftate
Antifungal agent, topically active, used
as an ointment or powder in the treatment of skin infections.

f   tolnaftate
d   Tolnaftat
i   tolnaftato
e   tolnaftato

3857 toloconium methylsulphate
Quaternary ammonium compound used as local disinfectant, being a cationic surface agent. It is marked by demulcent properties and is used in aqueous solutions of different concentration for the disinfection of hands and the sterilization of surgical instruments and syringes. It is also used for vaginal douches, cleaning infected wounds and disinfection of mouth and teeth.
f   méthylsulfate de toloconium
d   Tolokoniummethylsulfat
i   solfato di tolocomio
e   sulfato de tolocomio

3858 tolonium chloride
Heparin antagonist and antihaemorrhagic, antimenorrhagic agent mainly indicated in the treatment of menstrual disorders and of intoxications caused by agents capable of increasing methaemoglobin. A degree of methaemoglobinaemia may occur as side effect.
f   chlorure de tolonium
d   Toloniumchlorid
i   cloruro di tolonio
e   cloruro de tolonio

3859 tolperisone
Muscle relaxant and antispastic agent also marked by a peripheral vasodilator property and indicated for the treatment of multiple sclerosis, disseminated encephalomyelitis, spinal arachnoiditis, post encephalitis spasms, amyotrophic lateral sclerosis and all those diseases that are caused by a compression of the spinal cord. The drug is also used in thromboangiitis obliterans, varicose ulcers and in rheumatism. Gastro-intestinal disorders and general malaise are possible side effects.
f   tolpérisone
d   Tolperison
i   tolperisone
e   tolperisona

3860 tolpropamine
Antihistaminic agent and a mild topical anaesthetic and antiseptic. It is principally used for the treatment of allergic dermatoses, urticaria and stings.
f   tolpropamine
d   Tolpropamin
i   tolpropamina
e   tolpropamina

3861 tolu balsam tree
Tropical American tree from which a brown solid balsam is obtained, which is used as a stimulating expectorant and a flavouring for cough syrup.
f   baumier de Tolu
d   Tolubalsambaum
i   balsamino di Tolu
e   balsamero de Tolú

3862 toluene
Toluole, methylbenzene. Light mobile liquid aromatic hydrocarbon, resembling benzene but less volatile, less flammable and less toxic. It was originally obtained by distilling balsam of tolu. It is mainly used as a solvent and in the manufacture of several drugs.
f   toluène
d   Toluen
i   toluene
e   tolueno

3863 toluenesulphonate
Or tosylate. The salt of toluensulphonic acid. Toluensulphonates are more easily soluble in water than the acid from which they derive.
f   toluènesulfonate
d   Toluensulfonat
i   toluensolfonato
e   toluensulfonato

3864 toluensulphonic acid
Or tosic acid. Chemical compound obtained from toluene by substituting one or more hydrogen atoms with the same number of $-SO_3H$ groups. Toluensulphonic acids are used as catalyzers in many organic syntheses.
f   acide toluène-sulfonique; acide tosique
d   Toluensulfonsäure
i   acido toluensolfonico; acido tosico
e   ácido toluensulfónico; ácido tósico

3865 toluidine
Aromatic amine, a superior homologue of aniline, which exists in three isomeric forms, all of them soluble in organic solvents and in acids. They are used as intermediates in the production of several synthetic organic dyes, e.g. safranine and alizarin. Toluidine blue is a dyeing substance frequently used in histology as it is particularly suitable for the staining of nerve cells.
f   toluïdine
d   Toluidin
i   toluidina
e   toluidina

3866 toluidine blue
Basic dye or its chloride of the thiazine class intimately related to methylene blue. It is principally used as a biological stain and, therapeutically, for the treatment of some haemorrhages owing to its capacity to inhibit the anticoagulant effect of heparin.
f   bleu de toluïdine
d   Toluidinblau
i   blu di toluidina
e   azul de toluidina

3867 tolycaine
Local anaesthetic mainly used in dentistry. It is generally well tolerated and

does not affect the circulation. A local ischaemia may, albeit very rarely, occur as a side effect.
f tolycaïne
d Tolycain
i tolicaina
e tolicaina

3868 tolylantipirine
Toluene analogue of phenazone mainly used as an analgesic. Its salicylate is frequently used in the treatment of rheu matism.
f tolylantipirine
d Tolylantipirin
i tolilantipirina
e tolilantipirina

3869 tolylenediamine
Chemical compound principally used in the manufacture of dyestuffs, but also thought to have some cholagogue action.
f tolylènediamine
d Tolylendiamin
i tolilenediammina
e tolilenediamina

3870 tolynol
Choleretic agent capable of stimulating the production of bile in terms of volume without notably increasing the secretion of biliary pigments and salts in terms of percentage. It is mainly used in the treatment of hepatic disorders, biliary dyspepsia and cholangitis.
f tolynol
d Tolynol
i tolinolo
e tolinolo

3871 toothbrush tree
Salvadora persica. Tree growing mainly in African and West Asian steppes, whose twigs are sometimes bound in clus ters and used as toothbrushes.
f salvadore de la Perse; arac
d persische Salvadore; Zahnbürstenbaum
i salvadora di Persia
e salvadora pérsica

3872 tormentil root
The root of a yellow-flowered Eurasian herb, which contains an astringent agent.
f racine de tormentille
d Tormentilwurzel; Blutwurzel
i rizoma di tormentilla; rizoma di poten- tilla
e rizoma de potentila; rizoma de tormentila

3873 tosactide
Synthetic human corticotrophin, pituitary hormone having a trophic effect on the adrenal glands and mainly used in the treatment of their insufficiency, in cases of a prolonged therapy based on cortico- steroids and as a diagnostic test of the adrenal functions.
f tosactide
d Tosactid

i tosactide
e tosactida

3874 tramazoline
Sympathomimetic agent used topically as nasal decongestant, in particular in acute and chronic rhinitis, conjunctivitis and inflammation of the palpebral canal. The drug should not be used for child- ren under six years of age.
f tramazoline
d Tramazolin
i tramazolina
e tramazolina

3875 tranexamic acid
Antiplasmitic agent, i.e. an agent capa- ble of inhibiting the activity of the pro- teolytic enzyme derived from the activa- tion of plasminogen.
f acide tranexamique
d Tranexamsäure
i acido tranessamico
e ácido tranexámico

3876 tranylcypromine
(+)trans-2-phenylcyclopropylamine. Its sulphate is a mono-amine oxidase inhib- itor indicated in the treatment of depres sive states, including endogenous, reac- tive and psychoneurotic depression. Orthostatic hypotension, liver disorders, pallor, nausea and tremors may occur as side effects.
f tranylcypromine
d Tranylcypromin
i tranilcipromina
e tranilcipromina

3877 trazodone
Stimulant of the central nervous system. More specifically, an anxiolytic and antidepressant agent acting electively on the mechanism of emotive-affective integration of experience. It is indicated in the treatment of depression, not neces sarily accompanied by anxiety, psycho- somatic diseases, for the relief of some pains and in preanaesthesia.
f trazodone
d Trazodon
i trazodone
e trazodona

3878 treosulfan
Cytotoxic drug marked by an alkylating action and used in the treatment of ovarian cancer. It is metabolized by the liver to active, epoxide form. It acts by damaging the DNA thus affecting cell replication. Gastro-intestinal disorders, skin reactions, hair loss and bone mar- row depression may occur as side effects.
f tréosulfan
d Treosulfan
i treosolfan
e treosulfan

3879 tretamine
Antineoplastic agent acting on the nu-

cleic acids and used in the treatment of chronic lymphatic and myeloid leukaemia and Hodgkin's disease. Leukopenia, thrombocytopenia and anorexia may occur as side effects.
f trétamine
d Tretamin
i tretamina
e tretamina

3880 tretinoin
Vitamin A derivative capable of promoting irritation and desquamation of the outer protective layer of the skin and used in several dermatological treatments.
f trétinoïne
d Tretinoin
i tretinoina
e tretinoina

3881 tretoquinol
Beta-stimulant, bronchodilator marked by a very mild cardiovascular action and mainly used in the treatment of bronchial asthma, chronic bronchitis and pulmonary emphysema. Palpitations may occur as side effect.
f trétoquinol
d Tretochinol
i tretochinolo
e tretoquinolo

3882 triacetin
Oily liquid often used as a plasticizer. Strong disinfectant agent against mycetes.
f triacétine
d Triazetin
i triacetina
e triacetina

3883 triamcinolone
Synthetic steroid, one of the cortisonic drugs specifically marked by antiallergic activity. Owing to its corticosteroid action, it is used in the treatment of adrenal incompetence, hypopituitarism, Addison's disease and inflammatory and allergic dermatopathies, as well as severe allergic states or reactions. Dyspepsia, gastro-intestinal ulcers, osteoporosis and hyperglycaemia are among the most frequent adverse effects of the drug.
f triamcinolone
d Triamcinolon
i triamcinolone
e triamcinolono

3884 triamterene
Diuretic agent acting on the distal tubule and Henle's loop and capable of promoting the excretion of sodium and chloride without modifying the excretion of potassium. It is indicated in the treatment of oedema, e.g. hepatic or renal oedema, and of water deposits in serous cavities. Gastro-intestinal disorders may easily occur in hypersensitive patients.
f triamtérène

d Triamteren
i triamterene
e triamtereno

3885 triaziquone
Antineoplastic drug indicated in the treatment of lymphogranulomatosis, leuko sarcomatosis and mammary tumours. Leukocytopenia and thrombocytopenia may be observed as side effects.
f triaziquone
d Triazichon
i triaziquone
e triaziquona

3886 triazolam
Benzodiazepine tranquillizer and hypnotic used in the treatment of insomnia and states of anxiety.
f triazolam
d Triazolam
i triazolam
e triazolam

3887 tribromethanol
White crystalline substance used in solution of amylene hydrate as a basal anaesthetic.
f tribrométhanol
d Tribromethanol
i tribromoetanolo
e tribromoetanolo

3888 tribuzone
Antiphlogistic, analgesic and antipyretic agent indicated in the treatment of rheumatoid arthritis, hyperuricaemic syndrome and phlebitis. Nausea, diarrhoea and stomach aches may occur as side effects.
f tribuzone
d Tribuzon
i tribuzone
e tribuzona

3889 trichloracetic acid
Chemical compound. Colourless deliquescent crystals marked by a pungent odour, soluble in water. It is used in organic syntheses. It has astringent properties. The compound is also used as a herbicide.
f acide trichloroacétique
d Trichloressigsäure
i acido tricloroacetico
e ácido tricloroacético

3890 trichlorethylene
Colourless transparent liquid marked by a characteristic chloroform-like odour and a pungent sweetish taste. It causes depression of the central nervous system and is mainly used by inhalation as a general anaesthetic which is less potent but much less irritant than ether. The drug may slow the heart and cause irregular rhythms. Nausea, vomiting and headache may occur as side effects. High concentrations may depress liver and kidney function and cause severe

poisoning sometimes leading to death in coma.

f trichloréthylène
d Trichloräthylen
i tricloroetilene
e tricloroetileno

3891 trichlormethiazide
Diuretic agent more effective than chlorthiazide, acting on the proximal renal tubule. It increases the excretion of sodium and chloride. Its action is generally retarded but long lasting. It is mainly indicated in the treatment of cardiac, renal and hepatic oedemas. Hypotension may occur as side effect.

f trichlorméthiazide
d Trichlormethiazid
i triclormetiazide
e triclormetiazida

3892 trichlorofluoromethane
Aerosol propellant and refrigerant capable of producing intense cold by its rapid evaporation, a process which makes tissues insensitive to touch and pain. It is often used for the relief of muscle spasms.

f trichlorofluorométhane
d Trichlorfluoromethan
i triclorofluorometano
e triclorofluorometano

3893 trichlorophenol
Chemical compound existing in several isomeric forms. The derivative having the chlorine atoms in position 2, 4, 6 forms colourless, pungent-smelling crystals, not easily soluble in water, marked by antiseptic and disinfectant properties.

f trichlorophénol
d Trichlorphenol
i triclorofenolo
e triclorofenol

3894 triclocarban
Antiseptic and disinfectant agent mainly used in skin preparations for the treatment of bacterial and fungal infections.

f triclocarban
d Triclocarban
i triclorcarbano
e triclorocarbano

3895 triclofos
Non-barbituric hypnotic agent mainly indicated in the treatment of insomnia and also to induce sleep in electroencephalography. Drowsiness, general malaise, ataxia, gastro-intestinal disorders may occur as side effects.

f triclofos
d Triclofos
i triclofoso
e triclofoso

3896 tridihexethyl iodide
Parasympatholytic and anticholinergic agent especially used for the relief of gastro-intestinal spasms and the reduction of gastric secretions. Dizziness, a blurred vision and constipation may occur as side effects.

f iodure de trihexéthyle
d Trihexäthyljodid
i ioduro di tridiesetile
e yoduro de tridiesetilo

3897 trifluoperazine
Phenothiazine tranquillizer/antiemetic chiefly indicated in the treatment of psychic disorders caused by menopause and anxiety. For its side effects, see chlorpromazine. A symptomatology of extrapyramidal disorders must be added.

f trifluopérazine
d Trifluoperazin
i trifluoperazina
e trifluoperacina

3898 trifluorothymidin
Antiviral drug indicated in the treatment of herpetic and metaherpetic keratitis. It should not be used in the course of pregnancy.

f trifluorothymidine
d Trifluorothymidin
i trifluorotimidina
e trifluorotimidina

3899 trifluperidol
Butyrophenone tranquillizer marked by a neuroleptic action that is particularly effective in the treatment of schizophrenia. It is also used in the treatment of hebephrenia, catatonia and hallucinatory states. For side effects, see chlorpromazine. In addition, skin sensitization has been observed in some cases.

f triflupéridol
d Trifluperidol
i trifluperidolo
e trifluperidolo

3900 triflupromazine
Phenothiazine thymoleptic agent acting on the hypothalamus. An antipsychotic tranquillizer mainly indicated in the treatment of acute and chronic schizophrenia. Also used as a sedative in anxiety and psychic tension. The patient must be warned that some extrapyramidal reactions are merely a side effect of the drug. For other adverse effects, see chlorpromazine.

f triflupromazine
d Triflupromazin
i triflupromazina
e triflupromacina

3901 trigonelline
Alkaloid, methylbetaine of the nicotinic acid. It is found in various vegetables and in the seeds of hemp, strophantus etc. Colourless, watersoluble crystals which can be prepared by treating nicotinic acid with methyl iodide and silver hydrate. Medicinally it is used in the

treatment of metabolic disorders.
f trigonelline
d Trigonellin
i trigonellina
e trigonelina

3902 trihexyphenidyl
Anticholinergic agent acting like an atro‐
pine sulphate on the smooth muscles, in
particular the intestinal muscles. Owing
to its CNS depressant action, the drug
is indicated in the treatment of Parkin‐
son's disease, muscular spasms and
extrapyramidal motorial disorders. Also
used to treat torticollis.
f trihexyphénidyle
d Trihexyphenidyl
i triesifenidile
e trihexifenidilo

3903 tri-iodothyronine
Crystalline iodine-containing phenolic
amino acid occurring in small amount
with thyronine and regarded as being
formed from thyroxine by loss of one
iodine atom per molecule. Its action is
more rapid and more potent than thyro‐
xine, but considerably shorter. It is
generally used in the treatment of hypo‐
thyroidism and metabolic disorders.
f tri-iodothyronine
d Tri-Jodothyronin
i tri-iodotironina
e tri-yodotironina

3904 tri-isopropylphenoxy-polyethoxyethanol
Dispersant and emulsifying agent used
to disperse and repel spermatozoa in
order to prevent conception.
f tri-isopropylphénoxy-polyethoxyéthanol
d Tri-Isopropylphenoxy-Polyethoxyäthanol
i tri-isopropilfenossi-polietossietanolo
e tri-isopropilfenoxi-polietoxietanolo

3905 trilostane
Therapeutic agent used to inhibit the
synthesis of hormones in the adrenal
cortex while treating a pathological con‐
dition marked by obesity of the trunk,
purple striae atrophicae on the abdomen
and flanks, polycythaemia, hypertension,
osteoporosis and glycosuria (the Cushing
syndrome).
f trilostane
d Trilostan
i trilostan
e trilostan

3906 trimebutine
Therapeutic agent capable of promoting
the motility of the digestive tube, marked
by an elective affinity with Meissner's
and Auerbach's autonomous plexuses
with specifically regulating and syn‐
chronizing effect on the amplitude and
rhythm of the gastro-intestinal contrac‐
tions. The drug is indicated in the treat‐
ment of digestive disorders of various
nature but mostly characterized by an
altered motility of the system.

f trimébutine
d Trimebutin
i trimebutina
e trimebutina

3907 trimecaine
Or mesocaine, mesdicaine. Topical anaes‐
thetic mainly used in dentistry in asso‐
ciation with epinephrine and nor-epine‐
phrine.
f trimécaïne; mésocaïne
d Mesdikain; Mesokain
i trimecaina
e trimecaina; mesocaina

3908 trimeprazine
Chemical compound, a methyl derivative
of promazine, having antipruriginous
and dyspnoeic properties.
f triméprazine
d Trimeprazin
i trimeprazina
e trimepracina

3909 trimetaphan
Antihypertensive having a short duration
of action. It is mostly used to produce
controlled hypotension so as to reduce
blood losses in the course of surgery.
f trimétaphan
d Trimetaphan
i trimetafano
e trimetafano

3910 trimetaphan camsilate
Ganglionic blocking agent of short dura‐
tion. As a hypotensive, it is very useful
in plastic surgery and gynecology owing
to its prompt and short-lasting action.
A tendency to haemorrhages represents
its principal adverse effect.
f camsilate de trimétaphan
d Trimetaphankamsilat
i camsilato di trimetafano
e camsilato de trimetafano

3911 trimethadione
Anticonvulsant agent mainly indicated
in the treatment of petit mal and of
idiopathic epilepsy. It is frequently
used in conjunction with other antiepilep‐
tic drugs. Trimethadione is characterized
by a high toxicity; serious ophthalmolo‐
gical disorders and diseases of the op‐
tical nerve may occur as side effects.
f triméthadione
d Trimethadion
i trimetadione
e trimetadiona

3912 trimethobenzamide
Antiemetic indicated for the treatment of
nausea and vomiting, marked by a cen‐
tral mechanism of action. Some accommo‐
dation disorders may occur as side ef‐
fect.
f triméthobenzamide
d Trimethobenzamid
i trimetobenzamide
e trimetobenzamida

3913 trimethoprim
Antimicrobial agent capable of inhibiting the conversion of folic acid to folinic acid. It is mostly used against malaria, but also for the prevention of infections of the urinary tract. Combined with sulpha drugs, e.g. sulphamethoxazole, it is used in the treatment of respiratory infection, septicaemia, gonorrhoea, salmonellosis and various skin infections.
f triméthoprime
d Trimethoprim
i trimetoprim
e trimetoprim

3914 trimipramine
Tricyclic antidepressant, mainly psycho-therapeutic drug with a mechanism of action resembling that of imipramine. It is mostly used in the treatment of mental depression characterized by melancholia and involutionary neuroses. The often ensuing drowsiness appears to be heavier than that caused by imipramine.
f trimipramine
d Trimipramin
i trimipramina
e trimipramina

3915 trioxysalen
Therapeutic agent acting on the skin pigmentation as it produces the formation of melanin, thus promoting the oxidation of tyrosine by thyroxinase, a reaction which is activated by ultra-violet radiations. The drug concentrates on melanocytes and is indicated in order to increase skin pigmentation.
f trioxysalène
d Trioxysalen
i triossisalene
e trioxisaleno

3916 tripelennamine
A chemical compound whose derivatives, with nitric acid or hydrochloric acid, form colourless, bitter-tasting crystals, easily soluble in water, and acting as antihistaminic agents.
f tripélennamine
d Tripelennamin
i tripelennamina; tripelennammina
e tripelenamina

3917 triprolidine
Prompt- and long-lasting antihistamine mainly indicated in the treatment of allergic conditions, e.g. urticaria, intense pruritus, hay-asthma, Quinke's oedema and conjunctivitis, particularly when due to excessive production of histamine. Some drowsiness may occur as side effect.
f triprolidine
d Triprolidin
i triprolidina
e triprolidina

3918 trisodium edetate
Edetic acid. Chelating agent forming inactive complexes with Ca ions or with other bi- or tri-valent metals. It is used intravenously in hypercalcaemia and locally for lime burns in the eye. Also used in coronary angina and scleroderma and as an antidote in digitalis intoxications. Nausea, diarrhoeas, cramp and pains in the limbs may occur as side effects.
f édétate trisodique; acide tétracetique
d Trisodiumödetat
i acido edetico; acido etilendiaminotetracetico
e ácido edético

3919 troleandomycin
Oleandomycin-derived antimicrobial agent effective against streptococci, staphylococci, pneumococci and gonococci which are resistant to other antibiotics. Some nausea and, in case of prolonged therapy, liver disorders may occur as side effects.
f troléandomycine
d Troleandomycin
i troleandomicina
e troleandomicina

3920 trolnitrate
Organic nitrate used as vasodilator in the treatment of angina pectoris. Hot flushes, headache, dizziness and general weakness may occur as side effects.
f trolnitrate
d Trolnitrat
i trolnitrato
e trolnitrato

3921 tropicamide
Parasympatholytic agent mainly used as a mydriatic. Its action is similar to that of atropine. It has a rapid but short effect and is generally associated with epinephrine for the examination of the eye fundus and with prednisone against eye inflammation. Some accommodation disorders may occur as side effect.
f tropicamide
d Tropicamid
i tropicamide
e tropicamida

3922 tropine
Tertiary base and heterocyclic alcohol present as an ester in the atropine and hyoscyamine. Colourless crystals, soluble in water with an alkaline reaction. By hydrolyzing the atropine with soda or with acids, it is possible to obtain a molecule of optically inactive tropine and one of tropic acid. Pure tropine does not present the mydriatic properties which are characteristic of atropine.
f tropine; tropanol
d Tropin
i tropina; tropanolo
e tropina; tropanolo

3923 troxerutin

Vitamin derivative (P factor) used to improve the strength and reduce the permeability of blood vessels. It is also indicated for the treatment of venous incompetence, e.g. varices during pregnancy and haemorrhoids.

f troxérutine
d Troxerutin
i trosserutina
e troxerutina

3924 trypsin

Proteolytic digestive enzyme produced by the exocrine portion of the pancreas in the form of an almost inactive proenzyme (trypsinogen) which subsequently, once it reaches the intestine, is activated by another enzyme, the enteropeptidase. Medicinally, it is indicated in the treatment of abscesses, necrotic wounds, ulcers and burns. It is also used by instillation in haematomas, haemothorax, fistulae and pneumonia, orally in association with chymotripsin and by parenteral route in inflammatory and oedematous states.

f trypsine; tryptase
d Trypsin
i tripsina
e tripsina

3925 tryptophan

Or triptophane. Aromatic amino acid isolated among the products of the tryptic digestion. It is one of the constituent elements of the proteic molecules and has a considerable nutritional importance. It is one of the essential amino acids for human beings as it is indispensable for the biological synthesis of the nicotinic acid, of the prosthetic group of haemoglobins and of some pigments. Tryptophan is used medicinally as a psychostimulant and antidepressive agent in both endogenous and reactive depressions and in the depressive phase of manic-depressive psychoses.

f tryptophane
d Tryptophan
i triptofano
e triptófano

3926 tubocurarine

Relaxant of the skeletal muscles acting by blocking the passage of impulses at the neuromuscular junction. It is mainly used as an adjunct to anaesthesia and in shock therapy and muscle spasms. The drug is equally useful in the treatment of convulsions caused by infectious diseases and intoxications. Electrolytic imbalance, hypertension and release of histamine may occur as side effects.

f tubocurarine
d Tubokurarin
i tubocurarina
e tubocurarina

3927 tunicamycin

Antibiotic inhibitor. Hydrophobic analog of UDP (uridine diphosphate) -N-acetylglucosamine. It blocks the addition of N-acetylglucosamine to dolichol phosphate, this being the first step in the formation of the core oligosaccharide.

f tunicamycine
d Tunikamyzin
i tunicamicina
e tunicamicina

3928 tybamate

Tranquillizer mainly used in the treatment of anxiety, psychic tension, insomnia, hyperemotivity, senile psychosis and neuroses which are due to alcoholism. The drug must be administered with great caution, especially when it is taken with other psychodrugs.

f tybamate
d Tybamat
i tibamato
e tibamato

3929 tylocapol

Mucolytic agent administered by inhalation from a nebulizer in order to liquefy mucus and aiding expectoration especially where viscid mucus is very unpleasant, as in chronic bronchitis. Inflammation of the eyelids may occur as a side effect. The drug must be administered with great caution in asthmatic patients.

f tylocapol
d Tylocapol
i tilocapol
e tilocapol

3930 tymazoline

Vasoconstrictor and anticholinergic agent chiefly used as a nasal decongestant in the treatment of heavy colds, hay fever, and acute and/or chronic allergic rhinitis.

f tymazoline
d Tymazolin
i timazolina
e timazolina

3931 tyramine

A mydriatic used both as a therapeutic and diagnostic agent whenever a short action is required.

f tyramine
d Tyramin
i tiramina
e tiramina

3932 tyrotricin

Antibiotic produced by the Bacillus brevis. It has a polypeptidic chemical structure and is basically constituted by two substances endowed with antibiotic properties, i.e. gramicidin and tyrocidine. It is active against Gram-positive and some Gram-negative germs, but owing to its toxicity is only employed for topical treatment of infections of the skin and mucosae.

f tyrotricine
d Tyrotrizin
i tirotricina
e tirotricina

U

3933 ubidecarenone
Ubiquinone 10 or coenzyme q. Ubiquinone -derivative indicated in the treatment of mild myocardial insufficiency, oedema, pulmonary congestion and anginal symptoms. Nausea, swollen eyelids and a sensation of cold in the upper and lower limbs may occur as side effects.
f ubidécarénone; ubiquinone 10
d Ubidekarenon
i ubidecarenone; ubichinone 10
e ubidecarenona; ubiquinona 10

3934 ubiquinone
Term denoting a group of fat-soluble benzoquinones present in the majority of aerobic organisms, from bacteria to plants and superior animals. They act as electron transporters in the processes of respiration and oxidation.
f ubiquinone
d Ubichinon
i ubichinone
e ubiquinona

3935 undecenoic acid
Or undecylenic acid. Unsaturated acid produced by the heating of castor oil. It is used as an antifungal agent and in the treatment of athlete's foot (tinea pedis).
f acide undécénoïque; acide undécylénique
d Undecylenylsäure; Undecensäure
i acido undecilenico
e ácido undecilénico; ácido undecenoico

3936 uracil
Pyrimidin deoxyderivative which is obtained by the hydrolysis of nucleic acids. It is one of the constituents of ribonucleic acids.
f uracile
d Uracil
i uracile
e uracilo

* **uracil–chlorethamine   s.   uramustine**

* **uracil riboside   s.   uridine**

3937 uramustine
Cytotoxic agent used in the palliative treatment of chronic lymphatic leukaemia and malignant lymphoma. It is also indicated in the treatment of chronic myelocytic leukaemia, ovarian carcinoma, polycythaemia vera, multiple myeloma and thrombocytosis. Anorexia, nausea, vomiting, nervous depression and hepatic disorders may occur as side effects.
f uramustine
d Uramustin
i uramustina
e uramustina

3938 urate
Salt of uric acid that may be neutral or acid.
f urate
d harnsaures Salz
i urato
e urato

3939 urea
Chemical compound which can be considered the amide of the carbamic acid or the diamide of the carbonic acid. It is formed in animals as the final product of the metabolism of the nitrogenous compounds and represents the principal solid component of the urine of man and other mammals. It is present in small quantities in the blood and other body fluids as well as in the liver. It is a very weak base and can forms salts only with very strong acids. It is used in several chemical syntheses.
f urée
d Urea; Harnstoff
i urea
e urea

3940 urethane
Cytotoxic drug formerly widely used in the treatment of some neoplastic diseases, presently superseded by newer and more effective drugs. The drug is also characterized by mild hypnotic properties and is used as anaesthetic for small animals.
f uréthane
d Urethan
i uretano
e uretano

3941 uric acid
White crystalline substance soluble in alkalis and less so in water. It occurs in two tautomeric forms (keto and enol) and is a constituent of the urine of carnivorous animals including human urine as the result of the breakdown of the purines of the nucleic acids and nucleoproteins. In man and primates, uric acid represents the principal termi-

nal product of the purinic metabolism. In other animals, uric acid is subjected to a further oxidation by the uricase and transformed into allantoine which is more soluble and thus more easily eliminated.
f acide urique
d Harnsäure
i acido urico
e ácido úrico

3942 uricase
Uricolytic enzyme present in the liver of most mammals and the kidney of some of them. It oxidizes uric acid to allantoin and carbon dioxide. It is used in the treatment of severe primary or secondary hyperuricaemia, acute and chronic kidney disorders, haemodialysis and gout with urinary lithiasis.
f uricase
d Urikase
i uricasi
e uricasa

3943 uridine
Nucleoside obtained by hydrolysis of ribonucleic acid and uridylic acid. In the form of phosphate derivatives, it plays an important part in carbohydrate metabolism.
f uridine
d Uridin
i uridina
e uridina

3944 uridylic acid
Nucleotide of the ribonucleic acid constituted by the pairing of three molecules, one of phosphoric acid, one of ribose and one of uracil.
f acide uridylique
d Uridylsäure
i acido uridilico
e ácido uridílico

3945 urocanic acid
Heterocyclic carboxylic acid originally found in the urine of a dog and later in the sweat of man. It displays a protective action on the skin and is formed by the decomposition of histidine.
f acide urocanique
d Urokansäure
i acido urocanico
e ácido urocánico

3946 urogastrone
Antacid agent capable of selectively in-

hibiting gastric juices and marked by a cicatrizing action by stimulating fibro blastic proliferation and reepithelization. It is therefore indicated in the treatment of peptic ulcer and gastric hypersecretion.
f urogastrone
d Urogastron
i urogastrone
e urogastrona

3947 urokinase
Kidney-produced enzyme which is excreted in urine and which is capable of activating plasminogen and converting it into plasmin. It is used intravenously to break down recently formed clots of embolic or thrombotic origin, as well as in the treatment of thrombosis of retinal vessels, haemorrhage of the anterior chamber of the eye and pathological accumulation of fibrin.
f urokinase
d Urokinase
i urochinasi
e uroquinasa

3948 uronic acid
Any of a group of aldehyde-acids which are the product of oxidation of sugar. Uronic acids are found combined in many polysaccharides and in urine.
f acide uronique
d Uronsäure
i acido uronico
e ácido urónico

3949 ursodeoxycholic acid
Therapeutic agent used to dissolve cholesterol gall stones. Diarrhoea may occur as side effect.
f acide ursodéoxycholique
d Ursodeoxygallensäure
i acido deossicolico
e ácido deoxicólico

3950 usnic acid
Usninic acid, usnein. Colourless crystals, insoluble in water, optically active. It exists in the two optical antipodes and in the racemic form. The acid has antibacterial properties in skin diseases.
f acide usnique
d Flechtensäure; Usninsäure
i acido usnico
e ácido úsnico

V

3951 **vaccine**
Any preparation which is capable of producing active immunity. It may be viral or bacterial. Originally the term was used to denote the vesicle fluid of cowpox which was inoculated as immuniz ing agent for smallpox.
f vaccin
d Impfstoff
i vaccino
e vacuna

3952 **valerate**
Salt or ester of valeric acid. Several salts of the isovaleric (or isovalerianic) acid are of some importance in medicine owing to their sedative properties. Also known as valerianate.
f valérianate
d Valerianat
i valerianato
e valerianato

3953 **valerian**
Therapeutic agent consisting of the dried rhizome and roots of the garden heliotrope, i.e. Valeriana officinalis, used as a carminative and a sedative, especially in nervous conditions.
f valériane
d Baldrian; Valerian
i valeriana
e valeriana

3954 **valerianic acid**
Or valeric acid. Pentanoic acid which is found free and as ester in the roots of Valeriana officinalis and of Angelica archangelica. It is prepared by oxidation of normal amyl alcohol and valerianic aldehyde; a colourless, tasteless, bad-smelling oily liquid, used in organic synthesis and, in medicine, as sedative.
f acide valérianique; acide valérique
d Valeriansäure; Baldriansäure
i acido valerianico
e ácido valeriánico; ácido valérico

3955 **valerian root**
The root of the garden heliotrope from which an oily liquid is obtained, which is medicinally used as a carminative and sedative agent.
f racine de valériane
d Baldrianwurzel; Katzenwurzel
i radice di valeriana
e raíz de valeriana

3956 **valine**
Crystalline amino acid which in its dextrorotatory form is essential in the nutrition both of men and lower animals. It is obtained in the wanted form by the hydrolysis of proteins, e.g. casein. In its racemic form, it is obtained by synthesis.
f valine
d Valin
i valina
e valina

3957 **vanillic acid**
Colourless, not easily soluble in water crystals obtained from vanillin either by oxidation or directly in the extraction of vanillin from the bisulphitic leach liquors. The esters of the acid are used medicinally as disinfectant.
f acide vanillique
d Vanillinsäure
i acido vanillico
e ácido vainílico

3958 **vanillin**
Crystalline phenolic aldehyde, the principal fragrant component of vanilla. It is mostly used in flavouring.
f vanilline
d Vanillin
i vanillina
e vainillina

3959 **vanitiolide**
Choleretic agent without cholecystokinetic action indicated in the treatment of hepatobiliary disorders, dyspepsia and vomiting, in particular vomiting caused by acetone in infants, icterus, biliary lithiasis and biliary dyskinesia.
f vanitiolide
d Vanitiolid
i vanitiolide
e vanitiolido

3960 **vasoconstrictor**
Relating to vasoconstriction or inducing vasoconstriction, i.e. the narrowing of the lumen of blood vessels particular as a result of vasomotor nervous action.
f vasoconstricteur
d Vasokonstriktor; vasokonstruktorisches Mittel
i vasocostrittore

e vasoconstrictor

3961 vasodilator
Any agent, e.g. a drug or a parasympathetic nerve fibre, which causes or induces vasodilatation, i.e. the increase of the lumen of blood vessels.
f vasodilatateur
d Vasodilatator; vasodilatatorisches Mittel
i vasodilatatore
e vasodilatador

3962 vasopressin
Antidiuretic hormone (ADH). One of the hormones of the posterior lobe of the hypophysis, formed by 9 amino acids. Vasopressin is synthesized in the supra-optical and paraventricular nuclei of the hypothalamus in the form of a larger polypeptide from which it is liberated by proteolysis and is transported into the neurohypophysis by specific carriers of proteic nature.
f vasopressine; hormone antidiurétique; ADH
d Vasopressin; antidiuretisches Hormon; ADH
i vasopressina; ADH
e vasopresina; adiuretina

3963 vasopressor
Exerting a vasoconstricting effect or causing a rise in blood pressure.
f vasopresseur
d Vasopressor; blutdruckerhöhendes Mittel
i vasopressore
e vasopresor

3964 verapamil
Vasodilator and adrenergic blocking agent especially used in prevention of angina of effort. Although very active on the coronary arteries, it does not alter the rhythm and frequency of the heart.
f vérapamil
d Verapamil
i verapamile
e verapamil

3965 veratrum
Natural product capable of reducing the sympathetic tone and used in the past for the treatment of cardiac dysrhythmias. It is no longer administered owing to its rather severe side effect, e.g. respiratory depression and abnormal heart rhythms.
f vératre
d Veratrum
i veratro
e veratro

3966 verazide
Isoniazid-derived agent especially used as antiinfective in the treatment of primary pulmonary tuberculosis. Gastric pains and sometimes allergic reactions may occur as side effects.
f vérazide

d Verazid
i verazide
e verazida

3967 vidarabine
Antiviral agent used in the topical treatment of ophthalmic and cutaneous herpetic infections. Also, a cytotoxic agent used in the treatment of leukaemia and Hodgkin's disease. Bone marrow depression may occur as side effect.
f vidarabine
d Vidarabin
i vidarabina
e vidarabina

3968 villikinin
Hormone produced by the intestinal mucosa capable of activating the movement of the intestinal villi.
f villikinine
d Villikinin
i villichinina
e viliquinina

3969 viloxazine
Antidepressant agent marked by anticonvulsant properties and by supposed absence of anticholinergic and sedative properties. It is mainly used in the treatment of depression, especially when caused by drugs.
f viloxazine
d Viloxazin
i vilossazina
e viloxacina

3970 viminol
Analgesic agent marked by the absence of narcotic effect · Central mechanism of action. It is generally used for the symptomatic treatment of several painful conditions, in particular osteoarticular and neuritic pains. It has a sedative action but may cause epigastric disorders.
f viminol
d Viminol
i viminolo
e viminolo

3971 vinbarbital
Barbituric drug used as hypnotic and sedative and indicated in the treatment of asthma and cough, as well as for the relief of psychosomatic insomnia.
f vinbarbital
d Vinbarbital
i vinbarbital
e vinbarbital

3972 vinblastine
Cytotoxic alkaloid obtained from the West Indian periwinkle, indicated in the treatment of neoplastic diseases, e.g. reticulosarcoma, lymphosarcoma, mammary and stomach cancer, which have proved resistant to other cytotoxic drugs. Anaemia, bone marrow depression and neuropathy may occur as side effects.

f vinblastine
d Vinblastin
i vinblastina
e vinblastina

3973 vincamine
Cerebral vasodilator acting directly on the vascular smooth muscles so as to produce a considerable increase of blood flow without modifying the systemic arterial pressure. It is principally used to improve mental power in patients suffering from cerebrovascular disorders, in the treatment of cerebral sclerosis and senile cerebral involution.
f vincamine
d Vinkamin
i vincamina
e vincamina

3974 vincristine
See vinblastine for action and effects.
f vincristine
d Vinkristin
i vincristina
e vincristina

3975 vindesine
Semisynthetic cytotoxic agent derived from vinblastine, q.v. It has a broader spectrum of antitumoural activities and does not share cross-resistance with either vinblastine or vincristine. The drug is principally used in the treatment of leukaemia and malignant melanoma. Haematological, neurological and cutaneous disorders may occur as side effects.
f vindésine
d Vindesin
i vindesina
e vindesina

3976 vinylbital
Hypnotic agent of the barbituric series acting directly on the substantia reticularis and of the hypothalamus mesencephalon. It is indicated in the treatment of insomnia, psychic erethism and convulsive psychopathies.
f vinylbital
d Vinylbital
i vinilbitale
e vinilbitalo

3977 viomycin
Antibiotic displaying a bacteriostatic activity against Mycobacterium tuberculosis. It is also active against strains which have proved resistant to streptomycin. Hypokalaemia and a degree of deafness may be observed as side effects.
f viomycine
d Viomycin
i viomicina
e viomicina

3978 viquidil
Vasodilator mainly indicated for the symptomatic relief of cerebrovascular disorders. It acts directly on the vascular smooth muscles so as to produce a gradual progressive relaxation thus increasing the volume of the peripheral and cerebral blood flow. The drug has also proved useful in the treatment of senile cerebrovascular disorders and apoplexy.
f viquidil
d Vichidil
i vichidile
e viquidilo

3979 virginiamycin
Antibiotic isolated from cultures of Streptomyces virginiae, very active against staphylococci and thus used in the treatment of bacterial infections of the ear, nose, mouth, larynx and eyes.
f virginiamycine
d Virginiamycin
i virginiamicina
e virginiamicina

3980 visnadine
Coronary vasodilator used in small doses for the treatment of myocardial disorders caused by inadequate supply of oxygenated blood in elderly or greatly debilitated patients.
f visnadine
d Visnadin
i visnadina
e visnadina

3981 vitamin
Any of those important organic compounds that are present in natural food in extremely small concentration and totally without protein, fat, carbohydrate and inorganic salts. They are essential for the good health and the development of the organism and whenever they are deficient, outright absent or fail to be assimilated, various diseases may occur, generically called avitaminosis. Vitamins cannot be synthesized by the organism and must therefore by ingested with the aliments, a characteristic which distinguishes them from hormones.
f vitamine
d Vitamin
i vitamina
e vitamina

3982 vitamin A
Fat-soluble unsaturated alcohol, two molecules of which are formed by hydrolysis of one molecule of ß-carotene. It is especially present in butter, fat and the yolk of eggs, but the most copious source is the fish-liver oil. Deficiency of this vitamin may cause xeroma (keratinization and dryness of the conjunctiva) and night blindness. An excessive intake of vitamin A may prove toxic.
f vitamine A
d Vitamin A
i vitamina A

e vitamina A

3983 **vitamin B**
Each of many members of vitamin B complex, i.e. the B_1, also known as thiamine or aneurin having antineuritic activity and present in yeast, eggs, cereals and fruit; the B_2, or lactoflavine
which favours growth and development
and whose deficiency causes alterations
in the mucosae; the B_6, or pyridoxine
(adermine) that is present in all aliments and especially in the liver,
yeasts and milk. Its deficiency may
cause cutaneous diseases. The B_{12}, or
cobalamine contains some cobalt and
shows a considerable activity against
pernicious anaemia. It can be synthesized by microorganisms and fungi and is
. copiously present in the liver.
f vitamine B
d Vitamin B
i vitamina B

e vitamina B

* **vitamin C**   s. ascorbic acid

* **vitamin D**   s. calciferol

* **vitamin E**   s. tocopherol

* **vitamin H**   s. biotin

* **vitamin K**   s. phylloquinone

3984 **vitamin P**
The factor which regulates the permeability of the blood capillaries. Its nature
is reputed to be between hesperidin and
rutin, i.e. substances which are adminis
tered in the treatment of abnormal capil
lary fragility.
f vitamine P; vitamine d'antiperméabilité
d Vitamin P; Permeabilitätsfaktor
i vitamina P; fattore d'impermeabilità
e vitamina P; vitamina de impermeabilidad

W

3985 walnut
The seed of the Juglans regia. The distilled spirit is characterized by antispasmodic properties.
f noix
d Nuss
i noce
e nogal

3986 walnut oil
Very pale fatty oil obtained from wal-
nuts and used in several pharmaceutical
preparations.
f huile de noix
d Nussöl
i olio di noce
e aceite de nogal

3987 warfarin
Coumarin-derived anticoagulant capable
of lengthening the prothrombin time by
inhibiting the formation of Factor VII
owing to the antagonism of prothrombin
with vitamin K. In practice it interferes
with the synthesis of clotting factors by
the liver. Alopecia, dermatitis and urticaria may occur as side effects. The
drug should not be administered in the
course of pregnancy and in the presence
of hypertension.
f warfarin; warfarine
d Warfarin
i warfarin; warfarina
e warfarin; warfarina

*** wassermycine   s. metacycline**

3988 wood sugar
The xylose that is obtained from plant
sources; also a mixture of pentose and
hexose sugars obtained through the hydrolysis of pentosan and cellulose of
wood.
f sucre de bois
d Holzzucker
i zucchero di legno
e azúcar de madera

X

3989 xanthine
Chemical compound found in the sediments of urine. It belongs to the group of purine bases and is widely diffused in the animal and vegetable kingdoms. The colourless crystals which are obtained are almost insoluble in cold water and break down on heating developing carbon dioxide, ammonia and hydrocyanic acid. In man and other superior animals, xanthine is formed from guanine by the action of a deaminase and is transformed into uric acid by the action of an oxidase. In the purinic metabolism, it represents the precursor of uric acid. It also participates in the constitution of the nucleic acids.
f xanthine
d Xanthin
i xantina
e xantina

3990 xanthophyll palmitate
Oxygenizing agent used in the treatment of retinal disorders. It is a physiological precursor of sight pigments and is capable of increasing the reactivity of the retina by increasing the rate of regeneration of sight pigments and the oxygen consumption by the retinal tissues.
f palmitate de xanthophylle
d Xanthophyllpalmitat
i palmitato di xantofilla
e palmitato de xantofila

3991 xantinol nicotinate
Cardiocircular tonic chiefly indicated in the treatment of peripheral vascular diseases. It increases the rate of circulation. Flush and itching may occur as side effects.
f nicotinate de xantinol
d Xantinolnikotinat
i nicotinato di xantinolo
e nicotinato de xantinolo

3992 xenbucin
Hypocholesterolaemic agent capable of diminishing the endogenous synthesis of cholesterol in the blood. Its peak effect is attained after 20 to 40 days. It is mainly indicated in the treatment and prophylaxis of all hypercholesterolaemic conditions, e.g. generalized and localized arteriosclerosis.
f xenbucine
d Xenbucin
i xenbucina
e xenbucina

3993 xenytropium bromide
Atropine-like parasympatholytic agent mainly indicated in the treatment of psychosomatic and psychovegetative syndromes, as well as psychemotional lability. Gastro-intestinal disorders may occur as side effect.
f bromure de xénytropium
d Xenytropiumbromid
i bromuro di xenitropio
e bromuro de xenitropio

3994 xipamide
Diuretic agent characterized by an inhibiting effect on the reabsorption of sodium in the distal renal tubule. It is mainly indicated in the treatment of oedemas caused by cardiac or renal disorders, venous incompetence and hypertension. Hypokalaemia, hyperglycaemia and glycosuria may occur as side effects.
f xipamide
d Xipamid
i xipamide
e xipamida

3995 xylometazoline
Sympathomimetic agent marked by a peripheral mechanism and indicated as a nasal decongestant in the treatment of colds, vasomotor rhinitis and any congestive disorder of the pharyngeal mucosa. The systemic effects are negligible.
f xylométazoline
d Xylometazolin
i xilometazolina
e xilometazolina

3996 xylose
Pentose which is obtained from xylans by hydrolysis with diluted acid. Colourless, water-soluble sweet crystals. It exists in two optically active isomeric forms, as well as in the racemic one. It is sometimes used for the treatment of diabetes. It may, albeit rarely, be found in urine.
f xylose
d Xylose
i xilosio
e xilosa

Y

3997 yeast
Substance consisting of micro-organisms characterized by unicellular form and belonging to the fungi. The organisms secrete enzymes that are capable of converting sugar into alcohol and carbon dioxide. The term is also used to denote the cells and spores of Saccharomyces cerevisiae and other species of Saccharomyces which are cultivated and employed for the fermentation of grain. Yeast is also used in medicine for the treatment of several disorders deriving from deficiency of the B group vitamins.
f levure
d Hefe
i lievito
e levadura

3998 yohimbine
Plant extract having an alpha-adrenocep tor blocking action. It is a methyl ester of the yohimbic acid. The drug has been employed as an aphrodisiac but its results have never been satisfactorily proved. Vertigo, hot flushes and nervousness may occur as side effects.
f yohimbine
d Yohimbin
i johimbina; yohimbina
e yohimbina

Z

3999 zidovudine
Antiviral agent capable of preventing the replication of retroviruses, including the immunodeficiency virus in man (HIV) which is involved in AIDS. Although frequently in use, the drug cannot eliminate the infection and thus be regarded as a cure.
f zidovudine
d Zidovudin
i zidovudina
e zidovudina

4000 zinc chloride
White, granular, deliquescent powder marked by strong caustic and astringent properties and mainly indicated in the treatment of chronic ulcerated skin lesions. Also used as eye drops in a more diluted solution.
f chlorure de zinc
d Zinkchlorid
i cloruro di zinco
e cloruro de zinc

4001 zinc naphthenate
Powder or ointment used topically in the treatment of infections of the skin.
f naphténate de zinc
d Zinknaphthenat
i naftenato di zinco
e naftenato de zinc

4002 zinc oleate
Disinfectant and astringent used in dermatology, particularly for the treatment of eczema and excoriated skin.
f oléate de zinc
d Zinkoleat
i oleato di zinco
e oleato de zinc

4003 zinc oxide
White amorphous, odourless and tasteless powder used as an external application to the skin in the form of dusting powder, cream or ointment. It has a mild astringent and soothing effect and is indicated in the treatment of cutaneous disorders, e.g. eczema and excoriated skin. Also known as zinc white.
f oxyde de zinc; blanc de zinc
d Zinkoxyd; Zinkweiss
i ossido di zinco; bianco di zinco
e óxido de zinc; blanco de zinc

4004 zinc peroxide
Antiseptic, deodorant and astringent agent capable of liberating nascent oxygen when in contact with organic secretions. It is therefore indicated in the treatment of burns and lesions due to infections by anaerobic bacteria, diabetic gangrene and torpid wounds.
f peroxyde de zinc
d Zinkperoxyd
i perossido di zinco
e peróxido de zinc

4005 zinc sulphate
White vitriol. Astringent and mild antiseptic used in the treatment of disorders affecting the mucosa of the eye, pharynx and urethra. It is used topically in aqueous solutions. In conjunction with local anaesthetics, it is a component of some type of collyrium.
f sulfate de zinc
d Zinksulfat; weisser Vitriol; Bergbutter
i solfato di zinco
e sulfato de zinc

4006 zinc undecanoate
Or zinc undecylenate. Antifungal agent used topically in the treatment of fungal infections of the skin.
f undécylénate de zinc
d Zinkundecylenat
i undecilenato di zinco
e undecilenato de zinc

4007 zipeprol
Antitussive agent and bronchial antispasmodic mainly used in affections of the upper respiratory tract. Drowsiness and vertigo may occur as side effects whenever high doses are administered.
f zipéprol
d Zipeprol
i zipeprolo
e zipeprolo

4008 zolimidine
Antacid agent capable of restoring the secretion of gastroduodenal mucus with protective and regenerating effect on the natural mucous coat. It is used in the treatment of gastritis, duodenitis and any intestinal trouble caused by drugs.
f zolimidine
d Zolimidin
i zolimidina
e zolimidina

FRANÇAIS

DEUTSCH

ITALIANO

ESPAÑOL